Handbook of
Statistical Methods
and Analyses
in Sports

Chapman & Hall/CRC
Handbooks of Modern Statistical Methods

Series Editor

Garrett Fitzmaurice

Department of Biostatistics
Harvard School of Public Health
Boston, MA, U.S.A.

Aims and Scope

The objective of the series is to provide high-quality volumes covering the state-of-the-art in the theory and applications of statistical methodology. The books in the series are thoroughly edited and present comprehensive, coherent, and unified summaries of specific methodological topics from statistics. The chapters are written by the leading researchers in the field, and present a good balance of theory and application through a synthesis of the key methodological developments and examples and case studies using real data.

The scope of the series is wide, covering topics of statistical methodology that are well developed and find application in a range of scientific disciplines. The volumes are primarily of interest to researchers and graduate students from statistics and biostatistics, but also appeal to scientists from fields where the methodology is applied to real problems, including medical research, epidemiology and public health, engineering, biological science, environmental science, and the social sciences.

Published Titles

Handbook of Mixed Membership Models and Their Applications
Edited by Edoardo M. Airoldi, David M. Blei,
Elena A. Erosheva, and Stephen E. Fienberg

Handbook of Statistical Methods and Analyses in Sports
Edited by Jim Albert, Mark E. Glickman,
Tim B. Swartz, and Ruud H. Koning

Handbook of Markov Chain Monte Carlo
Edited by Steve Brooks, Andrew Gelman,
Galin L. Jones, and Xiao-Li Meng

Handbook of Big Data
Edited by Peter Bühlmann, Petros Drineas,
Michael Kane, and Mark van der Laan

Published Titles Continued

Handbook of Discrete-Valued Time Series
Edited by Richard A. Davis, Scott H. Holan,
Robert Lund, and Nalini Ravishanker

Handbook of Design and Analysis of Experiments
Edited by Angela Dean, Max Morris,
John Stufken, and Derek Bingham

Longitudinal Data Analysis
Edited by Garrett Fitzmaurice, Marie Davidian,
Geert Verbeke, and Geert Molenberghs

Handbook of Spatial Statistics
Edited by Alan E. Gelfand, Peter J. Diggle,
Montserrat Fuentes, and Peter Guttorp

Handbook of Cluster Analysis
Edited by Christian Hennig, Marina Meila,
Fionn Murtagh, and Roberto Rocci

Handbook of Survival Analysis
Edited by John P. Klein, Hans C. van Houwelingen,
Joseph G. Ibrahim, and Thomas H. Scheike

Handbook of Spatial Epidemiology
Edited by Andrew B. Lawson, Sudipto Banerjee,
Robert P. Haining, and María Dolores Ugarte

Handbook of Missing Data Methodology
Edited by Geert Molenberghs, Garrett Fitzmaurice,
Michael G. Kenward, Anastasios Tsiatis, and Geert Verbeke

Handbook of Neuroimaging Data Analysis
Edited by Hernando Ombao, Martin Lindquist,
Wesley Thompson, and John Aston

Chapman & Hall/CRC
Handbooks of Modern Statistical Methods

Handbook of Statistical Methods and Analyses in Sports

Edited by

Jim Albert

Bowling Green State University, Ohio, USA

Mark E. Glickman

Harvard University, Cambridge, Massachusetts, USA

Tim B. Swartz

Simon Fraser University, Burnaby, British Columbia, Canada

Ruud H. Koning

University of Groningen, The Netherlands

CRC Press
Taylor & Francis Group
Boca Raton London New York

CRC Press is an imprint of the
Taylor & Francis Group, an **informa** business

A CHAPMAN & HALL BOOK

CRC Press
Taylor & Francis Group
6000 Broken Sound Parkway NW, Suite 300
Boca Raton, FL 33487-2742

First issued in paperback 2019

© 2017 by Taylor & Francis Group, LLC
CRC Press is an imprint of Taylor & Francis Group, an Informa business

No claim to original U.S. Government works

ISBN-13: 978-1-4987-3736-4 (hbk)
ISBN-13: 978-0-367-33101-6 (pbk)

Visit the Taylor & Francis Web site at
http://www.taylorandfrancis.com

and the CRC Press Web site at
http://www.crcpress.com

Contents

Preface

A strong relationship has always existed between sports and the statistics that are used to measure player and team performance. Many interesting questions about sports have led to serious research in sports statistics. Comprehensive surveys of statistics in sports research have been provided by books such as *Management Science in Sports* (1976), *Optimal Strategies in Sports* (1977), and *Statistics in Sport* (1998). The American Statistical Association created a section on Statistics in Sports in 1992, the International Statistical Institute created a Sports Statistics Committee in 1993, and the journal *Chance* has devoted a regular column to sports statistics.

In the approximate 20 years since the publication of *Statistics in Sport*, there has been a remarkable change in both the accumulation of sports data and the opportunities to address sports questions using statistical methods. Researchers and sports professionals have seen a recent explosion in the proliferation of data collected on a variety of aspects of sports. Motion-tracking technology has permitted the accumulation of detailed information on player-level dynamics, and large data archives for sports have become much more accessible worldwide. For example, a baseball researcher has opportunities to explore season data by Lahman database (`www.seanlahman.com`), game and play data using Retrosheet (`www.retrosheet.org`), and pitch-by-pitch data through the PitchFX system (`www.sportvision.com/baseball/pitchfx`). This accumulation of data has led to the development of new statistical methodologies. As an example, the traditional measurement of fielding in baseball is the fielding percentage, the fraction of plays by a particular fielder that is successful. With the development of new tracking systems, one can now measure the movement of a fielder toward a ball that is hit in his direction and obtain a much better measure of fielding performance that accounts for the range of a player.

The opportunities for statistical research in sports have correspondingly grown immensely, as reflected in new dedicated journals to the subject area. Two examples are the *Journal of Quantitative Analysis in Sports* founded in 2006 and the *Journal of Sports Analytics* founded in 2014. In addition, meetings focusing on statistics in sports such as MathSport International and the New England Symposium on Statistics in Sports are regularly scheduled events. Also, opportunities exist to present research on statistics in sports at industry professional meetings such as the MIT Sloan Sports Analytics Conference and the SABR Analytics Conference.

Volumes such as *Statistical Thinking in Sports* (2007, ed. Albert and Koning) and *Anthology of Statistics in Sports* (2005, ed. Albert, Bennett, and Cochran) consist of general arrays of statistical articles but are not designed to provide the reader with a complete survey of the state-of-the-art methods in statistics in sports. The general aim of this handbook is to provide a basic reference for statistical researchers and students with an interest in sports applications to learn about the fundamental background, problems, and ongoing challenges in statistical methods in sports. The chapters in this book provide both overviews of statistical methods in sports and in-depth treatment of critical problems and challenges confronting statistical research in sports. This handbook intends to provide the reader the necessary background to conduct serious statistical analyses for sports applications and to appreciate scholarly work in this expanding area.

This handbook should be of interest for three types of readers. First, the handbook can serve as the basis for a graduate course or seminar in statistical methods in sports. Using the handbook in this fashion can take advantage of connections between the methods typically used in a sports context with the methods that graduate students are learning in theoretical coursework. Second, the handbook can serve as a reference for statistical practitioners in professional sports who may not be aware of the breadth of statistical issues and problems in their area, or who may simply want a refresher in the problem areas they are likely to encounter. Finally, the handbook can provide statistical researchers who are interested in delving more into sports applications the requisite background to produce sound scholarly work that is set in a proper context. The handbook is organized by major sport (baseball, American football, basketball, hockey, and soccer) followed by a section on other sports.

The four chapters on baseball provide a general description of measures of player performance and situational and streakiness effects. The chapter by Ben Baumer and Pamela Badian-Pessot describes the wide range of measures proposed for evaluating batters and base runners. Carson Sievert and Brian Mills discuss, in their chapter, traditional measures of pitching performance and explore the opportunities for further insight about pitching using pitch-level data from the PITCHf/x system. Measuring fielding performance has been one of the more challenging problems in baseball analysis. The chapter by Mitchel Lichtman describes a plethora of fielding measures that have been proposed and the use of modern technology systems such as Statcast and Fieldf/x that can construct improved defensive metrics. In sports, fans are fascinated with streaky and "clutch" performances of players and teams, and Jim Albert, in his chapter, describes the use of statistical models to detect and estimate the size of situational and streaky effects.

The topics of the American football chapters address the issues of evaluating college talent, measuring player and team abilities, and decision-making within a game. The section on football starts with a chapter by Mark Glickman and Hal Stern describing methods for estimating NFL team abilities based on game outcomes. This is followed by a chapter describing the methods of evaluating NFL quarterbacks and placekickers by Drew Pasteur and John David. The next chapter by Julian Wolson, Vittorio Addona, and Rob Schmicker is devoted to forecasting the success of NFL players based on college performance. The final chapter in this section by Keith Goldner presents a discussion of quantitative methods for making optimal strategic decisions within a football game.

The four chapters on basketball analytics cover a range of topics about player abilities and game outcomes. The chapter by James Lackritz describes the ongoing controversy surrounding streak shooting and methods for detecting the hot hand in NBA basketball. This is followed by a chapter by Jeremias Engelmann on the history and current usage of plus/minus methods for evaluating player contributions from possession-based data. Brian Skinner and Matthew Goldman present the basics of optimal strategy in basketball in their chapter, with a focus on optimizing when players should take shots, at what point in the shot clock a team should take a shot, and under what circumstances teams should try high-risk tactics. Finally, the basketball section is concluded by a chapter by Luke Bornn, Daniel Cervone, Alexander Franks, and Andrew Miller that explores the current state of the art of analyzing player tracking data to measure both offensive and defensive player abilities.

The next group of chapters concerns hockey analytics. A chapter by Andrew Thomas develops the structure for an NHL match simulator using Poisson/Exponential models. The approach is comprehensive in that it takes into account various game situations (e.g., penalties, score, etc.) that affect the play of the game. In the next chapter, Bobby Gramacy, Matt Taddy, and Sen Tian use regularized logistic regression to assess player performance. Their approach may be seen as a generalization and an improvement of traditional plus/minus methods. A chapter by Michael Schuckers surveys the statistics used to evaluate goaltending. Some of the statistics take into account the detailed aspects of goal-tending, including the type of shots and the location of shots. In the final hockey chapter, Peter Tingling looks at drafting with a specific focus on the nuances of the NHL draft.

Soccer is an example of a low-scoring sport. Compared to, say, baseball, relatively few performance measures are available, and for that reason, focus has been on models for outcomes and the effect of interventions on these outcomes. The chapter by Phil Scarf and Jose Rangel Sellitti discusses recent developments in the literature on score modeling, focusing on the dependence between the number of goals scored by the home team and the number scored by the away team. At a slightly more abstract level, information about scores results in information about team quality. Ruud Koning, in his chapter, discusses a range of models that have been proposed to measure team quality. At a more disaggregate level, the chapter by Ian McHale and Samuel Relton describes the analysis of individual player ratings using data that have become available only recently. The soccer section concludes with chapters on intervention issues in the quantitative analysis of soccer. The chapter by Martin van Tuijl discusses whether dismissing a coach midseason results in better performance. Another issue explored by the chapter by Rob Simmons is whether referees are biased or impartial judges of the game, a topic that is relevant to other sports as well.

The final section of the handbook contains three contributions related to other sports. The first chapter by Mark Broadie and Bill Hurley investigates the two major research areas in golf. They look at the use of detailed ShotLink data in golf analytics and the age-old problem of handicapping in golf. In the second chapter, Tim Swartz surveys statistical research in cricket, the second most popular sport in the world. One take-away from this chapter is that the game has been underexplored and that opportunities exist in cricket analytics. The final chapter by Elmer Sterken discusses the issue whether inequality in performance in Olympic sports between top athletes increases or decreases over time. As discussed in this chapter, a general observation is that the performance of athletes tends to improve, but once a sport has reached maturity, improvement can level off.

We thank the chapter writers for their important contributions and our many colleagues who have inspired us to create a handbook that we hope sets a standard for researchers and practitioners in statistics in sports.

Jim Albert
Mark E. Glickman
Tim B. Swartz
Ruud H. Koning

Contributors

Vittorio Addona
Department of Mathematics, Statistics, and
    Computer Science
Macalester College
Saint Paul, Minnesota

Jim Albert
Department of Mathematics and Statistics
Bowling Green State University
Bowling Green, Ohio

Pamela Badian-Pessot
School of Operations Research and
    Information Engineering
Cornell University
Ithaca, New York

Benjamin S. Baumer
Program in Statistical & Data Sciences
Smith College
Northampton, Massachusetts

Lucas M. Besters
Department of Economics
Tilburg University
Tilburg, the Netherlands

Luke Bornn
Department of Statistics and
    Actuarial Science
Simon Fraser University
Burnaby, British Columbia, Canada

Mark Broadie
Graduate School of Business
Columbia University
New York, New York

Babatunde Buraimo
Department of Economics, Finance, and
    Accounting
University of Liverpool Management
    School
Liverpool, United Kingdom

Daniel Cervone
Center for Data Science
New York University
New York, New York

John A. David
Department of Applied Mathematics
Virginia Military Institute
Lexington, Virginia

Jeremias Engelmann
Consultant
ESPN.com
Heidelberg, Germany

Alexander Franks
Department of Statistics
University of Washington
Seattle, Washington

Mark E. Glickman
Department of Statistics
Harvard University
Cambridge, Massachusetts

Matthew Goldman
Microsoft Research
Redmond, Washington

Keith Goldner
numberFire
New York, New York

Robert B. Gramacy
Department of Statistics
Virginia Tech
Blacksburg, Virginia

William J. Hurley
Department of Mathematics and Computer
    Science
Royal Military College of Canada
Kingston, Ontario, Canada

Ruud H. Koning
Department of Economics, Econometrics
   and Finance
University of Groningen
Groningen, the Netherlands

James Lackritz
MIS Department
College of Business Administration
San Diego State University
San Diego, California

Mitchel Lichtman
Baseball Analyst
Canandaigua, New York

Ian G. McHale
Centre for Sports Business
Salford Business School
University of Salford
Manchester, United Kingdom

Andrew Miller
Department of Computer Science
Harvard University
Cambridge, Massachusetts

Brian M. Mills
Department of Tourism, Recreation, and
   Sport Management
University of Florida
Gainesville, Florida

R. Drew Pasteur
Department of Mathematics and
   Computer Science
The College of Wooster
Wooster, Ohio

José Sellitti Rangel Jr.
Salford Business School
University of Salford
Manchester, United Kingdom

Samuel D. Relton
School of Mathematics
University of Manchester
Manchester, United Kingdom

Phil Scarf
Salford Business School
University of Salford
Manchester, United Kingdom

Michael E. Schuckers
Department of Mathematics,
   Computer Science, and Statistics
St. Lawrence University
Canton, New York

Robert Schmicker
Department of Biostatistics
University of Washington
Seattle, Washington

Dirk Semmelroth
Department of Management
University of Paderborn
Paderborn, Germany

Carson Sievert
Department of Statistics
Iowa State University
Ames, Iowa

Rob Simmons
Department of Economics
Lancaster University Management School
Lancaster, United Kingdom

Brian Skinner
Department of Physics
Massachusetts Institute of Technology
Cambridge, Massachusetts

Elmer Sterken
Institute of Economics, Econometrics,
   and Finance
University of Groningen
Groningen, the Netherlands

Hal S. Stern
Department of Statistics
University of California, Irvine
Irvine, California

Tim B. Swartz
Department of Statistics and
    Actuarial Science
Simon Fraser University
Burnaby, British Columbia, Canada

Matt Taddy
Microsoft Research New England
Cambridge, Massachusetts

and

Booth School of Business
University of Chicago
Chicago, Illinois

Andrew C. Thomas
Minnesota Wild
National Hockey League
Saint Paul, Minnesota

Sen Tian
Stern School of Business
New York University
New York, New York

Peter M. Tingling
Beedie School of Business
Simon Fraser University
Vancouver, British Columbia, Canada

Jan C. van Ours
Department of Applied Economics
Erasmus University Rotterdam
Rotterdam, the Netherlands

and

Department of Economics
University of Melbourne
Parkville, Australia

Martin A. van Tuijl
Department of Economics
Tilburg University
Tilburg, the Netherlands

Julian Wolfson
Division of Biostatistics
School of Public Health
University of Minnesota
Minneapolis, Minnesota

1

Evaluation of Batters and Base Runners

Benjamin S. Baumer and Pamela Badian-Pessot

CONTENTS

Since Henry Chadwick began publishing boxscores around the turn of the twentieth century (Schwarz, 2005a), people have been interested in evaluating the performance of baseball players. In particular, the offensive contributions made by position players—in the form of both batting and baserunning—are the most obviously varied and carefully studied contributions. In this chapter, we will catalog the most enduring sabermetric models for evaluating batters and base runners. Our approach is model centric, in that we will attempt to categorize metrics based on the type of model on which they are built. It should not be surprising that over time these models have become more sophisticated, both in terms of the model complexity and the rigor with which any parameters are estimated.

The fundamental challenge in evaluating batters and base runners is that run scoring in baseball is the result of interdependent actions among teammates and opponents. While separating the individual contributions of these team actions is considered easier in baseball than in many other sports, it is far from trivial. For example, consider an inning in which the first batter leads off with a walk, steals second base, advances to third base on a groundout, and then scores on a sacrifice fly. What is unambiguous is that the team scored one run. What is debatable is how much of that run is attributable to each player. Does the second batter make a positive contribution by advancing the runner, even though he made an out? How much credit did the first player accrue though baserunning? The tools developed in this chapter will help us answer these questions.

One common technique that will prime the search for models is

- To recognize that neither the runs scored statistic (R), which gives full credit for the team run to the player who crossed the plate, nor the runs batted in statistic (RBI), which gives full credit for the team run to the player who drove in the run, are reasonable ways to apportion the run
- To find a model for R that works well for teams
- To apply that model to individual players

Later in this chapter, we will show how this method can be used to evaluate the accuracy of offensive metrics.

In Section 1.1, we motivate this line of inquiry, outline some basic tools necessary to understand these models, and define our notation. In Section 1.2, we explore models that can be computed using data that are aggregated at the seasonal level. These are the simplest but most numerous and storied models for offensive performance. Models that require more detailed play-by-play data are explored in Section 1.3. Predictive models, including Bayesian models, are discussed in Section 1.4. We conclude the chapter in Section 1.5 by pointing toward some open problems.

1.1 Basic Tools

1.1.1 Overview

Several comprehensive assessments of offensive players have advanced the field of saber-metrics. Many have drawn inspiration from the work of Bill James interspersed in his books (James, 1986; James and Henzler, 2002). Perhaps the first book-length treatise was Cook (1964). This was followed 20 years later by Thorn and Palmer (1984), an important book that was reprinted in 2015 (Thorn and Palmer, 2015). The article-length analysis of Bennett and Flueck (1983) was greatly expanded by Albert and Bennett (2003) into what remains probably the best place to start reading about sabermetrics. The more recent book by Tango et al. (2007) not only focuses more on in-game strategy, but also includes some measures of offensive assessment. Methods for analyzing baseball data using the statistical computing environment R—as we do in this chapter—are explicated by Marchi and Albert (2013).

1.1.2 Expected Run Matrix

One of the fundamental tools in sabermetric analysis is the *expected run matrix*, which gives the expected number of runs scored in the remainder of an inning, given that the inning is currently in one of the 24 (*base, out*) states. Throughout this chapter, we will use the notation **R** to refer to this 8×3 matrix, and the notation **r** to refer to the corresponding vector of length 24. Specifically, the notation $\mathbf{R}_{23,2}$ indicates the value of the expected run matrix when runners are on second and third with two outs.* The complete expected run matrix for 2013 is presented in Table 1.1.

1.1.3 Notation

In this chapter, we denote the value of player i's batting and baserunning contributions as y_i. Typically, but not always, player value is measured in the units of runs. Since, as we discussed earlier, there is no way to *directly* measure the run value of these contributions for individual players, we consider y_i to be unknown. The goal of this chapter is to describe models for y_i in a coherent fashion. The estimates for y_i will be denoted by $\hat{y}_i$. Note that from the example at the beginning of this chapter, each of the three offensive players has a y_i—we just don't know what they are.

An example may make this clearer. It is an undisputed fact that the St. Louis Cardinals scored $y_{STL} = 798$ runs in 1987. This number represents the total batting and baserunning contributions of the entire team. However, we don't know how many of those 798 runs are attributable to each player. We know that Ozzie Smith scored 104 of those runs and drove in another 75, but neither of those numbers represents y_{Smith}. Furthermore, neither is

---

* There are several commonly used notations for referring to the configuration of base runners indicated by *base*. The most intuitive is a notation like 23 (or ×23), which indicates that there are runners on second and third. We will use this notation in the text of this chapter. However, this notation is very inconvenient computationally. In our computations, we use the following notation: imagine each of the three bases as a binary digit that can either be unoccupied (0) or occupied (1). Then the binary string 110 indicates that runners are on second and third. This has the decimal equivalent of 6. Thus, the notations x23, 110, and 6 all refer to having runners on second and third. We trust the reader will be able to keep this clear in context.

TABLE 1.1

The Expected Run Matrix for 2013

BaseCode\Outs	0	1	2
000	0.456	0.240	0.091
001	0.812	0.491	0.211
010	1.096	0.617	0.301
011	1.382	0.838	0.402
100	1.261	0.925	0.344
101	1.828	1.108	0.480
110	2.080	1.390	0.558
111	2.179	1.568	0.714

Source: MLBAM, GameDay files, http://gd2.mlb.com/components/game/mlb/, accessed July 1, 2016.

Note: The rows correspond to the configuration of the bases, while the columns correspond to the number of outs. The entry $(110, 2)$ implies that based on 2013 data, about 0.558 runs are expected to be scored in an inning in which there are two outs and runners on second and third.

TABLE 1.2

Some Basic Statistics for Eight Notable National League Players in 1987

Player	Team	PA	R	H	2B	3B	HR	RBI	BB	HBP	SB	CS
Ozzie Smith	SLN	706	104	182	40	4	0	75	89	1	43	9
Dale Murphy	ATL	693	115	167	27	1	44	105	115	7	16	6
Tony Gwynn	SDN	680	119	218	36	13	7	54	82	3	56	12
Andre Dawson	CHN	662	90	178	24	2	49	137	32	7	11	3
Darryl Strawberry	NYN	640	108	151	32	5	39	104	97	7	36	12
Tim Raines	MON	627	123	175	34	8	18	68	90	4	50	5
Eric Davis	CIN	562	120	139	23	4	37	100	84	1	50	6
Jack Clark	SLN	558	93	120	23	1	35	106	136	0	1	2

Note: Dawson won the MVP Award, with Smith and Clark placing next in the voting. Gwynn's season is considered the best by WAR, followed by Davis, Murphy, and Raines.

a particularly good *estimate* of y_{Smith}. We will see later in this chapter that our best estimate of y_{Smith} is closer to 98.

For seasonal metrics, we will follow a list of eight prominent National League players from the 1987 season to illuminate the metrics discussed herein. The list of players, along with their unambiguous counting statistics, is shown in Table 1.2. It is worth noting that Andre Dawson of the Chicago Cubs—largely by virtue of his NL-leading 49 home runs—won the NL Most Valuable Player (MVP) Award, becoming the first player from a last-place team to do so.*

---

* In hindsight, Tony Gwynn of the San Diego Padres produced the highest Wins Above Replacement (*rWAR*), according to Baseball-Reference.com. We do not discuss WAR in this chapter since it involves the evaluation of pitching and fielding.

1.2 Models Based on Seasonal Data

It should not be surprising that the most storied and widely used models for offensive performance in baseball are linear models. That is, models in which some number of discrete batting or baserunning events are associated with a specific value, and each player's value is quantified by summing these values over the number of these events attributed to that player.

Formally, let Ω be a set of commonly recorded offensive and baserunning events. Ω includes things like singles ($1B$), doubles ($2B$), triples ($3B$), and home runs (HR), as well as things like stolen bases (SBs) and caught stealings (CSs). To build a linear model, we might choose p of the events in Ω that seem important and tally the number of occurrences of these p events for player i over some fixed time period—say, a season. This becomes the vector $x_i \in \mathbb{N}^p$. Note that each element of x_i is a nonnegative integer, since it represents a count.

Next, we could assign a value to each of these p events. These values are most often in the units of *runs*, but they could be measured in any unit. Our coefficient vector becomes $\beta \in \mathbb{R}^p$, where here we allow negative values, since events like strikeouts (SO) and caught stealings are clearly detrimental to a team's offense.

Finally, we consider $y_i = x_i^T \beta$. The scalar value y_i is in the same units as the elements of β. If we have n players, then we can extend this framework to the matrix equation $y = \mathbf{X}\beta$, where $y \in \mathbb{R}^n$ and $\mathbf{X}$ is an $n \times p$ matrix with the ith row being x_i. That is, each row in $\mathbf{X}$ represents a player, and each column a variable. This expression is *linear*—in what follows we catalog the most significant models for offensive player value of this form. In each case, the primary differences in these models are

1. Which events to consider? That is, which elements of Ω make up the columns of $\mathbf{X}$ and the elements of β? We denote this choice by $\omega \subseteq \Omega$.

2. How to choose (or estimate) the elements of β? We will see that many early models simply asserted values for β or chose them based on natural elements of the game. More recently, sabermetricians have employed techniques for optimizing β according to some criteria.

The remaining differences are choices about units and about converting y to a *rate*. That is, as presented earlier y is a sum of weighted counts. As the playing time of players varies, it is often more useful to think about a player's value in terms of his value *per game*, or *per plate appearance*. In this case, we let $z_i \in \mathbb{N}^q$ be a vector of counts for q events $\psi \subseteq \Omega$ that record player i's playing time, and $\alpha \in \mathbb{R}^q$ a vector of corresponding weights. Then $z_i^T \alpha$ is a scalar that assesses player i's playing time, and y_i/z_i is a rate that measures player i's contributions on a per unit of playing time basis. We will reinforce this notation throughout this chapter.

1.2.1 Batting Models

We begin by considering models that incorporate batting statistics only.

1.2.1.1 The Triple-Slash Models

The so-called "triple-slash" statistics for batting are batting average (*AVG*), on-base percentage (*OBP*), and slugging percentage (*SLG*). These are the most commonly cited statistics in baseball and are ubiquitous on television broadcasts, the back of baseball cards, and web sites. We have also included on-base plus slugging (*OPS*) in this section, as it is the simple sum of *OBP* and *SLG*.

1.2.1.1.1 Batting Average

The most widely known offensive statistic is batting average, which dates back to the 1800s and Henry Chadwick (Schwarz, 2005a). Defined simply as hits over at-bats, the appeal of *AVG* is that it reflects the common sense perception that better offensive players will get hits in more of their opportunities.

$$AVG = \frac{H}{AB}$$

In our framework, we have the trivial decomposition where $\psi = \{AB\}$, $\omega = \{H\}$, and $\beta = 1$. Since, by definition, $H \leq AB$, the units here are a proportion.

While batting average remains popular, researchers have reached consensus that it is a relatively poor measure of offensive batting prowess, for both descriptive and predictive uses (Baumer, 2008; Albert, 2016) (see Table 1.11).

1.2.1.1.2 On-Base Percentage

While *AVG* only counts when a player reaches base by getting a hit, there are other ways of reaching base that also contribute to run scoring—most commonly through walks and hit by pitches. On-base percentage improves upon *AVG* in two ways: by including walks and hit by pitches in the numerator and by using plate appearances—excluding sacrifice hits—as a measure of opportunities in the denominator. *OBP* is essentially the rate a player reaches base, or—perhaps more importantly—the proportion of the time he doesn't make an out.

$$OBP = \frac{H + BB + HBP}{AB + BB + HBP + SF} = \frac{H + BB + HBP}{PA - SH},$$

where
  SF is sacrifice fly
  SH is sacrifice hit—more commonly known as sacrifice bunts

Here, we have $\psi = \{PA, SH\}$, $\omega = \{H, BB, HBP\}$, $\alpha = \begin{pmatrix} 1 & -1 \end{pmatrix}^T$, and $\beta = \begin{pmatrix} 1 & 1 & 1 \end{pmatrix}^T$.

1.2.1.1.3 Slugging Percentage

Slugging percentage is not a measure of the rate a player reaches base—it instead asks how far a player reaches, that is, how many bases he produces per at-bat.

$$SLG = \frac{TB}{AB} = \frac{1B + 2 \cdot 2B + 3 \cdot 3B + 4 \cdot HR}{AB}$$
$$= \frac{H + 2B + 2 \cdot 3B + 3 \cdot HR}{AB}$$
$$= AVG + \frac{2B + 2 \cdot 3B + 3 \cdot HR}{AB}$$

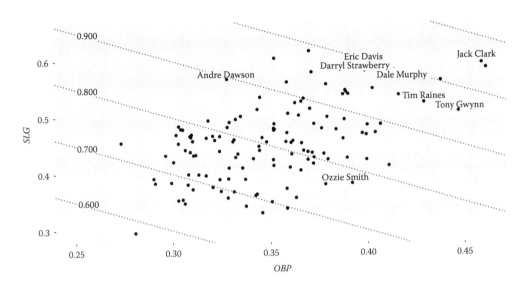

FIGURE 1.1
Scatterplot of *SLG* vs. *OBP* for all qualified (at least 502 plate appearances) NL batters in 1987. Dotted lines indicate *OPS* isometries.

In this manner, *SLG* differentiates between players who hit mostly singles and those who hit for power, unlike *AVG* and *OBP*. *SLG* weights hits by the number of total bases (*TBs*): one for singles, two for doubles, three for triples, and four for home runs. The term slugging percentage is itself a misnomer, in that it is the rate at which a player produces bases, not a percentage.

In our notation, $\omega = \{1B, 2B, 3B, HR\}$ and $\beta = \begin{pmatrix} 1 & 2 & 3 & 4 \end{pmatrix}^T$. The latter term in the final equation is known as *isolated power* (*ISO*) and accordingly measures a player's extra base power net of his batting average.

1.2.1.1.4 On-Base Plus Slugging

OBP improves upon *AVG* by including ways to reach base other than hits and *SLG* improves upon *AVG* by distinguishing between types of hits. On-base plus slugging (*OPS*) incorporates both of these ideas by simply adding these values together. Players with the highest *OPS* both reach base often and hit for power. While these skills are correlated, they are not the same. Figure 1.1 illustrates the relationship between *OBP* and *SLG* among qualified* NL players in 1987.

$$OPS = OBP + SLG$$

The triple-slash metrics for our eight notable players are shown in Table 1.3. We note that Gwynn's exceptionally high batting average does not carry the day according to *OPS*. Moreover, Dawson's low *OBP* and Smith's low *SLG* reveal the former's inability to draw walks and the latter's lack of power.

---

* To qualify for the batting title, a player needs 3.1 plate appearances per team game. For a full 162 game season, this translates to at least 502 plate appearances.

TABLE 1.3

Triple-Slash Statistics for Seven Notable National League Players in 1987

Player	Team	AVG	OBP	SLG	OPS
Jack Clark	SLN	0.286	0.459	0.597	1.055
Dale Murphy	ATL	0.295	0.417	0.580	0.997
Eric Davis	CIN	0.293	0.399	0.593	0.991
Darryl Strawberry	NYN	0.284	0.398	0.583	0.981
Tony Gwynn	SDN	0.370	0.447	0.511	0.958
Tim Raines	MON	0.330	0.429	0.526	0.955
Andre Dawson	CHN	0.287	0.328	0.568	0.896
Ozzie Smith	SLN	0.303	0.392	0.383	0.775

Since *OBP* is measured as a proportion, whereas *SLG* is an average, their units are incomparable, leading many to question whether the simple sum of the two measures is best. That is, if we generalize *OPS* and consider all models of the form

$$\hat{y}_i = a \cdot OBP_i + b \cdot SLG_i,$$

for constants a and b, what are the optimal values of a and b? In *Moneyball*, it is reported that the proper ratio a/b is closer to 3, as opposed to the ratio of 1 posited by *OPS* (Lewis, 2003). Wang (2006) estimates the value to be approximately 1.8 using trial and error—we confirmed this using data for all 1162 team seasons from 1960 to 2005 and estimating the optimal ratio using linear regression. The R^2 for the fit is 0.929, and the ratio of the coefficients is 1.781. While there is variability associated with this estimate, the conclusion of Wang (2006) seems to be in the right ballpark. That is, a modified *OPS* that weights *OBP* more heavily produces estimates that are "closer" to team runs scored than *OPS*.

1.2.1.2 Averaging Changes to the Expected Run Matrix

SLG posits the weights $\beta = (1,2,3,4)$ as values for singles, doubles, triples, and home runs, respectively, without considering how many runs, on average, a hit is actually worth. That is, is a double really twice as valuable as a single? How many runs is a double worth, on average? Using data from the 1959 to 1960 season, Lindsey (1963) estimated this figure to be 0.817. His idea was to tabulate the change in run expectancy for each situation in which a double could be hit and then take the weighted average according to how often those situations occur. Mathematically, the average value of a double β_{double} is given by

$$\beta_{double} = \sum_{(base,out)} \delta_{double}|(base,out) \cdot freq(base,out),$$

where
δ_{double} is the change in run expectancy for a double
$freq(base,out)$ is the likelihood of being in each $(base,out)$ state

More generally, the change in run expectancy $\delta_{f,i}$ from state i to state f is

$$\delta_{f,i} = \mathbf{R}_f + r_i - \mathbf{R}_i, \tag{1.1}$$

where
  $\mathbf{R}_i$ and $\mathbf{R}_f$ are the expected runs scored in the remainder of the inning before and after the play, respectively
  r_i is the number of runs scored as a result of the play

This value $\delta_{f,i}$ arises in other contexts later in this chapter.

For example, Lindsey found an expected run value of $\mathbf{R}_{12,2} = 0.403$ runs when a batter hits with two outs and men on first and second, but that he will bat in this situation only 3.3% of the time. If this batter hits a double, Lindsey supposed that half the time both runners will score and the other half one runner will score and the other will be left on third. The run value of a double in this situation is then

$$\hat{\delta}_{double}|(12, 2) = \frac{1}{2}(\mathbf{R}_{23,2} + 1) + \frac{1}{2}(\mathbf{R}_{2,2} + 2) - \mathbf{R}_{12,2}$$

$$= \frac{1}{2}(0.687 + 1) + \frac{1}{2}(0.297 + 2) - 0.403$$

$$= 1.589.$$

Thus, with two men on base and two outs, the value of hitting a double is quite high—nearly 1.6 runs. However, this situation is relatively rare, occurring only 3.3% of the time. Much more likely is that the batter will come up with no one on and no one out—24.2% of plate appearances occur in this configuration. The value of a double here is simply

$$\hat{\delta}_{double}|(0, 0) = (\mathbf{R}_{2,0} + 0) - \mathbf{R}_{0,0} = 1.194 - 0.461 = 0.733.$$

This is less than half of the value of the previous example, illustrating how the value of different events can change based on the situation. The *average* value of hitting a double $\hat{\beta}$ is the sum of the values of a double in all initial possible states, weighted by the frequency of each state.

Ultimately, Lindsey found the average value of a single to be 0.41 runs, and doubles, triples, and home runs to be worth 0.82, 1.06, and 1.42 runs, respectively, on average. With an eye toward measuring the value of hits relative to the average value of a single, Lindsey proposed altering the weights used for *SLG* from $1, 2, 3, 4$ to $1, 2, 2.5, 3.5$. Although this proposal was not adopted, it marks the beginning of using the expected run matrix to methodically determine optimal weights for different offensive plays. This idea has informed much of the subsequent analysis in all aspects of the game, as well as other sports, and is a truly enduring contribution.

Using the expected run matrix for the 2013 season shown in Table 1.1, we can use Lindsey's methodology to estimate the run values of any event. We show the results in Table 1.4. There are a number of interesting observations one can make based on these data—most of which were identified by Lindsey. These observations have informed sabermetric orthodoxy for decades. Among the most obvious are the following:

TABLE 1.4

Estimated Run Values for the 22 Most Common Events, 2013

Event	N	*Freq*	$\bar{r}$	$\hat{\beta}$
Home run	4,661	0.025	1.54	1.37
Triple	772	0.004	0.62	1.02
Double	8,185	0.044	0.40	0.75
Field error	1,516	0.008	0.20	0.49
Single	28,448	0.154	0.22	0.44
Hit by pitch	1,536	0.008	0.01	0.31
Walk	13,622	0.074	0.02	0.30
Intent walk	1,018	0.005	0.01	0.18
Sac fly	1,204	0.006	1.01	−0.00
Sac bunt	1,382	0.007	0.04	−0.10
Groundout	35,171	0.190	0.02	−0.20
Flyout	23,080	0.125	0.00	−0.23
Lineout	9,493	0.051	0.00	−0.23
Strikeout	36,573	0.197	0.00	−0.25
Popout	8,877	0.048	0.00	−0.26
Forceout	3,946	0.021	0.07	−0.31
Grounded into DP	3,731	0.020	0.03	−0.75

Source: MLBAM, GameDay files, http://gd2.mlb.com/components/game/mlb/, accessed July 1, 2016.

Note: The rightmost column shows the average change in run expectancy, which is a measure of the value of the play ($\hat{\beta}$), in runs. The column second from right ($\bar{r}$) shows the average number of runs that score on plays of each type.

- Walks are worth less than singles.
- A walk and a hit by pitch have the same run value, since they have the same result. However, an intentional walk is less valuable even though it too has the same result, because by design it is only employed in situations where adding the runner on first does not contribute as much to the run expectancy.
- A fielding error is worth about the same as a single, since it usually means that the batter reached first and the other runners moved up one or two bases.
- A sacrifice fly necessarily results in one run being scored, but because it also results in an out being recorded, it has a negligible effect in terms of the change in run expectancy.
- A sacrifice bunt is a marginally negative event for the offense.
- All other types of outs (e.g., groundouts, flyouts, popouts, lineouts, and strikeouts) are of approximately equal negative value.

1.2.1.2.1 Batting Runs (BRs)

Batting runs (Thorn and Palmer, 1984) use the same basic framework as Lindsey (1963) and consequently find nearly the same average run values for singles, doubles, triples, and home runs. However, Palmer expands on Lindsey's work by also including walks, hit by pitches, and the negative run value of an out. By Palmer's estimation, the value of an out changes yearly to account for changes in the league's run-scoring environment.

TABLE 1.5

Aggregated Linear Batting Models for Selected 1987 NL Players

Player	PA	OPS	BR	wOBA
Jack Clark	558	1.055	58.5	0.447
Eric Davis	562	0.991	43.9	0.421
Dale Murphy	693	0.997	57.6	0.417
Darryl Strawberry	640	0.981	48.6	0.415
Tony Gwynn	680	0.958	52.8	0.408
Tim Raines	627	0.955	46.9	0.405
Andre Dawson	662	0.896	26.9	0.382
Ozzie Smith	706	0.775	14.3	0.347

Note: We note that Clark leads by all three metrics.

BR measures the number of runs contributed beyond what an average player would have contributed while consuming the same number of outs. Accordingly, a league average player will have a BR of 0. The formula is

$$BR = 0.47 \cdot 1B + 0.85 \cdot 2B + 1.02 \cdot 3B + 1.40 \cdot HR + 0.33 \cdot (BB + HBP) - ABF \cdot (AB - H),$$

where ABF is calculated so that BR for the league is zero. A typical value of ABF is approximately 0.3. We show BR for selected players in Table 1.5.

1.2.1.2.2 Weighted On-Base Average (wOBA)

$wOBA$ is another metric that uses changes to the expected run matrix to determine the value of different events (Tango et al., 2007). However, $wOBA$ is designed so that the league average $wOBA$ is equal to the league average OBP, putting the former numbers on the familiar scale of the latter. It is necessary to inflate the run values by about 15% to make this happen. All weights are recalculated annually to adjust for changes to the run-scoring environment. The following equation defines $wOBA$ for the 2013 season:

$$wOBA = \frac{0.690 \cdot uBB + 0.722 \cdot HBP + 0.888 \cdot 1B + 1.271 \cdot 2B + 1.616 \cdot 3B + 2.101 \cdot HR}{AB + BB - IBB + SF + HBP}$$

$wOBA$ is presented as an improvement over OPS, as it incorporates both elements of reaching base and the value of the event on which the player reaches. Additionally, because $wOBA$ uses the same scale as OBP, many baseball fans already have a reference for what values are poor, good, or excellent. Weighted Runs Above Average ($wRAA$)—which is used for the hitting component of Fangraphs' WAR ($fWAR$)—uses $wOBA$ to calculate the number of runs a player contributes compared to an average player in the same number of plate appearances. In Figure 1.2, we illustrate how $wOBA$ compares to OBP for all NL players in 1987. Note that the distribution of $wOBA$ is similar to that of OBP.

In Table 1.5, we compare OPS, Batting Runs, and $wOBA$ for our selected 1987 NL players. While there is widespread agreement, there are also discrepancies. Note the nearly 14-run discrepancy between Eric Davis and Dale Murphy in Batting Runs, despite their being nearly identical in OPS and $wOBA$. This is the result of Batting Runs being a counting stat, while the others are rate stats.

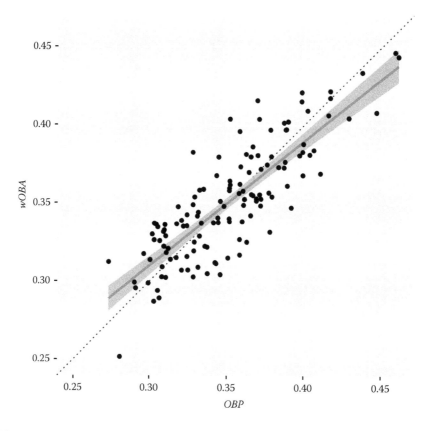

FIGURE 1.2
Comparison of *wOBA* to *OBP*. A regression line with error ranges has been added. Since *OBP* and *wOBA* are on the same scale, the dotted diagonal line allows a direct comparison of the values.

1.2.2 Baserunning Models

In this section, we discuss measures for baserunning alone. While thus far all the models covered use only batting statistics, a player's total offensive performance is clearly a combination of his hitting and baserunning skills. Both metrics covered in this section are offshoots of metrics from the previous section.

1.2.2.1 Basestealing Runs

Basestealing Runs are the basestealing counterpart of *batting runs*. Palmer used the same methods as in *BR* to find the average run value of stolen bases and caught stealing.

$$Basestealing\ Runs = 0.22 \cdot SB - 0.38 \cdot CS$$

1.2.2.2 Weighted Stolen Base Runs

wSB is the baserunning counterpart of *wOBA*. The values of stolen bases and caught stealings are found in the same way as *wOBA* and then compared to the league average. *wSB* measures runs created (RC) above (or below) the league average in the same number of opportunities. This statistic is used as part of the baserunning component of Fangraphs' *fWAR* (Fangraphs Staff, 2015b).

$$wSB = \beta_{SB} \cdot SB + \beta_{CS} \cdot CS - lgwSB \cdot (1B + BB + HBP - IBB),$$

where
$\beta_{SB} = 0.2$
$\beta_{CS} = -(2 \cdot RunsPerOut + 0.075)$, and *RunsPerOut* is the total number of runs scored by all teams divided by the outs in a season

The constant

$$lgwSB = \frac{\beta_{SB} \cdot SB + \beta_{CS} \cdot CS}{1B + BB + HBP - IBB}$$

provides a scaling factor. Its denominator is an approximation of the number of times that the player was on first base with second base potentially open. Thus, *wSB* measures the weighted run value of an individual player's contributions on the basepaths, above what a league average runner would have contributed the same number of times on first.

Clearly, there are shortcomings in *wSB* in that it only measures stolen base contributions and ignores contributions from "taking the extra base" (i.e., advancing from first to third on a single). These unmeasured contributions are known to be significant (Click, 2005; Fox, 2005). Second, the approximation for the number of times on first base is not a great measure of opportunities, since it ignores things like reaching first on a fielder's choice.

Table 1.6 displays baserunning metrics for our selected 1987 NL players.

1.2.3 Combined Models

We now turn our attention to measures that combine both batting and baserunning.

TABLE 1.6

Basestealing Runs for Selected 1987 NL Players

Player	PA	SB	CS	SB%	Base Runs	wSB
Tim Raines	627	50	5	0.91	9.10	7.57
Eric Davis	562	50	6	0.89	8.72	7.19
Tony Gwynn	680	56	12	0.82	7.76	5.72
Ozzie Smith	706	43	9	0.83	6.04	4.39
Darryl Strawberry	640	36	12	0.75	3.36	1.82
Andre Dawson	662	11	3	0.79	1.28	0.70
Dale Murphy	693	16	6	0.73	1.24	0.33
Jack Clark	558	1	2	0.33	−0.54	−0.96

Note: We note that the plodding Clark accrued a negative value by being caught stealing twice against only one successful stolen base.

1.2.3.1 Total Average

Total Average (*TA*) is most similar to the intuitive statistics covered at the beginning of this chapter. Indeed, Boswell (1982) introduced it as a revision to slugging percentage. As opposed to *SLG*—which is the rate of total bases per at-bat—*TA* is the total number of bases attributed to a player, either through hitting, walking, or baserunning, divided by total outs. The change from at-bats to outs in the denominator is important because unlike at-bats, outs are generally constant at 27 per game. This suggests that bases per out should be more closely related to runs per game than bases per at-bat.

$$TA = \frac{TB + BB + HBP + SB}{AB - H + CS + GIDP}$$

$$= \frac{1B + 2 \cdot 2B + 3 \cdot 3B + 4 \cdot HR + BB + HBP + SB}{AB - H + CS + GIDP}$$

1.2.3.2 Offensive Performance Average

Offensive Performance Average uses the weights computed by Lindsey (1963) for hits while additionally including walks, hit by pitches, and stolen bases (Pankin, 1978). Recall that the weights are relative to the value of single, that is, a stolen base is about half as valuable as single.

$$OPA = \frac{1B + 2 \cdot 2B + 2.5 \cdot 3B + 3.5 \cdot HR + 0.8 \cdot (BB + HBP) + 0.5 \cdot SB}{AB + BB + HBP}$$

1.2.3.3 Estimated Runs Produced

A response to Bill James' runs created (see Section 1.2.5), Johnson's *Estimated Runs Produced* (*ERP*) combines an intuitive model with weights found through trial and error to predict runs scored (Johnson and James, 1985). *ERP* aims to track the positive contributions of a player using *TB, BB, HBP, H, SB* and the negative contributions, that is, the number of outs made (*AB* − *H, CS, GIDP*) by a player. The weights were determined with the value of each play in mind but were estimated through trial and error.

$$ERP = (2 \cdot (TB + BB + HBP) + H + SB - (0.605 \cdot (AB + CS + GIDP - H))) \cdot 0.16$$

After a bit of algebra,

$$ERP = 0.48 \cdot 1B + 0.8 \cdot 2B + 1.12 \cdot 3B + 1.44 \cdot HR + 0.32 \cdot BB + 0.32 \cdot HBP$$
$$+ 0.16 \cdot SB - 0.0968 \cdot (AB - H) - 0.0968 \cdot CS - 0.0968 \cdot GIDP.$$

1.2.3.4 eXtrapolated Runs

Finally, in eXtrapolated Runs (*XR*), Furtado (1999) used regression analysis to determine the best-fit coefficients for a series of batting and baserunning events.* The unit of *XR* is runs.

---

* Furtado also created two simplified versions of *XR*, eXtrapolated Runs Reduced (*XRR*) and eXtrapolated Runs Basic (*XRB*), which simply omit terms that might be difficult to find in seasonal data.

$$XR = 0.50 \cdot 1B + 0.72 \cdot 2B + 1.04 \cdot 3B + 1.44 \cdot HR + 0.34 \cdot (HBP + TBB - IBB))$$

$$+ 0.25 \cdot IBB + 0.18 \cdot SB - 0.32 \cdot CS - 0.090 \cdot (AB - H - K) - 0.098 \cdot K$$

$$- 0.37 \cdot GIDP + 0.37 \cdot SF + 0.04 \cdot SH$$

XR represents one of the first attempts to *fit* a model to the data and estimate the coefficient vector β using formal statistical methods. It is not hard to approximate the weights specified in XR using regression. The coefficients we obtain are

Event	$\hat{\beta}$ for 1955–1997	$\hat{\beta}$ for 1969–2008
1B	0.5230	0.5327
2B	0.6617	0.6805
3B	1.0180	0.9993
HR	1.4756	1.4419
SB	0.1225	0.1358
CS	−0.2119	−0.1867
BIPO	−0.0967	−0.0998
SO	−0.0827	−0.0934
BB	0.3380	0.3425
IBB	−0.1491	−0.1854
GIDP	−0.3928	−0.3163
SF	0.6301	0.6438
SH	0.0630	−0.0141

It is important to understand that these coefficients provide a best fit only during the time intervals over which they were estimated, which in the case of XR was from 1955 to 1997. If instead, we fit the coefficients over the time period from 1969 (when the pitching mound was lowered and expansion occurred) to 2008 (when improved drug testing was implemented), we can see that most of the coefficients change only slightly. One notable exception is the coefficient for sacrifice bunts (SH), which flips its sign from positive to negative.

Since these coefficients can be interpreted as run values associated with each of these events, one can see evidence for many sabermetric insights—similar to those discussed earlier in the work of Lindsey—therein. To name a few, the small, and perhaps even negative coefficient for SH suggests that there is little value to sacrifice bunts, and that perhaps they are detrimental to offense. The similar coefficients for strikeouts (SO) and ball in play outs ($BIPO$s) suggest that there may be little difference in their value. The substantial negative coefficient on intentional walks (IBB) confirms that intentional walks are not nearly as helpful to the offense as a walk or hit by pitch. The ratio of the values on SB and CS provides an approximation for the "break-even" stolen base percentage. In XR, one needs $0.32/0.18 = 1.78$ stolen bases for every caught stealing just to positively impact run scoring, implying an effective stolen base percentage of 64%. Thus, runners who are stealing bases at less than a 64% success rate are actually costing their team runs.

One notable difference between the coefficients in XR and the output from the regression model discussed earlier is the weight on sacrifice flies (SFs). Here, collinearity is at work, since every sacrifice fly results in exactly one run being scored. Thus, Furtado has (somehow) adjusted the value downward. XR is often converted into a rate statistic by

TABLE 1.7

Summary of Combined Offensive Models for Selected 1987 NL Players

Player	PA	TA	OPA	ERP	XR	XR27
Jack Clark	558	1.265	0.615	113.3	114.3	10.09
Eric Davis	562	1.199	0.632	113.8	114.4	8.90
Tim Raines	627	1.146	0.587	119.6	118.7	8.68
Tony Gwynn	680	1.116	0.574	129.0	126.4	8.62
Dale Murphy	693	1.120	0.598	133.0	132.0	8.56
Darryl Strawberry	640	1.134	0.612	124.0	123.5	8.40
Andre Dawson	662	0.874	0.552	111.1	109.6	6.42
Ozzie Smith	706	0.833	0.466	96.2	97.9	6.06

Note: While Clark and Davis rank highly in terms of the rate metrics *TA*, *OPA*, and *XR27*, they did not play as often, leading to lower totals for the count metrics *ERP* and *XR*. Clark's *XR27* of 10.09 suggests that if Clark were to successively bat until he recorded 27 outs, his team would score 10.09 runs, on average.

dividing the total number of eXtrapolated Runs produced by the number of outs made and multiplying the result by 27 (the number of outs in a nine-inning game). This statistic, known as *XR27*, measures the expected number of runs the player would produce in a normal game if the lineup contained nine identical batters. We will return to this idea in the following section.

In Table 1.7, we summarize these combined offensive metrics for our 1987 NL players.

1.2.4 Comparison of Linear Models

Because of the linearity of the models discussed earlier, we can sensibly compare their coefficients. In Table 1.8, we summarize the value of each event considered by each model relative to a single. That is, for each event j, we report β_j / β_{single}. The similarities in these relative weights overwhelm the differences. It is easy to see, however, that home runs are overvalued by *SLG* relative to *wOBA* or *XR*, just as they are dramatically undervalued by *AVG* and *OBP*.

It should not be surprising then, that these metrics are highly correlated. In Table 1.9, we show the pairwise correlation matrix between these metrics for nearly 100,000 qualified player-seasons spanning 1871–2014.

The question of accuracy in matching team runs is mostly moot, since the regression-based results shown earlier provide the best fit among all possible linear models with least-squares criteria. Nevertheless, we discuss the accuracy of these metrics (and others) in Section 12.6.

1.2.5 Multiplicative Models

The linear models in the previous section were all predicated on the idea that each event has a fixed, average value, and that any accrual of those events would have a proportional impact upon scoring. Usually, this assumption is quite reasonable. However, it has been observed that the average run value of certain events will change based on the run-scoring environment. Moreover, one event might be worth a different amount depending on the value of another event. For example, consider the value of a walk. In the high run-scoring

TABLE 1.8

Comparison of Offensive Coefficients from Linear Models, Relative to the Value of a Single, that is, β_j/β_{single}

	BAVG	OBP	SLG	BR	WOBA	TAVG	OPA	ERP	XR
1B	1	1	1	1.00	1.000	1	1.0	1.0000	1.000
2B	1	1	2	1.81	1.431	2	2.0	1.6667	1.440
3B	1	1	3	2.17	1.820	3	2.5	2.3333	2.080
HR	1	1	4	2.98	2.366	4	3.5	3.0000	2.880
uBB	0	1	0	0.70	0.777	1	0.8	0.6667	0.680
IBB	0	1	0	0.70	0.000	1	0.8	0.6667	0.500
HBP	0	1	0	0.70	0.813	1	0.8	0.6667	0.680
SO	0	0	0	−0.64	0.000	0	0.0	−0.2017	−0.196
BIPOO	0	0	0	−0.64	0.000	0	0.0	−0.2017	−0.180
GIDP	0	0	0	−0.64	0.000	0	0.0	−0.4033	−0.920
SF	0	0	0	0.00	0.000	0	0.0	0.0000	0.740
SH	0	0	0	0.00	0.000	0	0.0	0.0000	0.080
SB	0	0	0	0.00	0.000	1	0.5	0.0000	0.360
CS	0	0	0	0.00	0.000	0	0.0	−0.2017	−0.640

TABLE 1.9

Correlation Matrix for Linear Models, Based on All Qualified Batters from 1871 to 2014

	BAVG	OBP	SLG	WOBA	BR	TAVG	OPA	ERP	XR
BAVG	1.00								
OBP	0.71	1.00							
SLG	0.59	0.65	1.00						
WOBA	0.70	0.87	0.94	1.00					
BR	0.70	0.89	0.92	1.00	1.00				
TAVG	0.62	0.87	0.89	0.97	0.97	1.00			
OPA	0.62	0.78	0.96	0.97	0.97	0.97	1.00		
ERP	0.67	0.83	0.96	0.99	0.99	0.96	0.98	1.00	
XR	0.68	0.84	0.94	0.98	0.98	0.98	0.98	0.99	1.00

environments of the late 1990s and early 2000s, it may have been the case that walks were quite valuable relative to their value in the low run-scoring environment in the early 2010s.* In the former era, home runs were comparatively likely. Thus, simply getting on base meant that your chances of scoring were much higher than in the current era, where a runner on first is more likely never to come around and score due to the lack of extra base power.

In linear models, the two separate skills of getting on-base (i.e., *OBP*) and moving runners around the bases (i.e., *SLG*) are not allowed to interact. That is, the benefit to increasing

---

* Running the regressions alluded to the aforementioned results in a run value of 0.35 for walks during the period 1995–2005, but a value of 0.31 for the period from 2006 to 2014.

slugging is the same whether on-base percentage is high or low. In the previous section, we considered the formula for a generic *OPS*:

$$OPS = a \cdot OBP + b \cdot SLG$$

Here, additional slugging power will result in b being contributed, but this is essentially independent of *OBP*.* Conversely, for the models discussed in the following text, additional slugging power could contribute more or less to overall offensive output based on the value of *OBP*. These models share a multiplicative functional form.

1.2.5.1 Runs Created

Bill James recognized linearity as a constraint and sought a metric that captured the interplay between getting on-base and moving runners around the bases (James, 1986). The basic notion of *Runs Created* (RC) was that slugging percentage was more valuable when *OBP* was also high, since there would be more runners on base to move around the bases with extra base hits. Thus, he modeled offensive production using a multiplicative model:

$$RC = \frac{(H + BB) \cdot TB}{AB + BB} = \frac{H + BB}{AB + BB} \cdot \frac{TB}{AB} \cdot AB \approx OBP \cdot SLG \cdot AB$$

Here, the contribution of *OBP* is nonlinear, since it depends on the value of *SLG*. Not so obviously, the value of *RC* is on the scale of runs, as we show in Figure 1.3.

1.2.5.2 Batter's Run Average

An even simpler multiplicative model is *batter's run average* (BRA), which was posited by Cramer and Palmer (1974). In its simplest version:

$$BRA = OBP \cdot SLG.$$

Subsequent versions incorporated stolen bases by expanding the formula to

$$BRA = \frac{H + BB + HBP + (1/2)(SB - CS)}{AB + BB + HBP} \cdot SLG$$

$$= \frac{H + BB + HBP + (1/2)(SB - CS)}{AB + BB + HBP} \cdot \frac{TB}{AB}.$$

While subsequent "improvements" to runs created have resulted in more complicated formulas that hew closer to the truth, the sabermetric value of runs created is the simplicity and insight of its original formulation. It is not surprising that linear weights estimators and other metrics that have been fit to data are more accurate than runs created. On the other hand, what is surprising is that runs created works as well as it does (see Figure 1.3). Specifically, the genius of creating a nonlinear model of the run creation process as an interaction between getting on-base and moving runners around is the lasting contribution of this line of inquiry.

---

* Of course, *SLG* and *OBP* are functionally dependent, since they both depend, they include hits, and thus the qualifier "essentially."

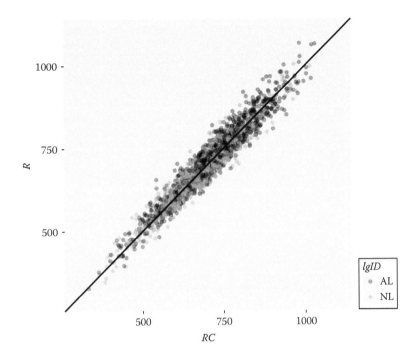

FIGURE 1.3
Scatterplot between runs created (*RC*) and runs scored (*R*) among all teams since 1915. The correlation is above 0.96.

1.2.5.3 Earnshaw Cook's Scoring Index

In his opus, Cook (1964) defined scoring index (*DX*) as

$$DX = SO \cdot \frac{H + BB + E + HBP - 2 \cdot SH - XBH}{PA} \cdot \frac{TB}{PA}.$$

The middle term in this equation is basically the "probability of reaching first base," and thus Cook's formula is very close in spirit to James' runs created, while preceding it by several years. However, unfortunately for Cook, Scoring Index was not adopted with the fanfare of Run Created.

In Table 1.10, we summarize the value of nonlinear models for our selected 1987 players.

1.2.6 Accuracy of Run Estimators

In Table 1.11 and Figure 1.4, we summarize the accuracy of the metrics discussed thus far on historical baseball data. As noted previously, it is not surprising that *XR* has the lowest root-mean-square error (*RMSE*), since it was constructed by mathematically optimizing for that criterion. It is more remarkable that simple, intuitive metrics such as *OPS* and *RC* perform so well.

The quest for a "best" model is quixotic. Nevertheless, this analysis confirms that eXtrap-olated Runs (*XR*) is a good general purpose run estimator. The concept (i.e., linear weights) is simple, the units (runs) are intuitive and the performance is excellent. The downside is

TABLE 1.10

Comparison of Nonlinear Models for Select 1987 NL Players

Player	PA	RC	BRA	DX
Dale Murphy	693	135.8	0.242	20.16
Tony Gwynn	680	134.6	0.228	5.54
Darryl Strawberry	640	122.2	0.232	16.53
Tim Raines	627	119.2	0.226	7.71
Jack Clark	558	115.3	0.274	21.99
Andre Dawson	662	113.5	0.186	11.78
Eric Davis	562	112.3	0.236	19.07
Ozzie Smith	706	90.5	0.150	3.39

TABLE 1.11

Summary of Accuracy of Various Run Estimators, 1954–1999

Metric	RMSE	R^2
XR_G	0.1339	0.9434
ERP_G	0.1408	0.9374
BRA	0.1566	0.9226
RC_G	0.1587	0.9205
wOBA	0.1592	0.9200
OPS	0.1597	0.9195
TA	0.1597	0.9195
OPA	0.1727	0.9059
SLG	0.2177	0.8505
OBP	0.2531	0.7980
AVG	0.3171	0.6829
BR_G	0.3760	0.5539
DX_G	0.4272	0.4243

Note: The *RMSE* is computed as the root mean squared error of the simple linear regression model between team runs per game and each metric. (Compare with page 230 of Albert and Bennett (2003).) R^2 is the coefficient of determination (i.e., the percentage of the variability in runs explained by the model) for the same regression model.

that the formula is long, difficult to remember, and painstaking to type in. If one is using a computer, then *XR* is hard to beat. Conversely, for back-of-the-envelope calculations, *BRA* and *OPS* can be quicker (given that *OBP* and *SLG* are available). *RC* is of similar complexity but has the advantage of intuitively meaningful units (runs). It is harder to see the practical virtues of *ERP*, *wOBA*,* and *OPA*.

---

* Proponents of *wOBA* often tout the scale of *OBP* as intuitively meaningful, but how is that more intuitive than runs per game?

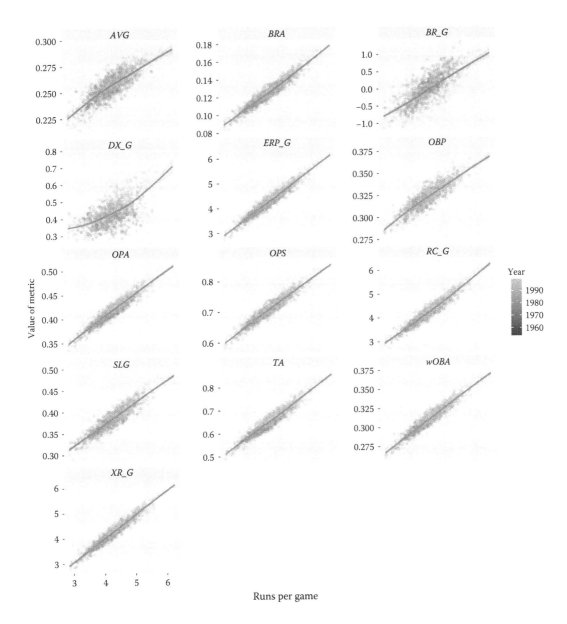

FIGURE 1.4
Comparison of accuracy of run estimators, 1954–1999. Note the tight fit for eXtraploted Runs (*XR*) and the relatively looser fit for Scoring Index (*DX*).

1.3 Models Based on Play-by-Play Data

The models in the previous section are computable using preaggregated data. That is, one only needs to know the counts of various events at the seasonal level to compute the metric. We now turn to models that use play-by-play data. These models are often situational or context-dependent. In this section, we will use examples from the 2013 major league season.

1.3.1 Markov Chain Models

A Markov chain is a *memoryless* stochastic process. That is, the behavior of the system in its present state does not depend on its previous state. Baseball is often assumed to have this property. For example, if a runner is on second and there is one out, it does not matter if the first batter hit a double and the next struck out, or if the first walked and moved up on ground out. Certainly, the run expectancy is the same in either case. Whether the psychological state of the pitcher is the same is debatable, but this simplifying assumption opens up a whole field of useful mathematical models (Trueman, 1977) (for a nice exposition, see D'Angelo (2010)).

Because baseball is naturally discrete and has 24 clearly defined (*base, out*) states, a Markov chain is a natural model to apply. These models can be useful in understanding the long-term probabilities associated with a fairly complex interdependent system.

Consider the 24 (*base, out*) states of an inning, and let's add a 25th state corresponding to having 3 outs. We denote the *state vector* of length 25 by **s**. Since every inning always begins in the state (0,0), the initial state vector $\mathbf{s}_0$ has a 1 as its first entry, followed by 24 zeros.

Next, consider the probability of moving from any one of the 25 initial states to any other of the 25 states. Clearly, some of these probabilities will be zero, since you can't go from two outs and nobody on to one out and a man on first. But many of these probabilities will be nonzero, and moreover they will be normalized so that the sum of the probabilities from any initial state is one. For example, in Table 1.12 we show the transition probabilities from the (0,0) state in which all innings begin. Only five outcomes are possible: either there is a runner on exactly one of the three bases, or the bases are empty and there is one out (if the first batter made an out) or no outs (if the first batter hit a home run). By far, the most likely outcome is that the first batter made an out (about 68% of the time), and the state is now (0,1).

Extending this idea, let the transition matrix **T** be the 25×25 matrix containing the probabilities of the inning moving from one state to the next. Formally, $\mathbf{T}_{i,j} = \Pr(s_f = j|s_i = i)$ for $i,j = 1,\dots,25$, where s_i and s_f are the states of the inning before and after the play, respectively.

The Markov process allows us to compute the probability of being in any given state as the inning evolves. Thus, the vector $\mathbf{s}_1 = \mathbf{T}\mathbf{s}_0$ gives the probabilities shown in Table 1.12—these are the probabilities of being in any state after one batter (the probability of being in any other state is zero). Similarly, the vector $\mathbf{s}_2 = \mathbf{T}\mathbf{s}_1 = \mathbf{T}^2\mathbf{s}_0$ gives the probabilities after

TABLE 1.12

Transition Probabilities from the Initial State of an Inning, 2013

Begin State	Begin Outs	End State	End Outs	Freq
0	0	000	0	0.0296
0	0	000	1	0.6799
0	0	001	0	0.2354
0	0	010	0	0.0497
0	0	100	0	0.0054

Source: MLBAM, GameDay files, http://gd2.mlb.com/components/ game/mlb/, accessed July 1, 2016.

Note: About 68% of the time, the following state is one out, no one on base.

two batters, and so on. After many batters have batted, the probability of being in the three out state approaches 1.

We learn a bit more with a little linear algebra. Let $\mathbf{Q}$ be the matrix obtained by deleting the three out row and column from $\mathbf{T}$. Inverting this 24×24 matrix gives us

$$\mathbf{E} = (\mathbf{I} - \mathbf{Q})^{-1},$$

where $\mathbf{I}$ is the identity matrix. It can be shown that the entries of $\mathbf{E}$ are the expected number of times that any given inning will be in each state. By summing over the rows, we arrive at the expected number of batters remaining in the inning (Albert, 2003). These values are shown in Table 1.13. Thus, the average number of batters per inning is about 4.2, but with two outs, it is always about 1.45, pretty much regardless of the configuration of the base runners. This makes sense, since the number of batters remaining in the inning has a lot more to do with the probability of the batter reaching base than it does with the configuration of the bases.

Another neat trick is that if we know the expected number of runs scored on a single play $\mathbf{q}$, we can use a little algebra to compute the expected run vector $\mathbf{r} = \mathbf{Eq}$. Table 1.14 displays the expected run matrix that results from the Markov chain model

TABLE 1.13

Expected Number of Batters Remaining in the Inning from Each of the States

BaseCode\Outs	0	1	2
000	4.244	2.857	1.458
001	3.992	2.685	1.452
010	4.295	2.890	1.505
011	4.005	2.678	1.437
100	4.336	2.951	1.518
101	4.106	2.733	1.483
110	4.400	3.072	1.543
111	3.967	2.704	1.421

TABLE 1.14

Expected Run Matrix Based on Markov Chain Model, 2013

BaseCode\Outs	0	1	2
000	0.464	0.245	0.093
001	0.819	0.491	0.213
010	1.088	0.634	0.301
011	1.394	0.870	0.411
100	1.340	0.929	0.349
101	1.700	1.131	0.488
110	1.986	1.384	0.558
111	2.143	1.535	0.725

Note: Compare these values to those presented in Table 1.1.

seeded by the transition matrix from the 2013 data. These values are very similar to the ones shown in Table 1.1, suggesting that the Markov chain model is a reasonable approximation to reality.

The transition matrix $\mathbf{T}$ described earlier is based on observations from the entire league. Suppose instead that we could estimate $\mathbf{T}_i$ for each player i. Then using the same equations outlined earlier, we could compute the vector $\mathbf{r}_i = \mathbf{E}_i\mathbf{q}$. The first entry in this vector is the expected number of runs scored in an inning with transition probabilities defined by $\mathbf{T}_i$, which are specific to player i. The interpretation is that if all nine batters were clones of player i, then $9\mathbf{r}_{i,1}$ is the expected number of runs scored per nine innings (e.g., per game) by a team of player i (Pankin, 1987, 2007). This is a useful measure of player i's offensive ability, comparable to $XR27$ as discussed earlier. A value of 8.48 runs per game, for example, would imply that the player was about twice as productive as the average batter.

However, the task of estimating $\mathbf{T}_i$ accurately is nontrivial. There are $25 \times 25 = 625$ entries in $\mathbf{T}_i$, many of which require observations to estimate. So even for players who play a full season and accrue 700 plate appearances, many entries in the transition matrix will have essentially no data. Thus, assumptions must be made. Most commonly, assumptions about baserunning are made to build transition matrices. Scoring index (Bukiet et al., 1997), offensive earned-run average (OERA) (Cover and Keilers, 1977), and modified offensive earned-run average (MOERA) (Katsunori, 2001) all use the same Markov model with slightly different baserunning assumptions.

1.3.2 *RE24*

Like the Markov models, $RE24$ is a context-dependent model that uses play-by-play data to evaluate batters (Fangraphs Staff; Appelman, 2008). Simply put, $RE24$ is the sum of the changes to the run expectancy matrix during a player's plate appearance. That is,

$$RE24 = \sum \delta_{f,i}$$

where $\delta_{f,i}$ is defined as in Equation 1.1. While the sum of the δ's over a long period of time should equal the number of runs scored, in each plate appearance a player may accrue either a negative or positive run value. Thus, the sum of these δ's over a period of time will provide a measure of the number of runs that were produced during each player's turns at-bat. $RE24$ forms the basis for the computation of *openWAR* (Baumer et al., 2015).

Skoog (1987) called this the "value-added" approach to runs created, while Albert (2001) considered this metric as well as the companion rate statistic $RE24/PA$ by dividing $RE24$ by the number of plate appearances. In Table 1.15 we report the top players in terms of $RE24$ for the 2013 season.

$RE24$ is an appealing concept that has arisen in multiple places over time. However, its context-dependent nature is an important feature (or bug, depending on your point-of-view). Because $RE24$ measures the actual changes in run expectancy, players with widely disparate opportunities could score higher or lower than their performance probably warrants. For this reason, $RE24$ is probably not a good basis for a prediction metric, such as those discussed in Section 1.4.

1.3.3 Baserunning Models

The idea behind $RE24$—that changes to the expected run matrix provide a measure of offensive performance—can be extended to baserunning as well as batting.

TABLE 1.15

RE24 Leaders for 2013

Player	PA	AB	H	HR	RE24	RE24_PA
Cabrera, M	651	554	193	44	81.33	0.125
Davis, C	673	584	167	53	75.53	0.112
Trout	716	589	190	27	69.78	0.097
Goldschmidt	710	603	182	36	64.91	0.091
Freeman, F	629	551	176	23	57.69	0.092
Ortiz, D	600	519	160	30	54.81	0.091
Cano	681	605	189	27	52.81	0.078
Votto	726	581	177	24	48.87	0.067
Carpenter, M	719	628	201	11	48.49	0.067
Encarnacion	621	530	144	36	48.21	0.078

Source: MLBAM, GameDay files, http://gd2.mlb.com/components/game/mlb/, accessed July 1, 2016.

In both Ultimate Base Running (UBR) and Base Runner Runs (BRRs), the baserunning value on an event e can be inferred from the difference

$$\delta_{f,i} - \bar{\delta}_{f,i}(e),$$

where

$\delta_{f,i}$ is again defined as in Equation 1.1

$\bar{\delta}_{f,i}(e)$ is the average value of the change in run expectancy on the batting event e

The difference between the two systems is how to measure $\bar{\delta}_{f,i}(e)$. Click (2005) estimates the value of $\bar{\delta}_{f,i}(e)$ by averaging δ's across the league, taking into account the state of the inning, how the ball is hit and where. For example, a ground ball to shortstop will result in a different value than a fly ball to rightfield. This notion of using play-by-play data to determine the expected position of base runners—as opposed to making assumptions about the movement of base runners based on the type of hit—is also employed in the baserunning component of *openWAR* (Baumer et al., 2015). However, in that system, a formal regression model is employed, unlike the empirical estimates of Click (2005). Alternatively, in Base Runner Runs, Fox (2005) estimates $\bar{\delta}_{f,i}(e)$ using conservative assumptions about baserunning. For example, a base runner is assumed to move up only one base on a single, and runners accrue additional value by advancing further.

1.3.4 Nonoutcome-Based Models

According to Lewis (2003), the Oakland A's worked with a company called AVM Systems on a model for the expected run value of any batted ball based on its location, but regardless of its outcome. That is, instead of giving batters credit for a bloop single just because the fielders couldn't catch it, why not give the batter credit for only the part of the transaction that he was involved in (namely, the weakly-hit fly ball)? If a ball was crushed, but a fielder happened to make an exceptional play, shouldn't the batter be credited for the hard-hit ball?

This type of modeling leads to a fundamentally different way of evaluating batters. Traditionally, for any batted ball i, we focus on the outcome ω_i (e.g., single, double, fly out), which is simply a designation from our list of categories Ω. Instead, the system alluded to assigns an expected run value $\mathbb{E}[r_i|x,y]$ to each batted ball, conditional upon its (x,y)-location on the field, but ignoring ω_i. One could imagine including additional covariates for batted ball speed, trajectory, etc.

In this section, we explore models based on this concept, but unfortunately very few details about these kinds of systems are publicly known. Nevertheless, new data sources that contain more accurate descriptions of the paths of batted balls are renewing interest in these types of models (Lindbergh, 2015).

1.3.4.1 DIBS

In 2015, Ben Jedlovec of Baseball Info Solutions presented a trajectory-based hitting evaluation metric called Defense-Independent Batting Statistics (DIBS). While the details of this proprietary system are unknown, it appears to be similar to the AVM system alluded to discussed earlier. It is claimed that DIBS has demonstrably greater predictive value than conventional batting statistics. DIBS includes the following covariates (Eddy, 2015):

- Batted ball location (e.g., (x,y)-coordinates)
- Hang time or ground ball velocity
- Batter handedness, speed, and power
- Park factors

1.3.4.2 HITf/x and Statcast

As we argued earlier, the outcomes ω_i are noisy—and this provides motivation for using the (x,y)-batted ball locations instead. But those locations are *also* noisy, suffering both measurement error and the vagaries of weather, altitude, etc. One further abstraction is to record the physical characteristics of the ball *as it leaves the bat*.

Since the late 2000s, Sportvision has been licensing HITf/x data, which contains launch angle, exit velocity, and trajectory information for batted balls in major and minor league parks (Sportvision, 2009). Unfortunately, these data are quite expensive, but many major league teams buy them. It is not hard to imagine how such data could be used to create a model for the expected run value of a batted ball. Note that in this case even the location of the batted ball is not important—rather we are using physical measurements from the milliseconds after impact to measure the strength of the hit. This is one further step removed from traditional metrics that focus on the outcome ω_i.

Statcast is a new player tracking system unveiled in 2014 and published in 2015 that uses Doppler radar to track the ball and video technology to track the players (Albert, 2016; Keri, 2014). These data provide unparalleled granularity about the locations of batted balls and players over fractions of seconds as each play unfolds. Undoubtedly, this will lead to new public domain models for batting and fielding evaluation.

1.3.5 Simulation-Based Models

Given the high degree of interaction within an inning of baseball, it is hard to understand how run scoring will change when more than one variable is changed. For example, how do changes in baserunning strategy affect run scoring for different profiles of offensive ability? These questions are not easily answered through parameter estimation but can be addressed using simulation.

The seeds of simulation-based study of baseball lie in the Strat-O-Matic tactile game that inspired so many sabermetricians (Guzzo, 2005). Since the late 1980s, Diamond Mind Baseball has produced a computer simulation game that its creator, Tom Tippett, has used for research purposes (Keene, 2016; Tippett). Here again, the program is proprietary, and Tippett has been an employee of the Boston Red Sox since the mid-2000s.

In two articles, Baumer (2009) and Baumer and Terlecky (2010) used a simulation-based approach to estimating base runners value. Baumer et al. (2012) used the same simulation engine to explore the relationship between baserunning and batting abilities. This approach enables, for example, the baserunning strategies of a particular base runners to be evaluated in the context of several different offensive profiles for the batter hitting behind him.

1.4 Predictive Models

While much of the early work in sabermetrics focused on developing methods for proper evaluation of the past performances of batters and base runners, in the past few decades interest has grown in predicting the future performance of these players. Indeed, while sportwriters and award voters may be primarily interested in the former problem, the question of how good a player *was* is almost irrelevant to the decisions that major league teams make. A good prediction of how good a player *will be* is far more germaine to the problem of constructing a winning team.

Thus, whereas so far in this chapter, we have focused on the problem of estimating y_t given $\mathbf{X}_t$, in this section, we consider the question of predicting y_{t+1} given $\mathbf{X}_t$.

As the ability to predict y_{t+1} accurately is of obvious importance for both general managers and fantasy players alike, it has become something of a sport unto itself. It would be impractical to review all of the publicly known projection systems here (for a larger list, see Fangraphs Staff (2015a), or Larson (2015a) for a collection of known projections). Instead, we focus on illuminating the guiding principles of a few of the better known projection systems and illustrating the key differences between them. The first characteristic distinction is between models that simply provide *point estimates* as opposed to those that provide *interval estimates*.

1.4.1 Models That Produce Point Estimates

1.4.1.1 Marcel

In comparing forecasting models, it is useful to have a benchmark. The most commonly used benchmark is the relatively simple Marcel projection system promulgated by Tango (2012). With great modesty, Tango describes Marcel as "most basic forecasting system you can have, that uses as little intelligence as possible." The system is not utterly trivial—it is a time-weighted 3-year average of each player's previous statistics but also takes into account each player's age and employs regression to the mean in the context of their playing time.

Let p_{ijt} be the rate at which player i achieves outcome j in season t. Since Marcel uses the previous three seasons of data, let $p'_{ijt} = \begin{pmatrix} p_{ij,t-1} & p_{ij,t-2} & p_{ij,t-3} \end{pmatrix}^T$ be the vector of those rates over the past three seasons. The corresponding 3×1 vector q'_{it} gives the number of plate appearances for player i in the three seasons preceding t.

Marcel uses a fixed time-discounting vector we define as $t = \begin{pmatrix} 5 & 4 & 3 \end{pmatrix}^T$ and a fixed regression to the mean vector we define as $\bar{q} = \begin{pmatrix} 100 & 100 & 100 \end{pmatrix}^T$ (for a fuller description, see Baumer (2014)).

In this notation, Marcel computes the predicted rate for each player as

$$\hat{p}_{ijt} = w_{it} \cdot A_{it} \cdot p'_{ijt} + (1 - w_{it}) \cdot A_{it} \cdot \bar{p}'_{jt}, \tag{1.2}$$

where

$\bar{p}'_{jt}$ is the corresponding vector of league average rates for event j over the past three seasons

$w_{it} = \frac{t * q'_{it}}{t * q'_{it} + t * \bar{q}}$

$A_{it} = \frac{t * diag(q'_{it})}{t * q'_{it}}$

$$diag(q'_{it}) = \begin{pmatrix} q_{i,t-1} & 0 & 0 \\ 0 & q_{i,t-2} & 0 \\ 0 & 0 & q_{i,t-3} \end{pmatrix}.$$

The quantity w_{it} depends only on player i's number of plate appearances over the past three seasons and is called *reliability*. Similarly, the quantity A_{it} depends on the same but acts as a scaling factor. Thus, by defining $\theta_{ijt} = A_{it}p'_{ijt}$ and $\bar{\theta}_{ijt} = A_{it}\bar{p}'_{jt}$, we have

$$\hat{p}_{ijt} = w_{it} \cdot \theta_{ijt} + (1 - w_{it}) \cdot \bar{\theta}_{ijt}.$$

This is immediately evocative of the well-known formula for the posterior mean for the frequency parameter in a beta-binomial Bayesian model. Here, θ_{ijt} is the time-discounted observed rate of event j for player i, and $\bar{\theta}_{ijt}$ is the corresponding league average rate, with the time-discounting depending on player i's playing time. Marcel's reliability w_{it} is apparent as a *shrinkage* factor.

The number of predicted plate appearances in season t is

$$\hat{q}_{it} = \frac{1}{2}q_{i,t-1} + \frac{1}{10}q_{i,t-2} + 200,$$

where this estimate is asserted without justification. The final computation is a piecewise linear age adjustment

$$f_i(t) = \begin{cases} -0.003 \cdot (29 - (t - birthYear_i)), & t - birthYear_i > 29, \\ 0.006 \cdot (29 - (t - birthYear_i)), & \text{otherwise,} \end{cases}$$

which has the effect of increasing the expected performance of those who have yet to reach their prime and decreasing the expected performance of those who are past their prime. While the intent is clear, no justification for this particular aging curve is given.

TABLE 1.16

Carlos Beltran's Recent History Preceding the 2004

Year	PA	HR	HR/PA
2001	680	24	0.0353
2002	722	29	0.0402
2003	602	26	0.0432

Consider the canonical example of predicting Carlos Beltran's home run count heading into the 2004 season. In Table 1.16, we present Beltran's relevant data. Beltran hit more home runs that the average player in each of these seasons. The home run rates for the average player are shown in Table 1.17.

Following Equation 1.2, for Beltran in 2004, we have

$$\hat{p}_{beltran,HR,2004}$$

$$= \underbrace{\frac{(3 \; 4 \; 5)\begin{pmatrix}680\\722\\602\end{pmatrix}}{(3 \; 4 \; 5)\begin{pmatrix}680\\722\\602\end{pmatrix}+(3 \; 4 \; 5)\begin{pmatrix}100\\100\\100\end{pmatrix}}}_{w_{beltran,2004}} \cdot \underbrace{\frac{(3 \; 4 \; 5)\begin{pmatrix}680 & 0 & 0\\0 & 722 & 0\\0 & 0 & 602\end{pmatrix}\begin{pmatrix}0.0353\\0.0402\\0.0432\end{pmatrix}}{(3 \; 4 \; 5)\begin{pmatrix}680\\722\\602\end{pmatrix}}}_{A_{beltran,2004} \cdot p'_{beltran,HR,2004}}$$

$$+ \underbrace{\frac{(3 \; 4 \; 5)\begin{pmatrix}100\\100\\100\end{pmatrix}}{(3 \; 4 \; 5)\begin{pmatrix}680\\722\\602\end{pmatrix}+(3 \; 4 \; 5)\begin{pmatrix}100\\100\\100\end{pmatrix}}}_{1-w_{beltran,2004}} \cdot \underbrace{\frac{(3 \; 4 \; 5)\begin{pmatrix}680 & 0 & 0\\0 & 722 & 0\\0 & 0 & 602\end{pmatrix}\begin{pmatrix}0.0300\\0.0278\\0.0286\end{pmatrix}}{(3 \; 4 \; 5)\begin{pmatrix}680\\722\\602\end{pmatrix}}}_{A_{beltran,2004} \cdot \bar{p}'_{HR,2004}}$$

$$= \underbrace{\frac{7938}{9138}}_{w_{beltran,2004}} \cdot \underbrace{\frac{318}{7938}}_{\theta_{beltran,HR,2004}} + \underbrace{\frac{1200}{9138}}_{1-w_{beltran,2004}} \cdot \underbrace{\frac{228}{7938}}_{\bar{\theta}_{beltran,HR,2004}}$$

$$= 0.8687 \cdot 0.0401 + 0.1313 \cdot 0.0287$$

$$= 0.0386$$

The number of plate appearances Beltran is expected to take in 2004 would be

$$\hat{q}_{beltran,2004} = \frac{1}{2}q_{beltran,2003} + \frac{1}{10}q_{beltran,2002} + 200$$

$$= \frac{1}{2} \cdot 602 + \frac{1}{10} \cdot 722 + 200$$

$$= 573.2$$

TABLE 1.17

MLB's Recent History Preceding the 2004 Season

Year	PA	HR	HR/PA
2001	177,941	5333	0.0300
2002	178,487	4958	0.0278
2003	177,376	5078	0.0286

Note: Pitchers are excluded by filtering players with at least 2 plate appearances per game played.

His predicted number of home runs is thus $573.2 \cdot 0.0386 = 22.1$. Since Beltran was 27 in 2004, this number gets adjusted up by $(29 - 27) \cdot 0.006 = 0.012$. Thus, our final estimate is that Beltran would have hit $1.012 \cdot 22.1 = 22.4$ home runs in 2004. (In reality, Beltran had a breakout year and hit 38 home runs.)

Despite its relative simplicity and idiosyncratic modeling choices, Marcel has proven to be a durable and effective projection system. Its primary contribution is to provide a benchmark against which all other projection systems can be compared.

1.4.1.2 ZiPS

ZiPS is a popular projection system created by Szymborski (2015). ZiPS is similar to Marcel in that is based on a time-discounted average of recent performance (Cockcroft, 2015). However, ZiPS makes several significant improvements over Marcel:

- It uses 4 years of data, rather than 3, with a $t = \begin{pmatrix} 8 & 5 & 4 & 3 \end{pmatrix}^T$ time-discount vector.
- Shrinkage toward the mean is different for different rate statistics. In particular, ZiPS uses McCracken (2001)'s Defensive-Independent Pitching Statistics (DIPS) theory as its starting point in determining shrinkage levels. This means that, for example, a batter's batting average on balls in play will be shrunk much further toward the mean than his strikeout rate.
- Age adjustments are based on a historical comparison with "similar" players. Thus, heavy-set sluggers may age differently than speedy leadoff hitters.
- Rookies and international players are projected based on major league equivalencies (MLEs).
- Park adjustments are included.

Of course, once season t is played, it is easy enough to measure the accuracy of any prediction system. The three obvious metrics are the mean absolute error ($MAE = \frac{1}{n}\sum_{i=1}^{n}|y_{it} - \hat{y}_{it}|$), root mean squared error ($RMSE = \frac{1}{n}\sum_{i=1}^{n}(y_{it} - \hat{y}_{it})^2$), and the Pearson correlation ($r(y_{it}, \hat{y}_{it})$). It is not surprising that by the later criteria, ZiPS tends to best Marcel (Meyer, 2014). What is perhaps more surprising is that Marcel occasionally bests ZiPS according to the *RMSE* criteria (Larson, 2015b). One explanation that has been proffered is that Marcel implicitly estimates an accurate run-scoring environment.

Another approach—common in machine learning—is to create an *ensemble* prediction based on weighted averages of existing projection systems (Highley et al., 2010).

1.4.2 Models That Produce Interval Estimates

Naturally, contientious forecasting of the future performance of baseball players involves uncertainty, and while the point estimates produced by the aforementioned models are certainly useful, general managers are better able to assess risk using forecasts that provide not just a point estimate for each player's future performance, but a probability distribution over that potential future performance.

1.4.2.1 Pecota

One of the first publicly available projection systems that published forecast distributions was Pecota (Schwarz, 2005b). Pecota is a proprietary system originally developed by Nate Silver, and now operated by Baseball Prospectus. In addition to the breakthrough that was publishing forecast distributions, Pecota is fundamentally different than most projection systems, in that it is based on *nearest neighbor* analysis (Silver, 2003). While the details of Pecota are proprietary and a full treatment of nearest neighbor algorithms is beyond what we can accomplish here, since this approach differs significantly from many of the other predictive models, we will describe the general framework upon which Pecota rests.

Note that in our notation, the matrix $\mathbf{X}$ has p columns and n rows, where each row corresponds to a player and each column to an attribute of that player. Each row is a vector in $\mathbb{R}^p$, and thus each player can be thought of as a point in a p-dimensional space. We can define a distance metric $d(s, t)$ for any two points $s, t \in \mathbb{R}^p$ (e.g., the Euclidean distance metric). Then given a desired number of comparison players k, for any candidate player x^*, we can find the k players who are "closest" to x^* using the distance metric d. The average performance of those k players provides an estimate of the performance of player x^*. Such k-NN algorithms are well documented in the machine learning literature (James et al., 2013).

In Pecota, we know that among the p attributes considered are

- Production metrics that measure past performance using typical sabermetric statistics
- Usage metrics that measure playing time
- Phenotypic attributes that are specific to the physical characteristics of the player
- Role metrics that focus on fielding position or starter vs. relief usage

We don't know much about the distance metric—a *similarity score* in the parlance of Bill James—used in Pecota (i.e., is it Euclidean?), but we do know that Pecota incorporates age and adjustments for ballpark and league.

1.4.2.2 Steamer

Steamer Projections are another commonly cited set of predictions. Like ZiPS, Steamer is a Marcel-rooted forecasting system with a host of additional complexities built-in. In particular, Steamer uses a DIPS-inspired framework with variable time-discounting.

More recently, Steamer has produced a Shiny app that allows a user to explore the forecast distribution while experimenting with parameters (Cross, 2014).

1.4.2.3 Bayesian Models

While all of the forecasting systems described earlier incorporate the notion of regression to the mean, and at least two (Marcel and Steamer) are evocative of a Bayesian framework, none of the models discussed earlier are explicitly Bayesian. However, the academic literature is rich with Bayesian models for all kinds of things, with batting and baserunning being no exceptions.

In a widely cited article, Efron and Morris (1975) motivated their use of Stein's estimator using an example from baseball. In particular, their goal was to use each player's first 45 at-bats of the 1970 season to predict their batting average over the remainder of the season. Thus, baseball has provided a natural context for Bayesian analysis for nearly as long as Bayesian analysis has been done.

More recently, Neal et al. (2010) use linear regression to predict second-half performance given first-half performance, improving upon the Bayesian models of Brown (2008). Jensen et al. (2009) formulate a rigorous hierarchical Bayesian model for home run rate that surpasses Marcel—but not Pecota—by the *RMSE* metric, while at the same time producing full posterior distributions.

It is common in baseball parlance to identify certain players as "home run hitters," while others are not identified as such. Statistical analysis confirms this intuition, as a batter's home run rate is relatively highly correlated across consecutive seasons. Yet while it is clear that home runs are an important characteristic for hitters, the question of which components of a batter's profile represent the strongest *signal* remains. From DIPS theory, we know that much of the *noise* in pitcher (and in turn hitter) performance is captured by the volatile statistic of batting average on balls in play. McShane et al. (2011) uses a hierarchical Bayesian variable selection model to tease out which other hitting statistics seem most persistent for hitters over time. They find that while many of the 50 commonly cited metrics were correlated, most of the signal is captured by just 5: strikeout rate, walk rate, ground ball rate, isolated power, and Bill James's speed score. This result—in conjunction with DIPS—provides a meaningful starting place for forecasting position players, in that it confirms that most of what we know about batters are their plate discipline, how fast they run, how often they hit for power, and how often they hit the ball on the ground. From the point of view of a forecaster, much of the rest may just be noise.

Rather than choosing specific rates to follow, in general, it is natural and convenient to consider the observations in $\mathbf{X}$ as realizations of a multinomial distribution with parameters n (i.e., the number of plate appearances) and ρ, a vector of length p giving the true probabilities for each of the p outcomes of a plate appearance (with $\sum_{i=1}^{p} \rho_i = 1$). Viewing the data as having this multinomial likelihood, a natural Bayesian approach is to employ the conjugate prior distribution: the Dirichlet, which has the hyperparameter vector κ. This framework is simply the multivariate extension of the beta-binomial framework suggested by Marcel.

Recognizing that this multinomial Dirichlet framework, however natural, has certain limitations (namely, it doesn't understand DIPS), Null (2009) constructed the *nested* Dirichlet distribution for extending this framework. This remains the current state of the art, although recent work by Albert (2016) provides a simpler model.

1.5 Conclusion

It is not a stretch to argue that the problem of evaluating batters (and base runners) has provided the primary motivation for sabermetrics, and in turn for sports analytics in general. This field has seen the development of fundamental tools (e.g., Lindsey's expected run matrix), brilliant insights (e.g., runs created), and sophisticated statistical models (e.g., Null's nested Dirichlet) that have collectively revolutionized the way we assess baseball players.

The problem of assessing the value of the contributions made by major league batters and base runners is largely solved. If the values of the expected run matrix can be estimated with reasonable accuracy, then *RE*24 answers the question of how the team's fortunes changed when a certain batter was at the plate, relative to a league average batter. The only remaining question is to separate the contributions of base runners from those of batter, for which we have discussed several reasonable attempts (Click, 2005; Baumer et al., 2015). While it is true that analysts continue to debate which particular metric is best, the variation in the estimates of competing metrics is generally small enough that it is unlikely that we will see major changes in the way that we evaluate batters and base runners in the future. This stands in stark contrast to, for example, the dramatic changes in the evaluation of fielders and catchers that have occurred in recent years.

On the other hand, the problem of predicting the future offensive performance of players is far from settled. Most of the widely cited projection systems (e.g., Pecota, ZiPS) are proprietary, and thus we don't really understand exactly how they work. Furthermore, while these projection systems do offer better accuracy than the relatively simple Marcel, the order of magnitude of the difference is not so large as to suggest that the problem has been "solved."

Moreover, these projection systems work well for major league players, less well for minor league players, and not at all for amateur players. Thus, there is still tremendous work to be done in predicting the future offensive performance of amateur players.

The most likely source of a breakthrough here is not a brilliant model, but rather a new data source—or the combination of a new model and a new data source. In what follows we outline some open problems.

1.5.1 Open Problems

Forecasting: The forecasting problem (i.e., given X_t, predict y_{t+1}) is particularly well defined, especially if we restrict X_t to include only information in the Lahman (or Retrosheet) database. This makes it a nice problem to tackle. Unfortunately, there is little consensus surrounding a "best" forecasting model. What is needed is a simple and/or elegant model that can consistently beat Marcel by a significant margin. It seems clear that we have models that can consistently beat Marcel, but they are significantly more complex (or proprietary—either way, we haven't learned much). It also seems likely that we have relatively simple models that can consistently beat Marcel but not by a significant margin (Albert, 2016). We suspect that bringing machine-learning tools to bear on this problem (e.g., random forests) might be a worthwhile approach.

Amateur forecasting: Forecasting the major league performance of amateur players is a fundamentally different problem, since performance data from colleges and high schools are considerably less reliable than minor league or major league performance data.

Here, we would like to see a model that incorporates traditional scouting information with phenotypic attributes and performance data. If a 6′2″ pitcher throws 90 mph in high school, and a scout describes him as having a "quick" arm, does he have a higher expected career WAR than a 5′11″ high school pitcher throwing 90 mph who is a "slinger?" Major league teams certainly have such models, but they are not known to the public.

Biometric models: Pursuant to the above, are there biometric indicators that are useful in predicting future performance, particularly that of amateurs? How does exit velocity—which is neither directly biometric nor directly a measure of performance—presage future performance?

Player allocation: There are only 9 positions on the field and 25 spots on the roster. If one assumes a free market, and that each player has a value (i.e., his expected WAR) and a cost (i.e., his salary), then what is the optimum allocation of players to positions given a fixed budget? Note that WAR is position dependent. Intuitively, it seems clear that signing the nine best hitters and forcing them to play the field is not the optimal solution, even though it would maximize offense. This is a variation on the multiple knapsack problem, which is known to be NP-hard, and thus, there is not likely to be a polynomial-time solution. Nevertheless, is the problem tractable for this size? Is there a better solution than trying all the possible combinations?

References

Albert, C. The metrics system. *Sports Illustrated*, pp. 45–48, August 22, 2016. URL http://www.si.com/mlb/2016/08/26/statcast-era-data-technology-statistics, accessed July 1, 2016.

Albert, J. Using play-by-play baseball data to develop a better measure of batting performance. Technical report, Bowling Green State University, Bowling Green, OH, September 2001. http://bayes.bgsu.edu/papers/rating_paper2, accessed July 1, 2016.

Albert, J. *Teaching Statistics Using Baseball*. MAA, Washington, DC, 2003.

Albert, J. Improved component predictions of batting and pitching measures. *Journal of Quantitative Analysis in Sports*, 12(2):73–85, 2016.

Albert, J. and J. Bennett. *Curve Ball: Baseball, Statistics, and the Role of Chance in the Game*. Copernicus Books, New York, NY, 2003.

Appelman, D. Get to know: Re24, March 2008. http://www.fangraphs.com/blogs/get-to-know-re24/.

Baumer, B. Marcel the matrix, June 2014. https://baseballwithr.wordpress.com/2014/06/25/marcel-the-matrix/, accessed July 1, 2016.

Baumer, B. and P. Terlecky. Improved estimates for the impact of baserunning in baseball. In *JSM Proceedings, Statistics in Sports, ASA*, Alexandria, VA, 2010.

Baumer, B. S. Why on-base percentage is a better indicator of future performance than batting average: An algebraic proof. *Journal of Quantitative Analysis in Sports*, 4(2):1–11, 2008.

Baumer, B. S. Using simulation to estimate the impact of baserunning ability in baseball. *Journal of Quantitative Analysis in Sports*, 5(2):1–16, 2009.

Baumer, B. S., S. T. Jensen, and G. J. Matthews. openWAR: An open source system for evaluating overall player performance in major league baseball. *Journal of Quantitative Analysis in Sports*, 11(2):69–84, 2015.

Baumer, B. S., J. Piette, and B. Null. Parsing the relationship between baserunning and batting abilities within lineups. *Journal of Quantitative Analysis in Sports*, 8(2):1–17, 2012.

Bennett, J. M. and J. A. Flueck. An evaluation of major league baseball offensive performance models. *The American Statistician*, 37(1):76–82, 1983.

Boswell, T. (ed.). Welcome to the world of total average where a walk is as good as a hit. In *How Life Imitates the World Series*, pp. 137–144. Penguin Books, New York, 1982.

Brown, L. D. In-season prediction of batting averages: A field test of empirical bayes and bayes methodologies. *The Annals of Applied Statistics*, 2(1):113–152, 2008.

Bukiet, B., E. Harold, and J. Palacios. A markov chain approach to baseball. *Operations Research*, 45(1):14–23, 1997.

Click, J. Station to station: The expensive art of baserunning. In *Baseball Prospectus 2005*, pp. 511–519. Workman Publishing, New York, 2005.

Cockcroft, T. H. Inside the projections process, February 2015. http://espn.go.com/fantasy/baseball/story/_/page/mlbdk2k15_projectionstalk/how-fantasy-baseball-projections-calculated-how-best-use-them, accessed July 1, 2016.

Cook, E. *Percentage Baseball*. Waverly Press, Baltimore, MD, 1964.

Cover, T. M. and C. W. Keilers. An offensive earned-run average for baseball. *Operations Research*, 25(5):729–740, 1977.

Cramer, R. D. and P. Palmer. The batter's run average (B.R.A.), 1974. http://research.sabr.org/journals/batter-run-average, accessed July 1, 2016.

Cross, J. Steamer percentile projections, August 2014. https://steamerprojections.shinyapps.io/steamer_error_bars/error_bars.Rmd, accessed July 1, 2016.

D'Angelo, J. P. Baseball and markov chains: Power hitting and power series. *Notices of the AMS*, 57(4):490–495, 2010.

Eddy, M. Sabr analytics: New takes on batted-ball profiles, pitch framing, March 2015. http://www.baseballamerica.com/majors/sabr-analytics-bis-offers-new-takes-batted-ball-profiles-pitch-framing-day-three/, accessed July 1, 2016.

Efron, B. and C. Morris. Data analysis using Stein's estimator and its generalizations. *Journal of the American Statistical Association*, 70(350):311–319, 1975.

Fangraphs Staff. Projection systems, 2015a. http://www.fangraphs.com/library/principles/projections/, accessed July 1, 2016.

Fangraphs Staff. wSB, 2015b. http://www.fangraphs.com/library/offense/wsb/, accessed July 1, 2016.

Fangraphs Staff. 2016. Re24. http://www.fangraphs.com/library/misc/re24/, accessed July 1, 2016.

Fox, D. Circle the wagons: Running the bases part iii, August 2005. http://www.hardballtimes.com/circle-the-wagons-running-the-bases-part-iii/, accessed July 1, 2016.

Furtado, J. Introducing XR, 1999. http://www.baseballthinkfactory.org/btf/scholars/furtado/articles/IntroducingXR.htm, accessed July 1, 2016.

Guzzo, G. *Strat-O-Matic Fanatics: The Unlikely Success Story of a Game That Became an American Passion*. ACTA Sports, Skokie, IL, 2005.

Highley, T., R. Gore, and C. Snapp. Granularity of weighted averages and use of rate statistics in AggPro. In *Proceedings of the 2010 Winter Simulation Conference (WSC 2010)*, Baltimore, MD, December 5–8, 2010, pp. 1318–1329. WSC, 2010.

James, B. *The Bill James Historical Baseball Abstract*. Random House Inc., New York, NY, 1986.

James, B. and J. Henzler. *Win Shares*. STATS Pub., Northbrook, IL, 2002.

James, G., D. Witten, T. Hastie, and R. Tibshirani. *An Introduction to Statistical Learning*. Springer, New York, NY, 2013. http://www-bcf.usc.edu/ gareth/ISL/, accessed July 1, 2016.

Jensen, S. T., B. B. McShane, A. J. Wyner, et al. Hierarchical Bayesian modeling of hitting performance in baseball. *Bayesian Analysis*, 4(4):631–652, 2009. http://projecteuclid.org/euclid.ba/1340369815, accessed July 1, 2016.

Johnson, P. and B. James. Estimated runs produced, 1985. http://www.baseballthinkfactory.org/btf/pages/essays/jameserp.htm, accessed July 1, 2016.

Katsunori, A. Modified offensive earned-run average with steal effect for baseball. *Applied Mathematics and Computation*, 120(1):279–288, 2001.

Keene, C. A. Statistician steps up to the plate for red sox. The Boston Globe, March 2016. https://www.bostonglobe.com/business/2016/03/04/baseball-numbers/rfdabpRE7z JObT7Ozvc8bP/story.html, accessed July 1, 2016.

Keri, J. Q&A: MLB advanced media's Bob Bowman discusses revolutionary new play-tracking system, March 2014. URL http://grantland.com/the-triangle/mlb-advanced-media-play-tracking-bob-bowman-interview/, accessed July 1, 2016.

Larson, W. The baseball projection project, 2015a. http://www.bbprojectionproject.com/, accessed July 1, 2016.

Larson, W. 2014 projection review (updated), March 2015b. http://www.fangraphs.com/community/2014-projection-review-updated/, accessed July 1, 2016.

Lewis, M. *Moneyball: The Art of Winning an Unfair Game*. WW Norton & Company, New York, NY, 2003.

Lindbergh, B. Before beane, July 2015. http://grantland.com/features/2015-mlb-avm-systems-ken-mauriello-jack-armbruster-moneyball-sabermetrics/, accessed July 1, 2016.

Lindsey, G. R. An investigation of strategies in baseball. *Operations Research*, 11(4):477–501, 1963.

Marchi, M. and J. Albert. *Analyzing Baseball Data with R*. CRC Press, Boca Raton, FL, 2013.

McCracken, V. Pitching and defense: How much control do hurlers have? 2001. http://baseballprospectus.com/article.php?articleid=878, accessed July 1, 2016.

McShane, B. B., A. Braunstein, J. Piette, and S. T. Jensen. A hierarchical Bayesian variable selection approach to major league baseball hitting metrics. *Journal of Quantitative Analysis in Sports*, 7(4):1–24, 2011.

Meyer, D. Evaluating the 2014 projection systems, December 2014. http://www.hardballtimes.com/evaluating-the-2014-projection-systems/, accessed July 1, 2016.

MLBAM, GameDay files. http://gd2.mlb.com/components/game/mlb/, accessed July 1, 2016.

Neal, D., J. Tan, F. Hao, and S. S. Wu. Simply better: Using regression models to estimate major league batting averages. *Journal of Quantitative Analysis in Sports*, 6(3):1–12, 2010. http://www.degruyter.com/view/j/jqas.2010.6.3/jqas.2010.6.3.1229/jqas.2010.6.3.1229.xml, accessed July 1, 2016.

Null, B. Modeling baseball player ability with a nested Dirichlet distribution. *Journal of Quantitative Analysis in Sports*, 5(2):1–36, 2009. http://www.degruyter.com/dg/viewarticle/j$002fjqas.2009.5.2$002fjqas.2009.5.2.1175$002fjqas.2009.5.2.1175.xml, accessed July 1, 2016.

Pankin, M. Evaluating offensive performance in baseball. *Operations Research*, 26(4):610–619, 1978.

Pankin, M. D. Baseball as a Markov chain. In *The Great American Stat Book*, James, B. (ed.), pp. 520–524. Ballantine Books, 1987.

Pankin, M. D. Markov chain models: Theoretical background, pp. 11–26, 2007. http://www.pankin.com/markov/theory.htm, accessed July 1, 2016.

Schwarz, A. *The Numbers Game: Baseball's Lifelong Fascination with Statistics*. Thomas Dunne Books, New York, NY, 2005a.

Schwarz, A. Predicting futures in baseball, and the downside of damon, November 2005b. http://www.nytimes.com/2005/11/13/sports/baseball/predicting-futures-in-baseball-and-the-downside-of-damon.html, accessed July 1, 2016.

Silver, N. Introducing pecota. In *Baseball Prospectus 2003*, pp. 507–514. Brassey's Publishers (Dulles, VA), 2003.

Skoog, G. R. Measuring runs created: The value added approach. In *The Bill James Baseball Abstract*, James, B. (ed.). Ballantine Books, 1987.

Sportvision. HiTf/x, 2009. https://www.sportvision.com/baseball/hitfx%C2%AE, accessed July 1, 2016.

Szymborski, D. Zips Q&A, 2015. http://www.baseballthinkfactory.org/szymborski/zipsqa.rtf, accessed July 1, 2016.

Tango, T. Marcel 2012, 2012. http://www.tangotiger.net/marcel/, accessed July 1, 2016.

Tango, T. M., Lichtman, and A. Dolphin. *The Book: Playing the Percentages in Baseball*. Potomac Books, Dulles, VA, 2007.

Thorn, J. and P. Palmer. *The Hidden Game of Baseball: A Revolutionary Approach to Baseball and Its Statistics*. Doubleday, Garden City, NY, 1984.

Thorn, J. and P. Palmer. *The Hidden Game of Baseball: A Revolutionary Approach to Baseball and Its Statistics*, 3rd edn. University of Chicago Press, Chicago, IL, 2015. ISBN 978-0226242484.

Tippett, T. 2016. Diamond mind baseball. http://diamond-mind.com/, accessed July 1, 2016.

Trueman, R. E. Analysis of baseball as a markov process. *Optimal Strategies in Sports*, 5: 68–76, 1977.

Wang, V. The OBP/SLG ratio: What does history say? *By the Numbers*, 16(3):3, 2006.

2

Using Publicly Available Baseball Data to Measure and Evaluate Pitching Performance

Carson Sievert and Brian M. Mills

CONTENTS

2.1 Introduction

In contrast to many other sports, the game of baseball can largely be described in discrete events, and player roles are well defined, making it relatively easy to measure player performance. Perhaps the most important individual role is the pitcher, since they have the most direct control over run prevention. However, directly measuring a pitcher's ability to prevent runs is difficult, since many relevant events depend on outside factors such as the team's ability to field balls hit into play. Sabermetricians have created a number of measures that attempt to isolate pitching from fielding contributions, but many do so by ignoring certain events and/or using subjective/incomplete information to distribute responsibility using probabilistic models.

With the advent of more granular, detailed, and accurate measurements of on-field activity, it is anticipated that our ability to measure and evaluate pitching performance will improve.

Many traditional measures of pitching performance, such as Earned Run Average (ERA) or pitcher Wins (W), confound pitching with other aspects of the game (e.g., fielding, baserunning, hitting, etc.). However, there are certain outcomes in the game that depend only on the skill of the pitcher (and batter): strikeouts, walks, and home runs. Fielding is irrelevant in these plate appearance outcomes since no ball is put in play.* This observation was first noted in Scully (1973) where a strikeout-to-walk ratio is used to measure pitching performance. But in 1999, Voros McCracken introduced and formalized the more robust Defense Independent Pitching Statistics (DIPS), made up of strikeouts, walks, and home runs allowed. While DIPS measures improved the understanding of pitcher contributions to game outcomes, it is still not well understood to what degree pitchers can control balls in play.

For years it was believed that pitchers had little to no control over batting average on balls in play (BABIP), making DIPS a reasonable measure of pitching performance, but this assumption breaks down when considering the type of ball hit into play (e.g., fly ball, ground ball, etc.). This finding should not shock those familiar with Simpson's paradox, as global trends rarely reflect local trends. If the goal is to measure overall performance of a pitcher, we believe it's best to avoid such assumptions and use all the information available. Moreover, pitchers can impact the game in more ways than delivering the ball to the batter (e.g., fielding, batting, baserunning), so an overall measure should respect these aspects of the game as well.

Inspired by James and Henzler (2002) and Jacques (2007), it is now commonplace to measure individual performance by estimating the number of wins contributed relative to "freely available talent" at that player's position. These win above replacement (WAR) measures are becoming increasingly popular since they are relatively easy to interpret and make it easy to compare player value. Unfortunately, most WAR measures use data/methods that are not freely available to the public. However, enough has been written to know that WAR measures are often a simple linear weight of season-level statistics. Moreover, it is common to see WAR measures designed specifically for pitchers use DIPS assumptions and consequently ignore batted ball information. As the public gains access

---

* We note that recent advances in understanding catcher skill as it relates to pitch framing are relevant here. We will address these in a later section.

to higher-resolution data, we expect to see more measures use at-bat and/or pitch-level data, which can improve our ability to measure individual performance.

openWAR (Baumer et al., 2015) is a completely open and reproducible WAR measure that uses publicly available at-bat level data. Since responses are defined on the at-bat level, openWAR can distribute defensive responsibility amongst pitcher and fielder(s) on batted balls in fashion similar to Jensen et al. (2009). openWAR does currently make the assumption that the value of each run scored counts equally for the offense and against the defense, but the underlying framework could be augmented to incorporate "clutchness" weights. It also assumes the change in expected runs of a plate appearance can be divided amongst four aspects of the game: (1) batting, (2) baserunning, (3) fielding, and (4) pitching.[*] In order to distribute value among fielding and pitching on balls hit in play, openWAR estimates the probability of an out based on locational data, and uses the estimated residuals to distribute value for good/bad fielding plays.

openWAR uses two-dimensional density estimation, a topic we cover and apply to pitch location data in Section 2.4, to estimate the probability of an out. This is almost certainly an oversimplified model that ignores useful information beyond the terminal location of the batted ball, but at the time, this was the only batted ball data available to the public.[†] Jensen et al. (2009) use more accurate and detailed (but proprietary) batted ball data to develop a more sophisticated model estimating this same probability. That study found that the location has a large effect on the likelihood of an out, but exit velocity also has a nonnegligible effect.[‡] More information (the trajectory of the ball, the starting location of the fielders, etc.), which will soon be available to public, could not only lead to more accurate division of run prevention responsibility on batted balls, but potentially new and better measures of individual performance in general. An overview of the new data that we anticipate, as well as more traditional baseball data sources, is covered in Section 2.2.

In Section 2.3, we provide a more detailed look at traditional measures of pitching performance and propose some ideas for future research directions in this area. In Section 2.4, we discuss research using pitch level (aka PITCHf/x) data. Section 2.4 also covers statistical methodology that has proven useful for research using PITCHf/x data (in particular, generalized additive models, density estimation, and cluster analysis), but these methods could also be useful for at-bat-level data. The chapter is concluded with a discussion of future challenges in the evaluation of pitching performance.

2.2 Data Usage and Availability

2.2.1 Traditional Statistics

Many traditional statistical measures of pitcher performance have been available for more than 100 years. For example, strikeouts, wins, earned run average, walks, batting average against, and other statistics are commonly found on many baseball web sites. Depending

---

[*] After estimating individual contributions in each category, these categories are added together, to obtain an overall value measured in runs. The run value of an appropriate replacement player is then subtracted, and the resulting quantity is divided by 10 (empirical evidence suggests that 10 runs roughly equals a win).

[†] Someone watching the game determines these terminal locations.

[‡] The velocities were generated by someone watching the game and were recorded as one of the following: {soft, medium, hard}. Having a more objective and granular measurement of velocity would improve the estimate of that effect.

on the analysis, one might need statistics on a career, seasonal, game, or even a play-by-play level. Serious analysts should consider obtaining play-by-play data from a reputable source such as http://www.retrosheet.org, since this data can always be aggregated to a desired level. Among play-by-play baseball data sources, Retrosheet has the richest set of play-by-play measurements spawning the greatest number of years. Play-by-play data is also available via the PITCHf/x system (discussed in Section 2.2.3), but it is only available for the 2007 season and later.

2.2.2 Balls in Play

The work of McCracken allowed us to separate two main areas of outcomes for pitchers: those pitches put in play, and those not put in play. While initial evaluation of batting average on balls in play (BABIP) indicated that pitchers had no control over BABIP, more recently analysts have argued that ball-in-play outcomes do in fact differ by pitcher. Strikeouts, walks, and home runs have been recorded for a large portion of baseball history, allowing one to make comparisons of the rate at which balls in play result in a hit across the history of professional baseball. For example, we can easily calculate BABIP for players like Cy Young in his 1890 rookie year with the Cleveland Spiders (0.269), as well as the entire careers of players like Nolan Ryan (0.265), Greg Maddux (0.281), and Clayton Kershaw (0.272). These data are readily available at web sites like Fangraphs or Baseball Reference but can also be calculated by anyone willing to delve into the play-by-play files at Retrosheet dating back to 1921.

More detailed data on balls in play, tracked by Baseball Info Solutions (BIS) classified as ground balls, fly balls, or line drives can be found on web sites like Fangraphs but are not available prior to the 2002 season. Information on in-play outcomes such as batted ball velocity, launch angle, and fly ball distance became publicly available in the 2015 season. This information can be obtained from http://www.baseballsavant.com/ or retrieved programmatically (Sievert, 2015a). These data may be particularly useful in developing measures of the quality of contact on pitches put in play, some of which may be attributed to pitcher skill.

2.2.3 PITCHf/x, Trackman, ChyronHego, and Statcast

In 2001, shortly after an umpire labor dispute and years of turmoil, Major League Baseball instituted the QuesTec system to monitor its umpire ball-strike calls. QuesTec revealed the location of each pitch as it crossed the plate, which could be matched with the call made by each umpire to assess the accuracy with which umpires were calling balls and strikes. The system was originally installed in only four MLB parks, but in 2004—after ratification by the umpire union—grew to be installed in half of the stadiums across the league. This data was not publicly available, though the league was able to use it extensively to understand the strike zone that was called by its umpires during this time.

While QuesTec was used through the 2008 season to monitor umpires, the league and its media subsidiary, Major League Baseball Advanced Media (MLBAM), began tracking pitches for display online and on broadcasts. The technology was developed by Sportvision, with three cameras set up to project the trajectory of pitches and where they would cross the front of the plate. This data is known more colloquially as PITCHf/x. In 2009, the league and its umpires agreed to allow the system to replace QuesTec for monitoring and evaluation, calling the new system Zone Evaluation. Fortunately for the baseball statistician, MLBAM publicly released all play-by-play and pitch location data from part of the 2007 season through to the present (2015 season). This data includes information on pitch trajectories and velocity, where pitches crossed the plate, pitch type, ball-strike count, base-out state, inning, the pitcher and the batter, runners on base, and the umpire(s) working the game. This data has been used extensively in modern baseball analysis and continues to evolve with more advanced integration of other data sources. As we demonstrate in Section 2.4.2.2, it is easy to acquire this data using the R package pitchRx (Sievert, 2014a). It can also be accessed through web sites such as http://www.brooksbaseball.net and http://www.baseballsavant.com.

In 2015, MLB began using what it has branded as Statcast data. This data uses the Trackman system that has been largely unreleased to the public, but used since at least 2011, paired with ChyronHego data. Trackman uses a radar system to reveal similar information to PITCHf/x such as velocity, movement, and where the pitch crossed the plate. It also adds the spin rate, angle of the ball, the angle off the bat, and information on the trajectory of a hit ball. ChyronHego tracks player movements, which has led to calculations of acceleration when stealing a base, measurements of baserunner leads, route efficiency for fielders, and top speed of baserunners and fielders. These can be used to evaluate quality of baserunning and fielding at a level that has not been reached with prior data. However, the large majority of this data is still not consistently made publicly available and is mostly used internally by teams, or for MLB Statcast broadcasts beginning in 2015. Most importantly, the data should lead to more precise measures of fielding ability, which will allow further identification of pitcher talent and performance independent of fielder quality.

2.3 Measures of Pitching Performance

2.3.1 Traditional Measures

Traditional measures of pitcher performance have serious flaws in both their ability to isolate a pitcher's contribution to run prevention and their ability to predict future performance of a pitcher. Traditional measures include Wins, which depend both on defense and the run scoring ability of a team's hitters, and ERA and batting average against (BAA), which are strongly dependent on defense. Other traditional statistics such as walks (BB), strikeouts (K), and home runs allowed (HRA) are more useful (Albert, 2006). As noted in Piette et al. (2010), this is particularly evident when normalized per inning or per nine innings pitched (K/9, BB/9, and HR/9) or as a rate statistic with the denominator being the total number of batters faced (K%, BB%, HR%).

As noted earlier, Voros McCracken and others have shown the importance of isolating pitcher performance from fielders and focusing on aspects of the game under control by the pitcher. For example, errors are dependent both on the quality of fielders behind the pitcher and on the subjective judgment of the scorekeeper for the game. By focusing on other outcomes and assuming BABIP is largely out of the control of the pitcher, the analyst can isolate performance in a more useful way. The advent of McCracken's observation has led to a plethora of measures focused on these other outcomes, though as noted earlier, there has been increased recent interest in balls in play with the advent of new ball tracking technologies.

2.3.2 Strikeout, Walk, and Home Run Rates across Eras

The availability of strikeout, walk, and home run rates makes yearly comparisons convenient for individual pitchers and the league as a whole. Strikeout rates have increased dramatically—part of a 50-plus-year trend. The scoring environment in MLB has decreased dramatically, particularly since the advent of performance-enhancing drug policies in MLB, making each home run allowed more important in the outcome of the game. In any analysis of the performance of pitchers across time, analysts can be misled without knowing the changes in base rates for strikeouts or run scoring. Figure 2.1 displays the time series of the overall strikeout rate, overall walk rate, and overall home run rate for the entire post–World War II era (1946–2015).

It is clear that a comparison of the strikeout rate of an average pitcher in 2015 to one just after WWII does not reveal particularly useful information in the context-specific skill level of a given pitcher. Because of the evolution of the game, standardizing pitcher statistics within season can be helpful to compare two pitchers from different eras. If this standardizing is not performed, it will be unclear whether or not the implied performance in a strikeout rate, for example, is a product of a changing environment in which the game is played, or attributed to innate talent in a given pitcher.

2.3.3 Defense Independent Pitching Statistics (DIPS)

Since McCracken's observation regarding balls in play, a number of statistics have been proposed that attempt measuring pitching performance independent of fielding performance. Basco and Davies (2010) summarize this development in SABR's *Baseball Research Journal*. Most DIPS measures derive from linear weights (Thorn and Palmer, 1985) that can be calculated through a Markov (or other) simulation using the various base-out states and associated transition probabilities. Marchi and Albert (2014) provide R scripts for most of these calculations using play-by-play Retrosheet data.

Fielding Independent Pitching (FIP), probably the most well-known DIPS measure, uses only home runs, walks, and strikeouts by the pitcher.[*] There are variations of FIP, with the general calculation as

$$FIP = C + \frac{\alpha * HR + \beta * BB - \gamma * K}{IP}.$$

In this notation, C is an arbitrary constant that is often identified as the league average ERA, allowing the measure to be interpretable on an ERA scale. The parameter α identifies the coefficient for home runs allowed (often between 11 and 13), β is the parameter for walks allowed (usually around 3), and γ is the parameter for strikeouts (often approximately 2). The parameters are estimated using relative linear weights run values for each outcome, and the measure is adjusted for the number of innings pitched by the pitcher. Typically the FIP measure is on a nine-inning scale, defined as expected runs allowed per nine innings pitched.[†]

---

[*] There are variations of this that include hit-by-pitches or remove intentional walks as well (Basco and Davies, 2010).

[†] Sabermetrician Tom Tango (2011) has a useful breakdown of how these parameters are estimated in the FIP formula. The reader is referred to this source for examination of the derivation of the measure: http://www.insidethebook.com/ee/index.php/site/comments/tangos_lab_deconstructing_fip/.

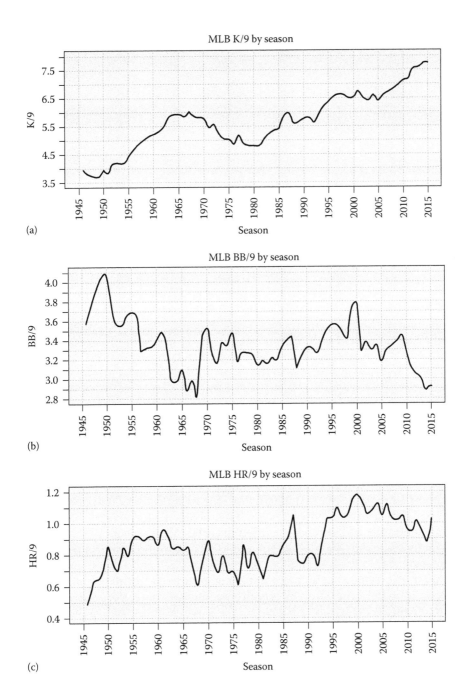

FIGURE 2.1
Plots of strikeouts per nine innings pitched (a), walks per nine innings pitched (b), and home runs per nine innings pitched (c) for the post–World War II era.

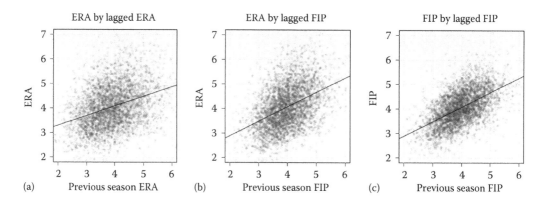

FIGURE 2.2

Plots of individual pitcher ERA in year t as a function of ERA in year $t-1$ (a; $r = 0.364$), ERA in year t as a function of by FIP in year $t-1$ (b; $r = 0.431$), and FIP in year t as a function of FIP in year $t-1$ (c; $r = 0.559$). Notice the stronger correlation for FIP in the rightmost panel. Data are from 1980 through 2015.

The FIP measure avoids the dependence on fielding performance and gives stable measures of pitching performance from year to year. Figure 2.2 constructs scatterplots between FIP in year t and FIP in year $t-1$, ERA in year t and FIP in year $t-1$, and ERA in year t and ERA in year $t-1$. The message from these plots is that the current FIP is a better predictor of the following season's ERA than the current ERA, and the current FIP is a relatively strong predictor of the following season's FIP. This is perhaps not surprising, given that we've already shown that strikeout, walk, and home run rates are more stable than other measures that depend on fielding or randomness of batted ball outcomes. In addition to being more stable, Piette et al. (2010) argue that FIP, HR/9, and BB/9 exhibit the most signal for starting pitchers, while ground ball percentage, fly ball percentage, and K/9 exhibit the most signal for relievers using a random effects model.

Variations of FIP include SIERA (Swartz and Seidman, 2010), xFIP (Studeman, 2005), and Component ERA (James et al., 2000). With the exception of Component ERA, each of these additional measures requires data on batted balls. Unfortunately, batted-ball data is often rather subjective and was not recorded in earlier MLB seasons.

Interestingly, measures such as SIERA and xFIP tend to be more stable over time than the FIP measure due to the repeatability of the ground ball and fly ball rates of individual pitchers, and FIP is influenced by a less stable home run rate. Ultimately, more precise data will improve upon these measures by identifying pitchers that reduce the quality of contact across batted balls. There is substantial room for research in this area, particularly with the advent of new data on balls in play.

2.3.4 Ground Ball and Fly Ball Rates

Identifiers for grounds balls and fly balls on balls in play have been available since 2002 on web sites such as Fangraphs. As with other outcomes like strikeouts, walks, and home runs, ground ball and fly ball rates are best presented as percentages. In this case, they are presented as a total percentage of all batted balls against a given pitcher. League average for ground ball, fly ball, and line drive rates in 2015 were 45.3%, 33.8%, and 20.9%, respectively. These rates show a little variation across seasons.

The importance of ground ball and fly ball rates relates to the relative value of inducing ground balls and fly balls, though this relationship is currently not well understood. Certain pitchers may be referred to as "ground ball pitchers" or "fly ball pitchers" if they have tendencies to allow pitches of one type. However, having ground ball or fly ball tendencies does not necessarily make one a high or low quality pitcher. For example, Dallas Kuechel found great success as a ground ball pitcher in 2015, with 66.3% of batted balls against him resulting in ground balls, while only 18.5% were fly balls. Alternatively, Max Scherzer was an early season Cy Young candidate in 2015 with only 36% ground balls and 45.4% fly balls. We know that fly balls are more likely to go for extra bases or over the fence for a home run, which might lead one to prefer the ground ball pitcher. Yet, Scherzer shows success can be had in the air. Further, there is some evidence that when ground ball pitchers do allow a fly ball, it is more likely to result in a home run (Swartz, 2011).

The type of pitcher preferred (ground ball or fly ball) depends somewhat on the ballpark in which they play regularly and the quality of infielders and outfielders behind them. For example, if a pitcher plays in a very large park like Citi Field—particularly with players like Juan Lagares roaming the outfield—it may be more advantageous to induce fly balls, as there is more space for outfielders to run down a ball that may otherwise go over the fence at smaller parks. Alternatively, at a stadium such as Coors Field, pitchers may want to avoid too many fly balls that could easily turn into home runs from the thin air.

The type of batted balls contributes to the probability that a given ball put in play will drop in for a hit. While we know that pitchers do not have full control over BABIP, batted ball data has revealed that there are some things that can be controlled by pitchers such as quality of contact (Swartz and Seidman, 2010). Further, it is important to note that ground ball and fly ball percentages include a substantial amount of subjectivity based on the judgment of those entering the data. The advent of more advanced data sets like Statcast will help reduce subjectivity in these classifications and add substantially to the literature on the value of certain types of batted balls. This is discussed in more detail in Section 2.3.6.

2.3.5 Velocity, Angle, and Hard Hit Balls

The availability of some Statcast data, beginning in 2015, has allowed the analyst to learn more about the patterns of balls hit in play. The data currently includes the velocity of balls off the hitters' bat and the angle at which the ball left the bat. This allows more precise and objective definition of the characteristics that define a ground ball, line drive, or fly ball.

One can model the probability of a hit, given the velocity and launch angle of the batted ball, to characterize balls that are "hard hit"—this is likely more accurate than a scorekeeper's judgment regarding what constitutes a line drive or fly ball. As a preliminary look, we model the probability of a hit conditional on batted ball velocity using the 2015 regular season balls in play from the two World Series contenders, the New York Mets and the Kansas City Royals. We use logistic regression to identify changes in hit probability as a function of this batted ball velocity. Here we are ignoring the importance of launch angle, which will play a role in where the ball lands and whether it is a ground ball or fly ball. Two models are fit, one with home runs included and one without.

TABLE 2.1

Logistic Regression of Hit Probability on Batted Ball Velocity

	No HR Included	HR Included
Constant	−4.676***	−5.327***
SE	(0.227)	(0.227)
Velocity (mph)	0.0436***	0.0518***
SE	(0.0025)	(0.0025)

Note: ***Refers to statistical significance at the 99% level.

From Table 2.1 one sees that an increase of 1 mph in batted ball velocity increases the probability of a hit—if the hit is in play—by about 4.5%.* If home runs are included in the model, one sees an increased hit probability of about 5.3%. This is not a surprising result, as we would expect balls that are hit harder would be more likely to drop in for hits.

Pitchers and pitching coaches can use this information to determine the best way to minimize batted ball velocity and therefore decrease hits. Making use of the locational information from PITCHf/x alongside the Statcast batted ball data—and using the methods in the following section—one can build models and visualize how hard batters hit pitches in different locations (Mills, 2015). In particular, if we assume that hard hit balls are more likely to fall in for hits, these visuals are informative about locations where the pitcher can minimize the probability of a hit by minimizing velocity off the bat. Generalized Additive Models (GAMs), discussed in Section 2.4.2, can be useful for this type of analysis.

2.3.6 Expected Run Values and Ball-Strike Count Progression

The ball-strike count, which is independent of fielding influence, can change the expected run value of an at-bat through strategic interaction from the pitcher. Increasing the number of strikes in the count relative to the number of balls could be seen as an independent skill set at a more granular level than even play-by-play data. The measured changes in run values can be used to measure that skill. Using techniques from Albert (2010), one can estimate that the average effect of changing a ball to a strike during the 2007–2010 seasons is about 0.146 runs (Mills, 2014a). Therefore, there is significant value in even individual pitches throughout the game.

Table 2.2 further breaks down values of pitches under various counts. These are likely to change depending on the sample used, as scoring has decreased rather substantially throughout the 2000s and 2010s from peak levels at the turn of the century.

Marchi and Albert (2014) further analyze ball-strike count effects, and how these result in changes in batter swings, noting that much of the data on batting outcomes after reaching a specific count can be found on Baseball Reference, with pitch sequencing information available through Retrosheet. Ultimately, Marchi and Albert show that the progression of the count is important, with a first pitch strike resulting in a decrease of offensive output of approximately 28%, while starting with an 0-2 count results in only 30% of the expected result at the beginning of the at bat. And one knows that the behavior of the batter will change with the count—for example, the batter is more likely to swing when he is behind in the pitch count.

---

* We exponentiate each coefficient for ease of interpretability.

TABLE 2.2[a]

Changes in Run Value by Pitch Count

Count	Run Value	Δ with Strike Call[a]	Δ with Ball Call	Difference[b]
0-0	−0.038	−0.043	0.038	0.081
0-1	−0.081	−0.052	0.025	0.077
0-2	−0.133	−0.300	0.013	0.313
1-0	0.000	−0.056	0.060	0.004
1-1	−0.056	−0.064	0.058	0.122
1-2	−0.120	−0.300	0.041	0.341
2-0	0.060	−0.058	0.107	0.049
2-1	0.002	−0.081	0.100	0.181
2-2	−0.079	−0.300	0.097	0.397
3-0	0.167	−0.065	0.330	0.265
3-1	0.102	−0.084	0.330	0.414
3-2	0.018	−0.300	0.330	0.630
Weighted Avg.[c]	—	—	—	*0.146*

Source: Modified from Albert, J., *J. Quant. Anal. Sports*, 6(4), 1, 2010; Mills, B.M., *Managerial and Decision Economics*, 35(6), 387, 2014a.

[a] This—as well as the difference after a ball call—is calculated as, for example, the difference between the run value of an 0-1 count and an 0-0 count: $(−0.081) − (−0.038) = −0.043$.

[b] Take the difference between a successive strike call and successive ball call in order to evaluate the net change from changing a strike to a ball or vice versa.

[c] Proportion of pitches thrown in the given count come from the data in Mills (2014a).

2.4 Tools for Analyzing PITCHf/x Data

In Section 2.4.1, techniques for visualizing pitch locations are described by the use of two-dimensional density estimation. Density estimation allows us to see patterns in a large amount of pitch locations, but does little to help *explain* those patterns or suggest what may have occurred under different circumstances. It is often more useful to have inferential tools to say, for instance, "Conditioned on this pitch location (and some other covariates), the estimated probability of event A is $P(A)$." GAMs are one attractive approach to making such inference, especially when the sample size is large.

GAMs and PITCHf/x data have allowed us to learn more about MLB pitching than just pitcher evaluation. Most noticeably in the academic literature, PITCHf/x has allowed researchers to better understand umpire decision-making with respect to the strike zone (Green and Daniels, 2015; Hamrick and Rasp, 2015; Kim and King, 2014; Mills, 2014, 2014b; Moskowitz and Wertheim, 2011; Parsons et al., 2011; Tainsky et al., 2015). Catchers have also been analyzed using this data (Fast, 2011; Judge et al., 2015; Marchi, 2013; Pavlidis and Brooks, 2014).

Publications on umpire decision-making have modeled the probability of a called strike as a function of pitch location (as it crosses the front of home plate), and other covariates, such as year, ball-strike count, and home field advantage. In Section 2.4.2, we provide a roadmap for building such models in R, from data collection, to model specification, model fitting and diagnosis, as well as visualization. A concise introduction to Generalized Additive Models (GAMs) is provided in Section 2.4.2.1 before diving into an application in Section 2.4.2.2. These methods are not restricted to umpire decision-making but also helpful in modeling a pitcher's ability to control pitches and make good decisions on the mound.

In Section 2.4.3, we shift focus from estimating densities and event probabilities to classification and clustering methods. The pitch type labels that come with publicly available PITCHf/x data are automatically created using a classification algorithm, which uses variables such as velocity and spin (Fast, 2008), but there is reason to believe the algorithm itself could be improved (Pane et al., 2013). In Section 2.4.4, some of the underlying physics is discussed on how these variables relate to pitch trajectory and classification.

2.4.1 Visualizing Pitch Location Frequencies

PITCHf/x data provides the vertical and horizontal locations of the ball the moment it crosses the front of home plate. There are some known accuracy issues with these locations, but previous work suggests that when the system is properly calibrated, they randomly deviate about a half of an inch away from the exact location (Nathan, 2008). Nevertheless, it is well known that pitch location is tied to pitcher performance (in general, it is more difficult to hit a baseball low in the zone as opposed to high in the zone); so pitch location summaries can offer insight into performance (as well as other things such as umpire decision-making). Sometimes it is useful to collect locations of hundreds or thousands of pitches, but a basic scatterplot of the pitch locations leads to an uninformative graphic, a problem known as overplotting.

There are many proposed graphical solutions to overplotting, each having its own drawbacks. Some approaches simply alter certain visual characteristics of the graphic (Few, 2008), while other methods are based on the estimation of the underlying probability density (so-called density estimation). In Figure 2.3a, a large number of pitches thrown by Yu Darvish from 2009 to 2012 are graphically displayed using a density estimate where the alpha transparency parameter is set such that 100 points would have to be overlaid on the same spot in order to appear fully opaque. This allows us to see where the highest density of pitches occurs, but it is difficult to infer just how many pitches were thrown in a particular location. There is no good rule of thumb for setting the transparency level, so this approach requires some trial and error to generate a decent-looking graphic.

Altering the visual characteristics of a graphic, such as the alpha transparency, can help reveal some structure, but it may not work well when most points are concentrated within a small area and the remaining points are scattered over a large area. A more robust approach is to use a binning and/or smoothing procedure to summarize the data into contiguous regions, thus eliminating the problem of overlaying points.

Two basic examples of binning are presented in Figure 2.3: one with rectangular binning and one with hexagonal binning. Binning algorithms are easy to understand as one simply counts the number of observations that fall within each region (absolute frequency), and (optionally) divide by the total number of observations to obtain a density estimate. In this case, the absolute frequencies are encoded using a linear color scale resulting in a graphic commonly called a heat map or level plot. Absolute frequencies are easier to interpret and are helpful when comparing multiple scenarios since they provide a sense of the total number of pitches.

Binning algorithms are simple, but it can be difficult to select an optimal number of bins. In the case of rectangular binning, traditional solutions depend on characteristics of the unknown underlying distribution and a default rule of thumb assumes a Gaussian

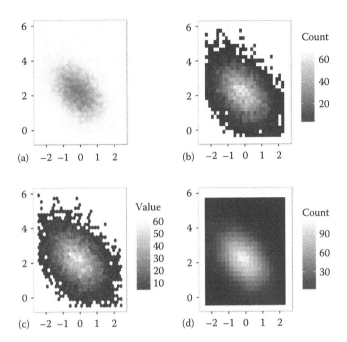

FIGURE 2.3
(See color insert.) Four plots of Yu Darvish's pitch locations from 2012 to 2014. These plots take the perspective of the umpire at the pitch crosses the front of home plate. (a) All pitches with alpha transparency, (b) 2D histogram with rectangular binning, (c) 2D histogram with hexagonal binning, and (d) bivariate normal kernel density estimate.

form (Scott, 2015). There are a number reasons to prefer hexagonal to rectangular binning (for one, hexagons are less visually distracting); however, there is relatively little literature on the choice of an optimal or sensible default for bin sizing.

Density estimation via binning can suffer from high variance. The same data and different anchor points for the bins can result in very different values for the binned frequencies. The high variance problem can be avoided by the use of kernel-based density estimation at the expense of higher bias if the bandwidth parameter is misspecified. The **MASS** package in R (Venables and Ripley, 2002) provides the **kde2d()** function for 2D density estimation with a bivariate normal kernel as well as a sensible routine for choosing the bandwidth of the kernel estimate (choosing the bandwidth is essentially the continuous version of choosing the number of bins in a 2D histogram). The density portrayed in Figure 2.3d was generated using **kde2d()**'s default bandwidth selection and the number of bins were set to match the 2D histogram in Figure 2.3b.

One can also apply some sort of smoothing technique to the binned values before visualizing. The **persp()** and **contourLines()** functions in the R package **graphics** both use interpolation techniques to approximate a smooth surface over an *xy*-plane, but the former creates a 3D plot whereas the latter projects the surface into two dimensions using contour lines to encode values of the *z*-axis. We do not recommend 3D surface plots since they can be visually deceptive and obscure/block interesting regions of the surface; however, interactive versions where users can alter the perspective can help alleviate these issues.

It's often useful to compare and contrast a number of density estimates across a variety of scenarios. Generally one should separate pitches thrown to left-handed batters from

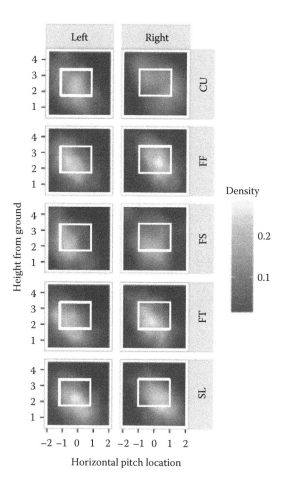

FIGURE 2.4
(See color insert.) Bivariate normal kernel density estimates of Yu Darvish's pitches for each combination of batter handedness and pitch type.

right-handed batters since these densities are typically very different. For this reason, the R package **pitchRx** provides the function `strikeFX()` to help quickly create many densities in a small amount of code (Sievert, 2014a). Figure 2.4 shows 10 different bivariate normal kernel density estimates of Yu Darvish's pitches: one for each combination of batter handedness and pitch type. Clearly, Darvish (as with most pitchers) tends to throw down and away to batters. Interestingly, Darvish appears to have thrown two-seam and sinking fastballs down and inside to right-handed batters.

A number of approaches to visualizing pitch locations using various 2D density estimation techniques have been described. For more details on how Figures 2.3 and 2.4 were actually generated, see the code provided at https://gist.github.com/cpsievert/fd83ec5516a07ab59c36. It is important to note that density estimation, in itself, doesn't allow for likelihood comparisons conditioned upon a pitch location. This type of inference is important, especially for studying umpire calls where, for instance, one would like to compare the probability of a called strike for various umpires. The next section covers one particular statistical method that allows us to make this type of inference.

2.4.2 Generalized Additive Models (GAMs)

A detailed overview of the Generalized Additive Models (GAMs) (Hastie and Tibshirani, 1986) framework is beyond the scope of this chapter. However, we introduce this method in Section 2.4.2.1, focusing on important properties, and apply them to PITCHf/x data in Section 2.4.2.2 using R (R Core Team, 2015). Clark (2013) provides a more detailed introduction to additive models in R and Wood (2006) gives a comprehensive introduction to the GAM framework, and its implementation in the R package **mgcv**.

2.4.2.1 A Brief Overview of GAMs

Those familiar with the Generalized Linear Models (GLMs) framework (Nelder and Wedderburn, 1972) will find many connections between GLMs and GAMs. The "generalized" term essentially means that in either framework, one can model the response distribution with any distribution belonging to the exponential family. The exponential family includes a number of common distributions such as normal, binomial, gamma, Poisson, etc. This flexibility is crucial for modeling PITCHf/x data, or any type of data, where the response of interest is a count, percentage, or rate observed over time. In cases where the response does not follow any one distributional form, one can indirectly select a distribution by modeling the relationship between its observed conditional mean and variance.

In addition to a distributional form, both frameworks require a one-to-one function, $g(\cdot)$, which links the conditional expectation, $E(Y|X = x)$, to the predictor space η, so that $g(E(Y|X = x)) = \eta$. This link function is necessary since there is no guarantee the range of $E(Y|X = x)$ will match the range of η. For example, if Y is a Poisson random variable, then $E(Y|X = x)$ is always positive (since $Y > 0$), but there is nothing to restrict η to be positive. In this case, we seek a function such as `log()` that can map the positive real numbers to all real numbers.

The key difference between generalized *linear* models and generalized *additive* models is in the assumed form of η. In the GLM setup, η is assumed to be a *linear* combination of independent variables multiplied by unknown coefficients $\beta_0 + \sum_{i=1}^{p} \beta_i X_i$. In the basic GAM setup, η is a sum of smooth functions of the independent variables $\beta_0 + \sum_{i=1}^{p} f_i(X_i)$, which is referred to as a nonparametric model. In practice, it is common to have a semiparametric model that contains both unknown coefficients and smooth functions; for example, $\eta = \beta_0 + \sum_{i=1}^{r} \beta_i X_i + \sum_{j=r}^{p} f_j(X_j)$.

A natural question at this point is "when are additive terms in a nonparametric model preferred over parametric terms in a linear model?" Parametric terms are generally more interpretable but require strong assumptions about the predictor space, whereas additive terms can automatically detect nonlinear effects. This flexibility is quite useful for modeling strike-zone event probabilities, as it's common to see a nonlinear relationship with pitch location. This flexibility comes at a computational price, however, as the "degree of smoothness" must be estimated using a cross-validation technique. Thankfully, recent advances have produced computationally efficient methods for estimating the level of smoothness (Wood, 2011).

2.4.2.2 Modeling Events over the Strike Zone with GAMs

In this section, we build a GAM to estimate the probability of a called strike as a function of pitch location and year. Although only binary outcomes over the strike zone are covered, remember the GAM framework supports any exponential family, so the response could even be multinomial. The multinomial case is more suitable for building more general models since a pitch can result in a number of outcomes.

Since the response distribution is binary, a natural choice for the link function is the logit(x) = log($x/(1 - x)$), which gives us a mapping from the domain of the response mean [0, 1] to the domain of the predictor space $(-\infty, \infty)$. Now let $Y_i \in \{strike, ball\}$ be the outcome of pitch i, X_{1i} be the corresponding vertical location, X_{2i} the horizontal location, $Z_{h(i)}$ be an indicator variable of whether the batter was left or right handed, and $Z_{t(i)}$ be an indicator of whether the pitch was thrown in 2008 or 2014. Then a sensible GAM model would be

$$\text{logit}(EY_i) = \beta_0 + \beta_1 Z_{h(i)} + \beta_2 Z_{t(i)} + \beta_3 Z_{h(i)} Z_{t(i)} + f_0(X_{1i}, X_{2i})$$
$$+ f_1(X_{1i}, X_{2i})Z_{h(i)} + f_2(X_{1i}, X_{2i})Z_{t(i)} + f_3(X_{1i}, X_{2i})Z_{h(i)}Z_{t(i)}$$

Note that with this model structure, we have one joint smooth function of horizontal and vertical location for each factor level. The reason why a *joint* function is preferred over marginal functions for each coordinate is that, contrary to the rulebook definition, strike zones are asymmetric Also, there is a different smooth for each factor level since the probability of a called strike on the edge of the rulebook strike zone is known to depend on batter stance and year.

The complete analysis, from data collection to model fitting, diagnostics and visualization, can be performed entirely within R. The R package **pitchRx** makes it easy to collect all available PITCHf/x data and store that data in a database (Sievert, 2014a). Storing PITCHf/x in a database is convenient (but not necessary) since there are many tables available that record information on various levels: pitches, at bats, hits in play, baserunning, games, players, umpires, coaches, etc. **pitchRx** works with any database connection, but it's particularly easy to create a SQLite database from R with **dplyr** which is illustrated here.

```
library(dplyr)
library(pitchRx)
db <- src_sqlite("pitchRx.sqlite3", create = TRUE)
scrape(start = "2008-01-01", end = Sys.Date(), connect =
db$con)
```

Now that a complete PITCHf/x database has been obtained, the package **dplyr** can be used to query variables of interest*:

- px: the horizontal location of the pitch (as it crossed the front of home plate)
- pz: the vertical location of the pitch (as it crossed the front of home plate)
- des: A description of the pitch outcome
- num: The order of at bat (within game). This is used to link pitch info to at-bat info.
- gameday_link: A unique identifier for game. This is used to link pitch info to at-bat info.

```
pitches <-tbl(db, "pitch") %>%
select(px, pz, des, num, gameday_link) %>%
filter(des %in%c("Called Strike", "Ball"))
```

---

* A more complete list of PITCHf/x variable descriptions can be found here: https://fastballs.wordpress. com/2007/08/02/glossary-of-the-gameday-pitch-fields/.

For this model, one also needs the stance of the batter and year, which is recorded on the at-bat level. For visualization purposes, the height of the batter is also used to draw an average rectangular strike zone for reference.

```
atbats <-tbl(db, "atbat") %>%
mutate(year =substr(date, 5L, -4L)) %>%
select(stand, b_height, num, year, gameday_link)
```

Now these tables are joined together, attention is restricted to 2008 and 2014 seasons, and an indicator variable for called strikes is created.

```
dat <- pitches %>%
left_join(atbats, by = c("gameday_link", "num")) %>%
filter(year %in% c("2008", "2014")) %>%
collect() %>%
mutate(strike = as.numeric(des == "Called Strike"))
```

Last, the model proposed at the start of this section is fit. Here the argument k is specified to increase the dimension of the basis allowing for smoother surfaces but reducing the computational efficiency. This increase in dimension was guided by diagnostics provided by the **gam.check**() function which suggested the initial fit using the default value for k was too smooth.

```
library(mgcv)
m <-bam(strike  ~interaction(stand, year) +
s(px, pz, by =interaction(stand, year), k = 50),
data = dat, family =binomial(link ='logit'))
```

The **strikeFX()** function in the **pitchRx** package is designed to work with **mgcv** for quick visualization of estimated probability surfaces over the strike zone. It also has support for visualizing the differences, as illustrated in Figure 2.5. In addition to visualizing a point estimate of difference, it can also be useful to compute the actual probabilities and associated measures of uncertainty. For quantifying this uncertainty, we recommend the bootstrapping procedure shown in Wood (2006). For an interactive visualization of all this information, see Sievert (2014b).

```
strikeFX(dat, model = m,
density1 =list(year ="2008"),
density2 =list(year ="2014")) +
facet_grid(. ~stand)
```

It is important to note that GAMs are not limited to strike probabilities. There are a number of other outcomes that are important to the pitcher, such as the probability of making contact, the quality of contact, whether balls in play become fly balls or ground balls, which pitches tend to be hit for home runs, and so on. The package used to fit these models in R is also not limited to the logit link function, and other appropriate links can be used. For example, when modeling exit velocity, the Gaussian link function may be used to model this continuous response data. Finally, these models are not restricted to pitching, as they could be used to further model fielding prowess with models of the probability that batted balls fall for hits, conditional on launch angle, batted ball velocity, and landing point.

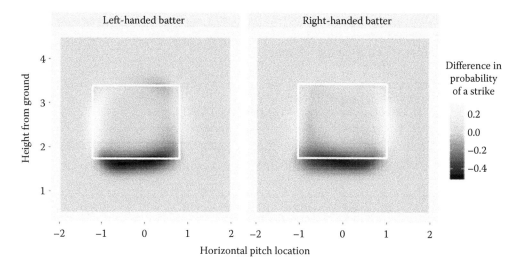

FIGURE 2.5
(See color insert.) The difference in the probability of a called strike in 2008 versus 2014 for both left- and right-handed batters. A pitch over the plate at the knees was about 40%–50% more likely to be called a strike in 2014. A pitch on the inner/outer portion of the plate (above the knees) was about 20%–30% more likely to be called a strike in 2008.

2.4.3 Cluster Analysis and Pitch Type Identification

While understanding the location of pitches, called strikes or balls, and which pitches will be hit well or poorly are key components for understanding pitchers, it is also helpful to identify the various types of pitches with different velocities and movements. Cluster analysis can be useful for the analyst in understanding these characteristics.

One of the key goals is the identification of the pitch types so that one may evaluate the success of each pitch in the context of the batting outcome and previous choices made. PITCHf/x data conveniently provide pitch classifications using a machine learning algorithm. However, the method used for this classification is not public, and it may be possible to improve upon this classification with statistical tools (Pane, 2013; Pane et al., 2013). This is ultimately an unsupervised learning problem as there is no training data with known membership to preface group assignment for the data of interest.

One can first think about the pitch type clustering problem by projecting pitches onto a two-dimensional space, inspecting characteristics of points, and look for separate groups. The goal of the clustering methods is to identify these pitches in their respective groups as closely as possible based on the given characteristics. It is desirable to identify every pitch exactly as it was intended with zero misclassifications, though the likelihood of perfect classification of each pitch is unlikely, due to variability around the spatial characteristics of human physical actions. Figure 2.6 shows the clusters of some randomly generated data first without assignment, and then with assignment denoted by color. The goal of the clustering methods is to find these clusters from the data all in black (Figure 2.6a), and identify them as separate groups as shown in red, green, and black (Figure 2.6b). Note that the clusters below are well separated; however, this is not usually the case with all pitch types gleaned from pitch data.

A number of clustering methods are potentially useful for this task, including hierarchical clustering (agglomerative or divisive), *k*-means clustering, and model-based clustering.

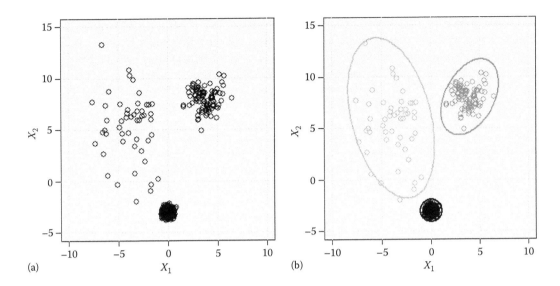

FIGURE 2.6
(See color insert.) Exhibition of unclassified data (a) and cluster assignment (b) of generic, randomly generated data.

The choice of clustering method depends largely on the structure and variance of the clusters, and the knowledge of the analyst about the number of different pitches a pitcher throws. If possible, clustering should always be done on the individual pitcher level, since differences across pitchers can confuse the clustering method.*

It is important to note that clustering does not require both a dependent and independent variable, but identifies cluster membership based on a group of variables (e.g., characteristics of each pitch) that the analyst finds appropriate for assigning pitches to different types. These features often include velocity, spin rate, spin direction, and trajectory of the pitch (specifically, trajectory relative to expectations for a four-seam fastball). The k-means and model-based clustering methods are described here, but it is worth mentioning that investigations into the appropriateness of hierarchical methods may also be worth examining.

2.4.3.1 k-Means Cluster Analysis

k-means clustering is a partitioning method requiring the researcher to choose a priori the number of pitch types, k, as well as the initial location of cluster centroids in a p-dimensional feature space. The initial locations are arbitrary and are commonly chosen at random or based on a previously developed hierarchical clustering method. To begin, the k-means algorithm assigns each pitch to its closest centroid in the feature space. The clustering method continues along this path, reassigning centroids and assigning data points to the closest centroids in an iterative fashion according to the Hartigan and Wong (1979) algorithm. The initial placement of these centroids may impact the quality of the clustering solution, and it is recommended that the initial placement of centroids be considered carefully, with possibly multiple starting points attempted. Further, while the k-means

---

* For example, Mark Buehrle's four-seam fastball averaged about 84 mph in 2013, while a change-up from a pitcher like Noah Syndergaard registers at about 88 mph.

iterations reach a stabilized solution relatively quickly, it is worth considering the number of iterations that balance a reasonable solution and computation time.

In the case of pitch clustering, characteristics such as horizontal and vertical movements, angle of the movement, and velocity of the pitch tend to be most useful. Work from physicist Alan Nathan (2008, 2012) discusses how to identify movement of pitches based on the provided variables in PITCHf/x data. But the data itself provide three very useful variables for clustering that are more intuitive before employing the tools of a physicist yourself: *start_speed*, *break_length*, and *break_angle*. The *start_speed* variable simply identifies the velocity of the pitch out of the pitcher's hand in miles per hour (mph). The *break_length* measure—as defined by Cory Schwartz, Vice President of Stats at Major League Baseball Advanced Media (MLBAM)—is "the measurement of the greatest distance between the trajectory of the pitch at any point between the release point and the front of home plate, and the straight line path from the release point and the front of home plate.*" Alternatively, the *break_angle* measure is "the difference between a pitch dropping perfectly perpendicular to the ground, and the actual trajectory of the pitch." The sign of *break_angle* encodes whether the pitch breaks toward the pitcher's handedness (positive if breaking across the pitcher's body, otherwise negative) and the magnitude encodes the actual break angle (0 would indicate the ball drops directly toward the ground).

Together these three PITCHf/x variables give us information about the traditional main characteristics that distinguish pitch types: the amount of movement, the direction of movement, and the velocity of each pitch. It is important to note that scaling variables can be helpful when variables are defined on different scales. This is clearly the case here, where the model includes characteristics of pitches like velocity, measured in miles per hour, break length, measured in inches), and break angle.

Figure 2.7a shows a scatterplot of *start_speed* and *break_angle* for pitches thrown by Mark Buehrle's during the 2013 season. A k-means classifier (k = 5) was fit to these pitches using *start_speed*, *break_length*, and *break_angle* for feature variables and the R function `kmeans()` in the **stats** package. The result of that fit is shown in Figure 2.7b. In this case, our feature space is 3D, so we *could* produce a 3D graph to view the model fit with respect to the entire feature space, but a method known as touring would be preferable since it allows us to view a high-dimensional feature space, which is a powerful tool for diagnosing statistical models such as k-means (Wickham et al., 2015).[†]

Understanding the relative quality of clustering solutions can be important in initially developing useful classifications of pitch types. The quality of the clusters using *k*-means methods can be measured using homogeneity (similarity within each cluster) as well as separation (difference across clusters). A popular measure of homogeneity is within-cluster sum of squared error (WSSE), which is defined as the sum of squared distances between each individual pitch and its cluster centroid. A popular measure of separation is between-cluster sum of squares (BSSE), which is defined as the sum of squared distances between each cluster centroid and the grand mean of the data.

The use of a silhouette coefficient and associated silhouette plot combines the homogeneity and separation considerations into a single measure. This silhouette coefficient is measured using information on the average distance a_i of each object i from each other

---

[*] From a discussion at The Book Blog (2007) at http://www.insidethebook.com/ee/index.php/site/comments/everything_you_ever_wanted_to_know_about_gameday/.

[†] Touring is a technique that requires dynamic interactive graphics software. For a simple video explanation and demonstration of touring PITCHf/x data, see Sievert (2015b).

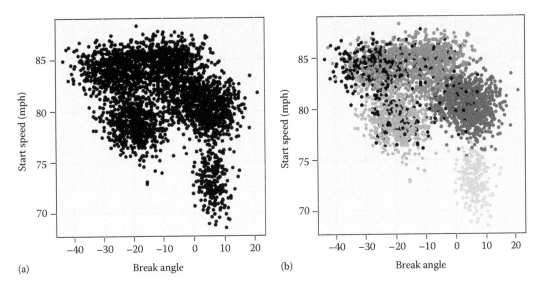

FIGURE 2.7
(See color insert.) Exhibition of unidentified pitches from Mark Buehrle's 2013 regular season (a) and pitch cluster assignment (b) using k-means clustering with five clusters.

object in its respective cluster, as well as the minimum average distance b_i of object i to objects in other clusters. The silhouette coefficient is defined as

$$s_i = \begin{cases} 1 - \frac{a_i}{b_i} & \text{if } a_i < b_i \\ \frac{a_i}{b_i} - 1 & \text{if } a_i \geq b_i \end{cases}$$

Typically, $s \in [0, 1]$, with values closer to 1 indicating better quality clustering. The average silhouette width can be calculated for a single cluster, as well as an entire clustering output by averaging the silhouette coefficient across objects (pitches). These coefficients can be plotted to visualize cluster assignment quality from the given solution.

There are, however, a number of drawbacks in using the k-means clustering method. First, the a priori selection of K requires the analyst to know how many pitches each pitcher intends to throw. Further, this specification may be too restrictive in that even if a pitcher is noted to throw a single curveball, the variability in that pitch may result in two different pitch types as perceived by the batter. Second, iterations of the centroids may result in either empty or very small clusters, in which case some outliers may be removed for better performance of the method. Even in the case where the sparse or empty issue does not arise, it is possible that iterations of cluster placement remain in a local optimum, rather than a global optimum for the clustering problem. Finally, the method does not handle differing size, density, and shape of clusters or other methods, such as model-based clustering.

2.4.3.2 Model-Based Clustering

Model-based clustering (Fraley and Raftery, 2002) is the least restrictive method of clustering pitch types—there is no need for the researcher to a priori choose the number of clusters (pitches) in a given pitcher's arsenal—and exploration of the data reveals that the various cluster densities (most pitchers have a primary pitch, which will occur much more often than others) and variance structures may be well addressed using probabilistic models. Model-based clustering uses population subgroup (cluster) densities to identify group membership based on maximum likelihood and the EM (Expectation-Minimization) with parameterized Gaussian mixture models (Fraley and Raftery, 1999). Further, the method estimates cluster membership probabilities for each observation (pitch) across each respective cluster (pitch type). Clusters are determined using the Bayesian Information Criterion (BIC), comparing across models with different parameterizations (number of clusters, size/shape, and orientation). Conveniently, R has its own package to implement this method (**mclust**; Fraley and Raftery, 1999, 2007; Fraley et al., 2012).

Fraley and Raftery (2007) exhibit that the clusters in this method come from a mixture density, $f(x) = \sum_{k=1}^{G} \tau_k f_k(x)$, where f_k is the probability density function of observations in group k, and τ_k is the probability that an observation comes from the kth mixture component. Components are largely modeled using a Gaussian distribution, characterized by mean μ_k and covariance matrix Σ_k, with probability density function

$$\phi(x_i; \mu_k, \Sigma_k) = \frac{\exp\left\{-(1/2)(x_i - \mu_k)^{\mathsf{T}}\Sigma_k^{-1}(x_i - \mu_k)\right\}}{\sqrt{\det(2\pi\Sigma_k)}}.$$

The likelihood for the data with n observations, with G components is

$$\prod_{i=1}^{n}\sum_{k=1}^{G}\tau_k \phi\,(x_i; \mu_k, \Sigma_k).$$

Fraley and Raftery (2007) further discuss the EM algorithm and geometric constraints on the G components, to which the reader is referred to both for the technical aspects of the clustering method as well as its implementation in R.

Pane et al. (2013) originally identified the usefulness of model-based clustering in the analysis of baseball data, and developed an adjusted Bayesian Information Criterion (BIC) for use with pitch data that avoids identifying too many sparse clusters. The traditional BIC method tends to result in too many identified clusters. The additional penalty from the adjusted BIC measure is shown to perform better with respect to this dimension of the method, and penalizes high intracluster correlations. Specifically, it reduces the number of chosen clusters, k, from a to b, $b < a$, when the intracluster correlation for $k = a$ is much higher than for $k = b$. The adjusted BIC measure is then calculated as (taken directly from Pane, 2013) follows:

$$BIC_{adj} = -2\log(f(Y|\hat{p})) - 2\lambda \sum_{i}\log(f(c_i)) + \left[k \cdot \left(j + \frac{j(j-1)}{2} + (k-1)\right)\right] \cdot \log(n).$$

As noted in Pane (2013), $f(Y|\hat{p})$ identifies the probability of the parameters given the data, $f(c_i)$ is the sum of the upper off diagonal elements of a correlation matrix, Σ_k, and j is the number of clustering variables. Further, the tuning parameter, λ, was determined using a training data set in Pane (2013), with $\lambda = 0.5$ being determined as the optimal tuning parameter choice for reducing pitch misclassifications.

Using this BIC_{adj} measure, we can visualize a solution to the clustering problem posed earlier for Mark Buehrle in Figure 2.8, with five pitch types identified.

Note the improvement in clusters going from k-means to model-based clustering, particularly with the black cluster that is likely to be two-seam fastballs. The k-means solution

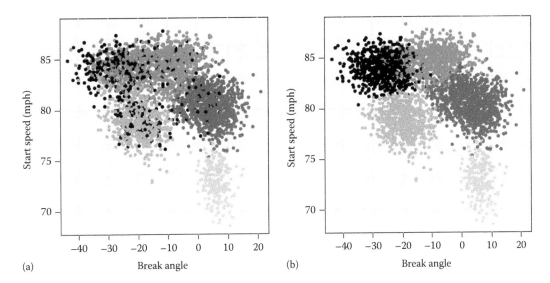

FIGURE 2.8
(See color insert.) Exhibition of Mark Buehrle's 2013 pitch cluster assignment using *k*-means clustering (a) and cluster assignment using model-based clustering (b). The latter shows more clear cluster assignment, particularly for the black-colored cluster.

resulted in a much more scattered cluster, while the model-based solution provides much tighter cluster assignment. The two-seam fastball classification difficulty was noted in Pane (2013), with the model-based solution providing more uniform clusters than other methods. It is likely that the darker blue cluster is a group of cutters, while the red cluster is four-seam fastballs. Buehrle's change-up is made up of the green cluster, while his curveball is the light blue cluster. Note that pitches to the left of the plot indicate "arm side movement," while those to the right of the plot indicate movement across the pitcher's body.

Ultimately, there are important limitations of clustering alone that relate to the inability for an outside researcher to know which pitch was *intended*. With any clustering method, there will exist misclassified pitches that could lead to incorrect conclusions regarding the strategy of the pitcher. Further knowledge of pitcher intention can help to identify shortcomings in approach, or provide a feedback loop for development of pitches.

For example, if there is substantial variation in the movement of a pitcher's curveball, then the use of clustering algorithms can tell us about the variation within this pitch type. Depending on the goals of the pitcher—developing two separate pitches, or increasing consistency in one pitch—the matching of cluster type and intention becomes paramount in this developmental process. Without knowledge of the intention, coaches and staff may not be able to intervene and give feedback on improvement. While this, of course, requires access to the players, analysts should be aware of the limitations of simplistic identification in the context of improving our understanding of pitching and pitch quality. However, a reasonably well-designed clustering method could reveal useful characteristics of strategic decision-making among pitchers, even for the outside analyst.

2.4.4 Pitch Movement and Trajectory

An important aspect of identifying pitch types after clustering includes understanding how the various measures used to cluster the pitches impact the trajectory of the pitch.

Without understanding this trajectory, identifying what each cluster represents in terms of pitch type would be rather difficult. Pitches are often classified casually by their respective trajectory and velocity—fastball, curveball, changeup, knuckleball. In this vein, Nathan (2008, 2012) lays the foundation for identifying pitch trajectories and defining the movement of a pitch as "the amount by which the trajectory deviates from a straight line with the effect of gravity removed." Relating the movement of a pitch—ultimately its trajectory deviation—back to expectations about the perceived movement of certain pitch types reveals the identity of each cluster in the absence of information on the pitcher's original intention with the pitch. While the physics behind pitch trajectories is beyond the scope of this chapter, Pane (2013, p. 18) includes a useful table to evaluate pitch types based on speed, vertical break, and horizontal break.

2.5 Current Extensions

2.5.1 Umpire Influences

There is a now relatively extensive academic literature making use of PITCHf/x data to evaluate umpire decision-making, and the impact that this has on the performance of players and general game play. The first paper to leverage this data in the academic literature is Parsons et al. (2011), who claimed to find evidence of race-based discrimination in umpire ball-strike calls. More recent work has called into question whether this is the case (Hamrick and Rasp, 2015; Tainsky et al., 2015), but the main crux of Parsons et al. (2011) remains: otherwise objective performance measures can be contaminated by subjective judgment of officials, and these must be accounted for when estimating specific skill. This is identical to the lesson of McCracken (1999) with respect to fielders.

Further, Moskowitz and Wertheim (2011), Green and Daniels (2015), and Mills (2014a) found that the size of the strike zone fluctuates depending on recent calls and the current ball-strike count, with calls favoring the pitcher (batter) when he is behind (ahead) in the count. And Mills (2014a) and Kim and King (2014) showed home plate umpires tend to give more favorable calls to both players with a higher status in the league as well as players who remain in close proximity to the umpire during the game. Further, Mills (2014b) and Roegele (2014) found evidence of strike-zone expansion from 2008 to 2014, which is believed to be a large contributor to the decrease in runs scored over the same period. Each of these considerations makes important contributions to the measurement and isolation of pitching performance.

2.5.2 Catcher Influences

The implications of the influence of umpires generalize to the influence of any other actor that could impact the measurement of pitcher performance. This, of course, includes the catcher. Work by a number of analysts (Fast, 2011; Judge et al., 2015; Marchi, 2013; Pavlidis and Brooks, 2014) has revealed not just catcher influences on pitching performance but have managed to use this information to evaluate catchers themselves. Many of these models are hierarchical in nature to identify catcher effects for pitchers when they throw to different catchers across a season or career.

In particular, Pavlidis and Brooks (2014) use PITCHf/x data and a GAM to estimate a probabilistic model of the strike zone—as discussed in Section 2.4—and attribute residuals to catchers. These are then applied to the run expectancy across each count to evaluate the total (retrospective) framing value of each catcher in the league (Tango et al., 2006). With these estimates in place, the impact of the pitcher's skill level can be isolated from this additional impact.

2.6 Future Challenges

2.6.1 Development and Minor League Projection

MLB teams are currently expanding their investments in technology for use in the development of players both in the minor leagues and when they reach MLB. Development using this technology requires a continuous feedback process, such as noting whether a curveball conforms to the pitch type cluster for an individual pitcher immediately after throwing the pitch. Or, release point data immediately available could result in pitchers developing better command more quickly as they progress. Full integration of data sources and communication with sports scientists should improve development as data availability and analysis techniques advance within franchise operations.

The intersection of statistical analysis using this advanced data and providing direct and actionable feedback is likely to be the most valuable future endeavor for pitching. Previously, pitcher development relied on the naked eye of pitching coaches and catchers to identify specific issues with pitches or approach. However, with new data—including physiological and biomechanical measurements—coaches will be able to provide more objective feedback related to shortcomings in a pitcher's game. While this seems to be a fruitful area of research, it clearly requires an insider's knowledge and access in order to bring it to its full potential.

2.6.2 Quality Pitches and Strategic Considerations

Understanding what makes a successful pitch, and of course how to throw it, has been a question long pondered by pitchers and pitching coaches alike. It is becoming more common to see higher-velocity pitches across the league, and given that this allows less reaction time for the batter, it seems clear that it would be more difficult to make solid contact with higher velocity pitches (all else equal). One might also assume that pitches that have less traditional movement would be harder to hit; however, there have been very few consistently successful knuckleball (perhaps the most nontraditional trajectory) pitchers in MLB. But we also know that these pitches with more movement tend to travel slower than a straight fastball. Contextual clues and recency effects—or the impact of seeing a fastball before a curveball—should play a role in future statistical evaluations of pitching success using the rich data available to the analyst.

While descriptive statistical analysis—as well as projections of future performance—has been widespread and accessible for many practitioners in sports, strategic interaction between pitchers and batters has been sorely under-researched. Surely, the problem at hand is much more complex than a simple description of past output; however, with the advent of PITCHf/x and Trackman/Statcast data, there is opportunity to test strategic theories within the sport. Marchi and Albert (2014) and Alamar et al. (2006) touch on this to some extent, noting some behavioral changes that take place with certain pitches and in certain counts. And in soccer, for example, economists have investigated the use of mini-max and other strategies on penalty kicks (Palacios-Huerta, 2003). There is little reason to believe that this sort of evaluation cannot be extended to pitching, though the task requires substantially more choices among players.

2.7 Conclusion

Measuring pitching performance is difficult, mostly due to the fact that run prevention depends on both pitching and fielding. Compared to traditional statistics like Wins or ERA, DIPS measures (functions of strikeouts, walks, home runs) are a better measure of performance but also ignore important information on balls hit into play. Probabilistic models are a sensible way to distribute run prevention contributions between the pitcher and fielders, and our ability to estimate the "true" distribution will only improve as more granular data becomes available.

Probabilistic models for categorical variables that can incorporate spatial and temporal components as covariates (e.g., Generalized Additive Models) will be important not only for estimating run prevention contribution but also for understanding the influence of confounding factors, such as bias in umpire decision-making or influences of catchers on ball-strike calls. The ability to integrate Statcast with PITCHf/x data will be a crucial first step to such analysis, which is not a trivial data management problem, and will require new tools to provide easy access to the data. There is ample opportunity for extension of these models in this space, which will continue to evolve in the coming years.

References

Alamar, B., Ma, J., Desjardins, G.M., and Ruprecht, L. (2006). Who controls the plate? Isolating the pitcher/batter subgame. *Journal of Quantitative Analysis in Sports*, 2(3), 1–8.

Albert, J. (2006). Pitching statistics, talent and luck, and the best strikeout seasons of all-time. *Journal of Quantitative Analysis in Sports*, 2(1), 1–30.

Albert, J. (2010). Using the count to measure pitching performance. *Journal of Quantitative Analysis in Sports*, 6(4), 1–28.

Basco, D. and Davies, M. (2010). The many flavors of DIPS: A history and an overview. *The Baseball Research Journal*, 39(2), 41–50.

Baumer, B.S., Jensen, S.T., and Matthews, G.J. (2015). openWAR: An open source system for evaluating overall player performance in Major League Baseball. *Journal of Quantitative Analysis in Sports*, 11(2), 69–84.

Clark, M. (2013). Getting started with additive models in R. Retrieved September 23, 2015 from: http://www3.nd.edu/~mclark19/learn/GAMS.pdf.

Fast, M. (2008). Drinking from a firehose. Retrieved September 23, 2015 from: http://www.hardballtimes.com/drinking-from-a-fire-hose/.

Fast, M. (2011). Spinning yarn: Removing the mask encore presentation. Baseball Prospectus. Retrieved October 25, 2015 from: http://www.baseballprospectus.com/article.php?articleid=15093.

Few, S. (2008). Solutions to the problem of over-plotting in graphs. Retrieved September 29, 2015 from: https://www.perceptualedge.com/articles/visual_business_intelligence/over-plotting_in_graphs.pdf.

Fraley, C. and Raftery, A.E. (1999). MCLUST: Software for model-based cluster analysis. *Journal of Classification, 16,* 297–306.

Fraley, C. and Raftery, A.E. (2002). Model-based clustering, discriminant analysis, and density estimation. *Journal of the American Statistical Association, 97,* 611–631.

Fraley, C. and Raftery, A.E. (2007). Model-based methods of classification: Using the mclust software in chemometrics. *Journal of Statistical Software, 18,* 1–13.

Fraley, C., Raftery, A.E., Murphy, T.B., and Scrucca, L. (2012). mclust Version 4 for R: Normal mixture modeling for model-based clustering, classification, and density estimation. Technical Report No. 597, Department of Statistics, University of Washington, Seattle, WA.

Green, E. and Daniels, D.P. (2015). Impact aversion in arbitrator decisions. http://papers.ssrn.com/sol3/papers.cfm?abstract_id=2391558.

Hamrick, J. and Rasp, J. (2015). The connection between race and called strikes and balls. *Journal of Sports Economics, 16,* 714–734.

Hartigan, J.A. and Wong, M.A. (1979). Algorithm AS 136: A k-means clustering algorithm. *Journal of the Royal Statistical Society, Series C (Applied Statistics), 28*(1), 100–108.

Hastie, T. and Tibshirani, R. (1986). Generalized additive models. *Statistical Science, 1*(3), 297–318.

Jacques, D. (2007). Value over replacement player. Retrieved December 31, 2015 from: http://www.baseballprospectus.com/article.php?articleid=6231.

James, B. and Henzler, J. (2002). *Win Shares.* San Francisco, CA: STATS Publishing.

James, B., Zminda, D., and Munro, N. (2000). *STATS All-Time Major League Handbook.* San Francisco, CA: STATS Publishing.

Jensen, S.T., Shirley, K.E., and Wyner, A.J. (2009). Bayesball: A Bayesian hierarchical model for evaluating fielding in major league baseball. *The Annals of Applied Statistics, 3*(2), 491–520.

Judge, J., Pavlidis, H., and Brooks, D. (2015). Moving beyond WOWY: A mixed approach to measuring catcher framing. Baseball Prospectus. Retrieved October 25, 2015 from: http://www.baseballprospectus.com/article.php?articleid=25514.

Kim, J.W. and King, B.G. (2014). Seeing stars: Matthew effects and status bias in Major League Baseball umpiring. *Management Science, 60,* 2619–2644.

Marchi, M. (2014a). The stats go marching in: Catcher framing before PITCHf/x. Baseball Prospectus. Retrieved October 25, 2015 from: http://www.baseballprospectus.com /article.php?articleid=20596.

Marchi, M. and Albert, J. (2014). *Analyzing Baseball Data with R.* CRC Press, Boca Raton, FL.

McCracken, V. (1999). Defense independent pitching statistics. Retrieved June 14, 2015 from: http://www.futilityinfielder.com/dips.html.

Mills, B.M. (2014a). Social pressure at the plate: Inequality aversion, status, and mere exposure. *Managerial and Decision Economics, 35*(6), 387–403.

Mills, B.M. (2014b). Expert workers, performance standards and on-the-job training: Evaluating major league baseball umpires. Retrieved September 30, 2015 from: http://ssrn.com/abstract=2478447.

Mills, B.M. (2015). Houston Astros whiffs and exit velocity. Exploring Baseball Data with R. Retrieved October 28, 2015 from: https://baseballwithr.wordpress.com/2015/06/30/ houston-astros-whiffs-and-exit-velocity/.

Moskowitz, T.J. and Wertheim, L.J. (2011). *Scorecasting: The Hidden Influences behind How Sports Are Played and Games Are Won.* Crown Archetype, New York.

Nathan, A.M. (2008). A statistical study of PITCHf/x pitched baseball trajectories. Retrieved September 29, 2015 from: http://baseball.physics.illinois.edu/MCAnalysis.pdf.

Nathan, A.M. (2012). Determining pitch movement from PITCHf/x data. Retrieved October 14, 2015 from: http://baseball.physics.illinois.edu/Movement.pdf.

Nelder, J.A. and Wedderburn, R.W.M. (1972). Generalized linear models. *Journal of the Royal Statistical Society: Series A, 135,* 370–384.

Palacios-Huerta, I. (2003). Professionals play minimax. *Review of Economic Studies, 70,* 395–415.

Pane, M.A. (2013). Trouble with the curve: Identifying clusters of MLB pitchers using improved pitch classification techniques. Dietrich College of Humanities and Social Sciences, Carnegie

Mellon University, Pittsburgh, PA.Retrieved September 1, 2015 from: http://repository.cmu.edu/hsshonors/190/.

Pane, M.A., Ventura, S.L., Steorts, R.C., and Thomas, A.C. (2013). Trouble with the curve: Improving MLB pitch classification. http://arxiv.org/pdf/1304.1756.pdf.

Parsons, C.A., Sulaeman, J., Yates, M.C., and Hamermesh, D.S. (2011). Strike three: Discrimination, incentives, and evaluation. *American Economic Review, 101*, 1410–1435.

Pavlidis, H. and Brooks, D. (2014). Framing and blocking pitches: A regressed, probabilistic model: A new method for measuring catcher defense. Baseball Prospectus. Retrieved October 25, 2015 from: http://www.baseballprospectus.com/article.php?articleid=22934.

Piette, J., Braunstein, A., McShane, B.B., and Jensen, S.T. (2010). A point-mass mixture random effects model for pitching metrics. *Journal of Quantitative Analysis in Sports, 6*(3), 1–15.

R Core Team (2015). R: A language and environment for statistical computing. R Foundation for Statistical Computing, Vienna, Austria. http://www.R-project.org/.

Roegele, J. (2014). The strike zone during the PITCHf/x era. The Hardball Times. Retrieved October 1, 2015 from: http://www.hardballtimes.com/the-strike-zone-during-the-pitchfx-era/.

Scott, D.W. (2015). *Multivariate Density Estimation: Theory, Practice, and Visualization*. Hoboken, NJ: John Wiley & Sons.

Scully, G. (1973). Pay and performance in Major League Baseball. *American Economic Review, 64*, 915–930.

Sievert, C. (2014a). Taming PITCHf/x data with pitchRx and XML2R. *The R Journal, 6*(1), 5–19. Retrieved September 22, 2015 from http://journal.r-project.org/archive/2014-1/sievert.pdf.

Sievert, C. (2014b). Interactive visualization of strike-zone expansion. Exploring Baseball Data with R. Retrieved October 20, 2015 from https://baseballwithr.wordpress.com/2014/11/11/interactive-visualization-of-strike-zone-expansion-5/.

Sievert, C. (2015a). Obtaining exit velocity and distance of batted balls. Exploring Baseball Data with R. Retrieved October 20, 2015 from https://baseballwithr.wordpress.com/2015/07/15/obtaining-exit-velocity-and-distance-of-batted-balls/.

Sievert, C. (2015b). The grand tour. Retrieved August 31, 2016 from https://vimeo.com/148050343.

Studeman, D. (2005). I'm batty for baseball stats. Hardball Times. Retrieved October 23, 2015 from: http://www.hardballtimes.com/im-batty-for-baseball-stats/.

Swartz, M. (2011). New SIERA, part two (of five): Unlocking underrated pitching skills. *Fangraphs*. Retrieved October 25, 2015 from: http://www.fangraphs.com/blogs/new-siera-part-two-of-five-unlocking-underrated-pitching-skills/.

Swartz, M. and Seidman, E. (2010). Introducing SIERA: Part 1. Baseball Prospectus. Retrieved October, 23, 2015 from: http://www.baseballprospectus.com/article.php?articleid=10027.

Tainsky, S., Mills, B.M., and Winfree, J.A. (2015). An examination of potential discrimination among MLB umpires. *Journal of Sports Economics, 16*, 353–374.

Tango, T. (2007). Everything you ever wanted to know about Gameday. Retrieved August 10, 2015 from: http://www.insidethebook.com/ee/index.php/site/comments/everything_you_ever_wanted_to_know_about_gameday/.

Tango, T. (2011). Tango's lab: Deconstructing FIP. Retrieved September 15, 2015 from: http://www.insidethebook.com/ee/index.php/site/comments/tangos_lab_deconstructing_fip/.

Tango, T.M., Lichtman, M.G., and Dolphin, A.E. (2006). *The Book: Playing the Percentages in Baseball*. Middletown, DE: TMA Press.

Thorn, J. and Palmer, P. (1985). *The Hidden Game of Baseball*. New York, NY: Doubleday/Dolphin.

Venables, W.N. and Ripley, B.D. (2002). *Modern Applied Statistics with S*. New York, NY: Springer.

Wickham, H., Cook, D., and Hofmann, H. (2015). Visualizing statistical models: Removing the blindfold. *Statistical Analysis and Data Mining: The ASA Data Science Journal, 8*, 203–225.

Wood, S.N. (2006). *Generalized Additive Models: An Introduction with R*. Boca Raton, FL: Chapman & Hall/CRC Press.

Wood, S.N. (2011). Fast stable restricted maximum likelihood and marginal likelihood estimation of semiparametric generalized linear models. *Journal of the Royal Statistical Society, Series B (Statistical Methodology)*, 73(1), 3–36.

3

Defensive Evaluation

Mitchel Lichtman

CONTENTS

3.1 Introduction

In baseball, *defense*, also known as *fielding* (technically, defense can include pitching), is one of the most controversial aspects of player evaluation. In this chapter, we will discuss exactly what we are attempting to measure when constructing a defensive metric, why it is important to any player or team evaluation system, and just how difficult it is to measure and quantify defense. We will also talk about why even the most robust modern metrics garner so little respect among some analysts and stat-friendly fans and commentators.

Attempts to measure defense have a storied history, from the traditional *fielding percentage* (FP) used in the early days of baseball (and somewhat remarkably, still used today) to the more modern and advanced play-by-play (PBP) metrics like *Defensive Runs Saved* (*DRS*) (Dewan, 2015), *Ultimate Zone Rating* (*UZR*) (Lichtman, 2010), and *Total Zone* (*TZ*) (Smith, 2008). In between we have metrics like Range Factor, Defensive Average, and simple Zone Rating (Basco and Zimmerman, 2010) which improved upon FP but were not quite as good as the modern, advanced PBP metrics. In the last section of this chapter, we will briefly discuss how we might use the most recent and technologically advanced data sets like *Statcast* (Jaffe, 2014) and *Field f/x* (Sportvision, 2014) to create the putative *holy grail* of defensive metrics.

We will also see how the various defensive metrics rate two recently retired shortstops. One is a future first-ballot Hall of Famer considered by many traditionalists and baseball insiders to be an excellent defensive shortstop throughout most of his career. The other is a much less-heralded player who, for a period of seven consecutive years, was one of the best defenders in baseball history.

Not too long ago defense was all but ignored when evaluating individual players. While most people in and out of the baseball analytic industry were generally aware that defense was an integral part of a player's skill set and that it significantly impacted a team's ability to prevent runs, good methods of evaluating defense were not widely available. It was also difficult to scale defensive metrics on the same level as offense and pitching. Additionally, we were not sure how to compare players across the fielding spectrum. For example, how would a below-average shortstop compare to an above-average left fielder—that is, which one had more *position-neutral* defensive value?

Eventually, with the advent of advanced fielding metrics that use the same currency (runs) as most modern offensive stats like *linear weights* (Thorn and Palmer, 1984) or offensive *WAR* (Baseball Reference, 2015), and an understanding and quantification of the concept of *positional adjustments* (MacAree, 2015), one became able to combine offense, defense, and base running in order to arrive at a *total player evaluation*. In addition, we are now better able to understand the role of defense in pitching and run prevention. For example, some analysts have estimated that team wins are a combination of around 50% offense (batting *and* base running), 25% pitching, and 25% defense. Compare that to the traditional (and wrong) saw that "Baseball is 90% pitching," or the careless assertion that baseball is half pitching and half hitting (with base running and defense completely left out of the equation).

3.2 Why Defense Is So Difficult to Evaluate?

Why is it that current defensive metrics, even the most advanced ones, are controversial and often mistrusted by fans, pundits, baseball insiders, and even the analysts who create them? In order to answer that, we can compare and contrast various methods of evaluating pitching, hitting, and defense. When that is done, we will see why evaluating defense is more difficult than, and in fact fundamentally different from, measuring offense and pitching, especially at the player level.

When a hitter completes a *plate appearance* (PA), he necessarily creates 1 of around 10 unique offensive events, defined by the rules of baseball and recorded by an *official scorer* (OS) (Major League Baseball, 2012). There is little or no ambiguity regarding the result of a plate appearance, at least as far as the record books and box scores are concerned. In order to evaluate a player with respect to a single PA, one can easily assign a theoretical run value to the resultant event. For example, using Pete Palmer's *linear weights* (Thorn and Palmer, 1984), the most widely accepted methodology for evaluating and quantifying offense, a single is worth around .47 more runs than a league-average PA, while a home run clocks in at 1.40 runs above average (the exact value of each offensive event depends, in part, on the run environment—that is, the average number of runs scored per game).

To quantify the theoretical offensive value of a player, notwithstanding context such as the timing of those events (generally thought to be largely out of a player's control), all one has to do is compute the *average* run value of all those events combined. In doing so, one arrives at a single number that represents that player's theoretical offensive contribution *in runs above or below average*. This kind of metric is fairly easy to understand and *credible*.

The exact same methodology can be used for evaluating pitchers. At the end of every PA, one can credit or debit a pitcher with the marginal run value of that PA (e.g., .47 runs for a single or *minus* .26 runs for an out), just as was done for the batter. This type of metric when prorated to 9 innings or 27 outs is often called *component ERA* (ERC). As with the batter, while this number represents *theoretical* rather than *actual* run prevention, it reflects actual events for which the pitcher was at least partially responsible and which can be easily quantified.

One can also evaluate pitchers based on their runs or earned runs allowed per some number of outs or innings—for example, the traditional and ubiquitous *earned run average* (ERA) and its cousin *runs allowed per 9 innings* (RA9). Whereas ERC describes *theoretical* run prevention based on individual offensive events, and ERA and RA9 reflect *actual* runs allowed, both cases represent an accounting of actual events that occurred on the field.

What about defense? On a team level, when a plate appearance is completed and a non-HR ball is put into play, there are only three primary outcomes, from a fielding perspective: the ball can be caught and turned into at least one out, it can fall for a hit, or it can be scored as an *error*. There are other more nuanced aspects to fielding, but these three primary outcomes are the basic events that can occur as a result of a PA and which are related to the skill and performance of the fielders. As with the hitters and pitchers, there is little ambiguity in this description, so why is evaluating defense so problematic? It isn't really, *at the team level*.

There are, in fact, team defensive metrics, like *defensive efficiency rating* (DER) (Baseball Prospectus, 2015), which measure the percentage of in-play batted balls turned into outs. Unfortunately, these metrics are not particularly robust. Among other things, they do not distinguish between singles and extra-base hits, nor do they account for the location, speed, and trajectory of each ball in play. These metrics also don't account for park effects, which can significantly affect the quality and distribution of hits, outs, and errors.

What about at the player level? After all, metrics are interesting mostly because they allow us to compare one player to another and to determine the identities of the most and least valuable players. How do we take those three basic fielding events, the hit, the out, and the error, and assign them to individual players? The outs and errors are easy of course. When a fly ball is caught, one player is assigned a *putout*. When a ground ball is fielded and turned into an out, one player (occasionally more) is typically credited with an *assist*, or a *putout* if he tags the runner or base by himself without a throw. Additionally when a routine batted ball is *not* converted into an out but *should* have been *with ordinary effort*, the offending player is charged with a *fielding error*. Again, there is no ambiguity in any of these results.

However, putouts, assists, and errors alone do not allow us to effectively evaluate defense. At best, they allow us to approximate a player's defensive skill or performance. In order to properly evaluate fielding, we *must* know the number of *opportunities* in which they occur, just like we *must* know how many singles or home runs a batter or team produces *per PA or AB* (offensive opportunities) in order to properly evaluate offense. Therein lies at least one of the problems associated with almost all fielding metrics, past and present. How do we assign those opportunities to individual fielders?

3.3 Early History of Defensive Metrics

Let's rewind the clock and see how baseball measured and evaluated defense *back in the day*, based on *what actually happened on the field* with respect to each fielder, bypassing this sticky issue of *responsibility* for batted balls that fall in for hits.

3.3.1 Fielding Percentage

As early as 1876, putouts, assists, and errors were recorded for fielders exactly as they are recorded today. The sum of those three outcomes was called *chances*, again, exactly as it is today. (Until 1887, wild pitches and passed balls were counted as errors.) So in the dawn of professional baseball, almost 150 years ago, the prevailing statistic for measuring fielding was exactly the same as it is today in the mainstream media, and among most fans, commentators, and baseball insiders. That enduring, time-honored, and crude metric is called *fielding percentage* (FP), or *putouts plus assists divided by chances* (Basco and Zimmerman, 2010).

It does not take a mathematical genius to figure out some of the flaws in that simple methodology. For example, infield putouts are awarded for catching pop flies as well as tagging runners and bases, and assists are given not only when a player fields a ground ball and retires the batter or a runner, but when he relays a throw from an outfielder, resulting in an out. A fielder is also credited with an assist when he touches a batted ball (or the ball touches him) that another fielder turns into an out.

A first baseman is awarded a putout every time he catches a throw from a fielder and the batter is out at first. Errors are given to a fielder when he muffs a play in which the batter or a runner *should have* been retired, or a batter or runner advances an extra base but for a bad (physical) play by the fielder. For outfielders, putouts are awarded on caught fly balls and assists on throws in which a runner or the batter is retired. Errors by outfielders are relatively rare and occur more often on muffed plays than by dropping a routine fly ball.

In other words, the denominator of fielding percentage is a *mishmash* of all kinds of defensive events, some of them requiring lots of skill and others not so much. The key to fielding percentage is errors. Essentially fielding percentage is the number of errors a fielder makes divided by some approximation of the number of plays in which he is involved—namely *chances*, or putouts plus assists plus errors (it is actually *one minus that number*).

For what fielding percentage actually does, which is to tell us the rate at which a fielder makes an error, it does a good job and relies on accurate information. It also only uses events that *actually occurred*, with no inferences, approximations, or subjectivity, other than whether a muffed play should be scored as a hit or an error by the official scorer. The advantage of fielding percentage is that it is easy to compute, it is simple to understand, it uses readily available data, and it is *believable* to the general public due to its simplicity and transparency.

The significant downside to fielding percentage is that it completely ignores one important aspect of fielding, *range*, which is the ability to *reach* batted balls and turn them

TABLE 3.1

Career Fielding Percentages of
Two Recently Retired Shortstops

Derek Jeter	.976
Adam Everett	.976
League-average SS (1995–2014)	.972

into outs. As it turns out, error rate, or fielding percentage, only constitutes approximately 25% of fielding talent for infielders and a much smaller percentage for outfielders. The remainder of fielding talent is the aforementioned *range* (excluding other less salient aspects of defense).

Some fielders, by virtue of their speed, agility, instincts, positioning, and jumps, are more adept at reaching and turning batted balls into outs than other less-talented fielders. For example, Andruw Jones in his prime in center field was spectacular at getting to balls hit far from his starting position, due to his speed and agility, while Manny Ramirez in left field had limited range. Yet, fielding percentage tells us almost nothing about the difference between the defensive talents of these two players. In fact, Manny averaged only one or two more errors per season than Andruw, and many of those were misplays in fielding balls off the Green Monster at Fenway, which Andruw never had to deal with. For outfielders, fielding percentage is a particularly poor measure of fielding talent.

Even for infielders, where fielding percentage represents *some* aspect of defensive skill, many players with low error totals are not very good fielders because of their limited range. Similarly, a player with a high error rate might be excellent at *getting to balls* and thus preventing hits. Good range can easily make up for a poor fielding percentage and vice versa. During the late 1990s and early twenty-first century, the two poster boys at the shortstop position for great and poor ranges (according to most analysts), Adam Everett and Derek Jeter, had identical career fielding percentages of .976 (see Table 3.1). Yet, the advanced defensive metrics, and even many of the less robust ones, suggest that Jeter was a below-average defender due to his limited range and that Everett was one of the best fielding shortstops in baseball history due to his exceptional range.

3.3.2 Range Factor

Despite its lasting popularity and ubiquitous nature, fielding percentage, which is just *error rate*, tells us almost nothing about the defensive talent of an outfielder and only a little about the run prevention ability of an infielder. Several baseball insiders recognized this as early as the nineteenth century. In the late 1880s, shortly after fielding percentage was popularized, two notable baseball figures, Henry Chadwick, historian, statistician, and perhaps the grandfather of sabermetrics, and Al Wright, a player for the Boston Braves and manager of the Philadelphia Athletics, invented what Wright called *fielding average* (rather than *percentage*). Fielding average (FA) was an early measurement of *range*. It ignored errors completely, categorizing them as hits. FA is simply "putouts plus assists divided by games played," equivalent to the modern *Range Factor* popularized by Bill James and Pete Palmer in the 1980s. Unfortunately, Wright and Chadwick's *fielding average* never gained any traction and it took 100 years or so before it was resurrected and renamed by James (Basco and Zimmerman, 2010).

James and Palmer's *Range Factor* (RF), like Wright's *fielding average*, is a relatively simple metric that tracks how often an individual fielder creates or participates in an out, and is the *sum of putouts and assists divided by innings or games played*. Like most of the early defensive metrics, RF reflects exactly what happened on the field with little ambiguity. For outfielders, a putout is awarded for catching a fly ball and an assist for throwing out a base runner or the batter; thus it is relatively straightforward. However, RF conflates two separate and largely independent skills, catching fly balls and throwing out runners. Also, since most OF errors are not dropped fly balls, by ignoring errors RF excludes bad defensive play by an outfielder, like muffing a hit or making a bad throw and allowing a base runner to advance an extra base.

Despite its weaknesses, a quick glance at the Range Factor leaderboards in the OF suggests that it is a pretty good measure of outfield range. For example, Paul Blair and Daren Erstadt, two of the best and quickest centerfielders of the modern era have a career RF of 3.046 and 3.033, while Matt Kemp and Carl Everett, not known for their defense, have career RFs of 2.286 and 2.217. Those numbers suggest that the speedy Blair and Erstadt made 3/4 more outs per game on defense than the less-skilled and slower Kemp and Everett.

For measuring infield defense, RF is a bit messier. As explained earlier, for infielders, assists and putouts are ambiguous. An assist can be a ground ball turned into an out or a relay throw from an outfielder that nails a runner. Occasionally it can be a ball that glances off an infielder's glove or body that is turned into an out by another fielder. It can also be awarded to a player or players involved in a successful rundown.

A putout is given to any infielder who retires a batter or runner without a throw by catching a popup or fielding a ground ball and tagging a runner or a base. Additionally, a putout is given to a fielder who receives a throw and tags a base or runner. Consequently a first baseman can amass hundreds of putouts per season with most of those plays requiring little skill. Pete Palmer at some point only used assists (i.e., he ignored putouts) for first basemen in his *Total Player Rating* (Thorn and Palmer, 1984).

For the rest of the infielders, there is still a good deal of noise in RF because of the various ways in which a fielder can be credited with a putout or an assist. Some infield plays have a large skill component, like fielding a ground ball, while other plays are relatively skill-free and routine like catching a pop fly on the infield area or receiving a throw from another fielder and tagging a base or a runner. Still, if one looks at a list of shortstop career RF, one finds players at the top of the list who were known for their outstanding defense, like Mark Belanger, Ozzie Smith, and Rey Sanchez, at 5.24, 5.22, and 5.14, per nine innings. At the bottom of the list, we see players like Derek Jeter, Tony Womack, and Hanley Ramirez, none of them known for their range, at 4.04, 4.05, and 4.17, respectively, more than one out per game worse than the great ones. Table 3.2 displays the career Range Factors for our two signature shortstops, Derek Jeter and Adam Everett.

TABLE 3.2

Career Range Factors for
Two Shortstops

Derek Jeter	4.04
Adam Everett	4.63
League-average SS (1995–2014)	4.46

As one can see, despite Jeter and Everett having exactly the same career fielding percentages, .976 (see Table 3.1), Everett was almost half a "play" (putouts plus assists) per game better than Jeter. That corresponds to a savings of around .31 runs per game (the difference between an out and a hit or error is approximately .73 runs). Interestingly, even though Jeter was a much better hitter than Everett, the difference in their batting linear weights (a precise measure of offensive value) was only .27 runs per game. According to Range Factor, Everett saved more runs on defense compared to Jeter (.31) than Jeter produced on offense compared to Everett (.27). In other words, once we combine offense, defense (using RF as our defensive metric of choice), and base running (both players were excellent base runners), they were roughly equivalent players *per game* (Jeter played more than four times as many games as Everett) throughout their careers. Most baseball fans, commentators, and those who played, coached, and managed the game would find that statement hard to swallow. Perhaps Range Factor is not accurately representing these players' defensive value. We will see shortly what the other, more advanced, metrics have to say about Jeter and Everett.

Two problems with Range Factor, for both outfielders and infielders, are that we don't know how many opportunities each fielder has had in order to compile his putouts and assists, and we don't know the difficulty of those opportunities. The latter issue tends to even out over large samples, especially in light of what we know about pitchers' balls in play (BIP)—most pitchers tend to allow around the same quality of batted balls, and any fluctuations that one observes in small samples tend to be as a result of chance. However, the number of opportunities per inning or per game can vary significantly from fielder to fielder at each position depending on the pitchers that each fielder plays behind (as well as other variables).

For example, over the course of a season or even several seasons, shortstop A might play behind predominantly ground-ball pitchers while shortstop B might play behind mostly fly-ball pitchers. In this scenario, shortstop A would get more opportunities to field ground balls than shortstop B, such that even if they had exactly the same skill at fielding those ground balls, shortstop A would necessarily have more assists and thus a better Range Factor than shortstop B. The same is true for outfielders in reverse.

Also *all* fielders will get more opportunities when "pitch to contact" pitchers are on the mound than when strikeout pitchers toe the hill. Pitcher handedness affects opportunities as well. For example, if a shortstop plays behind predominantly LH pitchers (thus more RH batters), he would get more ground balls hit in his direction than if he played behind mostly RH pitchers. If one can find a way to identify or even infer opportunities, one can refine RF using actual chances rather than innings or games as the "denominator," thus enabling analysts to create a more accurate defensive metric.

3.3.3 The First Play-By-Play or Batted Ball Defensive Metrics

Play-by-play (PBP) or *batted ball* data is typically comprised of the result of every plate appearance (as well as events that don't end a PA, like stolen bases, wild pitches, and passed balls), including the type and location of every ball in play (some PBP databases do not include type and/or location data). Prior to the advent of PBP databases, all of the defensive metrics, including those discussed earlier, only used information available in the standard box score or statistical compilation, namely, putouts, assists, and errors. In the mid-1980s, a group of *statheads* started a program called *Project Scoresheet*, whereby volunteer "stringers" recorded the result of every PA in every game including, eventually, the type and location of every batted ball. A few years later, a company called STATS also began recording PBP data from every game, selling and licensing this information to teams, the media, and occasionally the general public.

With the availability of this PBP data, it became relatively easy to create a metric that added an important piece to the defensive puzzle—opportunities. In fact, PBP data opened up a whole new world of offensive, defensive, and pitching metrics and greatly accelerated the pace of baseball analysis (sabermetrics) in general. Several analysts dove into this treasure trove of batted ball data, creating "zone-based" defensive metrics which, for the first time, included some semblance of *real opportunities* for fielders.

In the mid-1980s, Sheri Nichols and Pete DeCoursey invented *defensive average* (DA), which was defined as the number of ground balls for infielders and fly balls for outfielders turned into outs, divided by balls that were hit in the vicinity of each fielder and deemed "potentially catchable." A few years later, STATS devised a similar fielding measure and published the results in their annual *STATS Scoreboard*. They called their new metric *Zone Rating* (ZR), which was similarly defined as *ground balls or fly balls turned into outs divided by all balls hit within a predefined "area of responsibility" for each fielding position.* An "area of responsibility" was defined as every location in which at least 50% of the balls were turned into outs by a player at that position (Basco and Zimmerman, 2010). One can easily see why Nichols' *defensive average* and STATS' *zone rating* were called "zone-based" metrics. They are outs divided by opportunities, where the denominator is all balls hit within a "zone" or area of the field surrounding each fielder and defined by the creator of the metric.

These zone-based defensive metrics were very good at the time—much better than the previous attempts to measure defense such as *fielding percentage* and *Range Factor*. They represented a quantum leap in defensive evaluation, adding the missing element of true opportunities, and disentangling the *mishmash* of assists, putouts, and errors. Most of these metrics used only ground balls for infielders and "air balls" for outfielders.

Since the boundaries of these "zones" could vary from one metric to another, a player's ZR (e.g., .792) does not really mean anything unless one compared it to another player or to the league average at the same position. Using some simple math, one could even convert these ratios (outs divided by balls in zone) into runs saved or cost, much like the more modern defensive metrics that will be discussed later.

One issue with these metrics was what to do with balls fielded outside of a player's zone. On the average, an "out of zone" (OOZ) out was a very good or even a great play, yet in a simple zone rating system, this type of out wasn't counted at all—obviously a mistake. Eventually some of these metrics gave credit for a ball fielded OOZ by adding it to the numerator (sometimes the denominator as well) and perhaps giving it more weight than a ball fielded in zone (IZ). Still other metrics reported two numbers—an in-zone ratio and an out-of-zone one.

Using a metric that accounts for opportunities, for example, Revised Zone Rating (RZR), Table 3.3 shows that Everett is still a much better fielder than Jeter. It is easier to interpret these ratings by converting RZR ratios into runs saved or cost. Jeter averaged only 1.44 balls in zone (BIZ) per game while Everett averaged 2.14, which suggests that Range Factor really did exaggerate the difference between the two shortstops, since Jeter had many fewer opportunities than Everett. Jeter would therefore convert $(.816 - .792) * 1.44$, or .035 *fewer* balls per game into outs, as compared to a league-average SS. Everett fielded $(.871 - .816) * 2.14 = .12$ *more* balls than a league-average SS. According to RZR, Everett is only .155 outs per game better than Jeter, or .113 runs, which is only one-third of the difference one gets using Range Factor.

TABLE 3.3

Revised Zone Rating[a] 2003–2014 for
Two Shortstops

Derek Jeter	.792
Adam Everett	.871
League-average SS	.816

[a] Revised Zone Rating (RZR) is a single ratio
that combines in-zone and out-of-zone balls
fielded. It is reported only since 2003.

Although assigning "zones of responsibility" within these metrics is somewhat arbitrary, and the handling of OOZ plays is anything but elegant, these zone-based systems are still far better than those that do not count opportunities at all, like Range Factor. One can clearly see how RF greatly exaggerated the difference between Jeter and Everett's fielding skill by virtue of the fact that Everett had many more opportunities *per game* to field a ground ball than did Jeter. Zone Rating accounts for this while Range Factor does not.

3.3.4 Estimating Opportunities without Using Hit Location Data

There are several excellent defensive metrics that are able to estimate opportunities without using any batted ball data at all. Three of these are Michael Humphreys' *Defensive Regression Analysis* (DRA), Sean Smith's *Total Zone* (TZ), and Tom Tango's *With or Without You* (WOWY). Despite the absence of such granular information as the type, location, and speed of every ball in play, these metrics can be quite accurate, especially in large samples when these parameters tend to "even out," and the size of the data set enables the methodology to yield substantial statistical power.

3.3.4.1 *Defensive Regression Analysis*

DRA and TZ ratings are similar in that they essentially estimate fielder opportunities at each position using commonly available data, although they utilize different methodologies. After normalizing some of the data to disentangle cross-correlations, DRA uses a standard regression formula to determine which of the traditional statistics are helpful in explaining run prevention. Like the more modern PBP metrics (e.g., UZR and DRS) discussed later, DRA results in a number that represents runs saved or cost at each defensive position. In tests performed by Humphreys, DRA correlates well (close to a .9 correlation) with UZR for single season values, and the standard deviation of individual player ratings, which reflects the putative ability of the metric to identify small differences in skill, is similar for both methodologies. The advantage of DRA is that it can be used with historical data where no batted ball or PBP information is available. DRA, like some of the other excellent defensive metrics over the years, never gained much popularity (Humphreys, 2005).

Table 3.4 displays career DRA ratings for our two shortstops. For any defensive metric that is presented in runs above or below average, or saved/cost, such as DRA and most of those discussed later, the value of a league-average defender at any position is *zero* by definition. DRA, which uses a regression equation to account for parks, batters, pitchers, etc., has the difference between Jeter and Everett at .22 runs per game defensively, somewhere

TABLE 3.4

Defensive Regression Analysis (DRA) for
Two Shortstops

Derek Jeter (1995–2009)	−.13 runs per game
Adam Everett (2001–2009)	+.09 runs per game

TABLE 3.5

Total Zone (TZ) for Two Shortstops

Derek Jeter (1995–2010)	−.059 runs per game
Adam Everett (2001–2010)	.099 runs per game

between Range Factor and Revised Zone Rating. DRA also suggests that Jeter and Everett were much closer in *overall* talent than most people think.

3.3.4.2 *Total Zone*

Total Zone (TZ) also uses commonly available data and yields very good results. This measure is more transparent than DRA, as it does not use a multivariable regression formula. Basically, TZ uses league-wide PBP information when it is available to determine how often each batter normally makes an out on a ball hit to each position, such that it can estimate the defensive contribution of each fielder on a batter-by-batter basis. If this kind of batted ball information is not available, TZ estimates the number of outs that a batter makes to each position based on out-of-sample data using the handedness, batted ball rates, and ground/fly tendencies of the batter and pitcher. TZ is a good metric given large samples of data and can be used when no or limited PBP or batted ball data is available (Smith, 2008).

Table 3.5 displays career TZ ratings for our two shortstops. TZ is probably less accurate than DRA and tends to "shrink" extreme defensive performances toward zero, typical for a less robust metric. Consequently, TZ suggests that the difference between Jeter's and Everett's defensive performance per game is only .16 runs per game, around 25% less than DRA.

3.3.4.3 *With or Without You*

WOWY, or *With or Without You*, is an ingenious method devised by long-time sabermetric researcher and coauthor of *The Book: Playing the Percentages in Baseball* Tom Tango (Tango et al. 2006). WOWY basically accounts for the fact that not all fielders at each position get the same distribution of balls hit in their vicinity because their contextual parameters, such as parks, pitchers, batters, etc., are likely to be different, even in large samples. WOWY accounts for this by comparing the number of batted balls fielded by a particular fielder to the number fielded by all other players at that position, holding each of these parameters constant. Each data pair is weighted by the number of outs recorded by the player in question (Tango, 2008).

In Tango's article in the 2008 *The Hardball Times Annual*, it is reported that Derek Jeter played behind 124 different pitchers in his career and from 1993 to 2007 "his" pitchers pitched in front of a total of 308 different shortstops. For each of those pitchers, WOWY takes the number of batted balls turned into outs with Jeter at shortstop and compares that percentage with the number of balls turned into outs with all other players at shortstop (Tango, 2008).

Tango tells us that, for example, with Clemens on the mound and Jeter at short, 10.6% of all balls in play (anywhere in the park) were converted into outs by Jeter, while 12.2% were turned into outs by 22 other shortstops. So there is a 1.6% difference with Clemens on the mound, weighted by 1966, the number of BIP with Jeter on the field. The same calculation is made for every pitcher that Jeter plays behind and each difference is weighted by Jeter's number of BIP with that pitcher. A weighted average is then calculated for all of these pitchers. In Jeter's case, the final tally using more than 39,000 BIP while he is on the field, is 11.6% for Jeter and 12.5% for all other shortstops. That is around 38 fewer plays (or around 28 fewer runs prevented) for Jeter per season, a number that comports with the advanced defensive metrics like UZR and DRS, to be discussed later (Tango, 2008).

The same calculations are done for parks, batters, base runners, etc., and all the results are combined in a weighted average. In every case, Jeter is at the bottom of the list of short-stops, further cementing the fact that Jeter rates, defensively, as one of the worst everyday shortstops according to virtually every metric that incorporates range and not just error rate (Tango, 2008).

The attractiveness of WOWY is that it does not rely on inherently less-than-perfectly accurate and often biased estimates of the characteristics of batted balls and the inferred position of the fielders. For example, it assumes the distribution of batted balls allowed by each pitcher, as well as the position of the fielders, is essentially the same, especially in large samples, regardless of who is on the field defensively. The same assumption is made with regard to the batters, base runners, parks, and any other parameter that might affect a fielder's catch rate other than his defensive skill. It is also quite transparent and easy for a reader to understand and accept. For example, WOWY essentially says, "Here is a large sample of batted balls; Jeter turned 11.6% of them into outs. Here is another large sample of likely similar batted balls (since they were allowed by the same pitchers, hit by the same batters, occurred in the same parks, and with the same configuration of base runners and outs); all other shortstops turned 12.5% of them into outs." From that perspective, it is clear and believable that Jeter is probably a less mobile and thus less valuable shortstop than the rest of the field, in terms of turning batted balls into outs.

Using the WOWY methodology, we are able to quantify a fielder's defensive perfor-mance or skill in runs above/below average by taking the difference between the "with" percentage of balls fielded and the "without" percentage, multiplying it by the average number of BIP per game, and then again by the run value of the difference between a hit near the SS position and an out (around .73 runs). If we do that for our two shortstops, Jeter and Everett, we get the results presented in Table 3.6. The downside to WOWY is that it requires a large sample of both *with* and *without you* data (the latter type of data is often lacking) in order to be meaningful; thus it is generally only useful for long careers or many seasons worth of data.

WOWY may be quite accurate in large samples such as we have for our two shortstops. If that is true, one can argue that on a *rate or per game* basis at least, Everett was the *more*

TABLE 3.6

Career WOWY for Two Shortstops

Derek Jeter	−.19 runs per game
Adam Everett	.16 runs per game

valuable player overall once offense and defense are combined. Table 3.6 shows a .35 run difference per game in defense between our two shortstops. This can be contrasted with a difference in hitting of only .27 runs per game. According to these numbers, Everett has an overall .08 runs per game advantage in skill over Jeter. Keep in mind that Jeter played many more games than Everett and thus had far more career value.

3.4 Modern PBP Metrics

The principal weakness of metrics like DRA and TZ that do not directly use batted ball data is that they don't consider the *exact* location of each ball in play, as well as its type and speed. Additionally, these metrics don't consider the speed and power of the batter, the number of outs, or the location of any base runners, in order to infer fielder positioning. These parameters can significantly influence the chance of each ball being caught. This is especially problematic for small samples of data. In larger samples, many of these variables tend to "even out," especially with those metrics that adjust for pitcher and/or batter handedness, G/F tendencies, and in some cases, such as with WOWY, the exact identities of the pitchers, batters, and parks, and configuration of the base runners.

The most popular current advanced defensive metrics, like UZR and DRS, attempt to use all of these parameters. In order to compute the results of these metrics, very granular game-level data are required. Much of that information is provided by the nonprofit group *Retrosheet* (a progeny of *Project Scoresheet*) and companies like STATS, Inside Edge, BIS, and MLBAM, the media arm of Major League Baseball. The requisite data used by most of these modern *batted ball defensive metrics* include the type, location, and speed (or "hang time") of every batted ball in play, the identity and handedness of the pitcher and batter, the park, base runners, outs, and outcome of the play, including errors that allow base runners to advance. Not all advanced defensive metrics use every single one of these parameters, and some parameters are more important than others in the development of these metrics.

3.4.1 Ultimate Zone Rating (UZR)

One of the first attempts to move away from a single zone system like STATS Simple Zone Rating and Nichols' and DeCoursey's Defensive Average was also designed by STATS and was called *Ultimate Zone Rating*. One of STATS' founders, John Dewan, expanded the ZR methodology to account for the actual difficulty of each play. The difficulty of a play, and thus the amount of credit or debit given to a fielder, was based on how often *all* players at that position fielded a similar ball hit to the same location (Basco and Zimmerman, 2008). That seems now like an obvious upgrade, but it was revolutionary at the time and spawned the modern era of advanced PBP defensive metrics.

STATS and Dewan presented their results in the *2001 STATS Scoreboard*, but Dewan left STATS shortly thereafter and the original version of UZR went dormant. Years later, Dewan went on to develop a series of similar defensive metrics, Revised Zone Rating (RZR), Defensive Plus-Minus (PM), and Defensive Runs Saved (DRS), for his new company, Baseball Info Solutions (BIS) (Basco and Zimmerman, 2008). In 2002, Mitchell Lichtman developed his own version of Ultimate Zone Rating (UZR) using first Retrosheet data, then STATS data, and currently BIS data.

3.4.1.1 Parameters of Lichtman's UZR

The basic idea behind most of the batted ball defensive metrics is to determine the league-average catch rate at each defensive position for every batted ball that is put into play, given its type, location, and quality (e.g., speed or hang time), as well as the various parameters that can help us to infer the approximate initial position of each fielder and in some cases their ability to retire the batter based on other factors. Once that is done, for every ball put into play with Fielder A on the diamond, we can compare his result with that of an "average fielder" at his position, given the characteristics of that batted ball and the context in which it was hit.

UZR currently uses four classifications of batted balls, ground balls and bunt ground balls for infielders, and line drives and fly balls for outfielders. Note that line drives and popups on the infield are ignored, and *air balls* (any line drive, fly ball, pop fly, etc.) must be of a minimum distance in order to be included for outfielders. Keep in mind that these "type" classifications, as well as distances and locations, are according to the person or persons who record the data and are subject to human error and bias. In addition to the batted ball type, UZR uses the direction and location of every batted ball as well as a three-prong description of its relative speed (slow, medium, or fast). For direction and location, UZR splits the field up into sections rather than relying on the exact coordinates or vectors recorded by the video or "at-game" observers.

Since the initial position of each fielder affects his catch rate for any particular batted ball, one can improve the accuracy of the metric by attempting to infer that position *to some extent*. One of the primary drivers for fielder positioning is the side of the plate in which the batter stands. Most batters pull at least 2/3 of their ground balls and hit slightly more air balls to the opposite field. In addition, batted balls from RH and LH batters have different characteristics, including speed and spin, which can affect their chances of being caught at each fielding position. Thus, UZR treats left- and right-handed batters separately by creating two *buckets* for every batted ball type.

The UZR *engine* also uses outs and base runners to estimate fielding position. For example, with no runners on base or a runner on second base, we assume that, on the average, every first baseman is playing maybe 10 or 12 ft from the first base line. With a runner on first, and second base empty, however, the first baseman usually starts out on the bag and ends up at a position closer to the line. When the double play is in order, the middle infielders typically play shallower and closer to the second base bag. In a potential bunt situation, the first and/or third baseman may be playing up. With a runner on third and less than two out, some or all of the infielders may play in.

The speed of the batter is another variable that affects the infield "catch" rates on ground balls. For fast batters, the infielders must play a little shallower and are often forced to make a quicker and harder throw to first. For slow batters, they can play back, thus giving them more range, and take their time on throws to first, enabling them to turn more ground balls into outs. UZR creates three categories of batters—fast, medium, and slow—and calculates baseline catch rates separately for each.

For the outfield, batter power is another factor (besides handedness) that affects positioning. As with batter speed, UZR uses three power categories. Of course, one would prefer to know the average position of each fielder *for every batter and in every game situation* (or their actual starting position when the ball was put into play). Unfortunately, fielder positioning is not available within the traditional batted ball and PBP databases (some of them do include whether a "shift" occurred—UZR currently ignores all plays in which a shift influenced the result), so batter "hand," power, and speed must suffice.

The following "contextual" buckets are created—or adjustments are made—when determining the baseline catch rates in UZR:

- Batter hand
- Batter speed (slow, medium, fast)
- Batter power (small, medium, large)
- Base runners and outs
- Pitcher G/F ratio
- Park effects

These affect the positioning of the fielder and/or the speed, spin, etc., of the batted ball. These are the characteristics of each batted ball for which UZR creates separate buckets.

- Type (ground ball, bunt ground ball, fly ball, line drive)
- Ball speed (slow, medium, fast)
- Location ("slices" of the infield for ground balls and small sections of the OF for air balls)

3.4.1.2 UZR Nuts and Bolts

First, every batted ball is "bucketed" using some of the variables listed earlier. A ground ball is put into a bunt or nonbunt bucket and an air ball is put into a fly ball or line drive bucket. Those buckets are then subdivided into batter hand and batter speed (for infield GB) or power (for outfield air balls) buckets. We now have 36 possible ground ball buckets and an equal number of air ball buckets. In addition, there are eight buckets in the infield, representing the direction or vector of the ground ball, and several dozen outfield sections indicating the approximate landing zone of each air ball. In total, we have more than a thousand possible unique buckets. (Outs, base runners, parks, and pitcher G/F rates are not "bucketed." Mathematical adjustments are made to the baseline catch rates for these variables.)

The UZR algorithm looks at all batted balls that fall into each of those buckets over several years of data, in each league separately (both leagues can be combined), and computes the fraction that falls for a hit as well as the average value of that hit. The percentage of outs and errors is also computed, for each of eight fielding positions (catchers are excluded). For example, a fast ground ball hit by a speedy LHB in direction X may have resulted in a hit 60% of the time, an out by the shortstop 30% of the time, an out by the second baseman 8% of the time, an error by the shortstop 2% of the time, and an error by the second baseman 1% of the time. Adjustments are made to some or all of those numbers based on the base runners, outs, pitcher G/F tendencies, and park factors.

We now have baseline league-average numbers (hit rates and position-specific out and error rates) for each of those 1000-odd buckets as well as the adjustments explained earlier. We have to be careful with the number of buckets created. If the league-wide, multiyear sample size of any one bucket is small enough, we end up introducing lots of noise into the base-line rate for that bucket. Given so many buckets, even with relatively large numbers of batted balls in each bucket, our baseline rates tend to be a little noisy anyway. Statistically, it is likely that a few of our buckets will be very noisy. Our only solace is that within a large sample of opportunities for an individual player, the noisy buckets tend to cancel one another out, and a few spurious base-line rates (out of over 1000) won't significantly affect the final results. When we discuss SAFE, we'll introduce a powerful method developed by its creator, Shane Jensen, by which the noise associated with the location buckets can be significantly reduced by use of a *smoothing algorithm*.

Once these baseline numbers are created from several years of data (UZR uses six seasons prior to and including the season being evaluated, but it can be any number), the UZR engine goes through the database again season by season to create the results for each player or team (team UZR can be calculated using the same basic methodology, but with no regard for *which* position made an out or an error, only *whether a ball was fielded or not*). For every ball in play, it establishes the bucket to which it belongs, including of course, the most important parameters, its type and location. Then it notes the result, which can fall into three categories: a hit, an out, or the batter reaches on an error (ROE).

If the ball falls for a hit, we have to determine which fielders are going to be charged with some fraction of the run value of the difference between a hit and an out. To do that, UZR checks the baseline hit/out rate for that bucket. Every position that occasionally makes an out on an equivalent batted ball gets docked according to the proportion of outs that it makes in the league-wide database. For example, suppose a ground ball in bucket B (after applying all contextual adjustments) is normally converted into an out by all shortstops 27% of the time, and by all second basemen 9% of the time, based on the six seasons prior to and including the season in question. First, the UZR engine determines the average run value of a batted ball in bucket B by multiplying .36, the fraction that are turned into outs, by the average value of an out, which is normally around −.26 runs, and adding that to .64 (the hit fraction) times the average value of a hit in that bucket (depending on the proportion of singles, doubles and triples), in this case probably around .5 runs. That gives us an average *ball value*, in runs, for a batted ball in bucket B of .2264. (The calculations are displayed in Table 3.7.)

On a hit, all fielders combined must be charged with a total of .2736 runs, or .5 (the average run value of a *hit* in that bucket) minus .2264 (the average run value of all balls hit into that bucket). That .2736 is divided among all fielders based on the proportion of their "responsibility." In our example, the shortstop makes .27/.36, or 75% of the outs, and the second baseman, .9/.36, or 25% of the outs. So the shortstop gets charged with .75 ∗ .2736 runs and the second baseman, .25 ∗ .2736 runs. For that one play, that would reflect a UZR of −.2052 for the shortstop and −.0684 for the second baseman. (The calculations are summarized in Table 3.8.)

If that same batted ball were turned into an out—by the shortstop *or* the second baseman—the calculations are a bit more nuanced. First of all, when a ball is caught, no fielder is charged with negative runs, even if that fielder didn't actually make the catch. In our example, if the shortstop fields the GB, which it normally does 27% of the time, the

TABLE 3.7

Run Value of a GB in Bucket B in the
UZR Rating System

Out value	−.26
Hit value	.5
Fraction of balls that are hits	.36
Fraction of balls that are outs	.64
Ball value (−.26 ∗ .64 + .5 ∗ .36)	.2264

TABLE 3.8

UZR Calculations for the SS and 2B on a *Hit* in Bucket B

Hit value	.5
Ball value	.2264
All fielders combined must be charged with the difference (.5 − .2264)	.2736
SS fields 75% of balls in this bucket	
2B fields 25%	
SS UZR is *minus* (.75 ∗ .2736)	−.2052
2B UZR is *minus* (.25 ∗ .2736)	−.0684

TABLE 3.9

UZR Calculations for the SS and 2B on an *Out* in Bucket B

Out value	−.26
Ball value	.2264
SS fields the ball, he gets *plus* (−.26 − .2264)	.4864
2B fields the ball, he gets *plus* (−.26 − .2264)	.4864
The fielder who does not field the ball	0

second baseman is *not* debited any runs. The reasoning behind that decision is twofold: one, since we don't know the *exact* location and difficulty of each batted ball or the *exact* starting position of each fielder, if a ground ball in bucket B is caught by the shortstop, we must assume that the ball was probably closer to him and further from the second baseman than the average ball in that bucket (that is a *Bayesian inference*), and two, the second baseman *may* have also had an opportunity to catch the ball. In any case, we want individual UZRs to add up at the team level, and docking one player on a caught ball without giving the other player extra (underserved) credit would be problematic. If the shortstop does indeed catch our hypothetical ball in bucket B, he gets credit for −.26 runs minus the average value of that batted ball, .2264, or −.4864 runs. In other words, his catch created .4864 fewer runs on the average, than an average batted ball from bucket B. His UZR for that one play is +.4864 runs. These calculations are summarized in Table 3.9. (The *signs* of the numbers can be confusing. The normal convention is that *minus* is good for the defense and *plus* is good for the offense; however, UZR for a good fielder or a good play is always presented as *plus* and for a bad fielder/play it is presented as *minus*.)

One way to test whether these computations are correct is to make sure that, for average fielders, hits and outs sum to zero for each bucket. If 100 balls were hit from bucket B, remember that 27 would be fielded by the shortstop, 9 by the second base, and 64 would fall for hits. Of the 64 hits, the shortstop would be charged with −.2052 ∗ 64, or −13.1328 runs. The second base would be docked −.0684 ∗ 64, or −4.3776 runs. For the 27 balls fielded by the shortstop, he would get credit for .4864 ∗ 27, or 13.1328 runs. An average second baseman would field 9, for a UZR of .4864 ∗ 9, or 4.3776 runs. You can see that each fielder gets a total debit/credit or UZR of exactly zero.

Remember that in UZR a fielder never gets debited any runs when another fielder catches a ball in play. You may also notice that when a fielder catches a ball in a particular bucket, he gets exactly the same amount of credit as any other fielder, regardless of how often that position normally catches a ball in that bucket. In our example, both the shortstop and second baseman (one *or* the other) receive a credit of .4864 runs when they catch a ball in bucket B, even though the shortstop catches three times as many balls. Other similar batted ball metrics use a slightly different methodology.

A common method for crediting or debiting a fielder in many of the other advanced defensive metrics that use batted ball locations is to simply use a baseline catch rate for each bucket or location and apply that to whether a particular fielder at a position caught the ball or not. In these systems, errors are treated exactly the same as hits. In our example, if a batted ball results in a hit or error, the shortstop would be debited with .27 of a "catch" since he normally makes 27% of the outs in that bucket. The second baseman would be debited with .09 "catches." If the shortstop makes the play, he gets credit for .73 of a catch, the difference between how often he made the catch (1 of course) and how often a league-average fielder makes the catch (.27). The second baseman, meanwhile, gets debited .09 catches, even though the ball was caught by the shortstop—the same as if the ball fell for a hit. Docking the second baseman when the shortstop makes a catch in a bucket in which the second baseman also converts balls into outs is an example of where many of the other advanced batted ball metrics differ from UZR.

Using this alternative methodology, let's see if the numbers "add up." In this scheme, a "catch" is worth .76 runs—the difference between an average hit in that bucket (.5) and an out (−.26). So for the 64 hits, the shortstop is debited $64 * .27 * .76$, or 13.1328 runs, the same as in UZR. The second baseman gets charged with $64 * .09 * .76$, or −4.3776 runs, again, the same as in UZR. When the shortstop makes his 27 outs, he is credited with $27 * .73 * .76$, or 14.9796 runs. At the same time, the second baseman will get debited $27 * .76 * .09$, or −1.8468 runs. When the second baseman makes the play nine times, he receives a total of $.91 * .76 * 9$, or 6.2244 runs, and the shortstop gets subtracted 1.8468 runs.

In total, the shortstop's result is −13.1328 (for a hit), +14.9796 (for a catch), and −1.8468 when the second baseman makes the catch, for a total of zero runs. The second baseman gets $-4.3776 - 1.8468 + 6.2244$, or zero as well. So while the individual numbers vary slightly from UZR, the overall results are similar and everything adds up at the team level. Both methods are justifiable. The reason we don't have one single, optimal methodology is because we simply don't know the exact parameters of every batted ball, the precise starting location of each fielder, and the dynamics involved when one fielder makes a play and others do not on a ball hit in the vicinity of two or more players.

We mentioned that in some of these metrics, an error is treated in the same manner as a hit. In UZR, it is not. The reason is that we have more information on an error. A hit in a certain bucket implies that the ball was difficult to catch, on the average, regardless of the defensive prowess of the fielders and regardless of the average catch rate in that bucket. An error, on the other hand, is a ball that is deemed to be *catchable with ordinary effort*, according to the official scorer. While there is obviously some overlap between a hit and an error, and not every decision by a scorer is justifiable, clearly the average error is a much more catchable ball than the average hit. Treating a hit and an error the same is a mistake.

Table 3.10 shows that UZR, currently one of the more popular defensive metrics, is kind to Jeter compared to DRA and WOWY. The difference between Jeter and Everett, according to UZR, is once again only .14 runs per game, similar to less rigorous measures like RZR and TZ. Although UZR and many of the other modern metrics are complex and robust, they can only approximate defensive performance (and extreme values tend to over- or underrate performance), and that is why you will sometimes see fairly large differences in their outputs.

TABLE 3.10

UZR for Two Shortstops

Derek Jeter (2002–2014)	−.029 runs per game
Adam Everett (career)	.106 runs per game

3.4.2 Spatial Aggregate Fielding Evaluation

It was mentioned earlier that there is a defensive metric that uses smooth functions to model the probability of a "catch" given the various parameters of a batted ball, rather than using a discrete zone or vector-based methodology. This is an excellent solution to the problem of having relatively small samples in many of the buckets. Not only is sample size problematic with respect to these buckets but also is the notion of treating each bucket independently.

For example, let's say that the shortstop catch rate for hard hit ground balls is 60%, 50%, 40%, 30%, 35%, 20%, and 10%, for successive vectors on the field moving further and further away from the normal shortstop position. It is likely that the 35% is an outlier due to statistical noise in that "location bucket," particularly if the sample of opportunities is not especially large. If it is assumed that the further we go from the shortstop's normal starting position the harder it is to field a hard hit ground ball (which is a good assumption), then we probably want to find a way to "smooth out" these numbers such that they resemble a more reasonable sequence. Ideally, one might want to create a function that estimates the catch rate for each bucket, based on the angle of the batted ball vector relative to a line from home plate to the average or inferred starting position of each fielder. That is exactly what SAFE (Spatial Aggregate Fielding Evaluation) developed by Shane Jensen et al. does.

For outfielders and infielders, SAFE estimates the league-average starting position on the field for each fielder based on the maximum catch rates for each type and speed of batted ball. From there the system self-learns using historical batted ball data to create smooth model functions based on the angle of each ground ball for infielders, and the distance and direction (from the fielder's initial position) of air balls for outfielders (Jensen et al., 2009). This method can be adapted to any park and game situation (batter, outs, base runners, etc.) where the initial starting position of each fielder might be different from the standard one, and the characteristics of the batted balls might not be captured by the quality of the data. Like some of the other excellent metrics in their time, SAFE never gained much traction.

3.4.3 Other Advanced Batted Ball Metrics

Other batted ball metrics utilizing similar methodologies to UZR include Probabilistic Model of Range (PMR) (Pinto, 2003) and John Dewan's Defensive Runs Saved (DRS). Unlike UZR, which tracks "arm rating" and infield "GDP turned" separately, DRS *includes* those aspects of defense. The methodologies for evaluating outfield arms and infield double play defense used by DRS and UZR are quite similar. For outfield arms, run values are computed by recording how often outfielders throw out runners at the various bases or prevent/allow them to advance, as compared to league averages rates, given the type and location of the batted ball, the park, the base runners, and outs. The speed of the baserunner and even the game state, such as inning and score, can be used to fine tune these numbers. Infield GDP merely credits or debits infielders according to the number of double plays turned per double play opportunity, as compared to league averages.

TABLE 3.11

DRS for Two Shortstops

Derek Jeter (2003–2014)	−.066 runs per game
Adam Everett (2003–2011)	.162 runs per game

DRS also adds (to range/errors, outfield arms and infield GDP) what they call *good fielding plays* (GFP) and *defensive misplays* (DM). GFP and DM include things like outfielders making "over the wall" catches to prevent a home run, infielders making bad relay throws, first basemen handling difficult throws, and outfielders holding runners to a single on likely doubles, doubles on likely triples, or allowing runners to advance extra bases (e.g., slowly fielding a ball, misplaying a bounce off a wall, or overthrowing a cutoff man) without being given an error. As well, DRS uses more granular buckets for air balls (such as "fliners"—a combination of a fly ball and line drive) and in some versions uses batted ball "hang time" (for air balls, and "time through the infield" for ground balls) rather than simply three categories of speed, as UZR does. Generally, DRS is a very comprehensive defensive metric based on intensive video scouting (Dewan, 2015).

DRS in runs saved/cost per game for our two shortstops are displayed in Table 3.11. DRS sees Jeter as a worse defender than does UZR, but not nearly as poor as DRA or WOWY. It estimates the difference between the two shortstops at .23 runs per game, which once again suggests that they are equivalent players overall (combining offense and defense), a notion that is anathema to most baseball purists.

3.5 The Future of Defensive Metrics

All of the batted ball defensive metrics described here, even the most comprehensive ones like DRS and UZR, are limited in their accuracy due to the difficulty in identifying and recording precise, unbiased batted ball parameters and the exact starting location of each fielder. The latter is especially problematic with teams employing more and more infield shifts of varying degrees. The *holy grail* of defensive evaluation, at least with respect to fielding ground balls and air balls and turning them into outs, requires near-perfect information on essentially two things: one, the exact location and physical characteristics of every batted ball, and two, the starting position of every fielder who fielded or could have fielded the ball in question. With that, one can either create smooth models or functions, as SAFE does, or empirically compute the league-average probability of catching each type and location of batted ball from various discrete starting locations (for each fielding position), much like UZR, DRS, and PMR do now, but with more accurate data.

This kind of detailed and precise information is now available through MLB's advanced camera and tracking systems dubbed *Field f/x* (Jaffe, 2014) and *Statcast* (Casella, 2015). Unfortunately, at the time of this writing, the complete set of data generated from these complex and highly technological systems are not available to the general public or even nonteam affiliated researchers and writers. In addition to the location, speed, and trajectory of each batted ball, and the position of the fielders when the ball is put into play, *Statcast* provides precise measurements for such variables as the fielder's *first step, acceleration, top speed*, and *route efficiency*.

It is not exactly clear how to incorporate this kind of information into a defensive metric. For example, if fielder A, starting at position X, converts a particular ball in play into an out, and so does fielder B for an identical batted ball and from the same starting position, do we care which fielder had the better route, first step, or speed to the ball? An out is an out, given identical contexts. On the other hand, this kind of "data-driven scouting" might allow us to reduce the uncertainty associated with results based on small samples of data.

Perhaps this new technology will also enable us to treat misses (balls that fall for a hit) differently. For example, a player who misses a ball after taking an efficient route might be given "more credit" for a noncatch than a player who takes a poor route to the same ball, all other things being equal. In other words, a "near miss" with maximum effort and skill might be treated differently than a completely botched play.

There is also the philosophical question of how much credit to give a fielder for his initial positioning. For example, if fielders A and B turn identical batted balls into outs, but fielder A does it with a faster and more efficient route, while fielder B starts out closer to the ball, do we give them equal credit? How much is fielder positioning a function of the team's advance scouting and/or coaching skills and how much is it a function of the fielder's defensive prowess and instincts?

One thing is clear: Given the spectacular granularity and accuracy of the data that are now available, there will soon be (or perhaps there already is in some cutting-edge front office) a method of evaluating defense that may prove to be better than any metric that has ever come to light in the history of the game. This could transform defensive evaluation from a sabermetric stepchild to a veritable *wunderkind.*

References

Basco, D. and Zimmerman, J. (2010). Measuring defense: Entering the zones of fielding statistics. *Baseball Research Journal*, 39(1). http://sabr.org/research/measuring-defense-entering-zones-fielding-statistics.

Baseball Prospectus (2015). Glossary of terms. View details for defensive efficiency rating. Retrieved from https://www.baseballprospectus.com/glossary/index.php?search=DEF_EFF, accessed May 12, 2016.

Baseball Reference (2015). WAR explained. Retrieved from http://www.baseball-reference.com/about/war_explained.shtml, accessed May 12, 2016.

Casella, P. (2015). Statcast primer: Baseball will never be the same. Retrieved from http://m.mlb.com/news/article/119234412/statcast-primer-baseball-will-never-be-the-same, accessed May 12, 2016.

Dewan, J. (2015). The Fielding Bible Awards. Retrieved from http://fieldingbible.com/Fielding-Bible-FAQ.asp, accessed May 12, 2016.

Humphreys, M. (February 14, 2005). Defensive regression analysis: Complete series. *The Hardball Times*. Retrieved from http://www.hardballtimes.com/defensive-regression-analysis-complete-series/, accessed May 12, 2016.

Jaffe, J. (March 3, 2014). MLB unveils revolutionary next-level tracking technology on defense. *Sports Illustrated.com*. Retrieved from http://www.si.com/mlb/strike-zone/2014/03/03/mlb-advanced-media-defense-tracking-sloan-sports, accessed May 12, 2016.

Jensen, S., Piette, J., Shirley, K., and Wyner, A. (August 14, 2009). SAFE: Spatial Aggregate Fielding Evaluation. Retrieved from http://www-stat.wharton.upenn.edu/~stjensen/research/safe.html, accessed May 12, 2016.

Lichtman, M. (May 19, 2010). The FanGraphs UZR Primer. Retrieved from http://www.fangraphs. com/blogs/the-fangraphs-uzr-primer/, accessed May 12, 2016.

MacAree, G. (n.d.). Positional adjustment|FanGraphs Sabermetrics Library. Retrieved November 21, 2015, from http://www.fangraphs.com/library/misc/war/positional-adjustment/, accessed May 12, 2016.

Major League Baseball (2015). *The Official Rules of Major League Baseball* (2012). Chicago, IL: Triumph Books.

Pinto, D. (September 19, 2003). A probabilistic model of range. *Baseball Musings*. Retrieved from http://www.baseballmusings.com/archives/004765.php, accessed May 12, 2016.

Smith, S. (January 10, 2008). Measuring defense for players back to 1956. *The Hardball Times*. Retrieved from http://www.hardballtimes.com/measuring-defense-for-players-back-to-1956/, accessed May 12, 2016.

Sportvision (2014). Field f/x. Retrieved from https://www.sportvision.com/baseball/ fieldfx%C2%AE, accessed May 12, 2016.

Tango, T. (2008). With and Without...Derek Jeter. *The 2008 Hardball Times Annual* (p. 147).

Tango, T., Lichtman, M., and Dolphin, A. (2006). *The Book: Playing the Percentages in Baseball* (Chapter 1, p. 27). Florham Park, NJ: TMA Press.

Thorn, J. and Palmer, P. (1984). *The Hidden Game of Baseball: A Revolutionary Approach to Baseball and Its Statistics*. Garden City, NY: Doubleday.

4

Situational Statistics, Clutch Hitting, and Streakiness

Jim Albert

CONTENTS

4.1 Introduction

4.1.1 Situational Data

Baseball fans are familiar with splits, where player offensive or defensive statistics are broken down by different situations. For example, suppose one is interested in learning about the batting statistics of Bryce Harper who had a strong 2015 (MVP) season. Looking at the web site `fangraphs.com` and selecting the "Splits" tab for Harper, we see that

- Harper hits better against right-handed than left-handed pitchers.
- He hits better during home games compared to away games.
- He had a cool April, batting .286, but batted between .333 and .370 in the months May, June, and July.
- His batting average was .460 for fly ball hits, as opposed to .203 for grounders.
- His batting average was much higher when he hit to left field, as compared with center and right fields.
- He hit over .400 for situations with medium leverage (medium-pressure situations).
- His batting average was high for pitch counts that passed through "hitter's counts" such as 3-0 and 3-1 and low for "pitcher counts" such as 0-2 and 1-2.
- If one looks further in Harper's game log for the 2015 season, one notices that he hit six home runs in a three-game period. It appears that Harper had a short period of "hot" power hitting during these games.

This situational and game log data make for interesting reading. It is easy to find hitters who perform well in specific clutch situations, perform well against specific pitchers, or have experienced periods of cold or hot hitting.

These data are readily available for managers to create lineups and select replacements during the game. A manager may put a specific player in the lineup since he has good historical performance against the starting pitcher, or he may bench a player who is experiencing a hitting slump.

4.1.2 Outline

The purpose of this chapter is to provide a broad overview of situational data and streakiness in baseball. It is one thing to observe interesting situational or streaky behavior for players in a particular season. The key question is not whether situations affected past performance, but rather if this interesting data is predictive of situational behavior in a future season. A situational effect is "real" for one player if he has an underlying talent or ability to perform better in a particular situation not explained by chance or luck.

Section 4.2 reviews the literature on estimating hitting talents of baseball players. A useful representation for hitting data is a random effects model. When this model is fit, one obtains an estimate of the talent distribution of hitters. Moreover, this model gives improved estimates of players' hitting abilities by shrinking or adjusting the individual estimates toward a common value. With this background, Section 4.3 presents an overview of statistical models for situational data using hitting data against pitchers of the opposite and same side as an illustration. There are three broad classes of situational models, called "no effect," "bias," and "ability" models. The ability model is perhaps the most interesting since it implies that players have specific abilities to take advantage of particular situations. Section 4.3 presents a regression method for estimating situational effects for a group of players.

Sections 4.4 and 4.5 review particular studies on situational effects in Tango et al. (2007) and Silver in Keri and Baseball Prospectus (2007). Tango et al. (2007) explore platoon effects. They ask if there is indeed variation in batter skills in platoon situations. If the answer is yes, what is the size of this platoon skill variation and how can one measure the platoon skills of specific hitters? Silver talks about the general problem of evaluating clutch hitting—what is a reasonable way of quantifying a clutch situation? He proposes to measure a clutch contribution by seeing how one's contribution (measured in terms of wins) exceeds what would be expected based on his basic hitting statistics. Silver uses this clutch measure to see if a particular player such as David Ortiz is consistently clutch.

Section 4.6 reviews literature on the detection of true streakiness of players and teams. There are challenges in exploring streaky performance including the issue of looking at selected data, the notion of multiplicity, and the confounding of the streaky effect with the sample size and the success probability. This section reviews several measures of streakiness and describes some probability models for quantifying true streakiness. Some measures for testing streakiness are summarized; a simulation of predicted data is helpful for understanding the significance of observed streaky performance in the context of a large group of players. Section 4.7 summarizes some general observations about the analysis of situational data and streakiness and presents some problems for future study.

4.2 Estimating Hitting Talents

4.2.1 A Random Effects Model for Hitting Data

We begin by reviewing the general problem of estimating talents from a group of hitters. Efron and Morris (1975) was one of the first papers to illustrate the benefit of simultaneously estimating hitting talents of a group of baseball players. Albert (2002, 2006) illustrate the use of a beta-binomial random effects model for estimating hitting rates and pitching rates for players.

Suppose y_j is a measurement of the hitting performance of player j in n_j plate appearances in a given season. Our example uses the weighted on-base average (*wOBA*) as the measurement, although the ideas apply to any measurement of hitting performance. If one observes hitting data for N players, then we assume that $y_1, \ldots, y_N$ are independent, where y_j is normally distributed with mean θ_i and standard deviation $\sigma/\sqrt{n_j}$, where σ is an estimate of the standard deviation of the hitting measurements within a season. The normal means $\theta_1, \ldots, \theta_N$ represent the hitting talents of the N players.

One general assumption is that the hitting talents come from a single talent distribution. Specifically, one assumes $\theta_1, \ldots, \theta_N$ represent a random sample from a normal distribution with mean μ and standard deviation τ, where both parameters are unknown. In this random effects model, the mean parameter μ represents an average talent among the players and the standard deviation τ represents the spreads of the hitter talents.

To complete the model, the random effect parameters μ and τ are assigned a prior distribution that reflects one's knowledge about the location of these parameters before sampling. One common distribution saying little about the location of these parameters assumes that μ and τ^2 are independent, the average talent μ is assigned a uniform density, and the variance parameter τ^2 is assigned the proper density of the form

$$g(\tau^2) = \frac{1}{(1 + \tau^2)^2}, \quad \tau^2 > 0.$$

Gelman (2006) gives a general discussion on priors for variance parameters for random effects models.

4.2.2 Example

To illustrate fitting a random effects model, the *wOBA*s were collected for the 205 players in the 2014 season with at least 400 plate appearances. The sampling standard deviation was estimated from these data to be the value $\sigma = 0.521$. When the random effects model is fit to these data, one obtains the estimates $\hat{\mu} = 0.327$ and $\hat{\tau} = 0.025$. For this season hitting data, the *wOBA* talent curve for full-time nonpitchers in the 2014 season is represented by a normal curve with a mean of 0.327 and a standard deviation of 0.025.

4.2.3 Estimating Player Talents

If one explores the *wOBA* values for the 2014 season, one sees large variability. The values range from 0.243 to 0.412 with a standard deviation of 0.034. To perform good predictions of hitting performance for a future season, one wishes to estimate the player talents $\{\theta_j\}$. These talent estimates shrink or adjust the observed hitting performance toward an average value. In the Bayesian model, these estimates correspond to the posterior means of the hitting talents $\theta_1, \ldots, \theta_N$. The estimate of the jth hitting talent is given by the weighted average

$$\hat{\theta}_j = \frac{Pr_D}{Pr_D + Pr_P} y_j + \frac{Pr_P}{Pr_D + Pr_P} \hat{\mu},$$

where
  $Pr_D = n_j/\sigma^2$ is the precision of the data
  $Pr_P = 1/\hat{\tau}^2$ is the precision of the prior

For example, Jose Abreu had a strong 2014 season with $wOBA = 0.411$ in $PA = 607$ plate appearances. The data and prior precisions are given by $Pr_D = 607/0.521^2 = 2236$ and $Pr_P = 1/0.025^2 = 1600$. So the estimate of Abreu's hitting talent is given by

$$\hat{\theta} = \frac{2236}{2236 + 1600} \times 0.411 + \frac{1600}{2236 + 1600} \times 0.327 = 0.376.$$

Since $1600/(2236 + 1600) = 0.42$, the talent estimate shrinks Abreu's observed *wOBA* 42% toward the average. Assuming that the population of hitting talents does not change in successive seasons, a reasonable prediction of Abreu's *wOBA* for the 2015 season would be 0.376. It is interesting to note that Abreu's hitting performance in the 2015 season was 0.361, so he is demonstrating some regression from his strong 2014 hitting performance.

One can represent the talent estimate in terms of plate appearances. Write the prior precision as

$$Pr_P = \frac{1}{\hat{\tau}^2} = \frac{n_P}{\sigma^2}.$$

Then the estimate of Abreu's hitting talent can be expressed as

$$\hat{\theta} = \frac{PA \times y + n_P \times \hat{\mu}}{PA + n_P}.$$

Here, Abreu has $PA = 607$ plate appearances and one computes $n_P = 434$, and so the talent estimate is represented as

$$\hat{\theta} = \frac{607 \times 0.376 + 434 \times 0.327}{607 + 434}.$$

Essentially, one is combining Abreu's performance in 607 PA with 434 PA of an average hitter to get an estimate of his hitting ability.

4.3 Models for Situational Effects

4.3.1 Exploration

There is substantive literature on exploring situational data in baseball. One of the most interesting situations is the performance of hitters in clutch or high-leverage situations. Cramer (1977) and Palmer (1990) were two of the earlier papers to show little evidence for players to have a talent to perform better in clutch situations. Bennett (1993) explored Joe Jackson's performance during the 1919 World Series, and Albert (2002, 2007) explored the performance of hitters in different runner/outs situations. Hopkins and Magel (2008), Otten and Barrett (2013), and Yates (2008) are more recent papers exploring situational effects for a range of pitching and batting measures. Albert (1994) and Albert and Bennett (2007), in Chapter 4, provide overviews of probability models for situational baseball hitting data, distinguishing between no effect, bias, and ability models. Many of the current sabermetrics studies such as the chapters in Tango et al. (2007) and Keri and Baseball Prospectus (2007) focus on the interpretation of situational effects.

To illustrate situational data, batting data were collected for all players in the 2014 season. The analysis focused on right-handed and left-handed batters who had at least 150 plate appearances against pitchers of each arm. Figure 4.1 graphs the batter's *wOBA*

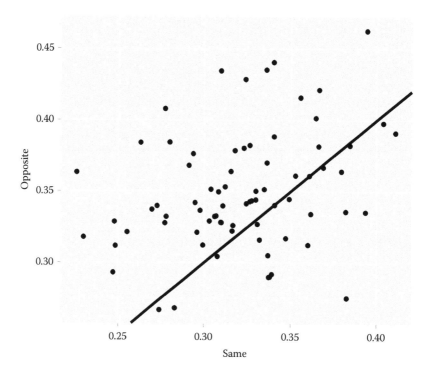

FIGURE 4.1
Scatterplot of batters' *wOBA* against pitchers of the same arm and the opposite arm.

against pitchers of the opposite arm versus the batter's *wOBA* against pitchers of the same arm. By comparing with the line $y = x$ in the graph, we see that hitters generally perform better against pitchers of the opposite arm. But there is great variability in the sizes of the observed situational effects.

Figure 4.2 plots the situational effect

$$EFFECT = wOBA_{opp\,arm} - wOBA_{same\,arm}$$

against the average *wOBA* for all hitters. It can be observed that several hitters hit 100 points better against pitchers of the opposite arm; other hitters actually hit 100 better against pitchers of the same arm. Due to the relatively small sample sizes when one breaks hitting data into categories, situational data such as the opposite-minus same-arm *wOBA*s will typically show this large variation.

4.3.2 No Effect, Bias, and Ability Models

It is clear that players have different batting talents, and Section 4.2 illustrated the use of a random effects model to estimate the talent distribution and the hitting talent of an individual player. What is less clear is whether players possess talents to perform well or worse

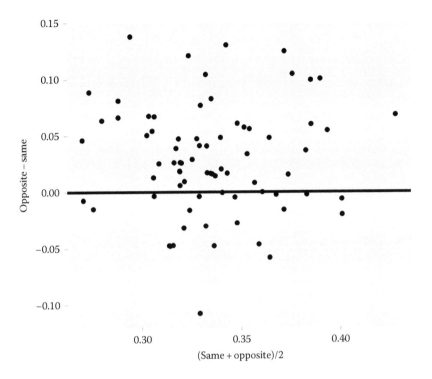

FIGURE 4.2
Scatterplot of batters' *wOBA* situational effect and their average *wOBA* for the opposite-arm/same-arm situation.

in particular situations. Suppose hitting data are collected for all data in two situations that will be called "opposite" (*O*) and "same" (*S*), although this can refer to any breakdown that has two categories. Let θ_j^O and θ_j^S denote the hitting talents in the *O* and *S* situations.

4.3.2.1 No-Effect Situations

Some situations will have no impact on the player's hitting ability. In this "no-effect" situation, for all players,

$$\theta_j^O = \theta_j^S, \quad j = 1, \dots, N.$$

One may think that particular situations such as day/night and pre-All Star Game/post-All Star Game fall in this particular scenario. There is no obvious reason why players should hit better in night games as opposed to day games. Likewise, there is no clear reason why players generally should have worse or better hitting talents during the second half of the season.

4.3.2.2 Bias Situations

In other situations, such as the home/away effect, one might believe that there is a situational effect that affects all hitters in a similar manner. This type of situational effect is commonly referred to as a bias. In our opposite/same scenario, this model states that the player's hitting ability in the opposite situation is larger than the hitting ability in the same situation, but this effect is a constant value, say B, for all batters:

$$\theta_j^O = \theta_j^S + B, \quad j = 1, \ldots, N.$$

Home versus away is a good illustration of a bias situation. There may be advantages for a player playing at home. For example, there is evidence (see Moskowitz and Wertheim, 2012) to suggest that the umpire may call balls and strikes favoring the batters from the home team (and disfavor batters from the visiting team). Although there may be an advantage, it is reasonable to believe that the size of this home advantage would be the same for all batters from the home team.

4.3.2.3 Ability Effect Situations

The most interesting model for situational data is the so-called "ability effect" where players have differing abilities to take advantage of a particular situation. Again considering the hypothetical opposite/same situation, if this is indeed an ability effect, the size of the advantage of a player batting in the opposite situation (as opposed to the same situation) is unique for each batter.

$$\theta_j^O = \theta_j^S + B_j, \quad j = 1, \ldots, N.$$

Some players might have small values of B_j, indicating a small situational effect, and other players may really take advantage of the situation and have a large value of B_j. One situation where one might plausibly see an ability effect is the "two strike"/"ahead in the count" breakdown. Some batters ahead in the count have a better chance of getting on-base and their batting talent is significantly greater in this situation. Other players with good batting control may have a small value of B_j since their batting abilities are similar in two-strike and ahead-of-the-count situations.

4.3.3 A Situational Random Effects Model

A more sophisticated random effects model can be used to represent situational hitting data. Using the same hypothetical opposite/same situation, let y_{jk} denote the measurement of hitting performance of the jth player in the kth situation, where $j = 1, \ldots, N$ and $k =$ "opposite side," "same side." As before we assume that

$$y_{jk} \sim N\left(\theta_{jk}, \frac{\sigma}{\sqrt{n_{jk}}}\right),$$

where
  σ is the sampling standard deviation
  n_{jk} is the number of plate appearances of the jth player in the kth situation

The key aspect of this model is the representation of the hitting talents. The talent of the jth player in the kth situation, θ_{jk} can be written as

$$\theta_{jk} = \begin{cases} \eta_j, & \text{if } k \text{ is "same side,"} \\ \eta_j + \gamma_j, & \text{if } k \text{ is "opposite side."} \end{cases}$$

The parameters $\eta_1, \ldots, \eta_N$ represent the hitting talents of N players that are distributed from a normal curve with mean μ_η and standard deviation σ_η. The parameter γ_j represents the advantage of player j in the opposite side situation. The situational effects for all players, $\gamma_1, \ldots, \gamma_N$ are assumed normal with mean μ_γ and standard deviation σ_γ.

The basic situational models described in Section 4.3.2 can be viewed as special cases of this situational random effects model.

1. The **no effects model** corresponds to the case where the situational parameters $\gamma_1, \ldots, \gamma_N$ are all equal to zero (when the random effects mean $\mu_\gamma = 0$ and the standard deviation $\sigma_\gamma = 0$).
2. The **bias model** corresponds to the case where $\mu_\gamma = B$. Since all players have the same situational effect, the standard deviation $\sigma_\gamma = 0$.
3. The **ability effect model** allows for different situational effects for N players. This model corresponds to a positive value of the random effects standard deviation σ_γ.

4.3.4 A Regression Random Effects Model

When one is primarily interested in learning about situational effects, the following two-step approach provides a simple method for estimating the situational effects. Gelman and Hill (2006) is a general resource for random effect models, describing different ways of fitting the models including the simple method described in this section.

1. *Fit N regression models*: Use the batting data for each player to estimate the situational effects. For the data for the jth player $(y_{j,same}, y_{j,opp})$, fit the regression model $y_{jk} \sim N(\theta_{jk}, \sigma/\sqrt{n_{jk}})$, where

$$\theta_{jk} = \begin{cases} \eta_j, & \text{if } k \text{ is "same side,"} \\ \eta_j + \gamma_j, & \text{if } k \text{ is "opposite side."} \end{cases}$$

From this fit, one gets the regression estimate $\hat{\gamma}_j$ with the corresponding standard error σ_j.

2. *Fit a random effects model on the individual regression coefficients*: The regression estimates $\hat{\gamma}_1, \ldots, \hat{\gamma}_N$ are assumed independent where $\hat{\gamma}_j \sim N(\gamma_j, \sigma_j)$. The situational effects $\gamma_1, \ldots, \gamma_N$ are assumed to be a random sample from a normal distribution with mean μ_γ and standard deviation σ_γ. The talent parameters $(\mu_\gamma, \sigma_\gamma)$ are assigned a weakly informative prior distribution of the form described in Section 4.2.1.

   The fit of the random effects model gives estimates $\hat{\mu}_\gamma$ and $\hat{\sigma}_\gamma$; the first estimate is the average situational effect and the second estimate is informative about the spread in the situational abilities. Following the general approach of Section 4.2.3, these estimates can be used to estimate the situational abilities $\gamma_1, \ldots, \gamma_N$ for N players.

4.3.5 Example

To illustrate this regression approach, we consider the opposite/same side platoon effects for all 2014 right-handed hitters with at least 150 plate appearances from each side. The random effects model estimates are given by $\hat{\mu}_\gamma = 0.0316$ and $\hat{\sigma}_\gamma = 0.0153$. On average, right-handed hitters have a *wOBA* talent that is 0.0316 higher for opposite-side pitchers. The size of the estimated standard deviation $\hat{\sigma}_\gamma$ is small (relative to the size of the average situational effect), indicating that players have similar talents for taking advantage of the opposite-side pitcher.

Let's return to the 2014 Jose Abreu who had a *wOBA* average of 0.462 in 152 PA against the opposite arm compared with a *wOBA* average of 0.394 in 455 PA against the same arm for a $0.462 - 0.394 = 0.068$ platoon advantage. In the regression fit for Abreu's data, one obtains the estimate $\hat{\gamma}_j = 0.068$ with the associated standard error $\sigma_j = 0.058$.

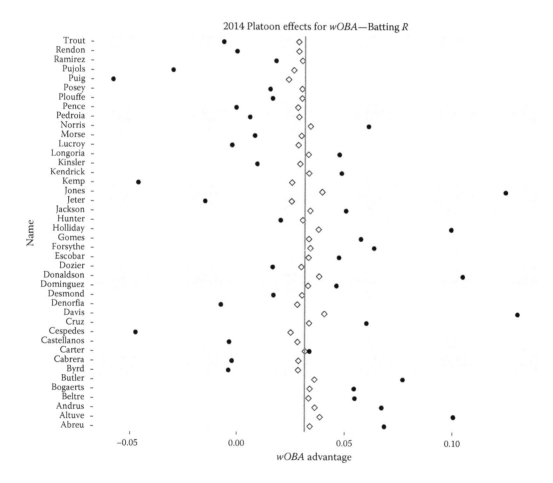

FIGURE 4.3

Dotplot of observed situational effects (black) and random-effects estimates (open circle) for all players in the 2014 season with at least 150 PA against opposite-arm and same-arm pitchers.

Here, the precision of the data is $Pr_D = 1/0.058^2 = 301$, the precision of the prior is $Pr_P = 1/0.0153^2 = 4259$, and the estimate of Abreu's situational talent is

$$\hat{\gamma}_j = \frac{301}{301 + 4259} \times 0.068 + \frac{4259}{301 + 4259} \times 0.0316 = 0.034.$$

Here, Abreu's observed situational effect is shrunk 93% toward the average situational effect of 0.0316.

As demonstrated earlier in Section 4.2.5, the estimates of the situational abilities $\gamma_1, \ldots, \gamma_N$ shrink the observed estimates toward the common estimate $\hat{\mu}_\gamma$. Figure 4.3 displays the situational estimates for the right-handed batters with at least 150 PA against each type of pitcher. As in Abreu's case, the ability estimates are clustered about the average situational effect. In particular, note that some of the observed situational estimates are negative, but none of the ability estimates are negative. The conclusion is that there are small differences in the situational abilities of the hitters in the opposite side/same side situation.

4.4 Understanding Platoon Effects in "The Book"

Chapter 6 of Tango et al. (2007) considers the same general issue in exploring situational data. Certainly, there is a general tendency for batters to hit better against pitchers of the opposite arm, but due to randomness and small samples, it is unclear if a player's extreme platoon split in one season is predictive of his platoon in the following season.

4.4.1 Do Platoon Skills Vary?

The authors begin by listing the left-handed batters who had the highest platoon split in weighted on-base percentage (*wOBA*) in favor of left-handed pitchers for individual seasons. When they collected the platoon split for the following season, they found that these players still had a higher than average platoon split. In other words, these players appeared to carry some of their extreme platoon split for the following season. The same phenomenon was observed for the 10 best left-handed batters who had extreme platoon effects against right-handers.

To make sense of these platoon splits, it might be the case that the conclusion depends on the batting measure that one uses. To check this, the authors also looked at the proportion of individual outcomes such as home runs, walks, hits, and strikeouts. From these data, they reached the same conclusion that there are indeed variations in real platoon splits among hitters, and this variation in platoon splits exists for all outcome types.

4.4.2 How Large Is the Platoon Skill Variation?

In Table 4.1, we reproduce part of Table 66 in Tango et al. (2007) that summarizes their analysis of platoon splits. This table addressed two questions. First, how large is the variability in platoon skills in contrast to the variability in player skills? Second, how does the variability in platoon skills contrast between batters and pitchers?

TABLE 4.1

Average Platoon Splits, 2000–2004 from Tango et al. (2007)

	Batters		Pitchers	
	RHB	**LHB**	**RHP**	**LHP**
Average OBP	.333	.349	.338	.342
OBP skill variation	.041	.040	.023	.025
Average OBP platoon split	.017	.019	.025	.011
Platoon skill variation	.014	.016	.021	.027
Average *wOBA*	.335	.349	.340	.342
wOBA skill variation	.046	.045	.024	.025
Average *wOBA* platoon split	.017	.027	.025	.019
Platoon skill variation	.013	.018	.022	.027

First, focus on the "Batters/OBP" section of the table. On average, right-handed hitters have on-base (OBP) talents around .333 and the standard deviation of these on-base talents is .041. (Left-handed hitters have a similar distribution of on-base talents.) In contrast, the average OBP platoon advantage for right-handed batters (opposite-arm compared to same-arm pitcher) is .017 and the standard deviation of these platoon skills is .014.

Next, look at the "Pitcher/OBP" section of the table, focusing on the comparison between batters and pitchers. The standard deviation of the OBP talents of the right-handed pitchers is .023 that is significantly smaller than the spread of the OBP talents of batters (a similar statement can be said for the left-handed pitchers.) The average size of the OBP platoon advantage for right-handed pitchers is .025—it is interesting that this average platoon split is higher than the average platoon split for the left-handed pitchers. Last, look at the standard deviation of the platoon skills for the pitchers—the values .021 and .027 for the two pitcher sides are significantly higher than the variation in platoon skills for batters.

The bottom half of Table 4.1 displays similar calculations for a different measure—the weighted on-base percentage *wOBA*. Focusing on the variation of platoon skills, we see that there exists significant platoon skill variation among batters and among pitchers. For example, the standard deviation of the platoon skills of the right-handed batters is .013, which is large relative to the average *wOBA* platoon split of .017. Furthermore, as we saw for OBP, there is larger player-to-player variation among pitchers than among batters.

Although this table confirms that players have varying platoon skills, it should be emphasized that the sizes of the platoon skills are small relative to the sizes of the batting skills. For example, a batter should not be used as a pinch hitter solely on the basis of his favorable platoon ability—a good hitter with a poor platoon split is still a better hitter than a bad hitter with a good platoon split.

4.4.3 Estimating Hitter's Platoon Skills

Now that the authors have established that players indeed have different platoon skills and have identified the mean and standard deviation of the platoon skill distribution, the next task is to estimate the platoon skills for individual batters and pitchers.

Recall from Section 4.2 that one combines his *wOBA* performance with the performance of an average batter to get a good estimate of the player's *wOBA* ability. In the case of

Bobby Abreu, the weights of the individual performance and the average were 607 and 434—these weights are essentially precisions (inverse of the variances) of the data and talent distribution. This same type of "regression to the mean" will be used to estimate a player's platoon skills—one takes a weighted average of the observed platoon effect and the average platoon effect.

Although the individual platoon skills are regressed toward the average, the degree of the shrinkage is different for platoon effects. We observed from Table 4.1 that the size of the platoon skill variation is relatively small, which implies a small talent standard deviation and a large prior precision. The implication is that the weights of the individual platoon effect and the average will be unbalanced where the average platoon effect will be given a much higher weight.

Specifically, Tango et al. (2007) find that the best estimate of a right-handed hitter's platoon skill is a weighted average of his actual platoon split and the league average where the weights are proportional to the number of lefties he has faced and 2200. Table 67 in their book displays some of these platoon skill estimates. For example, Brian Jordan who had a sizeable .101 *wOBA* split is estimated to have a platoon skill of only .037, and Derek Bell who had an unusual −.099 split is estimated to actually have a platoon skill of +.009. Since there is more variability among the platoon talents among pitchers, the weights are different—for example, for estimating the platoon ability of left-handed pitchers, the weight of the average is only 450.

One consequence of these estimates is that quite a large sample is needed to get reliable estimates of platoon skills. For example, Tango et al. state that a right-handed hitter needs at least 2000 plate appearances before his measured platoon split is more accurate than simply assuming that the player has an average platoon split.

Knowledge of the sizes of the platoon skills of batters and pitchers has strategic implications in the use of players during a game. Since the platoon advantage for pitchers is relatively large, there are clear advantages to the use of, for example, left-handed relief specialists during a game. One unorthodox strategy suggested by Tango et al. is the simultaneous use of two pitchers during the game, one pitching and one in the outfield, and rotate them depending on the handedness of the batter.

4.5 Detecting Clutch Ability (Is David Ortiz a Clutch Hitter?)

4.5.1 How to Evaluate Clutch Hitting?

Nate Silver in Chapter 1 of Keri and Baseball Prospectus (2007) describes the typical approach in the evaluation of clutch-hitting ability. One breaks down situations into two categories: clutch (say, a close game in the late innings) and other situations. One contrasts a hitting measure in the two situations and sees if there are specific hitters who have a propensity to perform better in clutch situations. Studies that have taken this approach have found little evidence of players having clutch ability.

There are problems with this approach. First, the choice of clutch and nonclutch situations is arbitrary. It is possible that plays that happen late in a game have little impact on the game's outcome, and other plays, say a home run in the second inning, actually are more important plays. A second problem is that the traditional approach for assessing clutch ability assumes that the value of a particular play, say a single, remains the same throughout a game. Actually, the value of a single depends dynamically on the game situation (inning, game score, runners on base, and outs).

Following this line of thought, imagine that a hitter is able to adjust his approach according to the game situation. He would attempt to hit a home run in situations where home runs were most valuable and perhaps try to hit a single in situations where any type of hit would be equally valuable toward the goal of winning the game. So Silver argues that the view of clutch hitting might be more accurately described as situational hitting.

A general way of quantifying a clutch situation is by means of win probabilities. At every state of the game (defined by inning, game score, runners on base, and outs), the home team has an associated probability of winning the game. A baseball play that changes the game state will change the win probability, and the change in win probabilities (called win probability added or WPA) quantifies the value of the play. Clutch situations can be associated with large potential values of WPA. Players can be evaluated in terms of their total contribution to WPA that Silver refers to as win expectancy or WinEx. This leads to a list of best hitters—the best clutch hitters tend also to be the best hitters in baseball.

But this is not what some people are thinking when they refer to clutch hitting. A player's conventional batting statistics may not accurately describe a player's value—a "clutch" hitter tends to get the hits at the right time, over and above what would be predicted based on his own traditional season statistics.

4.5.2 Players More Valuable than Their Regular Statistics

Silver proposes a method for determining the clutch contribution of a hitter. First, he estimates the "marginal lineup value" (MLV) of a player based on his season batting average, on-base percentage, and slugging percentage. The quantity MLV measures the number of additional runs that the player contributes if one position in a lineup of average hitters was replaced by this player.

Second, the value of MLV is converted to wins using a variation of Bill James' Pythagorean formula, which relates the ratio of wins and losses to the ratio of runs scored and runs allowed. This measure, called MLVWins, is similar to the popular WAR measure for evaluating players.

The third step of Silver's procedure is to adjust the value of MLVWins by the player's position in the batting lineup. Some players batting in the middle of the lineup may have a relatively large number of opportunities to produce runs in clutch situations, and other players in the bottom of the lineup will have fewer opportunities. Every plate appearance has an associated "leverage," and one averages leverage scores for all plate appearances for a player to get a measurement of how many clutch situations the player faced during the season.

Given a player's values of MLVRuns, MLVWIns, and Leverage, a regression model is used to predict a player's WinEx value—this prediction is called the "Leveraged MLV Wins" or LMLVW. A hitter's clutch performance is measured by the residual, the difference between the observed WinEx and the prediction:

$$Clutch = WinEx - LMLVW.$$

The value Clutch can be interpreted as the number of additional wins contributed by the player over and above the wins predicted by their season batting statistics, the leverage, and the run-scoring environment.

4.5.3 Exploration of "Clutch"

David Ortiz is widely viewed as a clutch player in baseball due to his batting performance in play-off game situations. Silver found that Ortiz had a clutch score of 3.6 in the 2005 season, indicating that he contributed 3.6 extra wins to his team's total wins due to his timely hitting. But this particular clutch score was unusual in Ortiz's career; indeed, his total clutch score for the nine seasons between 1997 and 2005 was only 5.2. Although one can construct a table of the players with the highest clutch career scores, there is little evidence to support that a player is consistently clutch through the seasons of his career.

Instead of focusing on specific players, Silver plots the number of clutch wins per 650 PA for players in even seasons against the number of clutch wins per 650 PA in odd seasons. There is a positive relationship in this graph with a correlation of .33. This graph suggests that roughly 10% of clutch hitting performance is explainable as skill.

Since clutch hitting ability appears to exist, the next obvious question is to explore the list of lifetime clutch hitters and try to figure out the batting qualities of the players who tend to be clutch. Silver looks at the correlation of different career hitting statistics, adjusted for league and park effects, against clutch rating per PA for players with at least 5000 plate appearances. From this analysis, he concludes that players who walk often and strike out rarely tend to perform well in clutch situations. Silver postulates that we are observing the effects of smart situational hitting. Hitters who are capable of adjusting their plan at the plate for the situation, say hit to right field to move a runner from first to third base, may be more successful in clutch situations.

4.6 Detecting True Streakiness

4.6.1 Streaky Performances by Players and Teams

Baseball fans are fascinated by streaky performances of players and teams. From a batting perspective, game hitting streaks are typically announced in broadcasts such as "Joe Smith has a 20-game hitting streak." Hitting slumps are often described in terms of "ofers" such as "Harry Taylor" is currently hitless in his last 25 at-bats—this is a "0 for 25" slump. Winning and losing streaks are recorded for teams. One of the streaks that motivated the book *Moneyball* (Lewis, 2004) was the 20-game winning streak of the Oakland Athletics during the 2002 season. This was perceived to be a long streak since there were no other winning streaks by baseball teams of length 20 or higher since the 1935 season.

Given the high interest in streaky performances, there is also large literature on the statistical interpretation of streaks. Albright (1993) uses a regression approach with a large selection of regression inputs to explore streakiness of hitting data. Albert (2008) and Albert and Bennett (2007), in Chapter 5, provide general discussions of methods for detecting "true" streakiness and Albert and Williamson (2001) show the use of simulation to perform inference about parameters of a streaky model. McCotter (2010) gives evidence of more hitting streaks that were predicted from "random variation." Thomas (2010) and Horiwitz and Lackritz (2013) explore specific types of hitting streaks. Albert (2004a,b) explore patterns of streaks in team performance and Albert (2013, 2014) focus on the use of spacings between occurrence of events in the search for true streakiness.

We focus on a particular streaky accomplishment that was noted in a recent game broadcast. During the 2015 season, it was noted that Ichiro Suzuki had 32 consecutive at-bats without striking out. Should one be impressed with this accomplishment?

At this point in the 2015 season, Ichiro had 38 strikeouts in 301 at-bats with a strikeout rate of $\hat{p} = 38/301 = 0.126$. Assuming independence of outcomes of individual at-bats, the probability of 32 consecutive at-bats without a strikeout would be the product:

$$(1 - 0.126)^{32} = 0.0134.$$

Since the probability of this event is small, this might appear to give sufficient evidence that Ichiro exhibited some "true" streakiness in his strikeout/nonstrikeout hitting during the 2015 season.

There are a number of concerns about this computation from a statistical perspective that are labeled in the following text as "selected data," "multiplicity," and "adjustment for sample size and probability."

4.6.1.1 Selected Data

The probability computation is flawed since we chose a particular subset of the data, namely, a streak of 32 nonstrikeouts, that looked interesting. The fact that we selected the interesting data will have a significant impact on the probability of the event. A more relevant probability to consider is the chance that in 301 at-bats, one would observe a streak of 32 (or more) consecutive nonstrikeouts. This is a more difficult probability to compute exactly, but it can be computed in a straightforward manner by simulation. If we assume that Ichiro's strikeout probability is 0.126, then a simulation with 10,000 iterations shows the probability of a streak of 32 or greater anywhere in the season is equal to 0.405. So actually observing a nonstrikeout run of 32 at-bats *sometime* during the season is relatively common for a batter like Ichiro (but not for a strikeout-prone hitter such as Ryan Howard).

4.6.1.2 Adjustment for Sample Size and Probability

A complication with the observation of streaky behavior is that the size of a streaky statistic, such as the length of the longest streak, is confounded with the number of opportunities and the probability of a success. It is more likely to observe long streaks from players with a large number of attempts. Likewise, poor teams with a small probability of winning a single game are more likely to experience long losing streaks, and good teams will tend to show more winning streaks. Any measure of streakiness has to make a reasonable adjustment for the number of opportunities (the sample size) and the probability of success.

4.6.1.3 Multiplicity

Another problem with this investigation of Ichiro's streak is that we focused on this player and the streak since it looked interesting. If one selects a random player, one might be interested in the probability that this particular player has a streak of a particular length or greater. A different question is whether *some* player in this particular season has a streak that exceeds a particular length.

This problem is analogous to the famous birthday problem. If a student is sitting in a class with 24 classmates, it is unlikely that he or she will find someone who shares his/her birthday (day and month). In contrast, it is relatively common that there will be two students in the class who share the same birthday. The reason why the second probability is high is that there many ways for two people to share common birthdays. Similarly, in the baseball setting, it is pretty common to observe one player (among the large number of regular players) with a long hitting streak during a season.

4.6.2 Measures of Streakiness

Given the statistical issues discussed in the previous section, there are a number of ways of measuring streakiness. We focus on streaky measures for the binary (0 and 1) sequence of hitting outcomes for a player during a season.

4.6.2.1 Runs

One can consider the pattern of *runs* or consecutive streaks of the same outcome. For example, if a player shows the following hit/out sequence

$$0\,0\,0\,1\,1\,1\,1\,0\,0\,0\,1\,0\,0\,0\,0\,0\,0\,1\,0\,0\,0\,0\,1\,1\,1\,0\,0\,0\,0\,0\,0,$$

then we see that he had a run of three outs, followed by a run of four hits and a run of three outs. One can summarize these runs in different ways. If one focuses on the runs of hits (1's), then one could compute

- The length of the longest run of hits
- The mean length of runs of hits
- The total number of runs of hits

One can also compute the longest run of outs (0's), the mean length of runs of outs, and the total number of runs of outs. A streaky player would likely have long runs of hits or long runs of outs.

4.6.2.2 Moving Averages

A different way to detect streakiness is to consider moving averages of outcomes in short time intervals. If the time series of batting measurements is $y_1, \ldots, y_n$, then the moving average of measurements from time j to time $j + w - 1$ is given by the average

$$m_j = \frac{\sum_{k=j}^{k=j+w-1} y_k}{w}.$$

If one plots the moving averages $\{m_j\}$ as a function of the time j, one sees the volatility of the players' measure of performance in short time intervals. Streaky players tend to display moving averages with a wide range. In contrast, a very consistent hitter would show a moving average plot that was nearly constant.

4.6.2.3 Measures of Dependence

A third way of measuring streakiness is to explore the association between a player's success and his success in previous at-bats. For example, if one focuses on individual at-bats, one can look at the association of success in the current at-bat with success in the previous at-bat. If one believes a hitter is streaky, then one would predict a positive relationship between the outcomes in the current and previous at-bats. A related idea would use a game to define the time interval and explore the association between a player's performance in the current game and the performance in the previous game.

4.6.2.4 Measures of Clumpiness

For sequences of binary measurements, Zhang et al. (2013) consider a class of measures of clumpiness and apply these measures to test for the "hot-hand" phenomenon. Let $w_1, \ldots, w_k$ denote the spacings or gaps between successive successes. Then one measure of clumpiness would be the second moment of these spacings.

$$L = \sum_{j=1}^{k} w_j^2.$$

A streaky hitter would have a "large" value of the clumpiness measure L.

4.6.3 Models

To get a better understanding of what it means for a hitter to have streaky ability, one defines statistical models. Suppose one focuses on a player who plays regularly and has 4 opportunities to hit during each of 160 games in a particular season. Let y_j denote the number of hits of the player in the jth game. We assume that y_j is distributed binomial with four trials and probability of success p_j. Here, p_j measures the player's hitting ability during the jth game.

4.6.3.1 The Consistent Model

If a player is truly consistent, then his hitting ability would not change across games, which indicates that his game hitting probabilities are equal to a constant value p.

$$p_1 = \cdots = p_{160} = p.$$

4.6.3.2 Streaky Models

In contrast, if a player is truly streaky, then the probabilities of hitting would not remain constant over the season. There are various ways to describe streaky models for hitting—each of the following models describes a pattern of variation of the hitting probabilities over the season.

One attractive model for streakiness (Albert, 1993) is based on the assumption that the batter is either "hot" with a hitting probability of p_{HOT} or "cold" with a hitting probability of p_{COLD}, where $p_{COLD} < p_{HOT}$. For the 160 games, the batter is moving between the two states "hot" and "cold." By use of a Markov switching model, one assumes that the batter moves between the two states by a Markov Chain, where the probability of remaining in the same state is given by γ, where $0 < \gamma < 1$. An assumption of $\gamma > 0.5$ would correspond to the common belief that if a player is hot in the previous game, he is likely to remain hot in the following game.

An alternative model assumes that the game-to-game probabilities $\{p_j\}$ are not constant but instead follow a probability distribution $g(p)$. For example, Albert (2008) assumes that the hitting probabilities, under the streaky model, follow a beta distribution with shape parameters a and b. The key parameter of the beta is the sum of parameters $a + b$ that controls the spread of the beta curve. A very streaky hitter would have a wide range of hitting probabilities and a small value of the parameter $a + b$.

4.6.4 Statistical Testing

4.6.4.1 Traditional Testing

If one observes a streaky event, say a run of 10 consecutive hits, then it is straightforward to test the hypothesis that the hitting probability is a constant value, say $p = 0.300$. One can repeatedly simulate Bernoulli outcomes with $p = 0.300$ and compute the p-value, the probability that the player gets 10 or more hits. If the p-value is sufficiently small, it can be concluded that the player is not truly consistent.

Although this testing method provides evidence against the consistent model hypothesis, it may be difficult to interpret. One does not believe that a player's hitting probability is exactly constant over the season—one would typically expect the hitting probability to exhibit some fluctuation. One is more likely interested in testing the hypothesis that the variation in hitting probabilities does not exceed a specific bound such as $\text{var}(p_j) \leq B$.

4.6.4.2 Bayesian Testing

If one is able to state two models, one for consistency and a second one for true streakiness, then the Bayes factor provides an approach for comparing the two models. For example, Albert (2008) considers the two models

$$M_C: p_1, \ldots, p_{160} = p, \quad M_S: p_1, \ldots, p_{160} \sim beta(a, b)$$

and then constructs a Bayes factor, which is a measure of the comparative supports of the streaky and consistent models. It is straightforward to compute the value of the Bayes factor. It is more challenging to construct interesting models on the underlying parameters that reflect the beliefs in consistency and streakiness. For example, in the specification of the streaky model M_S one needs to specify values of the beta shape parameters a and b that indicate the variability of the hitting probabilities when the player is truly streaky. If one believes that the Markov switching model is appropriate, then one needs to specify values of the hot and cold probabilities and the probability of moving from a hot state to a cold state in the next period.

4.6.5 Predictive Checking

4.6.5.1 Introduction

As mentioned earlier, one issue in the detection of interesting patterns of streakiness is multiplicity. When one observes the hitting patterns of many players over a season, one will likely observe some interesting streaky behavior by chance. It is analogous to flipping several hundred coins repeatedly. One will see interesting patterns of streakiness in the coin-flipping outcomes even though none of the coins are truly streaky.

A predictive simulation approach is helpful in understanding the patterns of streakiness among a collection of hitters. Suppose we have N hitters with respective constant hitting probabilities $P_1, \ldots, P_N$ for which we make reasonable estimates of the probabilities from the hitting data. (Different notation is used here to clearly distinguish these player hitting probabilities from the ones that can vary within a season.) This model assumes that each of the players is truly consistent with a constant hitting probability.

Given these N consistent hitters, what type of streaky hitting do we expect to find? To employ the predictive simulation approach, one simulates hitting data from Bernoulli distributions with these fixed hitting probabilities. For each player, some measure of streakiness S is computed, and the values of $S_1, \ldots, S_N$ are collected. If this simulation process is repeated K times, one obtains K sets of the streaky measures $\{S_1, \ldots, S_N\}$.

The predictive simulation gives us sets of plausible values for the streaky measures if indeed all of the N hitters were truly consistent with constant probabilities of hitting. Once the streaky measures for the N players, say $S_1^{OBS}, \ldots, S_N^{OBS}$ are observed, one compares this set of observations with the simulated sets of streaky measures. Specifically, suppose that the longest hitting streak among the N players is 25—is the value $S_{max}^{OBS} = 25$ special? This question is answered by computing the maximum hitting streak in each of the simulations and computing the predictive p-value

$$pred\text{-}p\text{-}value = P\left(S_{max} \geq S_{max}^{OBS}\right).$$

Suppose that the predictive p-value is equal to 0.03 for our example. This means that the observed hitting streak of 25 games is unusual. In our predictive simulation of hitting data from N consistent hitters, it was very unusual to see a hitting streak of 25 or longer. In this case, there would be some evidence that there is some true streakiness among our N players.

4.6.5.2 Example of Predictive Simulation

To illustrate the application of predictive simulation, suppose one wishes to explore the streakiness in home run hitting among all players with at least 300 at-bats for the 2014 season. The longest "ofer" or gap between consecutive home runs is used as the measure of streakiness.

To set up a model for consistent home run hitting, we initially fit a random effects model to the home run hitting data for the 2014 players. The count of home runs for the jth player (y_j in n_j at-bats) is assumed binomial (n_j, P_j). The home run probabilities for the N players, $P_1, \ldots, P_N$ are assumed to follow a beta distribution with shape parameters a and b. Fitting this model, one obtains the estimates $a = 3.4, b = 119.9$.

One performs a single predictive simulation from the consistent hitting model as follows:

1. One obtains home run probabilities $P_1, \ldots, P_N$ from the beta(3.4, 119.9) random effects distribution. It should be emphasized that each hitter is assumed to have a constant probability of hitting a home run during an at-bat during the season.

2. One simulates home run hitting data (sequences of 0's and 1's) for each player using the actual number of at-bats during the 2014 season.

3. For each home run hitting sequence, one computes the length of the longest ofer (gap between consecutive home runs).

When this single simulation is completed, one has a collection of streaky measures $S_1, \ldots, S_N$ for all hitters—these streaky measures were simulated assuming a consistent hitting model for all players. The question is how do these simulated streaky measures differ from the actual observed longest ofers $S_1^{OBS}, \ldots, S_N^{OBS}$ for the 2014 players?

Figure 4.4 illustrates this comparison. The top left panel displays a scatterplot of the number of home runs and the longest ofer for the 2014 hitters—we focus on hitters with 10–50 home runs. The remaining eight panels display scatterplots of simulated data from the consistent model. It is interesting to note that the pattern of streakiness (values of

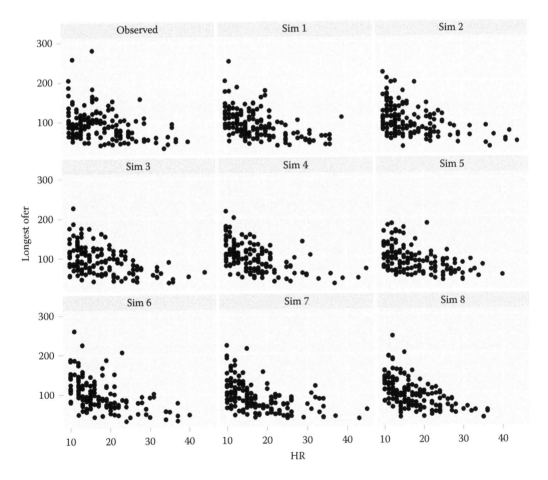

FIGURE 4.4

Scatterplots of number of home runs and the length of the longest ofer for hitters with at least 300 AB. The upper-left plot corresponds to the 2014 hitters and the remaining plots correspond to predictive simulations assuming all hitters are truly consistent with constant hitting probabilities.

the ofers) for the simulated data resembles the 2014 data. So in this case it is difficult to detect "true streakiness" using the longest gap between consecutive home runs as the streaky statistic.

4.7 Closing Comments

There is much interest in the sabermetrics literature on the existence of situational and streaky ability among ballplayers. James (2004) makes a list of nine statements including "clutch hitters don't exist," "batters have no individual tendency to hit well or poorly against left-handed pitching," and "batters don't get hot and cold" that have and will continue to motivate more statistical studies. Using James' terminology, one wishes to distinguish between transient and persistent phenomena. If a phenomenon such as clutch hitting is persistent, then players who are good in this characteristic one season will tend also to be good the following season. If players good in a characteristic one season *do not* tend to be better the following season, one says that the characteristic is transient.

The challenge in interpreting situational and streaky data is that, using James' terminology, there is much instability (chance variation) in this type of data and the sizes of the real situational or streaky effects are small relative to the variability in the observed data. This chapter has surveyed a model-based approach for handling situational data where each player is assigned individual parameters corresponding to his hitting and situational talents and the collections of talents are assigned random effects distributions. When these models are fit, they provide estimates of the situational talent distributions, and they also provide estimates of the situational effects for individual players.

There are some general conclusions that one can draw from the various studies described in this chapter. First, it can be challenging even to quantify what is meant by a situational effect. For example, Silver in Keri and Baseball Prospectus (2007) describes the challenge in measuring clutch play for a hitter since it is unclear what is really meant by the word "clutch." Second, due to the small sample sizes, situational and streaky data are inherently noisy and one can be deceived, say, by lists that give leaders in particular situational categories for a single season. One needs to look at many seasons of data for a player to see if he displays, say, a pattern of unusual hot and cold hitting over the seasons. Third, these studies indicate that players can indeed possess different talents to excel in a particular situation, although the sizes of these situational talents are smaller than the observed situational effects. Last, and perhaps most interesting, there is clearly a need for further research to better understand why players possess different situational talents. In the opposite-side/same-arm situation, what qualities of a hitter will tend to be associated with small situational effects? Given the common use of pitcher–batter matchups that exploit the platoon situation, a manager from the offensive team may wish to have more players with small situational splits. Perhaps players with good bat control have small platoon splits, and it might be advantageous to construct a team with more hitters with good bat control. In any event, it is clear that the interpretation of situational baseball data will continue to be a popular topic of research for many years to come.

References

Albert, J. (1993), Comment, *Journal of the American Statistical Association*, 88(424), 1184–1188.

Albert, J. (1994), Exploring baseball hitting data: What about those breakdown statistics? *Journal of the American Statistical Association*, 89(427), 1066–1074.

Albert, J. (2002), Hitting with runners in scoring position, *Chance*, 15(4), 8–16.

Albert, J. (2004a), A batting average: Does it represent ability or luck? Technical Report, http://www-math.bgsu.edu/~albert/papers/paper_bavg.pdf.

Albert, J. (2004b), Streakiness in team performance, *Chance*, 17(3), 37–43.

Albert, J. (2006), Pitching statistics, talent and luck, and the best strikeout seasons of all-time, *Journal of Quantitative Analysis in Sports*, 2(1).

Albert, J. (2007), Hitting in the pinch, *Statistical Thinking in Sports*, Chapman and Hall, Boca Raton, FL, pp. 111–133.

Albert, J. (2008), Streaky hitting in baseball, *Journal of Quantitative Analysis in Sports*, 4(1).

Albert, J. (2013), Looking at spacings to assess streakiness, *Journal of Quantitative Analysis in Sports*, 9(2), 151–163.

Albert, J. (2014), Streakiness in home run hitting, *Chance*, 27(3), 4–9.

Albert, J. and Bennett, J. (2007), *Curve Ball: Baseball, Statistics, and the Role of Chance in the Game*, Springer Science and Business Media, New York, NY.

Albert, J. and Williamson, P. (2001), Using model/data simulations to detect streakiness, *The American Statistician*, 55(1), 41–50.

Albright, S. C. (1993), A statistical analysis of hitting streaks in baseball, *Journal of the American Statistical Association*, 88(424), 1175–1183.

Bennett, J. (1993), Did shoeless Joe Jackson throw the 1919 World Series? *The American Statistician*, 47(4), 241–250.

Cramer, R. D. (1977), Do clutch hitters exist? *Baseball Research Journal*, 6, 74–79.

Efron, B. and Morris, C. (1975), Data analysis using Stein's estimator and its generalizations, *Journal of the American Statistical Association*, 70(350), 311–319.

Gelman, A. (2006), Prior distributions for variance parameters in hierarchical models, *Bayesian Analysis*, 1, 515–534.

Gelman, A. and Hill, J. (2006), *Data Analysis Using Regression and Multilevel/Hierarchical Models*, Cambridge University Press.

Hopkins, T. and Magel, R. C. (2008), Slugging percentage in differing baseball counts, *Journal of Quantitative Analysis in Sports*, 4(4).

Horowitz, I. and Lackritz, J. (2013), The value of s at which a Major League player with multiple hitting streaks of at least s Games might be termed streak prone, *Chance*, 26(3), 26–33.

James, B. (2004), Underestimating the fog, *Baseball Research Journal*, 33.

Keri, J. and Baseball Prospectus (2007), *Baseball Between the Numbers: Why Everything You Know about the Game is Wrong*, Basic Books, New York, NY.

Lewis, M. (2004), *Moneyball: The Art of Winning an Unfair Game*, W. W. Norton & Company, New York, NY.

McCotter, T. (2010), Hitting streaks don't obey your rules: Evidence that hitting streaks are not just byproducts of random variation, *Chance*, 23(4), 52–57.

Moskowitz, T. and Wertheim, L. (2012), *Scorecasting: The Hidden Influences Behind How Sports Are Played and Games Are Won*, Three Rivers Press, New York, NY.

Otten, M. P. and Barrett, M. E. (2013), Pitching and clutch hitting in Major League Baseball: What 109 years of statistics reveal, *Psychology of Sport and Exercise*, 14(4), 531–537.

Palmer, P. (1990), Clutch hitting one more time, *By the Numbers*, 2(2), 6–7.

Tango, T., Lichtman, M., and Dolphin, A. (2007), *The Book: Playing the Percentages in Baseball*, Potomac Books, Dulles, VA.

Thomas, A. C. (2010), That's the second-biggest hitting streak I've ever seen! Verifying simulated historical extremes in baseball, *Journal of Quantitative Analysis in Sports*, 6(4).

Yates, P. A. (2008), Estimating situational effects on OPS, *Journal of Quantitative Analysis in Sports*, 4(2).

Zhang, Y., Bradlow, E. T., and Small, D. S. (2013), New measures of clumpiness for incidence data, *Journal of Applied Statistics*, 40(11), 2533–2548.

5

Estimating Team Strength in the NFL

Mark E. Glickman and Hal S. Stern

CONTENTS

For NFL football fans, comparing team abilities and speculating on the outcomes of team matchups is one of the most enjoyable aspects of following the game. The comparison of teams can be a source of intense argument, with discussions commonly peppered with anecdotes and cherry-picked information to support the contention of one team being better than another. Various web sites promote a semblance of scientific rigor by computing power rankings, numerical team ratings, and other related quantities that are ostensibly reliable measures of team strength. But if interest centers on a serious study of NFL team strength, how can we appeal to a solid statistical foundation to measure ability and predict game outcomes?

This chapter covers basic statistical approaches to measure NFL team strength from historical game results. In Section 5.1, we formally lay out the problem and describe some basic procedures for estimating team strength. This is followed in Section 5.2 with a presentation of the statistical model for game outcomes along with common approaches to perform statistical inference of team strengths. The basic model is extended in Section 5.3 to account for the possibility that team strengths may be evolving over time. In Section 5.4, we explain the role of covariates in our models from the two previous sections. We then introduce some recent approaches to modeling game outcomes in Section 5.5. In Section 5.6, we demonstrate some of the methods in this chapter on NFL data, and we conclude the chapter in Section 5.7 with some discussion of future directions of NFL game outcome modeling.

5.1 Football Team Strength: The Basics

From a brief examination online, it is clear that many methods are in use to rate football teams. Massey (n.d.) regularly reports NCAA football team rankings from no fewer than 115 different methods of rating team strengths. As demonstrated on Massey's site, these methods produce rankings that are highly correlated despite the variety of methods implemented. Many of these approaches are principled and based on sound statistical reasoning, but plenty are ad hoc algorithms. We demonstrate in this section a particular principled approach to rate football teams.

The fundamental assumption for most models and rating methods for football team strengths is that the jth team's ability, $j = 1, \ldots, J$ ($J = 32$ currently for the NFL), can be represented as a scalar parameter θ_j, and that game outcomes between teams j and k depend on the team strengths only through their difference, $\theta_j - \theta_k$. The main goal of a statistical or rating procedure for football is to estimate $\theta = (\theta_1, \ldots, \theta_J)$ from game results. The resulting estimates can then be used as the basis for measuring relative strengths of teams, ranking teams, and predicting outcomes of new games.

Consider a set of n games to analyze, and let y_i be an outcome of game i played between teams j_i and k_i (the choice of assigning team labels within a pair is arbitrary). Let y denote the vector of all n game outcomes $(y_1, \ldots, y_n)$. The outcome y_i could be defined in a variety of ways. Common choices are the final score difference in the game from the perspective of team j_i, or the trinary outcome corresponding to 1 if team j_i wins, 0.5 if the game is a tie, and 0 if j_i loses. In the former case, a natural choice for estimating the θ_j is to minimize the sum of squared differences:

$$\mathrm{SS}(\theta|y) = \sum_{i=1}^{n} \left(y_i - (\theta_{j_i} - \theta_{k_i})\right)^2. \tag{5.1}$$

One way to interpret θ_j is through team j's mean margin of victory plus the mean strength of the opponents. Larger θ_j corresponds to better (higher relative scoring) teams and smaller θ_j corresponds to worse teams. The objective function in (5.1) is optimized solely based on game outcomes. We consider incorporating covariate information, such as which team is playing on its home field, in Section 5.4. This basic setup for football team strength was described in Stern (1995).

The objective function in (5.1) can be more conveniently represented in matrix form. Let X be an $n \times J$ matrix that encodes the identities of teams competing across the n games in the following manner. The ith row of X encodes information about game i. In row i, the j_ith element is set to 1, the k_ith element is set to -1, and the remaining elements are set to 0. An example of X might appear as

$$
\begin{pmatrix}
1 & 0 & -1 & 0 & 0 & \cdots & 0 \\
0 & 0 & 1 & 0 & -1 & \cdots & 0 \\
-1 & 0 & 0 & 0 & 1 & \cdots & 0 \\
\vdots & & & & & \ddots & \vdots \\
0 & -1 & 0 & 0 & 0 & \cdots & 1
\end{pmatrix}
$$

where the first game involves team 1 playing team 3, the second game involves team 3 playing team 5, and so on. Then the n-vector formed from $X\theta$ is the vector of ability parameter differences for each game, that is $(\theta_{j_1} - \theta_{k_1}, \theta_{j_2} - \theta_{k_2}, \ldots, \theta_{j_n} - \theta_{k_n})$. In the preceding example, $X\theta$ is given by

$$
\begin{pmatrix}
1 & 0 & -1 & 0 & 0 & \cdots & 0 \\
0 & 0 & 1 & 0 & -1 & \cdots & 0 \\
-1 & 0 & 0 & 0 & 1 & \cdots & 0 \\
\vdots & & & & & \ddots & \vdots \\
0 & -1 & 0 & 0 & 0 & \cdots & 1
\end{pmatrix}
\begin{pmatrix}
\theta_1 \\
\theta_2 \\
\vdots \\
\theta_J
\end{pmatrix}
=
\begin{pmatrix}
\theta_1 - \theta_3 \\
\theta_3 - \theta_5 \\
\theta_5 - \theta_1 \\
\vdots \\
\theta_J - \theta_2
\end{pmatrix}
$$

Equation 5.1 can then be reexpressed in matrix notation as

$$
\text{SS}(\theta|y) = (y - X\theta)'(y - X\theta). \tag{5.2}
$$

Recognizing that (5.2) is the sum of squared residuals for linear regression with design matrix X, the unique least-squares estimate $\hat{\theta}$ of θ is given by $\hat{\theta} = (X'X)^{-1}X'y$ only if $X'X$ is invertible. The problem is that the columns of X are linearly dependent because the sum of elements across every row is 0, so $X'X$ is not of full rank and therefore not invertible.

Before describing two common approaches to address the linear dependence in the columns of X, it is worth commenting on the structure of the $J \times J$ matrix $X'X$. Any least-squares estimate of θ, regardless of whether it is unique, satisfies the normal equations

$$
X'X\theta = X'y. \tag{5.3}
$$

The jth diagonal element counts the total number of times team j has competed among the n games in the data set. Each off-diagonal element, say the (j, k) element of $X'X$, is the negative of the number of times teams j and k have competed among the n games in the data set. The sum of elements in each row of $X'X$ therefore is 0, which also highlights that this square matrix is not invertible. Meanwhile, $X'y$ is the vector of J elements in which the jth element is the sum of the score differences for the games team j won less the sum of the score differences for the games team j lost. So, for the jth element of the left-hand and right-hand sides of (5.3), we have

$$
\left(\sum_{k \neq j} n_{jk} \right) \theta_j - \sum_{k \neq j} n_{jk}\theta_k = \sum_{i:\, j\,\text{wins}} |y_i| - \sum_{i:\, j\,\text{loses}} |y_i|, \tag{5.4}
$$

where n_{jk} is the number of times j and k have competed. Solving (5.4) for θ_j, we obtain

$$
\theta_j = \frac{\sum_{k \neq j} n_{jk}\theta_k}{\sum_{k \neq j} n_{jk}} + \frac{\sum_{i:\, j\,\text{wins}} |y_i| - \sum_{i:\, j\,\text{loses}} |y_i|}{\sum_{k \neq j} n_{jk}}, \tag{5.5}
$$

which demonstrates that θ_j can be understood as the average of the opponents' strengths plus the average margin of victory.

5.1.1 Incorporating Linear Constraints

An approach to address the linear dependence in X is to impose a linear constraint on θ. Such a constraint does not change the essential solution to the optimization problem; it only forces the selection of one solution among the infinite number in the unconstrained version. The most direct method to address the linear dependence is to reexpress $X\theta$ as $X^*\theta_{-J}$, where θ_{-J} is the vector of length $J-1$ of team strengths with θ_J removed, and X^* is the $n \times (J-1)$ design matrix that accounts for the linear constraint assumed on θ. This can be accomplished by creating a $J \times (J-1)$ contrast matrix W such that $X^* = XW$. This construction also implies that

$$X^*\theta_{-J} = XW\theta_{-J} = X\theta, \tag{5.6}$$

so that $W\theta_{-J} = \theta$.

For example, if the linear constraint assumed for identifiability was to set $\theta_J = 0$, then X^* would be simply the first $J-1$ columns of X. With $X^* = XW$, this is equivalent to assuming

$$W = \begin{pmatrix} 1 & 0 & 0 & \cdots & 0 \\ 0 & 1 & 0 & \cdots & 0 \\ 0 & 0 & 1 & \cdots & 0 \\ \vdots & \vdots & \vdots & \ddots & \vdots \\ 0 & 0 & 0 & \cdots & 1 \\ 0 & 0 & 0 & \cdots & 0 \end{pmatrix}.$$

If instead the linear constraint assumed was $\sum_{j=1}^{J} \theta_j = 0$, then X can be replaced with $X^* = XW$, where W is given by

$$W = \begin{pmatrix} 1 & 0 & 0 & \cdots & 0 \\ 0 & 1 & 0 & \cdots & 0 \\ 0 & 0 & 1 & \cdots & 0 \\ \vdots & \vdots & \vdots & \ddots & \vdots \\ 0 & 0 & 0 & \cdots & 1 \\ -1 & -1 & -1 & \cdots & -1 \end{pmatrix}.$$

This latter approach of obtaining the least-squares solution with the "sum contrast" linear constraint on θ is sometimes credited in the context of college football to Kenneth Massey (1997). In either case, the least-squares estimate $\hat{\theta}_{-J}$ of θ_{-J} is given by

$$\hat{\theta}_{-J} = (X^{*'}X^*)^{-1}X^{*'}y.$$

It is worth noting that the inclusion of a linear constraint does not itself guarantee a solution to the least-squares problem, let alone a unique solution. The extra requirement is that every team must be involved in at least one game in the data set so that the diagonal entries of $X'X$ are all positive. This is trivially satisfied by including team abilities in θ only if they have game results in the data set.

5.1.2 Regularization

An alternative approach to address the linear dependence among the columns of X is to modify the objective function in (5.1) so that it penalizes certain types of potential choices of θ. This approach to penalizing an objective function is sometimes known as regularization. A common type of penalty increases the objective function based on the sum of squared differences of the θ_j from a fixed set of values. A general form of the penalized sum of squared differences can be expressed as

$$\text{PSS}(\theta|y) = \sum_{i=1}^{n} \left(y_i - (\theta_{j_i} - \theta_{k_i})\right)^2 + \lambda \sum_{j=1}^{J}(\theta_j - \gamma_j)^2, \tag{5.7}$$

where $\lambda \geq 0$ and the γ_j are values set in advance of optimization.

The addition of the second term in (5.7) ensures that the optimizing values of the θ_j are centered at the γ_j. The value of λ attenuates the degree to which the θ_j are "shrunk" to the γ_j. For example, when $\lambda \to \infty$, the first term involving the score differences plays no role in the optimization and (5.7) is optimized at $\theta_j = \gamma_j$ for all j. When λ is set to a small value, such as $\lambda = 0.0001$, the impact of the second term is merely to break the tie among the infinite set of equally optimal solutions that satisfy optimizing only the first term. The method of adding a penalty term of this type for least-squares estimation is commonly known as ridge regression (Hoerl and Kennard, 1970).

It is straightforward to show from differential calculus that $\hat{\theta}$, the optimizing value of θ in (5.7), is equivalent to computing:

$$\hat{\theta} = (X'X + \lambda I)^{-1}(X'y + \lambda \gamma), \tag{5.8}$$

where $\gamma = (\gamma_1, \ldots, \gamma_J)$ and I is the identity matrix (of dimension J).

Adding λ to each diagonal element of the rank $J - 1$ matrix $X'X$ in (5.8) ensures the non-singularity of $(X'X + \lambda I)$ so that its inverse is uniquely defined.

The choice of values for λ and γ can be argued based on their role in (5.7). The parameter γ behaves as the center of θ. To ensure objectivity of the procedure, the elements of γ are commonly set to the same value. The specific choice of the common value is arbitrary and, for example, can be set to 0. The choice of λ on the other hand can be understood as a tuning parameter in a regularization problem. A principled procedure to choose λ is through cross validation. A particular form of cross-validation, namely K-fold cross-validation ($K = 10$ is a conventional choice), proceeds in the following manner.

- Divide the data set of n games at random into K mutually exclusive subsets. Let Y_k denote the kth subset of data and let Y_{-k} denote its complement. Assume γ is already selected.
- Suppose λ^* is a candidate value of λ.
  - For $k = 1, \ldots, K$, determine $\hat{\theta}$ from (5.8) with $\lambda = \lambda^*$ based on the data subset Y_{-k}. Then compute SS_k, the predicted sum of squared differences in (5.1) with $\theta = \hat{\theta}$ and restricting the sum to the validation data Y_k.
  - Compute $\text{SS}^* = \sum_{k=1}^{K} \text{SS}_k$. This is the sum of squared deviations of the predicted differences and the withheld observed score differences.

This process results in a predictive discrepancy for each candidate choice of λ. Optimization could proceed either by selecting a preset range of choices of λ, or preferably using an automated optimization procedure such as the Nelder–Mead algorithm (Nelder and Mead, 1965) to determine the optimizing value of λ. Once the optimized value of λ is chosen through cross-validation, it is fixed and then $\hat{\theta}$ can be determined from (5.8) using the entire data set.

An interesting special case of the penalized sums of squared differences is credited to Wesley Colley (2002). While not described as the solution to a regularized sums of squares problem, his approach involves applying (5.8) for particular choices of the variables. In the Colley setup, $\lambda = 2$, and $\gamma_j = 0.5$ for all j. Furthermore, for each $i = 1, \ldots, n$, $y_i = 0.5$ if team j_i wins and $y_i = -0.5$ if team k_i wins. Games in which teams tie in Colley's original system are ignored, but in fact they can be included in the penalized sum of squares framework letting $y_i = 0$ for such games.

5.2 Probabilistically Modeling Football Outcomes

The approaches described in Section 5.1 are limited in that they only provide point estimates of team abilities. Such methods do not acknowledge the uncertainty of the ability estimates and do not provide a mechanism to forecast game prediction distributions. They also do not provide a means to assess the significance or importance of the magnitude of differences in teams' strengths. In this section, we consider common probability models for football outcomes as a function of team strengths which permit addressing the aforementioned issues.

5.2.1 Modeling Wins versus Losses

A common approach to modeling football game outcomes is to ignore the final scores and record only which team won, or whether the game resulted in a tie. For the development to be discussed, we will ignore the possibility of a tie but will address this issue at the end of this section. Ties, which occur in football only when a game is tied at the end of regulation play and then if the game remains tied in overtime, occur typically about once per season. An argument for considering modeling game results as binary outcomes as opposed to actual scores is that teams are only incentivized to win games and not necessarily run up the score. For example, if a team consistently wins but plays conservatively so that it rarely wins by a large margin, then modeling the result of the game as a binary outcome would capture the team's winning tendencies without factoring in the margin of victory.

Assume teams j_i and k_i are involved in game i, with $i = 1, \ldots, n$. Let

$$y_i = \begin{cases} 1, & \text{if team } j_i \text{ wins,} \\ 0, & \text{if team } k_i \text{ wins.} \end{cases} \tag{5.9}$$

We can then model the probability of a game result as

$$p_i = P(y_i = 1) = F(\theta_{j_i} - \theta_{k_i}), \tag{5.10}$$

where θ_{j_i} and θ_{k_i} are the strengths of teams j_i and k_i, respectively and F is a monotonic function increasing from 0 to 1.

The most conventional choices of F are

$$F(x) \equiv \Phi(x) = \int_{-\infty}^{x} \frac{1}{\sqrt{2\pi}} \exp\left(-\frac{w^2}{2}\right) dw \tag{5.11}$$

and

$$F(x) \equiv \frac{1}{1 + \exp(-x)}. \tag{5.12}$$

The model in (5.11) is known as the Thurstone–Mosteller model (Mosteller, 1951; Thurstone, 1927), and the model in (5.12) is known as the Bradley–Terry model (Bradley and Terry, 1952). These models are often presented in the form of the distribution of a binary response (win versus loss) conditional on the strength parameters, but both models have an alternative representation as latent variable models. Suppose Z_{j_i} and Z_{k_i} are independent, unobserved continuous variables that can be understood as the latent performance for game i for teams j_i and k_i. If $Z_{j_i} > Z_{k_i}$, indicating that team j_i outperforms k_i, then $y_i = 1$ is observed; otherwise, $y_i = 0$ is observed. Under the assumption that $Z_{j_i} \sim N(\theta_{j_i}, 1/2)$ and $Z_{k_i} \sim N(\theta_{k_i}, 1/2)$ independently, the distribution of $Z_{j_i} - Z_{k_i}$ is $N(\theta_{j_i} - \theta_{k_i}, 1)$ and

$$P(y_i = 1) = P(Z_{j_i} - Z_{k_i} > 0) = \Phi(\theta_{j_i} - \theta_{k_i}), \tag{5.13}$$

which is the Thurstone–Mosteller model. Similarly, if Z_{j_i} follows a Gumbel distribution with location parameter θ_{j_i}, that is

$$p_Z(z) = \exp\left(-\left(z - \theta_{j_i} + e^{-(z-\theta_{j_i})}\right)\right), \tag{5.14}$$

and if independently Z_{k_i} has a Gumbel distribution with location parameter θ_{k_i}, then the distribution of the difference $Z_{j_i} - Z_{k_i}$ can be shown to follow a logistic distribution with density function

$$p_{Z_j - Z_k}(z) = \frac{\exp(z - (\theta_{j_i} - \theta_{k_i}))}{(1 + \exp(z - (\theta_{j_i} - \theta_{k_i})))^2} \tag{5.15}$$

so that

$$P(y_i = 1) = P(Z_{j_i} - Z_{k_i} > 0) = \int_0^\infty p_{Z_j - Z_k}(z)dz = \frac{1}{1 + \exp(-(\theta_{j_i} - \theta_{k_i}))}, \tag{5.16}$$

which is the Bradley–Terry model. The representations as latent variable models permit increased flexibility in modeling beyond the standard paired comparison models, such as modeling the latent performance as a function of game-specific covariates.

For the ordinary Bradley–Terry and Thurstone–Mosteller models, estimates of θ can be obtained through maximum likelihood by maximizing

$$L(\theta|y) \propto \prod_{i=1}^{n} p_i^{y_i}(1-p_i)^{1-y_i} \tag{5.17}$$

coupled with a linear constraint (as described in Section 5.1.1) on θ to ensure identifiability. The covariance of the maximum likelihood estimates, $\hat{\theta}$, that incorporate the linear constraint can be obtained through the usual method: evaluating the Hessian of the log-likelihood at the maximum likelihood estimate and then inverting. Standard errors are the square roots of the diagonal entries of this matrix. See the discussion at the end of Section 5.2.2 for computing the covariance matrix of the full vector of J parameter estimates.

To ensure identifiability of θ in (5.17), an additional condition, originally described by Ford (1957), must be met. For every possible partition of the J teams into two nonempty groups, some team in one group must have defeated a team in the other group. This condition can sometimes be difficult to satisfy in football, especially in the analysis of a partially completed regular season. The condition implies that no team has won every game or lost every game in the data set being analyzed. Thus, for example, analyzing the 2007 regular NFL season via maximum likelihood would not be possible because the Patriots won every game. In this situation, the likelihood in (5.17) would continue to increase as the ability parameter for the Patriots, θ_{NE}, increased.

Two other approaches are sometimes used in addressing concerns about uniquely identifying θ. One approach is to add ϵ to both y_i and $1-y_i$ in (5.17), where ϵ is a small positive constant set in advance. For example, the value $\epsilon = 0.2$ has been suggested (David, 1988). The effect of this change is twofold: First, because each game is essentially a win plus a fractional or partial loss, Ford's separability condition automatically holds. Second, the addition of ϵ to game outcomes has the effect of shrinking team abilities slightly toward the mean.

The other approach to uniquely identifying θ is to explicitly shrink the elements of θ to a constant via a penalty factor. For example, rather than maximizing the expression in (5.17), one could maximize

$$L(\theta|y) \propto \exp\left(-\frac{1}{2c}\theta'\theta\right) \prod_{i=1}^{n} p_i^{y_i}(1-p_i)^{1-y_i}, \tag{5.18}$$

which shrinks the θ to 0. This approach is identical to the Bayesian problem of finding the mode of the posterior distribution of θ for θ having a normal prior distribution with mean 0 and variance c. The shrinkage in this approach and the previous one can actually have improved predictive performance.

While ties are rare in football, they should be acknowledged in the context of a binary response model. Models for ties in the context of the Thurstone–Mosteller model include Glenn and David (1960), and for the Bradley–Terry model include Davidson (1970) and Rao and Kupper (1967). Rather than explicitly modeling the probability of a tie, which is rare, a more efficient approach is to act as though ties do not occur but incorporate them into the likelihood in (5.17) in a principled manner. Viewing a tie as equivalent information to half a win and half a loss, a factor in the likelihood for a tie could appear as

$$p^{1/2}(1-p)^{1/2},$$

where p is the probability of a win for the first team. Maximizing the likelihood with terms for tied games can proceed normally. This approach was used in Glickman (1999).

5.2.2 Modeling Score Differences

A more common approach to inferring NFL team strength is to model score differences rather than binary outcomes (with ties). Despite the arguments in Section 5.2.1 in favor of modeling outcomes as binary outcomes, the flip side of the coin is that score differences may convey greater detail about team strengths. In particular, games won by large margins generally indicate better team strength than games won by, say, an overtime scoring event.

Even though score differences are integer valued, the most common model for score differences is a normal distribution centered at the difference in team strength parameters. Choosing a normal distribution does not imply that the modeler believes that score differences are continuously distributed. Instead, this assumption is viewed as a continuous approximation to the true (discrete) probability distribution of score differences that might otherwise be too difficult to model in a convenient way. The normal model permits evaluating probabilities of ranges of score differences, including the probability one team would defeat another, using the normal distribution as a tool for probability calculations. Stern (1991) supports the use of a normal distribution around the expected score difference. While normal distributions are the most common continuous approximation for the true model of score differences, other continuous distributions could be explored, such as t distributions or other unimodal continuous and real-valued distributions.

Letting y_i be the score difference for game i involving teams j_i and k_i, the normal model assumes

$$y_i \sim \mathrm{N}(\theta_{j_i} - \theta_{k_i}, \ \sigma^2), \tag{5.19}$$

where σ^2 is the residual variance. The model can be expressed more compactly as

$$y \sim \mathrm{N}(X\theta, \ \sigma^2 I) \tag{5.20}$$

with X defined as in Section 5.1 and where I is the identity matrix of dimension n. Least-squares estimates $\hat{\theta}$ of θ can be obtained accounting for the collinearity in X using the adjustment described in Section 5.1.1. Once $\hat{\theta}$ is obtained, an unbiased estimate of σ^2 is given by

$$\hat{\sigma}^2 = \frac{1}{n - (J - 1)} (y - X\hat{\theta})'(y - X\hat{\theta}), \tag{5.21}$$

where J is the number of teams and $J - 1$ is the number of linearly independent columns of X. The estimated covariance matrix for J-dimensional $\hat{\theta}$ can be computed as

$$\widehat{\mathrm{Var}}(\hat{\theta}) = \widehat{\mathrm{Var}}(W\hat{\theta}_{-J}) = W\widehat{\mathrm{Var}}(\hat{\theta}_{-J})W' = \hat{\sigma}^2 W(X^{*\prime}X^*)^{-1}W' \tag{5.22}$$

with W defined in Section 5.1.1. Standard errors are the square roots of the diagonal elements of $\widehat{\text{Var}}(\hat{\theta})$ in (5.22). Note that the approach of obtaining the covariance matrix for $\hat{\theta}$ through the application of contrast matrices also applies to binary response models described in Section 5.2.1.

5.3 Dynamic Models for Score Differences

The approaches in Section 5.2 are appropriate if team abilities are not changing appreciably over time. We investigate in this section models for time-varying abilities. Because NFL teams typically make most of their player trades, acquire new players, make coaching staff changes, and lose players to retirement between seasons, it is natural to assume teams undergo ability changes mostly from season to season. The emphasis of this section is on state-space models for changes in ability, though we consider alternative approaches at the end of this section.

Let t index a regular season, where $t = 1, \ldots, T$. We now assume that team j during season t has a strength θ_{jt}, which may vary across seasons. As before, we assume that a score difference for game i during season t depends on current team strengths

$$y_{it} \sim \text{N}(\theta_{j_{it}t} - \theta_{k_{it}t}, \sigma^2) \tag{5.23}$$

for teams j_{it} and k_{it}. We further assume that teams' abilities evolve through an autoregressive process. For team j and all seasons t, we assume

$$\theta_{j,t+1} \sim \text{N}(\rho\theta_{jt}, \tau^2), \tag{5.24}$$

where τ^2 is an innovation variance and $|\rho| < 1$ is an autoregressive parameter.

The restriction on ρ is so that the stochastic process on θ_{jt} is stationary. This particular model assumes that the θ_{jt} for fixed t are distributed around 0 so that the impact of ρ is to regress the θ_{jt} toward the mean over time. This model is an example of a normal linear state-space model (West and Harrison, 1997) and is also an example of the Kalman filter (Kalman and Bucy, 1961; Meinhold and Singpurwalla, 1983). This approach has been applied to football score differences by Harville (1980) and Glickman and Stern (1998). The model can be extended to ensure that $\sum_{j=1}^{J} \theta_{jt} = 0$ for all t. For example, the model can be specified multivariately as

$$\theta_{t+1} \sim \text{N}\left(\rho\theta_t, \tau^2\left(\mathbf{I} - \frac{1}{J}\mathbf{11}'\right)\right), \tag{5.25}$$

where θ_t is the vector of J team strengths at time t, $\mathbf{I}$ is the $J \times J$ identity matrix, and $\mathbf{1}$ is a J-vector with 1 as each element.

The likelihood for the set of T vectors of team strength parameters $(\theta_1, \ldots, \theta_T)$ along with σ^2, τ^2, and ρ is an alternating product of multivariate normal densities for the game score differences, y_t, and multivariate normal densities for the θ_t that represent the innovations in strength over time. The full likelihood expression for the model in (5.23) and (5.24) can be written as

$$L(\theta_1, \ldots, \theta_T, \sigma^2, \tau^2, \rho | y_1, \ldots, y_T)$$

$$\propto N(y_1 | X_1\theta_1, \sigma^2 I_{n_1}) \prod_{t=2}^{T} N(\theta_t | \rho\theta_{t-1}, \tau^2 I_J) N(y_t | X_t\theta_t, \sigma^2 I_{n_t}), \tag{5.26}$$

where X_t is the design matrix of team pairings during season t, I_m is the identity matrix of dimension m, and the notation $N(\cdot | \mu, \Sigma)$ is the multivariate normal density of its argument with mean vector μ and covariance matrix Σ.

Inference for this model can be performed through maximum likelihood (Harville, 1980; Sallas and Harville, 1988), though our experience is that taking a Bayesian approach to inference permits greater flexibility in extending the basic model. To perform Bayesian inference for the model in (5.23) and (5.24), a prior distribution needs to be assumed on the two variance parameters, the autoregressive parameter, and the team strengths during the first season, θ_1. Particular choices are described in Glickman and Stern (1998). Inference may be implemented through Markov chain Monte Carlo (MCMC) simulation from the posterior distribution, a standard approach to inference in the Bayesian setting. The model can be implemented in a straightforward manner using standard Bayesian software that implements MCMC simulation, such as JAGS (Plummer, 2003).

The state-space model in (5.23) and (5.24) can be extended in a number of ways. For example, instead of assuming that team strength remains constant throughout a season, the model could be extended to allow for an autoregressive process for week-to-week variation in team abilities. Such a model was considered by Glickman and Stern (1998). Allowing for weekly variation can account for short-term effects such as injuries by franchise players. Furthermore, the evolution of team ability over time can be extended to more complex models including those applied in financial applications. Glickman (2001) considered a time component for the parameters in a model of binary football game outcomes that followed a stochastic volatility process (Jacquier et al., 2002; Kim et al., 1998). Instead of the autoregressive model in (5.24), the stochastic volatility extension assumes

$$\theta_{j,t+1} \sim N\left(\rho\theta_{jt}, \tau_{t+1}^2\right) \tag{5.27}$$

along with a stochastic process on the τ_t^2,

$$\log \tau_{t+1}^2 \sim N\left(\log \tau_t^2, \omega^2\right). \tag{5.28}$$

This model can account for sudden changes in team ability that are not well captured by the normal state-space model. Consideration of other processes for time variation is an open research question.

Assuming a stochastic process on team strengths is not the only approach to account for changes in team strength. In the context of binary outcomes, Baker and McHale (2014) used barycentric rational interpolation, a particular type of scatterplot smoother. Their approach assumes that a team's strength follows a nonparametric regression as a function of time.

While the focus of their work was on applications to binary outcomes, the approach can apply to score differences. In general, modeling team strengths as a nonparametric function of time is an underexplored avenue and is another opportunity for further development.

5.4 The Inclusion of Covariates

The models considered thus far rely only on the identity of the teams involved in a game. Incorporating game-specific or team-specific covariate information has the potential for substantially improving game predictions. We illustrate in this section an approach to incorporating covariate information into our football outcome models.

Covariate information can be divided into two types: factors that are endogenous to team strength, and those that are exogenous. Endogenous factors are ones that are intrinsic to describing a team's strength. These would include covariate data such as summaries of team performance, player-level factors, or any components of performance that describe some aspect of a team's ability. Exogenous factors are external to a team's make-up and might include location of the game (home versus away), weather-related information, time of the day a game was played, schedule-related factors (e.g., whether a team played after a "bye" week), and so on. The division of covariates into these two types has implications not so much for modeling but for summarizing team strengths as we describe in the following text.

Probably, the most compelling exogenous covariate information is whether a team played on its home field. Various authors have shown that playing on the home field conveys roughly a 3-point advantage in the final score (Acker, 1997; Glickman and Stern, 1998; Harville, 1980). The basic model in (5.19) can be extended to include a home field term. With teams j_i and k_i competing in game i, assume

$$y_i \sim N(\theta_{j_i} - \theta_{k_i} + h_i\delta, \ \sigma^2),$$ (5.29)

where

$$h_i = \begin{cases} -1, & \text{if team } k_i \text{ plays on its home field,} \\ 1, & \text{if team } j_i \text{ plays on its home field,} \\ 0, & \text{if the game is played on a neutral field} \end{cases}$$ (5.30)

and the parameter δ is the effect of a team playing on its home field. Glickman and Stern (1998) investigated an extension of (5.29) in the context of the dynamic normal linear model in which each team had its own home-field advantage. Thus, rather than considering a single δ, their model was parameterized with $J = 32$ parameters, $\delta = (\delta_1, \ldots, \delta_J)$.

A variety of covariates endogenous to overall team strength can be incorporated into the linear predictor for the score difference model. Typical information included in these models consists of season-to-date summaries (both offensive and defensive) of passing and rushing yards, rates of fumbles, and rates of interceptions. Ad hoc rules for including such measures early in a season have involved incorporating averages from the previous season. For game i involving teams j_i and k_i, it seems desirable to include separate variables for each measure specific to teams j_i and k_i, for example, season-to-date average rushing yards for team j_i as one variable and season-to-date average rushing yards for team k_i as a second variable. However, because the ordering of teams j_i and k_i is arbitrary, it would not be sensible to have effects of the two variables appear as the effect of teams j_i and k_i. Instead, a more principled approach is to transform the two variables into their difference and their average and use these two variables in the model. Thus, to include season-to-date average rushing yards for each team separately into a model, it would be more prudent to include the difference between the average rushing yards for team i and for team j, along with the average rushing yards for the two teams.

Let u_i be the vector of additional endogenous covariates to incorporate into the basic model, and let v_i be the vector of exogenous covariates. The u_i would include differences and averages of performance measures between the two teams involved in game i, and v_i would include h_i as a component if home field is recorded. Then, the model in (5.29) can be extended to

$$y_i \sim \mathrm{N}(\theta_{j_i} - \theta_{k_i} + u_i \beta_u + v_i \beta_v, \; \sigma^2), \tag{5.31}$$

where β_u and β_v are the vector of effects of u_i and v_i, respectively. Because all the parameters are included as linear terms, Equation (5.31) can be expressed more compactly in matrix form as

$$y \sim \mathrm{N}\left((X|U|V) \begin{pmatrix} \theta \\ \beta_u \\ \beta_v \end{pmatrix}, \; \sigma^2 \right), \tag{5.32}$$

where U is the matrix with rows u_i and V is the matrix with rows v_i.

Recognizing that (5.32) is a normal linear model, inference for the parameters can be obtained as usual through least-squares regression, accounting for the collinearity in X by replacing it with X^* as described in Section 5.1.1. This results in least-squares estimates $\hat{\theta}_{-J}$, $\hat{\beta}_u$, and $\hat{\beta}_v$, as well as the covariance matrix of this collection of estimates. To obtain the covariance matrix of estimates for $\hat{\theta}$, $\hat{\beta}_u$, and $\hat{\beta}_v$, similar to Equation (5.22), an analogous computation is required. Let

$$\hat{\eta} = \begin{pmatrix} \hat{\theta} \\ \hat{\beta}_u \\ \hat{\beta}_v \end{pmatrix} \quad \text{and} \quad \hat{\eta}_{-J} = \begin{pmatrix} \hat{\theta}_{-J} \\ \hat{\beta}_u \\ \hat{\beta}_v \end{pmatrix}. \tag{5.33}$$

Furthermore, let

$$\tilde{W} = \begin{pmatrix} W & 0 & 0 \\ 0 & I_u & 0 \\ 0 & 0 & I_v \end{pmatrix}, \tag{5.34}$$

where I_u and I_v are identity matrices having dimensions equal to the number of variables in u and v, respectively. Then

$$\widehat{\mathrm{Var}}(\hat{\eta}) = \widehat{\mathrm{Var}}(\tilde{W}\hat{\eta}_{-J}) = \tilde{W}\widehat{\mathrm{Var}}(\hat{\eta}_{-J})\tilde{W}', \tag{5.35}$$

and the standard errors can be determined as the square root of the diagonal entries.

Summarizing team strength in the presence of game-specific covariates is not as simple as summarizing inferences about the θ_j. Because the model in (5.32) adjusts the team parameters for endogenous covariate information, the θ_j have the interpretation of the effects beyond those of the covariates. If teams' covariate information correlates strongly with game score differences, then it is possible that the θ_j will be lower for stronger teams because the evidence of team strength will be encoded in $U\beta_u$. It would be more sensible to summarize inferences about

$$\theta_j + u^*\beta_u \qquad (5.36)$$

rather than θ_j alone, where u^* is a vector of specific covariate values that represent the team and a typical opponent. Because the variables v are exogenous to team strength, they may be omitted when describing team strength. The quantity in (5.36) can be estimated by replacing the parameters with their least-squares estimates. Note that this interpretation is specific to a particular set of covariate values u and is not an overall description of the strength of team j.

This approach can be generalized slightly. Rather than describing team strength in (5.36) by selecting particular values of u^*, one could consider a distribution of likely values of elements of u^* and summarize the distribution. For example, one could perform a Monte Carlo simulation of values of u^* from a specified joint distribution and then summarize the Monte Carlo distribution of $\hat{\theta}_j + u^*\hat{\beta}_u$ for team j. To account for the variability of the estimates $\hat{\theta}_j$ and $\hat{\beta}_u$, these parameters could also be simulated from their respective approximate normal distributions.

Another commonly used approach to address the difficulties in estimating team strength in the presence of endogenous covariates is simply to fit models that do not explicitly include strength parameters. Many authors, including David et al. (2011), Warner (2010), and Uzoma et al. (2015), analyze game outcomes as a function of covariate information but do not include team-specific parameters. This general approach assumes that team strength is sufficiently captured by the included covariates, and that game outcome predictions are sufficiently described by the information in these covariates and not on team identities. A disadvantage of this approach is that the evolution of team strength is difficult to capture only through covariate information.

The state-space model in (5.23) and (5.24) can be extended to include covariate information in a straightforward manner. As discussed earlier, the mean in (5.23) can include linear terms corresponding to covariate information, typically with coefficients that are nondynamic. For example, the inclusion of a home field variable as in (5.30) would likely have an effect that would not be expected to vary over time. In the context of an MCMC posterior simulation, nondynamic parameters can be updated via conditional posterior sampling as a block within one iteration of MCMC. Glickman and Stern (1998) demonstrate model fitting for the inclusion of nondynamic home-field advantage parameters (one per team) in their state-space model.

5.5 Other Approaches

Models that include team-specific parameters θ_j are not the only approach to predicting NFL game scores and as a method to assess team abilities. In this section, we describe several other approaches that have been considered for game prediction.

An active area of NFL game prediction is to model game outcomes using algorithmic approaches, such as through machine learning methods. Some examples include modeling game outcomes through neural network models (David et al., 2011), nearest-neighbor models (Uzoma et al., 2015), and Gaussian process models (Warner, 2010). In all instances, game outcomes (whether as a categorical outcome or as a score difference) are predicted from a possibly large set of covariate data. For example, in Warner (2010), 47 possible covariates were considered for game outcome prediction, including previous winning percentage, average points per game scored and allowed, total yards per game gained and allowed, rushing yards gained and allowed, temperature, and so on. Because team parameters are not included in these models, the distinction between endogenous and exogenous factors to team strength is not relevant.

In any of these algorithmic approaches, modeling game outcomes involves a combination of covariate selection along with constraining the model estimation from overfitting, the latter typically through cross validation. Because machine learning is a highly active area in the field of computer science and statistics, applications to NFL prediction and beating the Vegas point spread are likely to attract continued attention.

A novel approach to game score prediction was developed by Baker and McHale (2013). Their approach involved modeling the exact (discrete) game score outcome by viewing a game as a continuous-time stochastic process composed of a series of birth processes. Each "birth" event corresponds to one of ten different scoring events. These include for each of the home and away teams a touchdown without a conversion, a touchdown with a 1-point conversion, a touchdown with a 2-point conversion, a field goal, and a safety. The time to occurrence between events is modeled as a proportional hazards model that depends on team characteristics, home-field advantage, and an effect that depends on the type of scoring event. In their analysis, the authors worked with only final game scores, so they made simplifying assumptions in order to account for the multitude of ways in which the final score could be achieved from the point process. However, data on the exact timing of scoring events in NFL games are more easily accessible today, so analyses that recognize actual timings could be performed making use of some of the ideas in their paper. In particular, more precise modeling of interarrival times beyond proportional hazards may be worth exploring.

It is worth noting that neither of the aforementioned methods directly addresses measuring the ability of teams but rather focuses on modeling game outcomes. Ability measures could be derived quantities from these analyses, such as inserting team-specific covariate data as a proxy for estimating a team strength parameter. A potential area for future work is to extend the aforementioned approaches to include team strengths as explicitly defined parameters, though endogenous factors would need to be addressed by an approach analogous to that presented in Equation (5.36) to address the confounding.

5.6 Application to NFL Game Data

In this section, we demonstrate the application of some of the methods described in this chapter to NFL football outcomes. Game outcomes can be scraped directly from www.nfl.com, though we obtained our data from Australia Sports Betting (n.d.), which provided all NFL game data from the 2006 season onward conveniently in one spreadsheet.

We first demonstrate least-squares modeling from Section 5.2 to game outcome data from the 2014–2015 regular season. With $J = 32$ teams, each playing 16 games during the regular season, the number of games in our analyses is 256. We fit a least-squares model as in (5.19) with the inclusion of a home-field advantage parameter as in (5.29). Our data included information on which team played on their home field. Out of the 256 games, 4 were played on a neutral field so that $h_i = 0$ for these 4 games.

The results of the model fits are summarized in Table 5.1 with the teams' ranks ordered according to their fitted strengths. We incorporated the linear constraint $\sum_{j=1}^{J} \theta_j = 0$ for parameter identifiability using the methods in Section 5.2.2 and determined the $\widehat{\text{Var}}(\hat{\theta})$ using the computation in (5.22). The standard error of any $\hat{\theta}_j$ to two decimal places was 3.58. The standard error of the home field parameter was 0.886. The residual standard deviation, $\hat{\sigma}$, was about 14, suggesting large variation in score differences around the mean. Stern (1991) obtained a similar result.

The home-field advantage parameter was estimated as 2.443, suggesting that playing at home versus away conveyed roughly a 4.8-point advantage in the 2014–2015 regular season. The range in strengths spanned from -11.8 (Titans) to 10.9 (Patriots), suggesting that on a neutral field the Patriots would outscore the Titans on average by $10.9 - (-11.8) = 22.7$ points. The order of strength estimates in Table 5.1 arguably has face validity. The teams in the Super Bowl (the Patriots and the Seahawks) were two of the top three teams listed, and the teams that had poor regular season records were at the bottom of the rank order.

The variance of $\hat{\theta}_j - \hat{\theta}_k$ for teams j and k in the aforementioned analyses can differ due to the different covariances of the estimates. For example, because the Patriots and the Jets, who are in the same division, play twice during the regular season, the estimated variance of the difference in strength is lower than most pairs of teams. From the model fit, we have

$$\widehat{\text{Var}}\left(\hat{\theta}_{\text{NE}} - \hat{\theta}_{\text{NYJ}}\right) = \widehat{\text{Var}}\left(\hat{\theta}_{\text{NE}}\right) + \widehat{\text{Var}}\left(\hat{\theta}_{\text{NYJ}}\right) - 2\widehat{\text{Cov}}\left(\hat{\theta}_{\text{NE}}, \hat{\theta}_{\text{NYJ}}\right)$$
$$= 12.83 + 12.83 - 2(1.80) = 22.06$$

which corresponds to a standard error of $\sqrt{22.06} = 4.70$. In contrast, the same calculation to compute the estimated variance of the difference in strength between the Patriots and the Seahawks, who did not play during the regular season, is

$$\widehat{\text{Var}}\left(\hat{\theta}_{\text{NE}} - \hat{\theta}_{\text{SEA}}\right) = \widehat{\text{Var}}\left(\hat{\theta}_{\text{NE}}\right) + \widehat{\text{Var}}\left(\hat{\theta}_{\text{SEA}}\right) - 2\widehat{\text{Cov}}\left(\hat{\theta}_{\text{NE}}, \hat{\theta}_{\text{SEA}}\right)$$
$$= 12.83 + 12.83 - 2(-1.04) = 27.74$$

corresponding to a larger standard error of $\sqrt{27.74} = 5.27$. These standard errors are not appreciably different, and for the NFL, they do not vary considerably. However, in a larger less-connected league (e.g., NCAA college football) where teams mostly compete against each other in the same division, one might expect larger variation in the standard error of the team strength differences between divisions.

We also fit a dynamic model of team strengths described in Section 5.3 that includes a home-field advantage parameter as a nondynamic component of the model. Our model assumed that team j had a strength parameter θ_{jt} during season t, where $t = 2006, 2007, \ldots, 2014$, that evolved according to an autoregressive process presented in (5.24). We fit our model via MCMC simulation using the Bayesian software JAGS called from

TABLE 5.1

Least-Squares Estimates of θ for Score Difference Model
Based on NFL Game Outcomes during the 2014–2015
Regular Season

Team	Least-Squares Estimate
New England Patriots	10.92
Denver Broncos	9.60
Seattle Seahawks	9.54
Green Bay Packers	8.25
Kansas City Chiefs	5.66
Dallas Cowboys	5.23
Buffalo Bills	5.03
Baltimore Ravens	4.64
Indianapolis Colts	4.62
Philadelphia Eagles	3.91
Miami Dolphins	2.39
Pittsburgh Steelers	2.26
Arizona Cardinals	1.98
Detroit Lions	1.88
San Diego Chargers	1.86
Houston Texans	1.77
Cincinnati Bengals	0.77
St. Louis Rams	−0.84
San Francisco 49ers	−0.96
Minnesota Vikings	−1.69
New York Giants	−1.70
New Orleans Saints	−2.90
Carolina Panthers	−3.05
Atlanta Falcons	−3.70
Cleveland Browns	−3.86
New York Jets	−5.21
Chicago Bears	−6.76
Washington Redskins	−8.74
Oakland Raiders	−8.85
Tampa Bay Buccaneers	−9.77
Jacksonville Jaguars	−10.48
Tennessee Titans	−11.83
Residual standard deviation ($\hat{\sigma}$)	14.05
Home field ($\hat{\delta}$)	2.44

within R (R Core Team, 2015) using the Rjags package. We ran two parallel chains for 30,000 iterations, including a burn-in period of 10,000 iterations. Convergence diagnostics (Gelman et al., 2014) revealed that the chains converged, and parameter summaries were based on every fifth sampled value beyond the burn-in period to lessen the autocorrelation among successive draws but also to save on space resources. Thus, 8000 simulated values from the posterior distribution for each parameter were saved based on the MCMC simulation.

Summaries of the team strengths for the 2014–2015 season based on the dynamic model, along with other model parameters, are presented in Table 5.2. The team strengths are ordered according to the posterior means. The posterior mean strengths range from 8.42 (Seahawks) to –9.41 (Jaguars). The posterior mean of τ, the innovation standard deviation, equal to 4.28 indicates a nontrivial amount of change in team abilities from season to season, though with the autoregressive parameter $\rho \approx 0.67$ top teams and bottom teams are regressing substantially toward the mean team ability.

The comparison of the results in Table 5.2 with those in Table 5.1 reveals some interesting differences. First, the home-field advantage is inferred to be nearly the same for both sets of models, about 2.4 points relative to playing on a neutral field. The residual standard deviations are also of similar magnitude, with the dynamic model estimate of 13.1 compared to the least-squares estimate of 14.0. The Bayesian equivalents of the standard errors of the team strengths are all around 2.66, which is smaller than the corresponding value of 3.58 in the least-squares fit. The lower standard error is explainable by the use of historical game outcome data to aid in inferring 2014–2015 team strengths. The spread in estimated team strengths is slightly narrower for the dynamic model. This may again be due to incorporating previous seasons' game results, which recognize that teams' strengths regress toward the mean between seasons. The order of team strengths for the 2014–2015 season is similar between the two analyses (e.g., the top five and bottom five teams are the same, though in different orders), but some notable exceptions exist. For example, the Saints in the dynamic model are estimated to be nearly 3 points higher in strength than in the least-squares fit. This may reflect much better performances by the Saints in seasons previous to 2014, which factor more directly into the dynamic model. Similarly, the Bucs are inferred by the dynamic model to be a substantially better team at the end of the 2014–2015 season than through the least-squares estimates. Again, this may be due to the Bucs having a particularly poor performance during the 2014–2015 season. The Seahawks are inferred to be slightly better than the Patriots in the dynamic model compared to the least-squares model, and this may be a carryover from the superb 2013–2014 season in which the Seahawks won the Super Bowl.

Trajectories of team strengths over time can be understood by examining the posterior mean strengths across multiple seasons. The estimated strengths from all 32 teams are displayed in Figure 5.1 organized by division (each consisting of four teams who are in competition with each other to earn a play-off spot). Several features of these plots are worthy of mention. The AFC East plot demonstrates how dominant the Patriots have been over the last decade in their division, maintaining a posterior mean strength higher than the Bills, Jets, and Dolphins all 9 years. Only the Green Bay Packers come close to demonstrating this level of dominance in their division. Several teams have shown particularly quick improvement over the last few years. The Seahawks, for example, were a subpar team between 2008 and 2010, but, with the hiring of Pete Carroll in 2010 and the signing of quarterback Russell Wilson in 2012, they have improved to be best in their division by the 2013–2014 and 2014–2015 seasons. Similarly, the Denver Broncos were a below-average team through the end of the 2011–2012 season. With the acquisition of Peyton Manning from the Colts, the Broncos from 2012 onward were the strongest team in the AFC West and eventually won the Super Bowl at the end of the 2015–2016 season.

A closer examination of the Broncos team strength over time illustrates the comparison between inferences from the dynamic linear model and the least-squares estimates

TABLE 5.2

Posterior Means and Standard Deviations of 2014–2015
Team Strengths and Other Model Parameters Based on a
Normal Dynamic Linear Model for NFL Game Outcomes
Recorded for the Period 2006–2014

Team	Posterior Mean	Posterior Std Dev
Seattle Seahawks	8.42	2.65
New England Patriots	8.19	2.65
Denver Broncos	7.93	2.71
Green Bay Packers	5.79	2.68
Kansas City Chiefs	3.86	2.65
Dallas Cowboys	3.85	2.63
Baltimore Ravens	3.48	2.66
Indianapolis Colts	3.25	2.66
Philadelphia Eagles	2.91	2.66
Buffalo Bills	2.02	2.68
Pittsburgh Steelers	1.97	2.62
Cincinnati Bengals	1.94	2.65
Arizona Cardinals	1.78	2.68
Detroit Lions	1.28	2.70
San Diego Chargers	1.08	2.65
San Francisco 49ers	1.00	2.66
Miami Dolphins	0.77	2.70
Houston Texans	0.14	2.67
New Orleans Saints	−0.01	2.68
Carolina Panthers	−0.25	2.67
St. Louis Rams	−0.66	2.67
New York Giants	−1.59	2.68
Minnesota Vikings	−1.71	2.68
Atlanta Falcons	−2.14	2.61
Cleveland Browns	−3.36	2.67
Chicago Bears	−4.71	2.68
New York Jets	−5.19	2.68
Tampa Bay Buccaneers	−6.59	2.68
Washington Redskins	−6.98	2.66
Oakland Raiders	−8.23	2.69
Tennessee Titans	−8.79	2.65
Jacksonville Jaguars	−9.41	2.70
Residual standard deviation (σ)	13.12	0.20
Home field (δ)	2.41	0.27
Innovation standard deviation (τ)	4.28	0.36
Autoregressive factor (ρ)	0.67	0.07

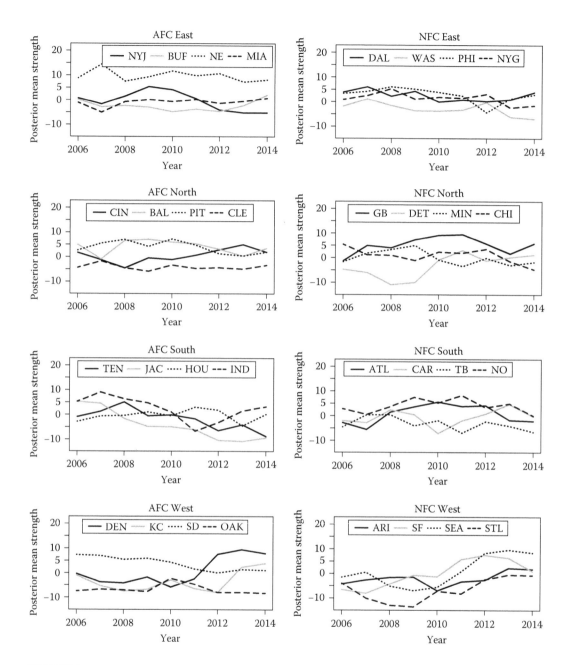

FIGURE 5.1
Posterior mean team strengths from the Bayesian dynamic linear model for all 32 NFL teams across nine seasons grouped according to division.

obtained on a season-by-season basis. In Figure 5.2, 95% central posterior intervals and 95% confidence intervals for $\theta_{\text{DEN},t}$ are plotted side by side for each season. Because the dynamic model pools information across all seasons, inferences about team strengths in any given year have less posterior uncertainty compared to the least-squares model that estimates strength for each season separately. This explains why the dynamic model interval

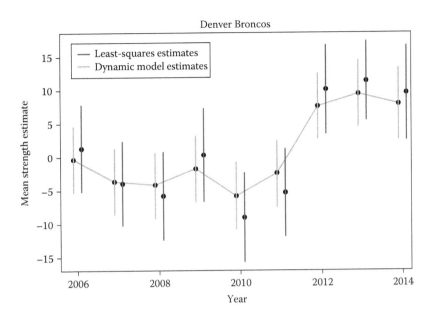

FIGURE 5.2
Point and interval estimates for the strength parameter of the Denver Broncos over time. The gray segments are the 95% central posterior intervals from the dynamic linear model; the black segments are the 95% confidence intervals based on least-squares estimates. The points in the center of each interval are the corresponding posterior means and least-squares estimates.

estimates are narrower than the corresponding least-squares intervals. Furthermore, the dynamic model estimates tend to shrink more to the adjacent years' estimates relative to the corresponding least-squares estimates. For example, in 2009 and 2011, the posterior mean estimates of the Broncos were −1.74 and −2.41. The least-squares estimate in 2010 was −9.03, much lower than the posterior means in 2009 and 2011. The dynamic model in effect averages the −9.03 least-squares estimate with the neighboring strengths in 2010 to produce a posterior mean estimate of −5.83, somewhat higher than the least-squares estimate. When the Broncos' results in the 2012–2013 season were a large improvement over the previous season, the least-squares estimate was 10.10, but the shrinkage due to the neighboring estimates pulled the dynamic model estimate down slightly to 7.58.

5.7 Conclusions

This chapter described the basics of estimating NFL team strength based on game outcomes within a season, game outcomes recorded across seasons, and the inclusion of potential covariate information to sharpen inferences. These basic approaches rely on establishing a relationship between game outcomes and differences in team strength parameters. This foundation has served the statistical community well for many years and is a solid basis for team rankings and predicting game outcomes.

Plenty of room exists for exploring and developing new ideas for measuring NFL team strength. For example, representing NFL team strength by a scalar parameter within a linear model is arguably a limiting approach. Such a simple linear model, which assumes a strict ordering in strengths, does not acknowledge the existence of circular triads in which team *A* is better than *B* which is better than *C*, but that team *C* is better than *A*. The development of such models, which surely require vectors of team strengths rather than scalar strengths, is an open area. Furthermore, now that detailed information is available for individual players, including player information from play-by-play data and the increased availability of player tracking data, possibilities exist for constructing team strength models based on player contributions that incorporate synergistic relationships among players (e.g., strength of quarterback/receiver pairs). Improvements in team strength estimation derived from these novel approaches may lead to new ways of conceptualizing ability measurement in NFL team strength and competitive ability in sports more generally.

References

Acker, J. C. (1997): Location variations in professional football, *Journal of Sport Behavior*, 20, 247.

Australia Sports Betting. (n.d.): NFL from 2006, www.aussportsbetting.com/data. Accessed: December 3, 2015.

Baker, R. D. and I. G. McHale (2013): Forecasting exact scores in National Football League games, *International Journal of Forecasting*, 29, 122–130.

Baker, R. D. and I. G. McHale (2014): A dynamic paired comparisons model: Who is the greatest tennis player? *European Journal of Operational Research*, 236, 677–684.

Bradley, R. A. and M. E. Terry (1952): Rank analysis of incomplete block designs: I. The method of paired comparisons, *Biometrika*, 39(3/4), 324–345.

Colley, W. N. (2002): *Colley's Bias Free College Football Ranking Method: The Colley Matrix Explained*, Princeton, NJ: Princeton University.

David, H. A. (1988): *The Method of Paired Comparisons*, New York: Oxford University Press.

David, J. A., R. D. Pasteur, M. S. Ahmad, and M. C. Janning (2011): NFL prediction using committees of artificial neural networks, *Journal of Quantitative Analysis in Sports*, 7(2), DOI: 10.2202/1559-0410.1327.

Davidson, R. R. (1970): On extending the Bradley–Terry model to accommodate ties in paired comparison experiments, *Journal of the American Statistical Association*, 65, 317–328.

Ford, L. R. (1957): Solution of a ranking problem from binary comparisons, *American Mathematical Monthly*, 64(8), 28–33.

Gelman, A., J. B. Carlin, H. S. Stern, and D. B. Rubin (2014): *Bayesian Data Analysis*, 3rd edn., Boca Raton, FL: CRC Press.

Glenn, W. and H. David (1960): Ties in paired-comparison experiments using a modified Thurstone–Mosteller model, *Biometrics*, 16, 86–109.

Glickman, M. E. (1999): Parameter estimation in large dynamic paired comparison experiments, *Journal of the Royal Statistical Society: Series C (Applied Statistics)*, 48, 377–394.

Glickman, M. E. (2001): Dynamic paired comparison models with stochastic variances, *Journal of Applied Statistics*, 28, 673–689.

Glickman, M. E. and H. S. Stern (1998): A state-space model for National Football League scores, *Journal of the American Statistical Association*, 93, 25–35.

Harville, D. (1980): Predictions for National Football League games via linear-model methodology, *Journal of the American Statistical Association*, 75, 516–524.

Hoerl, A. E. and R. W. Kennard (1970): Ridge regression: Biased estimation for nonorthogonal problems, *Technometrics*, 12, 55–67.

Jacquier, E., N. G. Polson, and P. E. Rossi (2002): Bayesian analysis of stochastic volatility models, *Journal of Business & Economic Statistics*, 20, 69–87.

Kalman, R. E. and R. S. Bucy (1961): New results in linear filtering and prediction theory, *Journal of Fluids Engineering*, 83, 95–108.

Kim, S., N. Shephard, and S. Chib (1998): Stochastic volatility: Likelihood inference and comparison with arch models, *The Review of Economic Studies*, 65, 361–393.

Massey, K. (n.d.): *College Football Ranking Composite*, Massey Ratings, masseyratings.com/cf/compare.htm. Accessed: April 13, 2016.

Massey, K. (1997): Statistical models applied to the rating of sports teams, Technical report, Bluefield, VA: Bluefield College.

Meinhold, R. J. and N. D. Singpurwalla (1983): Understanding the Kalman filter, *The American Statistician*, 37, 123–127.

Mosteller, F. (1951): Remarks on the method of paired comparisons: I. The least squares solution assuming equal standard deviations and equal correlations, *Psychometrika*, 16, 3–9.

Nelder, J. A. and R. Mead (1965): A simplex method for function minimization, *The Computer Journal*, 7, 308–313.

Plummer, M. (2003): JAGS: A program for analysis of Bayesian graphical models using Gibbs sampling, in *Proceedings of the Third International Workshop on Distributed Statistical Computing (DSC 2003)*, March 20–22, Vienna, Austria.

R Core Team (2015): *R: A Language and Environment for Statistical Computing*, R Foundation for Statistical Computing, Vienna, Austria.

Rao, P. and L. L. Kupper (1967): Ties in paired-comparison experiments: A generalization of the Bradley–Terry model, *Journal of the American Statistical Association*, 62, 194–204.

Sallas, W. M. and D. A. Harville (1988): Noninformative priors and restricted maximum likelihood estimation in the Kalman filter. *Bayesian Analysis of Time Series and Dynamic Models*, New York, NY: Marcel-Dekker, pp. 477–508.

Stern, H. (1991): On the probability of winning a football game, *The American Statistician*, 45, 179–183.

Stern, H. S. (1995): Who's number 1 in college football?... and how might we decide, *Chance*, 8, 7–14.

Thurstone, L. L. (1927): A law of comparative judgment. *Psychological Review*, 34, 273.

Uzoma, A. O., E. Nwachukwu et al. (2015): A hybrid prediction system for American NFL results, *International Journal of Computer Applications Technology and Research*, 4, 42–47.

Warner, J. (2010): Predicting margin of victory in NFL games: Machine learning vs. the las vegas line, Technical report, State College, PA: Pennsylvania State University.

West, M. and J. Harrison (1997): *Bayesian Forecasting and Dynamic Models*, 2nd edn., New York, NY: Springer-Verlag.

6

Forecasting the Performance of College Prospects Selected in the National Football League Draft

Julian Wolfson, Vittorio Addona, and Robert Schmicker

CONTENTS

> Without question, one of the primary factors affecting success in the National Football League is having talented players ... [T]he most important step in securing the players a team needs is evaluating the available talent pool...
>
> **—Bill Walsh (Walsh, Billick, and Peterson, 1997, p. 113)**

6.1 Introduction

Many factors contribute to the success of a professional football organization. Quality coaching, excellent game planning for opponents, and good fortune with regard to the health of players are all important components of a winning team. Arguably paramount among these factors is the effective evaluation of on-field talent; a team that can find "diamonds in the rough" and avoid highly touted "busts" will position themselves well for success. This has perhaps never been more true than in today's salary cap era, where the league seeks to achieve competitive balance among franchises.

Although players in the National Football League (NFL) can be traded or signed in free agency, player movement via these routes is relatively limited. Most player procurement occurs via the amateur draft, a mechanism by which NFL teams select from a pool of eligible college players. The selection order for teams that did not make the play-offs is based on their previous year's winning percentage (in reverse order). Teams that made the play-offs then select based on the order in which they were eliminated, with the Super Bowl champion selecting last. Drafting a player gives a team exclusive rights to negotiate a contract with that player. Though players may be selected earlier or later in the draft for a variety of reasons (e.g., positional needs, off-field marketing potential, hometown popularity), a player's draft position can generally be viewed as a proxy for the team's assessment of their overall skill level.

In this chapter, we address whether NFL franchises have shown an ability to predict the success of players available in the amateur draft and discuss the issues surrounding this objective. While we frame some of the following discussion in terms of predicting success for players at all positions, we pay particular attention to the quarterback position. Quarterbacks play a central role in the on-field success of a franchise; hence, there is a strong incentive for teams selecting near the top of the draft to find their "quarterback of the future." As a result, quarterbacks drafted within the first few picks sign large guaranteed contracts and face tremendous scrutiny from the fan base.

Due to the high cost of signing drafted players as well as the opportunity cost of failing to acquire the rights to a future star, organizations have vital interest in predicting how successful college players will be in the NFL. Nevertheless, in spite of the enormous volume of information available about draft-eligible prospects and the hundreds of person-hours spent assessing each player's abilities, it is not uncommon for players to perform dramatically better or worse than anticipated. For example, quarterbacks who have over- or under-performed relative to their draft position are frequently highlighted in the popular media. Several current or recent NFL starting quarterbacks (e.g., Tom Brady, Russell Wilson, and Matt Hasselbeck) were drafted in the third round or later, meaning that at least 60 players, including a number of quarterbacks, were selected before them. Others (e.g., Kurt Warner and Tony Romo) went undrafted entirely. Moreover, several quarterbacks selected with one of the first five picks overall (e.g., Mark Sanchez, Vince Young, JaMarcus

Russell, Tim Couch, Akili Smith, Ryan Leaf, and Heath Shuler) have played poorly in the NFL. Why does this happen? Do NFL teams lack the ability to discern future success from the information available on draft day? Or, is the task of projecting future performance in the NFL an inherently difficult one?

6.1.1 Previous Research by Position

Some attempts have been made to assess the ability of NFL teams at predicting the future performance of players at various positions. Inextricably linked to this question is the choice of one or more metrics for measuring the success of players in the NFL. This section reviews some previous research on these matters for different positions.

One way to evaluate performance, which seeks to evade many difficult issues, merely tallies the degree to which a player has participated in the NFL during their career. For example, in developing an alternative NFL draft pick value chart, Schuckers (2011) uses the number of games played, number of games started, and number of Pro Bowl appearances to measure career success. Moreover, Boulier et al. (2010) use career length in seasons played as a measure of performance for quarterbacks and wide receivers. One appeal of such general metrics is that they are applicable across different positions. The simplicity of these performance measures, however, often leads researchers to search for more intricate position-specific alternatives.

6.1.1.1 Running Backs

Literature on offensive "skill positions" (i.e., running backs, wide receivers, and tight ends) is limited. For running backs, we could find only two attempts at predicting NFL success:

1. Kuzmits and Adams (2008) find that better sprint ability is related to greater success in the NFL, as measured by games played and average yards per carry in the first 3 years of a running back's career, and
2. Blees (2011) defines a score (*RBScore*) for NFL running backs as follows:

$$RBScore = Yards - 3 * Plays - 30 * Fumbles + 18 * TDs \qquad (6.1)$$

where
  Yards = rushing yards + receiving yards
  Plays = rushing attempts + receptions
  Fumbles = number of fumbles that the running back loses to the opposing team
  TDs = rushing touchdowns + receiving touchdowns

Blees (2011) concludes that no aspect of a running back's pre-draft characteristics is associated with NFL RBScore.

6.1.1.2 Tight Ends

Mulholland and Jensen (2014) investigate predictors of NFL success for tight ends, using three different measures of success: games started, tight end score, and tight end score per game. Here, tight end score (*TEScore*) is defined as

$$TEScore = Yards + 19.3 * TDs \qquad (6.2)$$

where
  Yards = receiving yards
  TDs = receiving touchdowns

The authors find only two factors consistently related to tight end career success: broad jump ability and number of college receptions. They also conclude that size measures (BMI, weight, and height) are overemphasized by NFL teams.

6.1.1.3 Wide Receivers

For wide receivers, the literature seems to arrive at two main findings:

1. Tests performed at the NFL Combine are not good predictors of professional success, as measured by games played, average yards per reception, and a wide receiver score similar to (6.2) (Kuzmits and Adams, 2008; Dhar, 2011).
2. Draft position is associated with professional success (as measured by seasons played and receiving yards) indicating that NFL teams are effective at evaluating the wide receiver talent pool (Boulier et al., 2010).

6.1.1.4 Field Goal Kickers

Kickers are typically evaluated based on their field goal success rate, but even the seemingly simple question "Which placekickers are the best?" can get very complicated. Field goal success can vary with a multitude of factors: distance, weather, altitude, opposing defense, quality of the snap and hold, and stress level, to name a few. Pasteur and Cunningham-Rhodes (2014) develop a logistic regression model for the success probability of a field goal attempt based on eight explanatory variables: distance, temperature, wind metric, Denver indicator, play-off indicator, defense quality, an indicator for whether the kick taker was the starter, and fatigue. The authors then evaluate kickers by comparing them to a "replacement-level" kicker, arriving at a per-kick percentage improvement metric and a sum total of points above replacement kicker metric.

6.1.1.5 Offensive Line

Offensive linemen compete at a position where even rudimentary statistics on individual performance are lacking, if not nonexistent. Alamar and Weinstein-Gould (2008) pioneered work in this area by collecting data to help demonstrate that the longer a lineman can hold a block on a designed pass, the greater is the chance of a successful completion. Alamar and Goldner (2011) follow up on this research by creating a metric to quantify the total number of yards a lineman contributed to their team's passing game. Although these are exciting preliminary steps, there is still a long way to go before we can provide a comprehensive appraisal of the achievements of an individual lineman.

6.1.1.6 Defensive Players

The paucity of research on the evaluation of defensive players is even more pronounced. Of course, one could use very simple statistics, for example

1. Interceptions or passes-defensed for defensive backs
2. Sacks or "Hurries" for defensive linemen and linebackers
3. Tackles for linebackers and all defensive players
4. Forced fumbles for all defensive players

But the aforementioned metrics are unsatisfactory. For example, Richard Sherman may have very few interceptions, or passes-defensed, because no throws are attempted in his direction, and J.J. Watt's contribution may come in the form of drawing the attention of two offensive linemen. The complexity of the problem is evidenced by the fact that we could not find any work attempting to evaluate defensive player performance in a more sophisticated manner.

6.1.1.7 *Quarterbacks*

Presumably due to the prominence of the position, the preponderance of football-related research has focused on quarterbacks. We present a brief overview of some quarterback research here and provide a more in-depth examination of the quarterback prediction problem in Section 6.3. Two common threads in the literature on evaluating quarterbacks are that measures should

1. Link common statistics to game outcomes (e.g., how many wins does 4000 passing yards translate to?) and/or
2. Account for context (down, distance, field position, etc.) when assessing statistics (e.g., a 4-yard completion on 3rd and 3 is worth much more to a team than a 4-yard completion on 3rd and 10)

One metric that addresses context is Expected Points Added, which credits the quarterback with changes in the number of points their team can expect to score from one play to the next (see Burke (2014) for a more detailed explanation). Alternatively, Wins Produced uses multiple regression to determine weights for different quarterback statistics in order to arrive at an estimate of the number of wins they produce per 100 offensive plays (Berri, 2007; Gerrard, 2007). Of course, some statistics do not produce the wins they purport to because they are earned in low-leverage situations, for example, throwing a 50-yard touchdown pass when your team is losing by 28 points with 3 minutes remaining in the game. To address the shortcoming of neglecting score and time remaining, one could use a Win Probability Added metric. This is very similar to Expected Points Added, except using changes in the chance of winning a game, instead of changes to expected points, from play to play. It is important to point out that no single measure of success is flawless, for instance, Win Probability Added can be heavily influenced by a team's defense which might provide a quarterback with more, or fewer, opportunities to win a game.

Berri and Simmons (2011) quantify a quarterback's NFL performance using a variety of *per-play* metrics (e.g., Wins Produced per play and the related Net Points per play). Using these per-play statistics, they concluded that the draft position of a quarterback was not associated with how well they performed in the NFL. Quinn et al. (2007) arrive at similar conclusions using different performance metrics (e.g., the NFL's quarterback rating [QBR] and a metric called Defense-adjusted Value Over Average). Brian Burke (of *Advanced Football Analytics*) takes an interesting approach to circumvent the potential selection bias created by setting aside a class of non-qualifying quarterbacks for per-play statistics (a point we cover in detail later): Impute an expected performance value, so that all quarterbacks can be included in the subsequent analysis. Burke (2010) explores a few simple ways to carry out the imputation and finds evidence of an "unmistakable relationship between draft order and career per-play performance." Similarly, Boulier et al. (2010) use cumulative statistics to conclude that "although their ability to rank future performance of players is less than perfect, football executives are very successful at evaluating the talents of athletes."

In Section 6.2, we discuss several challenges that researchers encounter in attempting to quantify success in professional football and confront the player prediction problem. In particular, in Section 6.2.1, we address the issue of available data sources needed to perform any analysis of competencies of NFL teams. Section 6.2.4 introduces the dilemma of the numerous drafted players who never partake in an NFL snap and thus may have undefined success metrics. In Section 6.3, we turn our attention to a thorough consideration of the quarterback prediction problem: the challenge of identifying college quarterbacks who will succeed in the NFL. Section 6.4 concludes with a general discussion and some potential for future research.

6.2 Challenges to Predicting Player Performance

Quantifying the success of professional football players is not as simple a task as it may first seem. Data linking both college and NFL performances are not readily available, so datasets must be assembled manually. Furthermore, these datasets may have incomplete information on important predictor variables. Even with high-quality data in hand, there are several conceptual questions to grapple with. Players have vastly distinct roles on a team, can accomplish the goals of their position via different styles, and interact very closely with the players (teammates and opponents) around them. And, as we discuss in the following, the choice of whether to use cumulative or per-play metrics to quantify "success" can have a major impact on the conclusions of the analysis.

In this section, we discuss these key challenges in depth. Of course, there are many additional challenges that we do not address here. For example, we do not even begin to broach the issue of era effects, which may make comparison of the same metric difficult over time.

6.2.1 Data Availability

There is no single data source that provides comprehensive information on college and NFL performance for current and former players. Hence, the first step in analyzing the association between college statistics and NFL success is to form a dataset by combining multiple data sources. Particularly at the college level, there are major challenges involved in collecting and harmonizing performance data. First, the scope of the data collection effort is daunting: With 85 scholarship spots per team in the Football Bowl Subdivision (FBS) and well over 100 teams, there are 10,000 or more players to track (compare this to the NFL, with 53 players across 32 teams for approximately 1,700 players in total). Furthermore, while most play outcomes are objective (e.g., touchdowns, interceptions, yards gained, etc.), outcomes such as sacks and fumbles are more subjective. While it is possible to train professional scorers at 32 NFL stadiums to ensure that these outcomes are recorded consistently, the task of harmonizing outcome definitions across the many dozens of college football venues is much more challenging.

The NFL Scouting Combine (henceforth referred to as the "Combine") is another problematic source of data, chiefly because of missing information. Combine data are only sporadically available prior to the mid-2000s, and even with improved data availability some drafted players are not invited to the Combine. Furthermore, players (particularly elite skill-position players) have become increasingly selective about the events that they participate in, both to minimize injury risk and showcase their most outstanding skills. In statistical terms, missing Combine measurements are likely to be *informatively missing* as the population of players skipping a particular Combine event are likely to perform worse, as a group, than those who elect to participate in it.

Beyond the imperfections of the available data sources, lurks another fundamental problem: that many potentially useful predictors of NFL success are simply not made publicly available. NFL executives conduct wide-ranging interviews with prospects during the Combine, invite them for private workouts, and have access to the results of aptitude tests such as the Wonderlic Cognitive Ability Test (henceforth referred to as the "Wonderlic"). While Wonderlic scores sometimes leak to the public, the results of interviews and workouts are typically described in very vague terms. At the college level, several teams are experimenting with wearable devices which track player movements to predict game performance and injury risk (Feldman, 2015; Loh, 2015; Schroeder, 2015). These biometric measurements may (or may not) be useful for predicting future success. But the NCAA (which is relatively decentralized) may lack the will and authority to compel individual schools to report specific statistics about individual performance, particularly if schools see a competitive advantage in not publicly disclosing them.

As teams search for increasingly sophisticated techniques to divine which player is likely to be the next NFL star, the data environment for sports statisticians not affiliated with collegiate or professional teams may start to resemble an iceberg, with the vast majority of information hidden from view.

6.2.2 Defining "Success"

Measuring the success of an NFL football team is relatively straightforward:

> The bottom line in professional sports is winning. Everything has to focus on that product: winning football games. Other offshoots — the public relations, the merchandising, the high-sounding philosophical approach — mean little compared with being successful on the playing field.
>
> **—Bill Walsh (Harvard Business Review, 1993)**

Measuring the success of individual players who make up a team is a much more difficult problem. In spite of the common refrain in the media on the importance of players being "winners," surely we should not evaluate an individual purely by the outcomes of their team. Somewhat unique to football is the compartmentalization of roles on the playing field. In basketball and hockey, for example, all players are responsible (at least in part) for the offensive and defensive components that contribute to winning. In football, essentially all players are solely responsible for either the offensive or defensive facet of the game. Moreover, even among offensive players, different positions have such vastly different roles that it is difficult to compare, say, a running back to a tight end. The fundamental numerical quantities recorded for running backs, receivers, and quarterbacks (e.g., rushing yards, receiving yards, passing yards, touchdowns, touchdown passes, interceptions) are vastly different, making comparisons across position extremely problematic.

Although more achievable, within-position comparisons are still nontrivial. For example, different players who play the same position can contribute to a team's success in very distinct ways: one wide receiver might be more of a "possession" receiver while another could be a "deep threat"; one quarterback might be a "pocket passer" while another might be a "scrambler," and yet another could be a "game manager." How should one balance, for instance, a high completion percentage on very safe throws versus a higher yards per attempt value that comes at the expense of more interceptions?

One solution to differing within-position playing styles is the creation of aggregate metrics aimed at combining various individual statistics into a composite measure of a player's performance. For example, the most pervasive measure of a quarterback's performance is the NFL's QBR (Wikipedia, 2015):

$$QB\ rating = \left[5 \times \left(\frac{Comp}{PassAtt} - 0.3 \right) + 0.25 * \left(\frac{PassYds}{PassAtt} - 3 \right) \right.$$
$$\left. + 20 * \left(\frac{PassTD}{PassAtt} \right) + \left(2.375 - 25 * \frac{INT}{PassAtt} \right) \right] \times \frac{100}{6}$$

where
  Comp = number of completions
  PassYds = passing yards
  PassTD = passing touchdowns
  INT = interceptions
  PassAtt = passing attempts

This rating was developed by the league's statistical committee in 1971, at the request of commissioner Pete Rozelle, and was adopted by the NFL in 1973 (National Football League, 2015). The idea behind the NFL's quarterback rating is not dissimilar from the use of linear weights (Thorn and Palmer, 1984) in baseball's sabermetric community, that is, to identify factors which are correlated with success and assign weights to these factors according to the strength of their association.

As Berri and Simmons (2011) point out, however, the weights adopted in the NFL's quarterback rating are arbitrary. Moreover, quarterback rating ignores entirely contributions made via running, a facet of the game which has recently seen growing prominence with the emergence of quarterbacks such as Cam Newton, Russell Wilson, and Colin Kaepernick. To address the issues with the NFL's quarterback rating, some authors (Berri et al. 2006; Berri 2007) have defined derivative metrics that:

1. Incorporate a quarterback's rushing yards
2. Derive weights for the factors considered from regression coefficients

In particular, consider the *Net Points* statistic, which quantifies how many points a quarterback contributes to their team:

$$Net\ Points = 0.08 * Yards - 0.21 * Plays - 2.7 * INT - 2.9 * Fumbles, \qquad (6.3)$$

where

Yards = passing yards + rushing yards
Plays = passing attempts + rushing attempts + sacks
Fumbles = number of fumbles that the quarterback loses to the opposing team

A more general purpose metric to quantify performance at any offensive position, *Defense-adjusted Value Over Average* (*DVOA*) was developed by *Football Outsiders* (Football Outsiders Inc., 2015). A key feature of *DVOA* is that it attempts to contextualize the outcomes of individual plays while accounting for the opposing defense quality. For example, a 3-yard gain by a running back is a much more successful outcome if it occurred on a 3rd-and-2 against a high-quality defense than if it occurred on a 3rd-and-5 against a porous defense.

While aggregate statistics may put players with different playing styles on similar footing, they do have some drawbacks. First, since they combine multiple statistics, they cannot be computed if the value of one or more of these statistics is not available (or uncertain; for instance, we found some inconsistencies across data sources in how fumbles and fumbles lost were reported). Furthermore, they are dependent on the performance of teammates and opponents, over which the player in question has little, or no, control. For example, *DVOA* adjusts for the quality of a team's opponent but does not attempt to divide credit for a particular play's outcome between the offensive teammates.

6.2.3 Isolating the Impact of Individual Players

In certain sports, such as baseball, the task of isolating the contribution of each player is made easier by the intrinsically more discrete nature of the game. The continuous flow, and interactions between teammates and opponents, makes quantifying an individual's performance challenging in many other team sports, including basketball, hockey, soccer, and football.

Consider a wide receiver who plays on a team with (1) a poor quarterback, and/or (2) a poor offensive line, and/or (3) a coach who favors running over passing plays. The contribution that such a receiver makes to his team may be severely underestimated by any of the traditional metrics used to evaluate receivers (e.g., receptions, receiving yards, touchdown receptions). At some positions, like offensive line, the task of disentangling an individual's contribution is made even more difficult because there are so few individual performance statistics that are publicly available.

ESPN recently created a new rating system for quarterbacks, *TotalQBR*, which is based on the expected points added by the quarterback on each play. *TotalQBR* adjusts for how critical the play is to a game's outcome (i.e., a "clutch" factor for high-leverage situations) and divides credit among certain offensive players (Oliver, 2011). For example,

1. Passes that are deemed to be "drops" penalize the quarterback less than the receiver
2. Yards after the catch are credited more to the receiver than the quarterback
3. Some sacks penalize a quarterback less than their offensive line

The details of *TotalQBR*'s calculation are, however, proprietary, and its values are only available for quarterbacks dating back to the 2006 NFL season. Moreover, *TotalQBR* does not adjust for the strength of the opposing defense.

6.2.4 Cumulative versus Per-Play Metrics

Cumulative statistics (e.g., games played, passing/rushing/receiving yards, touchdowns) are an important marker of success in the NFL; after all, the leaders in these categories are typically elite players. However, cumulative statistics are closely related to the number of opportunities given to a player. For example, a quarterback who plays for a weak team that often finds itself behind may throw many more passes and hence accumulate more passing yards and touchdowns. Similarly, a running back playing for a better team which often finds itself in the lead may carry the ball more often and hence accumulate more rushing yards and touchdowns.

In analyses which use cumulative statistics as success metric, it is reasonable to assign players who never play in the NFL a value of 0. This assignment is predicated on the idea that the number of playing opportunities reflects the quality of the player, so that players who do not play would, on average, have had low cumulative statistic totals if given more playing time. However, if playing opportunities are awarded based on factors other than on-field performance, then cumulative statistics can be a poor reflection of the quality of a player's career. For instance, teams may be more reluctant to replace a player who is performing poorly if that player was drafted early, and hence highly paid. This phenomenon, known as the *sunk cost fallacy* or *escalation of commitment*, occurs when organizations who have committed resources to a project (player) maintain their commitment, even when the costs exceed the benefits (Staw, 1976). Staw and Hoang (1995) and Camerer and Weber (1999) found evidence for escalation of commitment in the National Basketball Association (NBA). In the context of the NFL, players (especially quarterbacks) drafted with the first few picks in the first round typically receive large contracts, which may lead to organizations awarding them more playing time than their on-field performance merits. At the opposite end of the spectrum, undrafted players may be undervalued by cumulative statistics if they are awarded playing opportunities only in cases where their skill level clearly exceeds that of their more highly paid teammates.

Per-play metrics (e.g., yards per attempt/carry/reception) avoid this potential problem with cumulative statistics by leveling the "opportunity" playing field. By reducing the dependence on team characteristics and playing style, per-play metrics may more accurately reflect a player's individual contribution. But the analyst wishing to use per-play metrics faces a key question: how to handle the large number of drafted players who have accumulated no playing time in the NFL. Many drafted players never actually participate in an NFL play: of the 3819 players drafted between 1997 and 2013, 408 (11%) never played an NFL game; for quarterbacks, 42 of 191 (22%) never saw the field. Per-play statistics are undefined for these players since their "denominator" (number of plays/attempts) is zero. Similarly, as pointed out by Burke (2010), those with very limited playing time can have per-play statistics that are highly variable.

To avoid these problems, several authors have proposed to restrict analyses to players whose career playing time exceeds a fixed threshold. For example, in their analysis of quarterbacks, Berri and Simmons (2011) exclude all quarterbacks who played fewer than 100 NFL snaps. Excluding players with less than some minimum threshold of playing experience (henceforth referred to as *non-qualifying* players) is problematic: the validity of subsequent analyses depends on the assumption that data from non-qualifying players are missing completely at random (MCAR), that is, that non-qualifying players would have performed similarly (if given similar playing time) to those above the minimum threshold. But the MCAR assumption seems tenuous, as once a college player has been drafted onto an NFL team, that team's coaches can observe their performance in training camp, team practices, and exhibition games before deciding whether or not to allow them to play in a regular season game. In other words, there is a reasonable alternative explanation for the surprising findings of, for example, Berri and Simmons (2011) and Quinn et al. (2007) that draft position is not related to future success: perhaps, player performance is unrelated to draft status *conditional* on an NFL coach deeming a player sufficiently skilled to play professionally, but earlier draft picks are more likely to possess this minimum skill level.

The debate over whether non-qualifying players can plausibly be assumed to be as good as qualifying players enjoyed some mainstream media attention when the bestselling author Malcolm Gladwell and Harvard evolutionary psychologist Steven Pinker found themselves on opposite sides of the dispute (Gladwell, 2009; Pinker, 2009). Arguing that we have no reason to believe that non-qualifying quarterbacks are any worse than qualifying quarterbacks, Gladwell is quoted as saying in jest (Bialik, 2009):

> Coaches and GMs turn out to be good decision-makers when it comes to drafting quarterbacks when you consider the fact that quarterbacks who never played aren't any good. And how do we know that the quarterbacks who never play aren't any good? Because coaches and GMs are good decision-makers!

Our goal here is not to resolve the debate, but rather to point out that neither cumulative nor per-play metrics give a complete and unbiased picture of player performance.

6.3 Case Study: Which Factors Predict NFL Quarterback Success?

We now present a case study that outlines an analysis to identify factors predicting the future success of college quarterbacks drafted into the NFL. The analysis is similar to that presented in Wolfson et al. (2011), with the only major change being that the data were updated up to include the 2013 NFL season. Throughout the case study, we highlight several analytic techniques which can be employed to (at least partially) address the challenges laid out in the previous section. We also emphasize the importance of explicitly estimating the predictive ability of models used with the goal of quantifying the inherent difficulty of the underlying prediction problem. Note that while this analysis involves only quarterbacks, the same principles and approaches could be applied to identify predictors of NFL success at other positions.

6.3.1 Background

Quarterback is widely regarded as the most important position on a football team. Accordingly, quarterbacks have traditionally commanded some of the largest contracts in the NFL. Quantifying quarterback performance suffers many of the same difficulties as exist for other positions. Perhaps due to the prominence of the position, however, a great deal more effort has gone into measuring the quality of a quarterback's play. The challenge of identifying college quarterback prospects who are most likely to succeed at the professional level is among the "Hilbert Problems" for football identified by Schatz (2005). We will henceforth refer to this challenge as the *quarterback prediction problem*.

Given the success of the analytics movement in other major sports (e.g., baseball and basketball), it is certainly possible that football executives are incorrectly valuing player attributes when deciding who to select in the NFL draft. If teams are poor at identifying college prospects who are likely to succeed as NFL quarterbacks, two possible explanations are

1. *The "Moneyball" theory*: NFL teams aggregate available information suboptimally, emphasizing some attributes which do not correlate with NFL performance and de-emphasizing other attributes which are more predictive of NFL success.
2. *The "Statistical Noise" theory*: The variability in individual performance due to random, unmeasured factors may make prediction inherently difficult, even if all available information is used optimally.

In what follows, we consider both of these explanations and attempt to quantify how much each contributes to the quarterback prediction problem faced by NFL teams. Important features of our analysis include

- We base our analyses on the cumulative statistics of all quarterbacks drafted into the NFL, not only on those whose career playing time exceeds some minimum threshold. As a result, we employ *zero-inflated regression models* to account for the excess zeroes introduced into the dataset by players who do not play in the NFL.

- We consider *several variants of our outcome*, including binary outcomes defined by thresholding the underlying "continuous" cumulative statistics.

- We perform *sensitivity analyses* by excluding first-round picks to explore the "additional opportunity" hypothesis which might bias results based on cumulative statistics.

- We report not only associations between college and NFL performance and their associated p-values but also *explicitly estimate the predictive ability of our models*. Estimating the prediction error of various models allows us to quantify the "predictive value added" by various combinations of college, Combine, and draft statistics. This provides insight into the inherent difficulty of the quarterback prediction problem.

6.3.2 Data

To perform the analyses in this case study, one of us (Schmicker) assembled a dataset of 208 quarterbacks drafted between 1997 and 2013, using information from multiple sources:

- College statistics going back to the 2000–2001 season were obtained from NCAA.com.

- Career college statistics for quarterbacks who played before 2000–2001 were obtained from several other sources including school web sites.

TABLE 6.1

Sample Summary Statistics

Draft years spanned: **1997–2013**
Number of quarterbacks: **208**
Number from FBS: **184**
Number of colleges: **116**
Number by round:

1st	2nd	3rd	4th	5th	6th	7th
45	19	21	20	27	36	40

NFL games played:

Total number	Median	Range
6995	14.5	[0, 256]

- Physical evaluations from the NFL Scouting Combine were obtained from nfl-draftscout.com.
- Draft position and most NFL statistics were obtained from Pro-Football-Reference.com.
- Number of sacks and fumbles lost (which are not consistently reported on Pro-Football-Reference.com) were obtained from NFL.com.

Drafted players who played a substantial amount at the quarterback position in college were omitted from our analysis if they played almost exclusively at other positions in the NFL (e.g., Antwaan Randle El, Brad Smith, Michael Robinson, and Isaiah Stanback). Some basic summaries of the data are presented in Table 6.1.

6.3.3 Outcomes and Predictors

For our analyses, we considered two cumulative statistics quantifying NFL performance:

1. *Games played*: Counts the total number of NFL games in which a quarterback has appeared for at least one play. In our analyses, we treated games played as an integer-valued random variable and also considered three binary variants. Letting G be the number of games played, we define

$$G^{(K)} = \begin{cases} 1 & \text{if } G \geq K \\ 0 & \text{if } G < K \end{cases} \tag{6.4}$$

for $K = 1$, 16, and 48. These cutoffs correspond, informally, to a minimal, moderate, and substantial degree of NFL success. Quarterbacks with $G \geq 1$ (i.e., $G^{(1)} = 1$) can be thought of as having reached a minimum competence threshold: their team's coaching staff has judged them good enough to play in an NFL game. Similarly, quarterbacks with $G^{(48)} = 1$ are generally considered very good to excellent, as few poor quarterbacks are allowed to play in three complete seasons' worth of games.

2. *Net Points*: Recall from (6.3) that the Net Points metric attempts to quantify the number of points a quarterback contributes to their team (Berri et al., 2006; Berri,

2007). Fractional Net Points are rounded to the nearest integer. Berri and Simmons (2011) computed Net Points only for quarterbacks who had accumulated statistics at the NFL level; for our analysis, we assigned zero Net Points to quarterbacks who have not played in the NFL, since they have not accumulated any of its component statistics. Thirty quarterbacks had small negative Net Points values, which we set to zero.

In Figure 6.1, we present a scatterplot of Net Points versus games played in the NFL, with some notable quarterbacks labeled. Note that our outcome measures are defined for all drafted quarterbacks and may be affected by the number of playing opportunities that a quarterback is afforded. The degree to which playing opportunities depend on factors other than on-field performance is unknown, but in Section 6.3.6, we present analyses contradicting the view that these factors play a major role in determining playing time for quarterbacks.

We considered the following predictors of NFL performance in our regression models: draft position (Pick), year drafted (DraftYear), passing statistics compiled during a quarterback's college career, and measurements from the Combine (including Height and Weight). Table 6.2 presents summary statistics of the predictors in our analysis. For the reasons

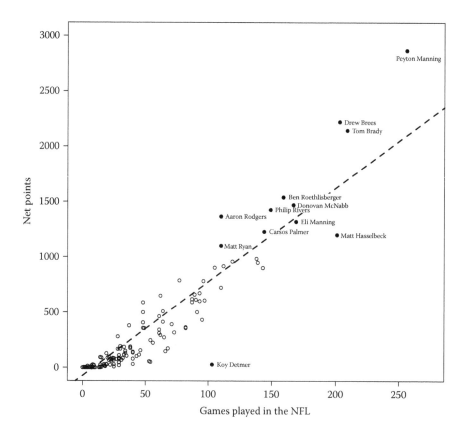

FIGURE 6.1
Net points vs. games played for all quarterbacks in our dataset.

TABLE 6.2

Summary Statistics for Predictors

Predictor	Median [Min, Max]	# Missing
ColGames	40 [12, 55]	19
Number of games played		
CompPerc	59.8 [40.9, 70.4]	13
Completion percentage = (# completions)/(# pass attempts)		
YPA	7.6 [5.7, 10.2]	13
Yards per pass attempt = (pass yards)/(# pass attempts)		
Int	28 [1, 64]	15
Number of interceptions		
TD	56 [0, 131]	11
Number of touchdowns		
Height	75 [70, 80]	54
Height (in inches)		
Weight	225 [192, 265]	54
Weight (in lbs)		
40-yard dash	48.2 [43.3, 53.7]	56
Time to run 40 yards (in 0.1 s of a second)		
Vertical jump	31.5 [21.5, 39.0]	93
Vertical leap height from a standing position (in inches)		
Cone drill	71.1 [66.6, 78.0]	99
Time to run a course marked by cones (in 0.1 s of a second)		

noted previously, a relatively large proportion of quarterbacks have missing Combine measurements.

6.3.4 Statistical Models

The binary variables $G^{(1)}$, $G^{(16)}$, and $G^{(48)}$ were modeled via logistic regression. Games played (G) was modeled via negative binomial (NB) regression (Agresti, 2002). Suppose that, given $\lambda > 0$, Y has a Poisson distribution with mean λ and that $\lambda \sim Gamma(k, \mu)$, then the marginal probability function of Y is negative binomial, taking the form

$$P(Y = y; k, \mu) = \frac{\Gamma(y + k)}{\Gamma(k)\Gamma(y + 1)} \left(\frac{k}{\mu + k}\right)^k \left(1 - \frac{k}{\mu + k}\right)^y$$

with $E(Y) = \mu$ and $Var(Y) = \mu + \mu^2/k$. $\theta = 1/k$ reflects the degree of overdispersion of the counts; as $\theta \to 0$, the negative binomial distribution converges to the usual Poisson distribution. In our case, both games played and Net Points showed evidence of overdispersion: the negative binomial regression models we fit generally estimated $\theta \approx 2$, with standard errors less than 0.4.

As noted, quarterbacks could have zero Net Points either because they did not play in the NFL and were assigned zero points by definition, or because they did play and had their Net Points rounded to zero. Since zero values for this outcome can be viewed as having been generated by two separate processes, we modeled the Net Points outcome as a zero-inflated negative binomial (ZINB) random variable (Greene, 1994; Yau et al., 2003). The ZINB model extends the NB model by allowing extra probability mass to be placed on the value zero, with the probability that an observation is a structural or "excess" zero modeled by logistic regression. ZINB regression is a generalization of Poisson regression appropriate for data which may be overdispersed (i.e., given predictors X, $Var(Y|X) > E(Y|X)$). It may also be the case that the assumed Poisson mean–variance relationship $Var(Y|X) = E(Y|X)$ for the nonzero observations is incorrect. The ZINB model explicitly models the excess zeros via a logistic model and accounts for possible overdispersion in the nonzero observations by estimating an additional parameter. Although it is possible to use different predictors for the two components of a ZINB model, we used the same sets of predictors for both components in our analysis.

For each outcome, we considered two primary models. The first model (Base) contained the college predictors (ColGames, CompPerc, YPA, Int, and TD) listed in Table 6.2, along with DraftYear; the second model contained all the Base predictors plus log(Pick), a term accounting for where a player was selected in the NFL draft.

6.3.4.1 Sensitivity Analyses

Due to the financial investment required to sign first-round draft selections, one could reasonably argue that the playing opportunities for these quarterbacks are most heavily influenced by external factors unrelated to their on-field performance. An analysis that excludes these players may indicate whether the predictors of success differ for more "disposable" quarterbacks who were selected later in the draft and did not command a large contract. To address this issue, we considered two secondary models with the same predictors as the two primary models, but excluding quarterbacks selected in the first round. We refit these four models using the Combine measurements (Height, Weight, 40-yard dash, Vertical jump, and Cone drill) from Table 6.2 in place of college statistics, leading to a total of eight regression models for each outcome.

6.3.5 Predictive Accuracy

We compared the predictive performance of eight distinct models, fitted to data from all quarterbacks. The models are

1. *DraftYear*: A model including draft year as the only predictor.
2. *+ log(Pick)*: A model including draft year and log(Pick).
3. *+ College Stats*: A model including draft year and the college statistics listed in Table 6.2.
4. *+ Combine Stats*: A model including draft year and the Combine statistics listed in Table 6.2.
5. *+ College + Combine*: A model including draft year, and college and Combine statistics.
6. *+ log(Pick) + College*: A model including draft year, log(Pick), and college statistics.
7. *+ log(Pick) + Combine*: A model including draft year, log(Pick), and Combine statistics.
8. *+ log(Pick) + College + Combine*: A model including all of the available predictors.

6.3.5.1 Quantifying Prediction Error

For the binary outcomes modeled using logistic regressions, we computed $\hat{\pi}_i^{(K)}$, the esti-mated probability that $G_i^{(K)} = 1$. Predictive accuracy was compared between models incorporating various predictors via the receiver operating characteristic (ROC) curve, which plots the true positive vs. false positive rate as the value of $\hat{\pi}_i^{(K)}$ beyond which one predicts $G_i^{(K)} = 1$ ranges from 1 to 0. Overall model accuracy was summarized by comput-ing the area under the ROC curve (AUC) using fivefold cross-validation (Efron and Gong, 1983). The AUC can be interpreted as the probability that $\hat{\pi}_i^{(K)} \geq \hat{\pi}_j^{(K)}$ for a randomly chosen pair of observations i and j with $G_i^{(K)} = 1$ and $G_j^{(K)} = 0$. Models which predict "success" no better than chance have an AUC of 0.5; an infallible prediction model has an AUC of 1. Uncertainty for the AUC estimate was quantified by bootstrapping the model-fitting and cross-validation procedure 100 times.

For the integer-valued outcomes, we label our predictions as $\hat{Y}_i$, referring either to predicted games played (NB models) or Net Points (ZINB models).

- In the NB regressions, predictions $\hat{Y}_i$ were obtained from the fitted values for each individual i.
- In the ZINB regressions, predictions were obtained for each individual i via

$$\hat{Y}_i = \begin{cases} 0 & \text{if } \hat{\phi}_i < 0.5 \\ \hat{\theta}_i & \text{if } \hat{\phi}_i \geq 0.5 \end{cases}$$

where
$\hat{\phi}_i$ is the estimated probability that individual i represents a structural zero
$\hat{\theta}_i$ is the estimated mean for individual i given that they are not a structural zero

Predictive accuracy for integer-valued outcomes was quantified via the absolute prediction error

$$APE = \frac{1}{n} \sum_i |\hat{Y}_i - Y_i|,$$

where Y_i refers to either games played or Net Points. As with the AUC, the APE was estimated via fivefold cross validation with uncertainty quantified by bootstrapping the original data.

6.3.6 Results

6.3.6.1 Games Played

Tables 6.3 through 6.6 report the results of regression models investigating the association between games played and college and Combine statistics. The values in Table 6.3 represent the percent increase in mean games played (and corresponding 95% confidence intervals)

TABLE 6.3

Percent Change in Number of NFL Games Played (with 95% Confidence
Intervals) Associated with One-Unit Differences in College and Combine
Statistics, Year Drafted, and Draft Position

Variable	All Quarterbacks		Rounds 2–7 Only	
	Base	Base + Pick	Base	Base + Pick
Games	1 (−2, 4)	1 (−2, 4)	3 (−2, 8)	2 (−3, 6)
Comp Pct	**8 (2, 14)**	**5 (0, 10)**	1 (−6, 9)	3 (−4, 10)
YPA	−6 (−29, 25)	**−29 (−45, −8)**	**−43 (−64, −8)**	**−37 (−58, −5)**
Int	0 (−2, 3)	0 (−3, 2)	−2 (−5, 1)	−1 (−4, 2)
TD	0 (−1, 1)	0 (−1, 1)	1 (−1, 3)	1 (−1, 2)
Draft year	**−8 (−13, −4)**	**−8 (−12, −4)**	**−13 (−20, −6)**	**−10 (−16, −3)**
log(Pick)	—	**−29 (−36, −21)**	—	−35 (−58, 2)
Height	−4 (−19, 15)	−3 (−17, 14)	−6 (−25, 18)	−3 (−23, 21)
Weight	2 (−1, 5)	0 (−2, 3)	0 (−3, 4)	0 (−3, 3)
40-Yard dash	−3 (−90, 896)	0 (−87, 690)	−7 (−98, 3489)	6 (−96, 2965)
Vertical jump	−1 (−13, 11)	−3 (−13, 7)	−3 (−19, 16)	−3 (−18, 14)
Cone drill	−48 (−84, 70)	17 (−62, 263)	6 (−79, 438)	49 (−70, 633)
Draft year	**−9 (−16, −2)**	**−10 (−16, −3)**	−10 (−19, 1)	**−10 (−19, −1)**
log(Pick)	—	**−29 (−41, −13)**	—	**−43 (−67, −4)**

TABLE 6.4

Percent Change in Odds of Playing ≥1 NFL Game (with 95% Confidence
Intervals) Associated with One-Unit Differences in College and Combine
Statistics, Year Drafted, and Draft Position

Variable	All Quarterbacks		Rounds 2–7 Only	
	Base	Base + Pick	Base	Base + Pick
Games	1 (−4, 7)	1 (−5, 7)	2 (−4, 8)	1 (−5, 8)
Comp Pct	4 (−6, 15)	1 (−10, 13)	2 (−8, 13)	1 (−11, 14)
YPA	41 (−21, 161)	37 (−27, 163)	22 (−35, 134)	36 (−33, 187)
Int	0 (−5, 4)	1 (−4, 5)	−1 (−5, 4)	1 (−4, 6)
TD	0 (−2, 3)	−1 (−3, 2)	0 (−2, 3)	−1 (−4, 3)
Draft year	**−14 (−23, −5)**	**−16 (−26, −5)**	**−15 (−24, −5)**	**−16 (−28, −4)**
log(Pick)	—	**−89 (−97, −73)**	—	**−89 (−97, −64)**
Height	0 (−24, 47)	3 (−19, 51)	4 (−19, 49)	4 (−21, 57)
Weight	4 (−2, 11)	2 (−5, 9)	2 (−5, 9)	2 (−5, 9)
40-Yard dash	−82 (−99, 542)	−62 (−99, 1584)	−83 (−100, 725)	−65 (−99, 2186)
Vertical jump	4 (−15, 27)	−1 (−19, 21)	−1 (−20, 22)	−1 (−20, 23)
Cone drill	0 (−91, 1072)	103 (−82, 2372)	71 (−85, 2035)	109 (−85, 3117)
Draft year	−11 (−23, 1)	−12 (−23, 1)	−11 (−23, 1)	−12 (−24, 2)
log(Pick)	—	**−82 (−95, −54)**	—	**−80 (−95, −36)**

TABLE 6.5

Percent Change in Odds of Playing ≥16 NFL Games (with 95% Confidence Intervals) Associated with One-Unit Differences in College and Combine Statistics, Year Drafted, and Draft Position

Variable	All Quarterbacks		Rounds 2–7 Only	
	Base	Base + Pick	Base	Base + Pick
Games	0 (−5, 4)	0 (−5, 5)	2 (−3, 8)	0 (−6, 7)
Comp Pct	6 (−3, 16)	1 (−8, 12)	3 (−7, 14)	2 (−9, 14)
YPA	6 (−34, 70)	−23 (−57, 36)	−21 (−57, 41)	−17 (−56, 57)
Int	−1 (−4, 3)	−1 (−5, 4)	−1 (−5, 3)	0 (−5, 5)
TD	2 (0, 4)	2 (−1, 4)	2 (−1, 4)	1 (−1, 4)
Draft year	**−10 (−18, −2)**	**−13 (−23, −4)**	**−12 (−22, −3)**	**−12 (−22, −2)**
log(Pick)	—	**−82 (−91, −70)**	—	**−80 (−91, −57)**
Height	16 (−16, 63)	33 (−4, 101)	23 (−7, 81)	40 (−2, 122)
Weight	3 (−3, 8)	−3 (−9, 3)	−1 (−7, 5)	−3 (−10, 4)
40-Yard dash	−80 (−99, 384)	−66 (−99, 1,975)	−91 (−100, 327)	−83 (−100, 1,527)
Vertical jump	9 (−9, 31)	2 (−16, 25)	1 (−17, 24)	1 (−19, 26)
Cone drill	29 (−84, 937)	756 (−21, 11,193)	421 (−47, 5782)	1,193 (−3, 23,109)
Draft year	−11 (−22, 0)	**−15 (−28, −2)**	−13 (−25, 0)	−14 (−28, 0)
log(Pick)	—	**−85 (−94, −67)**	—	**−86 (−96, −59)**

TABLE 6.6

Percent Change in Odds of Playing ≥48 NFL Games (with 95% Confidence Intervals) Associated with One-Unit Differences in College and Combine Statistics, Year Drafted, and Draft Position

Variable	All Quarterbacks		Rounds 2–7 Only	
	Base	Base + Pick	Base	Base + Pick
Games	1 (−4, 6)	3 (−4, 10)	5 (−3, 14)	4 (−5, 13)
Comp Pct	10 (−1, 22)	9 (−4, 24)	2 (−11, 16)	0 (−13, 16)
YPA	−1 (−42, 69)	−48 (−75, 3)	−50 (−79, 13)	−51 (−81, 18)
Int	0 (−5, 4)	0 (−6, 5)	−2 (−8, 4)	−2 (−8, 4)
TD	0 (−2, 2)	−1 (−4, 2)	1 (−2, 5)	1 (−3, 4)
Draft year	**−11 (−20, −2)**	**−12 (−23, −1)**	−13 (−26, 0)	−12 (−26, 2)
log(Pick)	—	**−61 (−73, −48)**	—	**−68 (−88, −19)**
Height	−8 (−32, 34)	0 (−23, 46)	−11 (−33, 27)	−8 (−32, 36)
Weight	6 (0, 13)	2 (−5, 9)	4 (−3, 13)	3 (−4, 12)
40-Yard dash	−75 (−100, 906)	−63 (−99, 2116)	−54 (−100, 5273)	−17 (−100, 12,610)
Vertical jump	−1 (−18, 21)	−7 (−26, 15)	−1 (−22, 26)	−1 (−23, 28)
Cone drill	−60 (−97, 311)	85 (−87, 2486)	−40 (−97, 872)	3 (−95, 1,792)
Draft year	−11 (−23, 2)	−14 (−28, 0)	−8 (−23, 8)	−7 (−23, 10)
log(Pick)	—	**−67 (−83, −44)**	—	**−70 (−90, −15)**

associated with one-unit increases in each predictor. Tables 6.4 through 6.6 give the percent increases in the odds of playing in ≥ 1, ≥ 16, and ≥ 48 games. Confidence intervals that exclude zero are highlighted in bold.

Overall, the models fitted to quarterbacks drafted after the first round yielded similar results to models fitted to all drafted quarterbacks. Unsurprisingly, draft year was negatively associated with games played in most models; more years in the league generally lead to more games played. The only other predictor that was consistently associated with games played was log(Pick), with quarterbacks drafted in the later rounds playing fewer games. The influence of draft status was relatively consistent across models: one log differences in draft pick number (e.g., the difference between the first overall selection and the 3rd, or the 10th overall selection and the 27th) were associated with 30–40% fewer games played, and 60–90% decreases in the odds of achieving the 1, 16, and 48 games played thresholds. Neither college statistics nor Combine measures were associated with playing in ≥ 1, ≥ 16, or ≥ 48 NFL games.

College completion percentage and yards per attempt were significantly associated with the integer-valued games played outcome. Completion percentage had a modest effect (5%–8% increase in games played for every additional percentage point of completed passes) in the model including all drafted quarterbacks. The effect was not significant in the models excluding first round picks. Yards per attempt had a strong association with games played in three of the four models which included college statistics, but the direction of the relationship is contrary to conventional wisdom: when considering all but the "Base" model including all quarterbacks, one additional YPA was associated with a 29–43% *decrease* in the number of NFL games played. For more discussion of this counterintuitive result, see Wolfson et al. (2011).

6.3.6.2 Net Points

Table 6.7 summarizes the results of the NB count portions of the ZINB models for Net Points, as before, reporting percent increases in the mean for a one-unit increase in each predictor, along with 95% confidence intervals. For the sake of brevity, we do not report coefficient estimates from the "excess zeros" portions of the ZINB models.

The conclusions for Net Points mirror those for the number of NFL games played. Models for Net Points fitted to all quarterbacks did not differ greatly from models fitted to quarterbacks drafted in rounds 2–7. Draft year and draft position were negatively associated with Net Points (i.e., quarterbacks drafted more recently and later in the draft produced fewer Net Points). Moreover, college/Combine statistics were mostly not associated with this outcome, with the exception of completion percentage and the unexpected negative association between YPA and Net Points.

6.3.6.3 Predictive Accuracy

Figures 6.2 and 6.3 summarize the cross-validation estimates (based on 100 runs of fivefold cross-validation) of the absolute prediction error for games played and Net Points, respectively. The model with the smallest prediction error included draft year, pick, and college statistics; however, this model offered at best a modest improvement (approximately 10%–15% for games played, and no improvement for Net Points) over the simpler model which included only draft year and pick. Models that did not include pick had the highest prediction errors. Incorporating college and Combine statistics offered some improvement over

TABLE 6.7

Percent Change in NFL Net Points (with 95% Confidence Intervals) Associated with One-Unit Differences in College and Combine Statistics, Year Drafted, and Draft Position

Variable	All Quarterbacks		Rounds 2–7 Only	
	Base	**Base + Pick**	**Base**	**Base + Pick**
Games	2 (−2, 6)	1 (−2, 5)	4 (−3, 11)	3 (−3, 11)
Comp Pct	**10 (3, 19)**	**8 (2, 15)**	5 (−3, 15)	5 (−3, 14)
YPA	1 (−29, 45)	**−32 (−53, −4)**	**−53 (−74, −14)**	**−49 (−72, −6)**
Int	0 (−3, 4)	−1 (−4, 2)	−4 (−8, 1)	−3 (−7, 1)
TD	0 (−2, 1)	0 (−2, 2)	2 (−1, 5)	1 (−1, 4)
Draft year	**−10 (−15, −4)**	**−11 (−16, −6)**	**−16 (−24, −8)**	**−16 (−24, −8)**
log(Pick)	—	**−30 (−39, −19)**	—	−29 (−61, 30)
Height	−18 (−34, 3)	−16 (−32, 3)	−21 (−39, 2)	−19 (−37, 5)
Weight	**4 (0, 9)**	3 (−1, 6)	4 (−1, 9)	3 (−1, 7)
40-Yard dash	28 (−96, 3903)	35 (−93, 2695)	683 (−97, 203,638)	1,470 (−91, 263,087)
Vertical jump	−9 (−24, 9)	−11 (−24, 3)	−6 (−27, 23)	−4 (−24, 22)
Cone drill	−72 (−94, 21)	−28 (−83, 207)	−77 (−97, 110)	−65 (−96, 177)
Draft year	**−11 (−19, −1)**	**−12 (−19, −3)**	−8 (−20, 6)	−10 (−21, 3)
log(Pick)	—	**−32 (−46, −13)**	—	**−54 (−77, −8)**

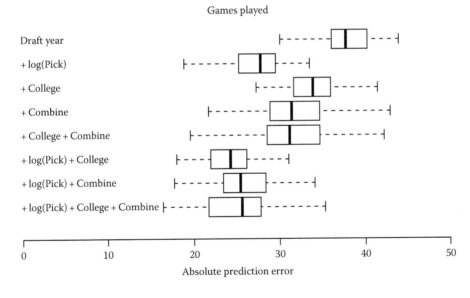

FIGURE 6.2

Cross-validated absolute prediction error estimates for NFL games played. Boxplots show prediction error estimates from 100 bootstrap replicates.

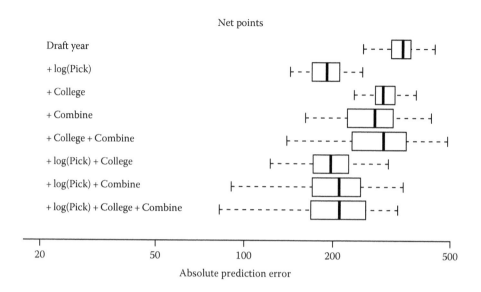

FIGURE 6.3
Cross-validated absolute prediction error estimates for Net Points. Boxplots show absolute prediction error estimates from 100 bootstrap replicates.

the draft year only model, but these models were substantially worse than models which only used information on draft pick.

Figure 6.4 displays ROC curves for four models applied to the three binary games played outcomes $G^{(1)}, G^{(16)}$, and $G^{(48)}$: draft year only; draft year and log(Pick); draft year and college statistics; and draft year, log(Pick), and college statistics. While the models incorporating log(Pick) (indicated by solid black and gray lines) clearly outperform those which exclude it, the gap in predictive accuracy is largest for the indicator of ≥ 16 NFL games

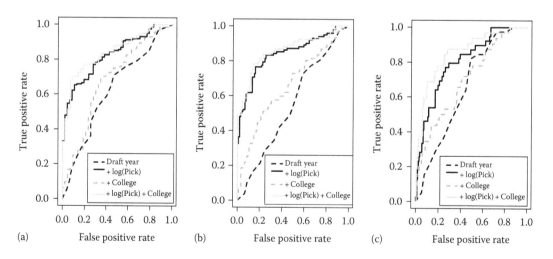

FIGURE 6.4
ROC curves for four logistic regression models predicting (a) ≥ 1, (b) ≥ 16, and (c) ≥ 64 games played.

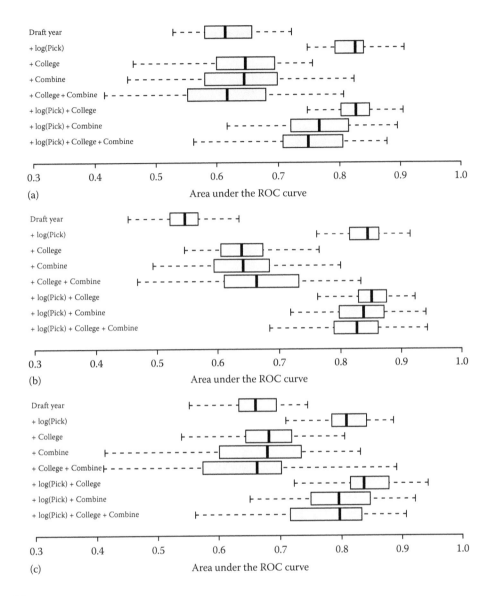

FIGURE 6.5
Cross-validated AUCs for eight prediction models applied to three binary games played variables: (a) ≥ 1, (b) ≥ 16, (c) ≥ 64. Boxplots show AUC values from 100 bootstrap replicates.

played, suggesting that draft position carries more information about whether or not a quarterback will be moderately successful, and less about whether they will play at all or become an established starter.

Figure 6.5 summarizes the cross-validated AUC for the eight predictive models enumerated at the beginning of this section. Note that, in contrast to the figures for integer-valued games played and Net Points, models with boxplots falling farther to the right in Figure 6.5 have better predictive accuracy. The overall conclusions are similar to the integer-valued games played and Net Points: models incorporating information on draft pick performed best, with college and Combine statistics providing little improvement in predictive accuracy.

Of particular note is the poor performance of models incorporating college and/or Combine statistics (but not draft pick) for the $G^{(1)}$ and $G^{(64)}$ outcomes. These models appear to offer virtually no gain over the basic model using only draft year; indeed, certain models (e.g., + *College* + *Combine* for the $G^{(64)}$ outcome) actually predict worse than the draft year only model. Across all outcome types, the results were similar when first-round picks were excluded from the analysis.

6.3.7 Conclusions and Limitations

Because quarterbacks with no NFL appearances are included, an analysis of this type could be expanded to include all college quarterbacks, not only those drafted to play in the NFL. While this idea is appealing, we chose to limit our analysis to drafted quarterbacks for two main reasons. First, it would be very challenging to obtain performance data for all NCAA quarterbacks. And second, teams and fans are generally most interested in assessing the future potential of college quarterbacks who have been deemed good enough to be drafted, with a particular focus on quarterbacks drafted in the first few rounds. These quarterbacks are closely scrutinized and endlessly discussed, and hence the lack of quantitative predictors of NFL success for this group is notable.

One alternative explanation for these results is that NFL teams systematically deny or award playing time to players based on their draft position without regard to on-field performance. Previous work has focused on this possibility, but as we have argued, the resulting approach (considering per-play data, and thereby excluding quarterbacks who have not played in the NFL) is vulnerable to selection bias, which may be severe. In the case study, we chose to consider outcomes that are dependent on the amount of playing time a quarterback is given. We investigated the plausibility of the hypothesis that playing time is awarded to highly drafted quarterbacks without regard to performance by fitting models which excluded quarterbacks drafted in the first round, precisely those one would expect to benefit most from a policy of awarding opportunities based on draft position rather than merit. Neither the effect of draft position nor any of the other predictors we considered was appreciably different in the analyses of this subset of quarterbacks. Though this finding does not rule out the possibility that external factors influence playing time decisions, it suggests that the role of such factors may be exaggerated.

6.4 Discussion and Future Research

Our overall conclusion is that college and Combine statistics for drafted players are not reliably associated with, or strongly predictive of, success in the NFL. In our own analyses, college and Combine performance statistics provided very little additional predictive ability beyond year drafted and draft position. Indeed, in some cases, including college/Combine measurements actually degraded predictive performance, suggesting that the amount of statistical noise in these predictors overwhelms most of the predictive value they might have. Publicly available information and statistics do not reliably indicate which players are most likely to find success in the NFL. And while it appears that NFL teams may be able to use more closely guarded (or less easily quantifiable) information to discriminate between players who are likely to be successful in the NFL and those who are not, there remains substantial uncertainty in predicting the future performance of college prospects.

Several additional complicating factors that we have not addressed here make it difficult to characterize the association between college and future NFL performance. These include

- Teams sometimes draft for "need," selecting a player earlier or later than his talent level might dictate because of a dearth of NFL-level players on the roster at a given position. Drafting for need makes draft position a noisier measurement of NFL teams' assessment of a player's ability, which will tend to degrade the accuracy of models which use draft position as a predictor. However, this phenomenon is difficult to identify based on the available data and is not typically incorporated into analyses.

- Though Combine data availability is improving, players are being increasingly selective about the events they participate in, often choosing only the ones that showcase their strengths. Analyses using Combine measurements to predict NFL success are therefore vulnerable to the selection bias induced by this selective participation.

- The notion of player "success" is not static but changes as the game evolves. For example, the ability of a running back to shoulder a heavy workload has been de-emphasized in recent years as teams rotate players more frequently. Even if pre-draft information is predictive of "success" as defined at the time that a player is drafted, that same information is likely to be less predictive if the concept of "success" shifts in the later years of that player's career. Combining data across different eras compounds this problem.

There are many opportunities for sports statistics researchers to improve upon the state of knowledge in this area. Most analyses in the published literature, as well as those presented in our case study, consider a relatively small number of individual statistics which are easily tabulated. The effects of a player's environment are often neglected; future analyses might consider adjusting college performance statistics for conference, strength of schedule, and even offensive scheme.

NFL teams still rely heavily on "semi-quantitative" predictors including scouting reports and video analysis; the relative ability of these to predict future success is an open question. Furthermore, with player tracking technology improving rapidly, the set of possible predictors could be expanded substantially. Systems like Hit f/x, Pitch f/x, and Statcast in professional baseball are not yet widely deployed in college or professional football, but the ability to accurately track quantities like throw velocity, acceleration, and response time on a play-by-play basis in college games could allow for more reliable predictions of future NFL success.

However, given the poor predictive performance of models incorporating a variety of quantitative measures, it seems unlikely that collecting more statistics on the performance of college players will yield a dramatically more accurate assessment of their likelihood of success in the NFL. Though NFL draft "experts" at major sports networks may object, it appears that factors which are inherently unmeasurable and/or random play a major role in determining whether a quarterback will succeed at the professional level.

References

Agresti, A. *Categorical Data Analysis*, 2nd ed. Wiley-Interscience, Hoboken, NJ, 2002.

Alamar, B. and K. Goldner. The blindside project: Measuring the impact of individual offensive linemen. *Chance*, 24(4):25–29, 2011.

Alamar, B.C. and J. Weinstein-Gould. Isolating the effect of individual linemen on the passing game in the National Football League. *Journal of Quantitative Analysis in Sports*, 4(2), January 2008. doi: 10.2202/1559-0410.1113.

Berri, D., M. Schmidt, and S. Brook. *The Wages of Wins: Taking Measure of the Many Myths in Modern Sport*. Stanford University Press, Stanford, CA, 2006.

Berri, D.J. Back to back evaluation on the gridiron. In J.H. Albert and R.H. Koning, eds., *Statistical Thinking in Sport*. Chapman & Hall/CRC Press, Boca Raton, FL, 2007.

Berri, D.J. and R. Simmons. Catching a draft: On the process of selecting quarterbacks in the National Football League amateur draft. *Journal of Productivity Analysis*, 35(1):37–49, 2011.

Bialik, C. The count: The art and science of drafting QBs, December 2009. http://blogs.wsj.com/dailyfix/2009/12/30/the-count-the-art-and-science-of-drafting-qbs/. Date accessed: January 21, 2016.

Blees, C. Running backs in the NFL draft and NFL combine: Can performance be predicted. Claremont McKenna College Senior Theses. Paper 127, 2011. http://scholarship.claremont.edu/cmc_theses/127. Date accessed: January 21, 2016.

Boulier, B.L., H.O. Stekler, J. Coburn, and T. Rankins. Evaluating National Football League draft choices: The passing game. *International Journal of Forecasting*, 26(3):589–605, 2010.

Burke, B. Steven Pinker vs. Malcolm Gladwell and drafting QBs, 2010. http://archive.advancedfootballanalytics.com/2010/04/steven-pinker-vs-malcolm-gladwell-and.html. Date accessed: January 21, 2016.

Burke, B. Expected points and expected points added explained, 2014. http://www.advancedfootballanalytics.com/index.php/home/stats/stats-explained/expected-points-and-epa-explained. Date accessed: January 21, 2016.

Camerer, C.F. and R.A. Weber. The econometrics and behavioral economics of escalation of commitment: A re-examination of Staw and Hoang's NBA data. *Journal of Economic Behavior & Organization*, 39(1):59–82, 1999.

Dhar, A. Drafting NFL wide receivers: Hit or miss? Technical report, Department of Statistics, University of California at Berkeley, Berkeley, CA, 2011.

Efron, B. and G. Gong. A leisurely look at the bootstrap, the jackknife, and cross-validation. *The American Statistician*, 37(1):36–48, 1983.

Feldman, B. 'I was blown away': Welcome to football's quarterback revolution, 2015. http://www.foxsports.com/college-football/story/stanford-cardinal-nfl-virtual-reality-qb-training-031115. Date accessed: January 21, 2016.

Football Outsiders Inc. Football Outsiders Information ... because you need to know, 2015. http://www.footballoutsiders.com/info/methods. Date accessed: August 25, 2015.

Gerrard, B. Is the moneyball approach transferable to complex invasion team sports? *International Journal of Sport Finance*, 2(4):214, 2007.

Gladwell, M. Let's go to the tape, November 2009. http://www.nytimes.com/2009/11/29/books/review/Letters-t-LETSGOTOTHET_LETTERS.html. Date accessed: January 21, 2016.

Greene, W.H. Accounting for excess zeros and sample selection in Poisson and Negative Binomial regression models, March 1994. http://papers.ssrn.com/abstract=1293115. Date accessed: January 21, 2016.

Kuzmits, F.E. and A.J. Adams. The NFL combine: Does it predict performance in the National Football League? *Journal of Strength and Conditioning Research*, 22(6):1721–1727, 2008. Date accessed: January 21, 2016.

Loh, S. Virtual reality QB trainer a "game changer", May 2015. http://www.sandiegouniontribune. com/news/2015/may/21/virtual-reality-quarterback-trainer-derek-belch/. Date accessed: January 21, 2016.

Mulholland, J. and S.T. Jensen. Predicting the draft and career success of tight ends in the National Football League. *Journal of Quantitative Analysis in Sports*, 10(4):381–396, 2014.

National Football League. NFL quarterback rating formula, 2015. http://www.nfl.com/help/ quarterbackratingformula. Date accessed: August 25, 2015.

Oliver, D. Guide to the total quarterback rating, 2011. http://espn.go.com/nfl/story/_/id/6833215/ explaining-statistics-total-quarterback-rating. Date accessed: January 21, 2016.

Pasteur, R.D. and K. Cunningham-Rhodes. An expectation-based metric for NFL field goal kickers. *Journal of Quantitative Analysis in Sports*, 10(1):49–66, 2014.

Pinker, S. Malcolm Gladwell, eclectic detective, November 2009. http://www.nytimes.com/2009/ 11/15/books/review/Pinker-t.html. Date accessed: January 21, 2016.

Quinn, K.G., M. Geier, and A. Berkovitz. Passing on success? Productivity outcomes for quarterbacks chosen in the 1999–2004 National Football League player entry drafts. Technical Report 0711, International Association of Sports Economists, Dayton, Ohio, (07), 2007. http://ideas.repec. org/p/spe/wpaper/0711.html. Date accessed: January 21, 2016.

Schatz, A. Football's Hilbert problems. *Journal of Quantitative Analysis in Sports*, 1(1), January 2005. doi:10.2202/1559-0410.1010.

Schroeder, G. Virtual reality becomes a reality for college football, June 2015. http://www.usatoday. com/story/sports/ncaaf/2015/06/09/strivr-eon-football-virtual-reality-training-college- quarterbacks/28725797. Date accessed: January 21, 2016.

Schuckers, M. An alternative to the NFL draft pick value chart based upon player performance. *Journal of Quantitative Analysis in Sports*, 7(2), 2011.

Staw, B.M. Knee-deep in the big muddy: A study of escalating commitment to a chosen course of action. *Organizational Behavior and Human Performance*, 16(1):27–44, 1976.

Staw, B.M. and H. Hoang. Sunk costs in the NBA: Why draft order affects playing time and survival in professional basketball. *Administrative Science Quarterly*, 40(3):474–494, 1995.

Thorn, J. and P. Palmer. *The Hidden Game of Baseball: A Revolutionary Approach to Baseball and Its Statistics*. Doubleday, New York, NY, 1984.

Walsh, B., Billick, B., and Peterson, J.A. *Finding the Winning Edge*. Champaign, IL: Sagamore Publishing, 1997. https://hbr.org/1993/01/to-build-a-winning-team-an-interview-with-head- coach-bill-walsh.

Wikipedia. Passer rating, 2015. https://en.wikipedia.org/wiki/Passer_rating. Date accessed: August 25, 2015.

Wolfson, J., V. Addona, and R.H. Schmicker. The quarterback prediction problem: Forecasting the performance of college quarterbacks selected in the NFL draft. *Journal of Quantitative Analysis in Sports*, 7(3), 2011.

Yau, K.K.W., K. Wang, and A.H. Lee. zero-inflated negative binomial mixed regression modeling of over-dispersed count data with extra zeros. *Biometrical Journal*, 45(4):437–452, 2003.

7

Evaluation of Quarterbacks and Kickers

R. Drew Pasteur and John A. David

CONTENTS

7.1 Evaluation of Quarterbacks and Kickers

Detailed measures of player performance arose first in baseball, due to the sport's structure as a series of discrete events, most of which involve only one or two players on each team. The credit for a home run goes exclusively to the batter, and the blame solely to the pitcher; a ground ball out may involve the batter, pitcher, shortstop, and (perhaps minimally) the first baseman. While basketball is more structurally complex due to continuous free movement of the players, possessions can be considered discrete events and the result of a shot depends largely on the shooter and his primary defender. By comparison, it is more difficult to assign credit or blame in football, due to the interconnectedness of the players' actions. On a play-action pass, in which the offensive team fakes running the ball before throwing it, most of the 22 players on the field may have some nontrivial role in the outcome. Gerrard classifies football in the category of "invasion sports," those which "involve a group of players co-operating to move an object (e.g., a ball or puck) to a particular location defended by opponents (e.g., across a line or between goalposts)" (Gerrard, 2007). Among the complexities of such sports noted by Gerrard, the most relevant for football is that of multiple players working directly together for the same action: two players jointly blocking a single pass rusher, multiple tacklers bringing down a running back, or a cornerback and a safety jointly responsible for covering a receiver. Soccer is even more difficult to assess, given the free-flowing spatial movement of competitors, the multiple possible offensive and defensive actions by a given player (sometimes in rapid succession), and the lack of easily discretized events.

At some positions in football, for example, cornerbacks and offensive linemen, impressions regarding the relative quality of players are largely formed by subjective analysis of their actions on each play, typically after the game is over, by independent analysts (such as those at profootballfocus.com) and/or coaches. While such information from film analysis can be useful for all players, there are more objective measures available for some players. The actions of a quarterback (QB) in his role as passer, quickly and accurately delivering the ball to an open receiver, can be measured reasonably well by information contained in game-total or play-by-play statistics. Similarly, the performance of a kicker or punter can be sufficiently isolated from the actions of other players to allow for assessment using readily available game statistics. In this chapter, we will review the history of evaluating players at these positions, along with discussing current challenges and related open questions in quantifying their performance.

7.1.1 Evaluation of NFL Quarterbacks

7.1.1.1 History

Attempting to quantify the performance of NFL players, in particular quarterbacks (QBs), is not a new phenomenon. It began in 1932, as a tool to sell more tickets, by promoting that a visiting player was the league leader in some particular category (Carroll et al., 1989). For nearly the next 40 years, measuring QBs took various forms, and the interested reader is referred to Carroll et al. (1989) for more on the evolution of early quarterback measures. In the early 1970s, NFL commissioner Pete Rozelle tasked the league's statistical committee with standardizing the way official statistics were kept (Steinberg, 2001). In creating an overall quarterback measure that would come to be known as the "passer rating," this group looked at 13 years of data and decided to focus on four categories: percentage of completions, percentage of touchdowns, percentage of interceptions, and average gain per pass attempt. They then decided that average performance in each category would be worth one point, record-level performance two points, and very poor performance zero points. They then summed these points and normalized the scores, so that the average rating would be around 67, with a great season scoring near 100 or above.

The equation describing these ideas is

$$PR = \left(\frac{\frac{COMP}{ATT} - 0.3}{0.2} + \frac{\frac{YDS}{ATT} - 3}{4} + \frac{\frac{TD}{ATT} - 0.3}{0.05} + \frac{0.095 - \frac{INT}{ATT}}{0.4} \right) \frac{100}{6}$$

where
  PR is the passer rating
  COMP is the number of completions
  YDS is the number of passing yards
  TD is the number of passing touchdowns
  ATT is the number of passing attempts
  INT is the number of interceptions

There have been many criticisms of this metric, from both inside and outside the NFL, notably including von Dohlen (2011), Berri (2007), and Carroll et al. (1989), but perhaps Berri sums it up best: "In addition to being both complex and non-intuitive, the measure also ignores rushing attempts, rushing yards, sacks, yards lost from sacks, and fumbles lost. Furthermore, it is not clear that the factors that are included are actually weighted correctly in terms of team outcomes. In sum, this measure is quite complex, clearly incomplete, and of questionable accuracy" (Berri, 2007). There have since been several attempts to improve this system, each with its own advantages and disadvantages.

7.1.1.2 Wins Produced

Primarily based on the work of Gerrard (2007) and Berri (2007), the goal of a *wins produced* rating is to use multiple regressions to link QB-related statistics to the outcomes of games. Ultimately, based on these regressions, one can use a QB's stats to estimate the number of wins he produces per 100 offensive plays. This method uses a solid theoretical and statistical approach, especially when compared to passer rating. Additionally, it only requires either game or cumulative season statistics. However, it does suffer a few drawbacks. The first is that it assumes the QB deserves full credit for all statistics credited to them. The second is that it credits a given play equally, whether it occurred during a close game or late in a blowout.

7.1.1.3 Expected Points Added

Passing yards gained are significant in many QB rating systems; however, not all yards of field position are equally valuable. Most defenses will happily give up 13 yards on third and 15, and it is generally much easier to gain 5 yards at midfield than inside the opponent's 10-yard line. In general, the combination of down, distance, and field position can affect the value of a particular yardage gain. A concept that can help determine the value of a particular play relative to how many points it contributes is the idea of expected points (EPs), in which field position has value based on future scoring potential. In computing EP, consider the following example. If we take all first-quarter instances in which an NFL team has a first down at midfield, then record the change in score differential for the remainder of the half (which could be either positive or negative), then take an average over all such instances, we have estimated the value of having an offensive first and ten at midfield. The team with a first and ten at midfield will, on average, outscore their opponents by two points, so this situation is given +2.0 EP (Burke, 2009b).

In 1971, Carter and Machol (1971) were the first to publish this idea, using data from the first few weeks of the 1969 season to estimate EP values of first downs at all field positions. Carrol, Palmer, and Thorn constructed EP values from the 1986 NFL Season (Carroll et al., 1989). In 2009, the now-defunct web site Advanced NFL Stats (archived at advancedfootballanalytics.com), run by future ESPN analyst Brian Burke, published estimated EP values for all common down and distance situations. While EP values often have been used to help determine optimal decisions, for example, in Romer (2006), Carter and Machol (1978), and Burke (2009a), our focus is on the application of EP to QB rating. Consider the aforementioned first and ten at midfield, worth 2 EP. If the QB completes a 20-yard pass, his team will now have a first and ten at the opponent's 30-yard line, worth 3 EP. Assuming the QB is credited for this play, he has added 1 EP, which we would characterize as +1 expected point added (EPA). This statistic could be used to measure a QB's contribution either in absolute (total EPA over a season) or relative (EPA/play) terms. Unlike regression-based approaches, EPA helps determine how particular statistics and plays are contributing to points being scored, allowing for comparisons across positions. A downside to this measure is that it requires play-by-play data to compute, and it does not consider the game situation, with regard to time remaining and current score margin.

7.1.1.4 Success Rate

Success rate (SR) was originally proposed in 1989 by Carroll, Palmer, and Thorn (Carroll et al., 1989) and grades each play as a success or failure. A simple rule of thumb is used for offensive successes: on first down a gain of at least 4 yards, on second a gain of at least half the distance to go, and on third a conversion to first down. Advanced NFL Stats refined the idea to define a success as any play that improved a team's net point expectancy, that is, one with a positive EPA. This new definition allows for the metric to consider not only down and distance but also field position; for example, a gain on third down that does not result in a first down but does put the team into field goal range may be considered a success (Berri and Burke, 2012).

7.1.1.5 Win Probability Added

Win probability (WP) addresses the primary shortcoming of all of the aforementioned metrics, the fact that none considers the game situation, that is, the score margin and time remaining. Passing for many yards or touchdowns late in a one-sided game may improve a player's raw statistics but will not significantly improve the likelihood of the passer's team winning. The fundamental idea of WP is that based on score, field position, down, distance, and time remaining, it is possible (with a sufficiently large data set) to estimate the probability of a team winning, in any situation.

WP models are very common in baseball, due to the limited number of game states (inning, number of outs, and which bases are occupied) and huge amounts of historical data to estimate the probabilities. Berri and Burke estimate that "there are over one billion potential combinations of score, field position, down and distance and time remaining in an NFL football game" (Berri and Simmons, 2011), and there are fewer than 300 NFL games played annually, so building such models in football was more difficult. While there were other efforts to apply the WP concept, Advanced NFL Stats in 2008 built the first widely used, public WP model. It estimated first down probabilities using regression and estimated values for other downs from Markov model interpolations (Berri and Simmons, 2011).

While we will focus on using WP to rate QBs, it has many other applications, including the same ones as EP, but it also has been used to rank the most influential plays in NFL history (Healy, 2015). For example, consider a game in which professional oddsmakers consider the teams equally likely to win, with one team having first and ten at their own 20, with the score tied in the first quarter. In this situation, Pro Football Reference estimates that the offensive team has a 50.8% chance of winning. The QB completes a 10-yard pass, setting up first and ten from their own 30, increasing the chance of winning to 52.8%. This play added a 2% chance of winning to the offensive team. This addition is termed win probability added (WPA). One can sum the plays the QB was significantly involved in (passes, QB runs, etc.) and divide by this number of plays to estimate a QB's WPA per play.

7.1.1.6 Total Quarterback Rating

Total quarterback rating (QBR) is a proprietary rating system developed by ESPN in 2011. It is perhaps the most data-intensive and analyst-intensive metric discussed in this chapter. Dean Oliver described it as follows:

> [QBR is a] statistical measure that incorporates the contexts and details of those throws (throws in different down, distance, scoring margins, and contribution of players around the QB) and what they mean for wins. It's built from the team level down to the quarterback, where we understand first what each play means to the team, then give credit to the quarterback for what happened on that play based on what he contributed.
>
> **—Oliver (2011)**

It was presented at the *2012 MIT Sloan Sports Analytics Conference* and in a television special as a part of the network's "Year of the Quarterback" promotion. Based on ESPN's relationship with Football Outsiders, this metric has been calculated for all seasons from 2006 to the present. QBR can be thought of as starting from EPA, but mitigating its shortcoming of assuming a QB to be completely responsible for any statistic assigned to him, by incorporating the situational importance of a play using WPA, and dividing credit (or blame) on a play among multiple involved players.

WPA helps contextualize a play, in terms of understanding the potential influence over the game's outcome. Clearly, a fourth-quarter play in a close game has more impact on the outcome than one in which a team has built a four-touchdown lead. QBR uses this concept together with empirical data, to create a "clutch index" that describes whether a QB is producing EP in situations that significantly affect the outcome of the game. The use of such a "clutch index" and/or WPA does not come without challenges. Dean Oliver notes, "It is a nice intuitive concept, but it has strange consequences. Single plays can dominate the rest of the game" (Oliver, 2013b). For example, the Patriots went from 0.12 WP to a 0.99 WP in a single play at the end of Super Bowl XLIX (http://www.advancedfootballanalytics.com/). Thus, "clutch index" was originally added in order to "minimize the value of plays made in games already decided" (Oliver, 2013b). Reweighting was done in 2013 to further emphasize this goal (Oliver, 2013a).

Another key component of QBR builds on the work of Ben Alamar (Alamar and Weinstein-Gould, 2008), (Alamar et al., 2006) to divide the credit for QB-related statistics between the QB and his blockers and receivers. This is arguably the most complex part of this model, and perhaps its biggest contribution to evaluation of quarterbacks. Anyone who has collected data in any sport recognizes that there is a degree of subjectivity in many sports statistics. Assists in basketball are an excellent example; the definition of an assist is very vague, leading to a good deal of subjectivity in crediting assists. Tommy Craggs describes some such issues in telling the story of an ex-NBA scorekeeper (Craggs, 2009), while Blake Murphy found that, over six recent seasons, the home team's assist percentage was, on average, 2.7% higher than that of the visiting team (Murphy, 2013). Despite the potential subjectivity of particular statistics, to truly look at QB performance, particularly at the single-game level, one must understand whether receivers are dropping passes or the QB is over- or under-throwing receivers. QBR attempts to mitigate the subjective nature of certain statistics using strict guidelines for how their video trackers chart these items and each game is analyzed at least twice to reconcile inconsistencies (Oliver, 2013b).

Some important attributes of QBR are as follows:

- It was developed in consultation with analysts from ESPN and NFL Films, including Ron Jaworski and Trent Dilfer (Oliver, 2011).
- It considers all aspects of a QB's performance: passing, rushing, avoiding sacks, and penalties. Jason Lisk did an analysis to suggest that QBs are "more responsible for sacks than we believe" (Lisk, 2009), in one of many studies that suggest the importance of considering multiple facets of a QB's play.
- It does not adjust for strength of opponents (but the NCAA version of QBR does, Oliver, 2013a). This seems reasonable, because the variation in team strengths is much smaller at the professional level.
- The team with the higher QBR score wins the game 86% of the time, a higher percentage than the team committing fewer turnovers or the team with the higher Passer Rating.
- A 2013 update de-emphasized very clutch situations, so that a few key plays cannot overshadow game-long performance. Handling of a few very specific situations was also included in this revision (Oliver, 2013a).
- It was updated again in 2015, to consider the expected (rather than actual) length of an interception return, because the QB has little control over this. QBs are now credited for playing well under pressure, and the QB blame for a sack was decreased to give more fault to receivers who fail to get open. Credit for yards after catch is now based on expected values given the throw, decreasing the effect of wide receiver quality on a QB's rating (Katz, 2015).

QBR is a data-intensive, complicated, ambitious, and fascinating quantitative endeavor. Dean Oliver summed it up as follows:

> As we neared the end of the development of QBR, we talked to Ron Jaworski and Greg Cosell at NFL Films about its evolution. Cosell said at one point, 'Football is not complex, but it is very detailed.' I realized then that QBR is like that. It is very detailed, accounting for a lot of different situations, but it is not particularly complex. It really does try to see the game the way we have gotten used to seeing it in its elegant simplicity. We hope you, the fan, appreciate it, as well.

> **—Oliver (2011)**

7.1.1.7 Current Challenges

There have been significant improvements to QB statistics and ratings in recent decades. Each of the metrics described in this chapter has added understanding and insight into how we understand what is arguably the most important position on the football field. There are many ways forward in furthering this understanding, and we will discuss only a few. Regarding QBR, there remain several questions:

- Does QBR reward good QBs with bad defenses, in that they get more "clutch opportunities"? (Oliver, 2013b)
- Are there additional ways QBR can/should be updated, after recent adjustments?

- Is there a simpler approach that captures the same information as QBR, but without such daunting time and data requirements?
- QBR has been applied to NCAA QBs (Oliver, 2013a). Does this metric predict NFL success, and if so what can be learned from it?
- How predictive is QBR of other aspects of football, in particular wins? Chase Stuart (2014) has begun to look at this question, but significant opportunities for using this metric remain, given that nearly a decade of NFL QBR data are now available.
- The specific methods and intermediate data associated with QBR remain proprietary to ESPN. With the repeatability of many scientific findings (in a variety of fields) being questioned, there is value in independent validation of such results.

Other general QB rating issues include the following:

- Are there other paradigms for evaluating QBs? Wins Above Replacement (WAR), which is related to WPA, is a common baseball metric but is difficult to apply to the NFL, given the much smaller number of games per season. Hughes, Koedel, and Price used this metric to look at the positional WAR in the NFL, finding not surprisingly that QB is the most valuable position in the NFL (Hughes et al., 2015). Is there a way to leverage this methodology to learn more about the QB position or QBs in general?
- NFL QB success is difficult to predict (Berri and Simmons, 2011). Improvements in this area offer significant competitive and economic advantages for teams.

All of the methods listed take a top-down approach, that is, they use data and game situations in attempting to quantitatively describe what QBs do. An alternative approach is an agent-based modeling approach, attempting to characterize each of the underlying agents in a model, then running many simulations to see if these individual characterizations produce overall system-level performance that matches the overall real system. The NFL has a 4-billion-dollar (total franchise sales) agent-based model with roughly 5 million annual testers, the Madden video game franchise (Paine, 2015). Neil Paine details the intricate process by which all players are rated in 43 categories. As far as the authors know, this particular rating system has not been as quantitatively evaluated in terms of its predictive and descriptive ability in real games. These data and approach have a large amount of potential for understanding NFL QBs.

7.1.1.8 Analysis of Recent QBs by Current Metrics

In order to better illustrate the previously discussed metrics, we have compiled and analyzed the data for individual QBs since 2006, the first year for which QBR data are available. We used all QBs that ESPN.com lists for season-long QBR (usually the top 30) and all QBs that advancedfootballanalytics.com lists numbers for (usually the top 40 in WPA/G). This gave us a data set of 268 player-seasons. All techniques are tools with their own purposes and trade-offs (Berri and Burke, 2012). PR is simple, requiring only box score information but omits many key factors to being a QB, notably including rushing, and ignores any information about the game situation (down, field position, score margin, etc.). QBR considers many of the actions of a QB but requires not only play-by-play data but also analysis of each play (e.g., whether an incompletion was caused by an underthrow versus being dropped by

TABLE 7.1

Correlation between Various QB Metrics

	PR	QBR	WPA/G	EPA/P
QBR	0.86			
WPA/G	0.75	0.82		
EPA/P	0.87	0.88	0.88	
SR	0.76	0.81	0.78	0.87

the receiver). All metrics make some trade-offs in complexity, data availability, and game context sensitivity. Ultimately, these metrics correlate strongly with each other, as can be seen in Table 7.1. QBR tends to correlate most strongly with all other metrics, while PR and SR correlate most weakly with the others. The weakest correlations between any metric pairs are between PR and WPA/G and between PR and SR. These reduced correlations can be explained by differences in sensitivity to game context.

Typically, we expect that a metric that accurately assesses a skill will show some degree of consistency over time. Berri and Burke address this question for all of these metrics except QBR in (Berri and Burke, 2012). We considered all season pairs where a QB played for the same team in two consecutive seasons and computed the year-to-year correlation for each statistic. In line with the findings of Berri and Burke, we found that these correlations are not strong: 0.43 for PR, 0.46 for WPA/G, 0.53 for EPA/P, and 0.52 for QBR. We had hypothesized that QBR might have a higher correlation, given that one of its stated goals is to divide credit between the QB and other offensive players, but this was not the case for our data set. We believe that this returns us to the fundamental problem of evaluating the performance of a position so deeply linked to the performance of their teammates and opponents.

Another interesting line of analysis is to consider the top QB-seasons according to QBR, as they relate to age, winning percentage, and the other metrics, as can be seen in Table 7.2. All but two of the top 20 performances were by QBs aged 28 or older. Overall, there is a positive correlation of 0.26 between QBR and age, which may reflect a survivorship bias, but also the considerable time it takes to develop an elite NFL QB. Additionally, all but one of the top 20 performances occurred as part of a team having a winning season. Among the metrics considered, QBR has the highest correlation with win percentage, 0.64.

It can be argued that in the era of QBR data (since 2006), the elite NFL quarterbacks have been Peyton Manning, Tom Brady, Aaron Rodgers, Drew Brees, Ben Roethlisberger, Tony Romo, Eli Manning, and Phillip Rivers. The seasonal QBR ratings for these eight QBs are shown in Table 7.3. By QBR, Peyton Manning has consistently been the best QB in the NFL, followed by the group of Brady, Rodgers, and Brees, all of whom have been comparable, except during Rodgers QBR-record-setting 2011 season. The next tier appears to be Romo, Roethlisberger, and Rivers. As Eli Manning and Philip Rivers were traded for each other on the day both were drafted, it is interesting to compare their career statistics, and it could be argued that San Diego got the better of this deal.

Finally, we compared each of these eight QBs' career average ranking in each category in Table 7.4 and found these rankings to be fairly consistent with one another across all the metrics, a result that is not surprising given the strong positive correlations among these metrics. Peyton Manning led three categories and Aaron Rodgers the remaining two, with the other QBs in nearly the same order for all metrics, over the last nine seasons.

TABLE 7.2

Top 20 Performances by QBR

Player	Team	Age	Year	QBR	PR	Win (%)	WPA/G	EPA/P	SR (%)
Rodgers	GB	28	2011	86.2	122.5	0.93	0.39	0.40	55.5
Manning, P	IND	30	2006	85.4	101.0	0.75	0.36	0.31	57.2
Brady	NE	30	2007	84.5	117.2	1.00	0.39	0.39	58.7
Brees	NO	32	2011	84.0	110.6	0.81	0.37	0.31	56.3
Romo	DAL	34	2014	83.6	113.2	0.80	0.28	0.28	53.7
Manning, P	DEN	36	2012	83.2	105.8	0.81	0.24	0.27	53.6
Manning, P	IND	33	2009	81.2	99.9	0.88	0.38	0.30	54.9
Manning, P	DEN	37	2013	81.0	115.1	0.81	0.36	0.33	55.4
Manning, P	IND	32	2008	79.0	95.0	0.75	0.31	0.24	51.5
Rodgers	GB	31	2014	77.4	112.2	0.75	0.32	0.34	53.6
Brady	NE	35	2012	76.7	98.7	0.75	0.29	0.27	55.3
Rivers	SD	28	2009	76.1	104.4	0.81	0.32	0.33	52.7
Griffin	WAS	22	2012	75.6	102.4	0.60	0.21	0.23	50.4
Brees	NO	30	2009	75.5	109.6	0.87	0.27	0.25	54.5
Manning, P	DEN	38	2014	75.1	101.5	0.75	0.25	0.20	51.3
Rivers	SD	32	2013	74.9	105.5	0.56	0.27	0.23	53.7
Brady	NE	37	2014	74.6	97.4	0.75	0.17	0.20	50.3
Foles	PHI	24	2013	73.8	119.2	0.80	0.17	0.30	48.0
Rodgers	GB	29	2012	73.5	108.0	0.69	0.31	0.22	53.2
Brees	NO	35	2014	73.4	97.0	0.44	0.18	0.19	53.3

7.1.2 Evaluation of Kickers and Punters

7.1.2.1 Introduction to Placekicking

Compared with quarterbacks, the performance of NFL placekickers is even more isolated. A kickoff depends solely on one player, although if the kick is returned, then that portion of the play involves many players. The success of field goal attempts depends primarily on the kicker, although a small percentage will be hindered or abandoned because of late/poor placement of the ball by the holder, or due to a defensive player quickly breaching the kicking team's protection scheme. In the vast majority of field goal (and point after touchdown, or PAT) attempts and all kickoffs, credit or blame for the outcome can be attributed fully to the kicker. The failure rate for PAT attempts, which were equivalent to very short field goals until a 2015 rule change, was routinely below 1% in the seasons leading up to the change, at which point the distance was substantially increased.

The performance of kickers has improved dramatically over the last several decades (Stapleton, 2010), (Pasteur and Cunningham-Rhoads, 2014). Jan Stenerud, the only player enshrined in the Pro Football Hall of Fame solely as a placekicker, converted 67% of his field goals in a career that spanned from 1967 to 1985, and 27% of attempts from 50 yards are more. By comparison, in the 2014 season, 84% of field goal attempts league-wide were converted, including 61% of those from 50+ yards, based on data from pro-football-reference.com. Possible reasons for the improvement include soccer-style techniques, early specialization, strength training, and better turf quality.

Traditional measurements for a kicker's contribution include the number of field goals made or points scored (a team gains three points for a field goal and one for a PAT), along

TABLE 7.3

Comparison of Selected QBR Ratings by QB Age, 2006–2014

Avg.	Age	Manning, P	Brady	Rodgers	Brees	Roethlisberger	Romo	Manning, E	Rivers
46	24					46			
59	25			61		69		44	62
51	26			67		39	66	39	45
65	27			62	67	65	67	60	67
66	28			86	61	62	48	62	76
62	29		58	74	64	63	58	56	63
71	30	85	85	62	76	69		59	64
64	31	73		77	64	57	66	66	46
67	32	79	65		84	72	60	34	75
69	33	81	72		65		59	66	71
74	34	73	68		73		84		
75	35		77		73				
72	36	83	60						
78	37	81	75						
75	38	75							
	Avg	79	70	70	70	60	63	54	63
	Rank	1	3	2	4	7	5	8	6

Note: Darker shading indicates a higher QBR value. The upward trend in QBR with age may reflect both survivorship bias (only very good QBs are still playing by age 35) and a steep learning curve for young QBs. At the bottom of the chart is each QB's average QBR over the included seasons. Note that QBR is designed so that 50 is average.

TABLE 7.4

Career Rankings in Various Categories

	PR	QBR	WPA/G	EPA/P	SR	Avg.
Manning, P	2	1	2	1	1	1.4
Rodgers	1	2	1	2	4	2.0
Brady	3	3	3	3	2	2.8
Brees	4	4	4	4	3	3.8
Romo	5	5	5	6	5	5.2
Rivers	6	6	7	5	6	6.0
Roethlisberger	7	7	6	7	7	6.8
Manning, E	8	8	8	8	8	8.0

with the percentage of a kicker's attempts that succeeded (field goal percentage, denoted FG%). As in basketball, countable volume-of-productivity measures seem to influence perception of top players, including the kickers who appear on NFL "all-star" teams (Pasteur and Cunningham-Rhoads, 2014). However, because the number of attempts varies considerably among the league's kickers, FG% better quantifies performance and has long been a common metric for kicker quality.

While weather conditions, situational factors, etc. could have an influence on the difficulty of a field goal attempt, the key factor is distance. The longer a kick attempt, the more difficult it is to convert it. Recording the distance of field goal attempts has been universal only since the 1970 season.

7.1.2.2 Review of Kicking Literature

In 1985, Berry and Berry (1985) published the first related paper, fitting a logistic model to the distances of each player's made and missed attempts, including PAT attempts as 20-yard kicks. The logistic models are used to create two metrics by which kickers are compared: the estimated distance at which a kicker would make half of his attempts, and the probability of making a generic field goal (i.e., one randomly selected from the league-wide distribution of distances.)

With play-by-play accounts of games increasingly available by the 1990s, Bilder and Loughin (1998) constructed a more comprehensive logistic model, seeking to determine which factors affected kick success. Among those considered were variables related to weather (temperature, wind speed, precipitation), the stadium (domed or not, playing surface, altitude), and several regarding the game situation. Their final model included distance and three binary variables, representing windy conditions (15+ mph), the kick being a PAT (rather than a field goal), and whether the kicker's team could take the lead with a successful kick (possibly indicative of psychological pressure.) One additional variable, the time remaining in the half, was also found to be significant yet was not included in the final model. Interestingly, several factors commonly believed to be important were not determined to be statistically significant in this study: kicking in a domed stadium, on grass (versus artificial turf), precipitation, and altitude. The last two of these may have been affected by small sample sizes for the wet and high-altitude cases, respectively.

Using data from the 2002 and 2003 seasons, Berry and Wood (2004) considered distance (in a slightly different manner than the aforementioned papers) plus nine binary variables. Several factors were found to be individually significant: domed stadium, precipitation, wind (15+ mph), and grass surface. The authors also constructed a full model with distance plus all nine factors. Kicking performance was found to be overall unaffected in pressure situations, defined as attempts in the last 3 min (or in overtime) when the kicking team could tie the score or take the lead. However, if the opposing team called a time-out before the kick in such pressure situations (in an attempt to make the kicker become nervous while waiting for play to resume, a common tactic known as "icing" the kicker), there was evidence, despite a small sample size, that the kicker was less likely to successfully convert the field goal.

Subsequent work by Goldschmeid et al. (2010) used six seasons of data to more deeply investigate high-pressure situations, using a narrower set of circumstances (attempts in the last 1 min or overtime, with the potential to tie/lead/win.) Using a hierarchical linear model, and controlling for distance, this study found stronger evidence that icing does indeed negatively affect kickers. However, there was no such effect if the kicker's team called a time-out before a pressure kick. Because an icing time-out typically is called just seconds before a kick would otherwise be attempted, it is theorized that the kicker is already mentally into his pre-kick routine at that point and must redo his preparation after the delay; by comparison, a time-out called by the kicker's team typically would occur much sooner, usually shortly after the previous play ends.

The work of Clark et al. (2013) was a major step forward, using a larger data set of nearly 12,000 field goal attempts over a dozen seasons (2000–2011). Along with distance, six binary variables for environmental conditions were studied: cold (below 50°F), altitude (above 4000 ft, i.e., Denver vs. all others), surface (turf vs. grass), precipitation, wind (10+ mph), and humid (60%+ relative humidity); all except for humidity were highly significant. Four other situational factors were considered: post-season vs. regular season, pressure situations (via multiple measures), home vs. away, and an "icing" time-out before the kick, but none of these was found to be significant. The authors then combined the significant factors into a full model, by which the probability of an average NFL kicker making a particular kick could be estimated. With such a model, a player's aggregate performance can be compared to the sum of the probabilities, giving a quality measurement that takes into account the distances, weather conditions, etc. of a kicker's attempts. In line with contemporary analysis of quarterbacks and players at other positions, the authors used the number of expected points added (either per attempt or the sum total over a season) as a unit for measuring kicker performance. In applying the model to stadiums, the toughest cities in which to kick were Green Bay, Pittsburgh, Cleveland, and Chicago, all outdoor venues known for cold, windy conditions; by comparison, Denver would be the easiest, followed by the domes.

Concurrently to Clark et al. (2013), Pasteur and Cunningham-Rhoads (2014) constructed a similar model for three seasons of field goal attempts (2008–2010), but with hourly weather data and several continuous variables, including temperature, wind speed, precipitation rate, score margin, etc. Stadium orientation was used to estimate the crosswind and headwind/tailwind speeds, but information to distinguish headwinds from tailwinds was unavailable. In factor selection, the authors used a cross-validation technique (Picard and Cook, 1984) to find the predictive error with and without particular variables, rather than a Wald statistic (Wald, 1943). The final model included distance, temperature, a quadratic wind speed measure, defense quality (based on points per game allowed), and binary variables for Denver (high altitude), playoff vs. regular season, starting kicker vs. replacement (distinguished based on the number of attempts for the season), and fatigue (4+ prior field goal attempts in the game.) The effects of these factors are noted to effectively lengthen or shorten the distance of a kick (e.g., 3 yards shorter in Denver, 7 yards longer with a 25 mph crosswind, etc.). Lists of the toughest and easiest venues in which to kick are very similar to those of (Clark et al., 2013). The version of expected points added uses a hypothetical "replacement-level" kicker as a baseline, as opposed to an average kicker in Clark's work, but both single out Rob Bironas over multiple seasons and Sebastian Janikowski in 2009 as top performers. The authors note that kickers selected to league-wide "all-star" teams in the seasons considered were often mediocre by EPA metrics but made many field goals and played on winning teams. Season-to-season correlations of individual kickers' EPA were near zero, casting doubt on whether differences among NFL kickers are significant; see Petti (2011) and section 8.4 of Berri and Burke (2012) regarding this concept. Differences in the EPA contributions of high-salary and near-minimum salary kickers were found to be small.

7.1.2.3 Contemporary Questions Regarding Kicking

The large sample size in Clark et al. (2013) makes that study rather authoritative regarding the importance of temperature, altitude, precipitation, and playing surface. However, there remain significant questions regarding these factors. Does the observed temperature dependence reflect only a negative impact of cold conditions, or are hot conditions beneficial? Is the effect nonlinear across the range of temperatures observed? Are there ways to better measure the effects of altitude in Denver, given that there are fewer than 50 field goal attempts there annually, and half of those are taken by the same kicker, who might be better or worse than average? Regarding wind, knowing at which goalpost a player attempted a kick, together with wind direction, could allow for determining whether the wind was in the kicker's face or at his back, offering hope for new insight into the magnitude of headwind/tailwind effects. Along the same lines, are there ways to study air flow patterns inside stadiums (perhaps including domes), which might be substantially different from those recorded at an airport miles away? To what degree does precipitation that fell before the game have an impact? Does the rate or type (rain versus snow) of precipitation during the game matter? Is the turf quality of a natural grass field different early in the season than in a December or January game, to the point that it affects kicking success? Factors such as fatigue, defense quality, and pressure may warrant further investigation, as well.

Given the steady increase in kicking quality over the last few decades, a long-term study with a non-constant baseline measure and even a few key environmental factors would be interesting to spotlight the best field goal kickers across eras. Further economic analysis and study of season-to-season consistency also would enrich the literature. None of the studies to this point consider the value of field position losses associated with missed kicks; in such cases, the defense takes possession at the spot of the kick (not the previous line of scrimmage), provided that the spot is outside of the 20-yard line. This is quite different than for a punt that goes into or beyond the end zone (which results in a touchback, if not returned) and discourages long field goal attempts, except at the end of a half.

Strength and accuracy are often conceptualized as being separate desirable qualities in a kicker; over a career, can these be characterized, based on over- or underperformance in various distance ranges? Knowledge about whether missed kicks were short or wide would be helpful. In an era of automated optical tracking in professional sports, data about the initial speed and angle of the kicked football might soon become available and could shed light on this issue.

Proposed or enacted rule changes may lead to other opportunities for timely analysis. For the 2015 season, the line of scrimmage for PAT kicks was moved back from the 2-yard line to the 15-yard line, effectively lengthening PAT attempts from 20 to 33 yards. Kicks from this longer distance have been successful more than 90% of the time in recent years; however, this may still encourage more two-point conversion attempts (by run or pass from the two-yard line), which succeed about half the time (145 of 289 were converted during the 2010–2014 seasons). During the Pro Bowl following the 2014 season, substantially narrower uprights (14 ft apart) were used as an experiment, along with the longer PAT distance, and veteran kicker Adam Vinateri missed two 33-yard PATs. It might be possible to construct a theoretical model of field goal attempts on narrower uprights, estimating the impact if they were adopted.

Converting field goals and PATs is only half of the job of an NFL placekicker. In most cases, the same player also takes a team's kickoffs, after each score and at the start of a half. Except in unusual circumstances, the goal on a kickoff is simply to kick the ball far enough that it cannot be returned, forcing a touchback (in which case the receiving team takes possession of the ball on the 20-yard line.) Strength is far more important than accuracy; a kickoff needs only to stay on 160-ft-wide field, compared to splitting uprights 18.6 feet apart on a field goal attempt. From 1994 to 2010, kickoffs were taken from the (kicking team's own) 30-yard line, that is, 70 yards from the receiving team's goal line. At least partially due to the frequency of injuries in high-speed collisions when blocking for kick returners (Vida, 2011), kickoffs were moved up to the 35-yard line in 2011 (as previously was the case from 1974 to 1993). With kickoffs moved up 5 yards, the percentage resulting in touchbacks increased from 17% in 2010 to 45% in 2011, according to pro-football-reference.com, with the percentage returned dropping accordingly. The average line of scrimmage after kickoffs moved back from the 26.5-yard line to the 22.2-yard line with the rule change, based on offensive drive stats from Football Outsiders. Kacsmar (2014) used data through 2013 to study the longer-term impacts, including possibly unintended consequences for offensive production and game strategy.

7.1.2.4 Evaluation of Punters

Statistics for punters are readily available, with the traditional measure being (gross) punt yardage, the downfield distance from the line of scrimmage that the ball travels before being fielded, rolling to a stop, or going out of bounds. This does not include the roughly 15 yards the ball has to travel to reach the line of scrimmage, as the punter stands well behind the line. When possible, the receiving team will try to return a punt, and potential tacklers are later getting downfield than on a kickoff because of the need to protect the punter, avoiding a blocked punt. For this reason, it is preferable to punt the ball high (with a long *hang time* in the air), to give teammates the opportunity to get near the returner before he catches the ball. A low, "line drive" punt may go farther but also will allow a returner the chance to be running at full speed before tacklers reach him, increasing the chance of a long return, so this type of punt is rarely used. Net punting yardage is computed by subtracting any return yardage from the punt distance and is considered a better measure of a punter.

If a punt goes into the end zone and is not returned, then a touchback occurs, effectively reducing the punt distance by 20 yards. To avoid this, punters will aim to land the ball around the 5- or 10-yard line, when in range to do so, intentionally reducing the length of the punt. Punters playing for good teams more frequently find themselves in such situations, kicking from near-midfield or beyond, as opposed to deep in their own territory, so the highest gross punting averages tend to come from players on weak teams. If the net result of a punt is that the opponent takes over inside their own 20-yard line, this is considered a good punt, and the number (or percentage) of such kicks is another measure of a punter; this one favors punters on good teams.

Burke (2010) illustrated that net punting distance decreases as field position improves (i.e., as the maximum distance without a touchback decreases). The effect becomes substantial at roughly the punting team's 45-yard line (55 yards from a touchback). Any advanced metric for punters would need to take into account the field position from which punts are taken. Because there is a nonlinear relationship between field position and net punting distance, some information would be lost if only a punter's average field position was available. The marginal value of a yard of field position is not constant as a team moves down the field; this is reflected in the Expected Points computations (on first down) of Romer (2006) and Burke (2009b). Tymins (2014) took this into account, when studying a limited subset of punts from several recent seasons, using EPA as a natural unit for comparison with players at other positions.

In analyzing decades of Pro Bowl selections, Lisk (2008) found that players from teams with below-average offenses are overrepresented (perhaps because of reliance on gross punting averages) and those from cold-weather cities substantially underrepresented, indicating that environmental conditions may have an impact. Using data from 2002 to 2011, Beouy (2012) documented punter improvement in recent years, so a punting metric applied across many seasons would need to use a non-constant basis for comparison.

Along with field position and temperature, other environmental factors (e.g., wind, altitude) also might be considered in a comprehensive model. Situational factors (score, time, playoffs, etc.) could also be studied, perhaps along with measures of the coverage (tacklers) and return (punt returner and blockers) units. The methodology necessarily would be different than for field goals, because of the continuously valued outcome of a punt, as opposed to the binary outcome variable for field goals. Along with evaluation of individual punters (and analysis of their salaries), such a model could be applied to fourth down decision theory, along the lines of the New York Times 4th Down Bot (The New York Times, 2015), which was an outgrowth of the results from Burke (2009a). Even more than for field goals, future availability of spatial-temporal data from automated cameras has the potential to transform analysis of punts and punt returns.

7.1.3 Conclusion

Bradbury notes that performance metrics should correlate with winning, measure skill as opposed to luck, and be intuitively reasonable (Bradbury, 2011). The metrics described in this chapter fit the first and third of these criteria, by rewarding scoring plays and (for quarterbacks) advancement of the ball downfield, while penalizing turnovers and missed opportunities. A typical measure of the second criterion is season-to-season correlation of a metric for the same individual. Because true ability is somewhat stable over time, while luck is not, measures of skill tend to have significant positive correlations across consecutive seasons, but measures influenced heavily by random effects typically have near-zero correlations (Petti, 2011).

The interconnectedness of players' actions in team sports raises challenges to objectively rating the performance of individual athletes. However, the increasing availability of detailed play-by-play statistics has made it easier to better characterize individual performance, and high-frequency data gathered from multiple fixed cameras promise much more insight. Even with the most detailed data, the nature of American football has components that defy simple classification, such as two-on-one blocks, zone pass coverage, and tackles involving multiple defenders. For these reasons, there remains a place for film analysis by experts in a particular sport (or even experts in a particular aspect of a sport) as an assessment tool. The frequency (including over- or under-use) and game situations of a player's opportunities often need to be considered as well. In contrast to other positions, the discrete and relatively isolated nature of a placekicker's job lends itself well to assessment based on play-by-play information, and QB play is so central to a team's offense that it can be reasonably well evaluated by the team's offensive performance.

References

Alamar, B., Ma, J., Desjardins, G. M., and Ruprecht, L. (2006). Who controls the plate? Isolating the pitcher/batter subgame. *Journal of Quantitative Analysis in Sports*, 2(3).

Alamar, B. C. and Weinstein-Gould, J. (2008). Isolating the effect of individual linemen on the passing game in the National Football League. *Journal of Quantitative Analysis in Sports*, 4(2).

Beouy, M. (2012). Are punters getting better? http://community.advancednflstats.com/2012/11/are-punters-getting-better.html. Accessed: September 29, 2015.

Berri, D. J. (2007). Back to back evaluations on the gridiron. In: *Statistical Thinking in Sports*, Albert, J. and Koning, R. H. (eds.), pp. 241–262. CRC Press, Boca Raton, FL.

Berri, D. J. and Burke, B. (2012). Measuring productivity of NFL players. In: *The Economics of the National Football League*, Quinn, K.G. (ed.), pp. 137–158. Springer, New York, NY.

Berri, D. J. and Simmons, R. (2011). Catching a draft: On the process of selecting quarterbacks in the National Football League amateur draft. *Journal of Productivity Analysis*, 35(1):37–49.

Berry, D. A. and Berry, T. D. (1985). The probability of a field goal: Rating kickers. *The American Statistician*, 39(2):152–155.

Berry, S. M. and Wood, C. (2004). A statistician reads the sports pages: The cold-foot effect. *Chance*, 17(4):47–51.

Bilder, C. R. and Loughin, T. M. (1998). "It's good!" An analysis of the probability of success for placekicks. *Chance*, 11(2):20–30.

Bradbury, J. (2011). *Hot Stove Economics*. Copernicus Books, New York.

Burke, B. (2009a). The fourth down study—Part 2. http://archive.advancedfootballanalytics.com/2009/09/4th-down-study-part-2.html. Accessed: September 29, 2015.

Burke, B. (2009b). Expected points and expected points added explained. http://archive.advancedfootballanalytics.com/2009/12/expected-point-values.html. Accessed: September 29, 2015.

Burke, B. (2010). Shane Lechler is overrated...or is he? http://archive.advancedfootballanalytics.com/2010/01/shane-lechler-is-overratedor-is-he.html. Accessed: September 29, 2015.

Carroll, B., Palmer, P., and Thorn, J. (1989). *The Hidden Game of Football*. Warner Books, New York.

Carter, V. and Machol, R. E. (1971). Operations research on football. *Operations Research*, 19(2):541–544.

Carter, V. and Machol, R. E. (1978). Optimal strategies on fourth down. *Management Science*, 24(16):1758–1762.

Clark, T., Johnson, A., and Stimpson, A. (2013). Going for three: Predicting the likelihood of field goal success with logistic regression. http://www.sloansportsconference.com/?p=10200. Accessed: September 29, 2015.

Craggs, T. (2009). The confessions of an NBA scorekeeper. http://deadspin.com/5345287/the-confessions-of-an-nba-scorekeeper. Accessed: September 29, 2015.

Gerrard, B. (2007). Is the Moneyball approach transferable to complex invasion team sports? *International Journal of Sport Finance*, 2(4):214.

Goldschmeid, N., Nankin, M., and Cafri, G. (2010). Pressure kicks in the NFL: An archival exploration into the deployment of timeouts and other environmental correlates. *Sport Psychologist*, 24(3):300.

Healy, A. (2015). One play away: The updated list of the most influential plays in NFL history. http://www.footballperspective.com/one-play-away-the-updated-list-of-the-most-influential-plays-in-nfl-history/. Accessed: September 29, 2015.

Hughes, A., Koedel, C., and Price, J. (2015). Positional WAR in the National Football League. *Journal of Sports Economics*, 16(6):597–613.

Kacsmar, S. (2014). Impact of the NFL's kickoff rule change. http://www.footballoutsiders.com/stat-analysis/2014/impact-nfls-kickoff-rule-change. Accessed: September 29, 2015.

Katz, S. (2015). Total QBR updates for 2015. http://espn.go.com/nfl/story/_/id/13302797/total-qbr-updates-2015-nfl. Accessed: September 29, 2015.

Lisk, J. (2008). Pro Bowl punters. http://www.pro-football-reference.com/blog/?p=563. Accessed: September 29, 2015.

Lisk, J. (2009). What quarterback rate stats stay most consistent when a quarterback changes teams. http://www.pro-football-reference.com/blog/?p=4152. Accessed: September 29, 2015.

Murphy, B. (2013). Introducing arena-adjusted assists. http://www.hoopdata.com/blogengine/post/2013/01/06/Introducing-Arena-Adjusted-Assists.aspx. Accessed: September 29, 2015.

Oliver, D. (2011). Total QBR FAQ. http://espn.go.com/nfl/story/_/id/6909058/nfl-total-qbr-faq. Accessed: September 29, 2015.

Oliver, D. (2013a). Total QBR updates for 2013. http://espn.go.com/nfl/story/_/id/9634173/nfl-total-qbr-updates-2013. Accessed: September 29, 2015.

Oliver, D. (2013b). Understanding Total QBR. http://espn.go.com/college-football/story/_/id/9612585/total-quarterback-rating-college-football. Accessed: September 29, 2015.

Paine, N. (2015). How Madden ratings are made: The process that turns NFL players into digital gods. http://fivethirtyeight.com/features/madden/. Accessed: September 29, 2015.

Pasteur, R. D. and Cunningham-Rhoads, K. (2014). An expectation-based metric for NFL field goal kickers. *Journal of Quantitative Analysis in Sports*, 10(1):49–66.

Petti, B. (2011). What hitting metrics correlate year-to-year? http://www.beyondtheboxscore.com/2011/9/1/2393318/what-hitting-metrics-are-consistent-year-to-year. Accessed: September 29, 2015.

Picard, R. R. and Cook, R. D. (1984). Cross-validation of regression models. *Journal of the American Statistical Association*, 79(387):575–583.

Romer, D. (2006). Do firms maximize? Evidence from professional football. *Journal of Political Economy*, 114(2):340–365.

Stapleton, A. (2010). NFL kickers dramatically improved over the years. http://www.washingtontimes.com/news/2010/nov/12/nfl-kickers-dramatically-improved-over-the-years/?page=all. Accessed: September 29, 2015.

Steinberg, D. (2001). How I learned to stop worrying and love the bomb: A survival guide to the NFL's quarterback rating system. http://www.bluedonut.com/qbrating.htm. Accessed: September 29, 2015.

Stuart, C. (2014). Is ESPN's QBR the best measure of quarterback play? http://www.footballperspective.com/is-espns-qbr-the-best-measure-of-quarterback-play/. Accessed: September 29, 2015.

The New York Times. (2015). 4th down bot. http://nyt4thdownbot.com/. Accessed: September 29, 2015.

Tymins, A. (2014). Assessing the value of NFL punters. http://harvardsportsanalysis.org/2014/10/assessing-the-value-of-nfl-punters/. Accessed: September 29, 2015.

Vida, J. (2011). Historical, statistical look at kickoffs. http://espn.go.com/blog/statsinfo/post/_/id/17417/historical-statistical-look-at-kickoffs. Accessed: September 29, 2015.

von Dohlen, P. (2011). Tweaking the NFL's quarterback passer rating for better results. *Journal of Quantitative Analysis in Sports*, 7(3):1–14.

Wald, A. (1943). Tests of statistical hypotheses concerning several parameters when the number of observations is large. *Transactions of the American Mathematical Society*, 54(3):426–482.

8

Situational Success: Evaluating Decision-Making in Football

Keith Goldner

CONTENTS

Up 34-28 with just over 2 minutes remaining, Bill Belichick—arguably the best NFL coach in the modern era—calls a time-out after an incompletion targeted at slot-receiver Wes Welker. It is fourth down, 2 yards-to-go from the New England 28-yard line. Any normal coach would punt the ball. Let your defense hold on to your six-point lead.

Standing on the far sideline is no ordinary player. Peyton Manning, 6′5″, laser rocket arm, is just itching to get one last chance. The Indianapolis Colts are the best team in football with an undefeated 8-0 record coming into the game. Do you really want to give that high-powered offense an opportunity to score the game-winning touchdown?

Belichick decides to gamble. Lined up in shotgun formation with three receivers to the left and Randy Moss split out to the right, running back Kevin Faulk motions out into the slot between Moss and the right tackle. Indy shows blitz and sends five guys after Tom Brady. Brady lets the ball go quickly, hitting Kevin Faulk on a quick out route. Melvin Bullitt meets Faulk immediately, driving him back behind the line-to-gain. But, rather than being given forward progress, the official marks Faulk down just before the first down due to his bobbling the ball. Turnover on downs.

It does not take Peyton Manning long to find the end zone with such good field position, hooking up with Reggie Wayne four plays later. The Colts win 35-34 en route to a 14-0 start and ultimately a Super Bowl loss to the New Orleans Saints.

Belichick surely gambled, but did he make a bad decision? Because the outcome was unsuccessful, does that make the play call the wrong choice? How can we effectively measure the success of a play, and is that the same as grading the success of the decision?

Football games are best broken down into drives, which are made up of plays. Drives consist of one team on offense, the other on defense, with special teams tying it all together. The essential building block of a football game is the play. Each play is a transition between game states. A 7-yard gain on 2nd-and-10 is the transition from 2nd-and-10 to 3rd-and-3. States can be defined using down, distance-to-go, yard line, score, time remaining, and more, but the actual play—the in-game action—is always a transition between states.

In this chapter, we will start with the history of situational football analytics and how innovation has led to improved decision-making. We then dive into the primary metrics used for evaluating plays, players, and teams—expected points, net expected points (NEP), and win probability. In examining the existing models, we will look at Markov chains in particular. Finally, we will discuss how to use these metrics for smarter decision-making in football.

8.1 History of Situational Football Analytics

Brigham Young has a long line of historically great quarterbacks: Jim McMahon, Steve Young, and Ty Detmer to name a few. But it is a relatively unknown quarterback who started this chain of greatness: Virgil Carter. Carter set 24 school records and 6 NCAA records as an academic all-American. He also helped Brigham Young win their first Western Athletic Conference title in 1965. Carter also holds the distinction of being the first quarterback to ever run Bill Walsh's West Coast Offense (Barras 2013).

After Greg Cook suffered a shoulder injury, Carter was the next man up for the Cincinnati Bengals in 1970. Carter did not quite have the arm of Cook, but he was smart and nimble. Walsh had to come up with a system in which Carter could be successful. So, he came up with a quick-hitting, precision passing attack that stretched the field horizontally, rather than vertically, to compensate for his quarterback. Before Carter helped set in motion one of the great developments in the history in football, he actually started another field: Football Analytics.

Carter studied statistics while he was a quarterback for the Brigham Young Cougars, and in 1967 he was selected in the sixth round at 142 overall by the Chicago Bears. In Carter's first off-season, the Bears ownership said they would subsidize an MBA from Northwestern University if he interned with the Bears front office. It took Carter 3 years to complete a 2-year program due to his football career, but he finished with a quantitative focus. One professor with a particular quantitative focus, Dr. Robert Machol, taught classes only during the Fall quarter (when Carter was in season). Carter and Machol worked on a 2-year project to analyze a football game, ultimately coming up with something called an Expected Points model (Levin 2003).

The paper, *Operations Research on Football*, looked at "8373 plays from the first 56 games of the 1969 schedule." The analysis was fairly simple: look at every play and determine

TABLE 8.1

Expected Point Values from Virgil Carter's *Operations Research on Football*

Yards from Opponent Goal Line	Expected Point Value
95	−1.245
85	−0.637
75	+0.236
65	+0.923
55	+1.538
45	+2.392
35	+3.167
25	+3.681
15	+4.572
5	+6.041

the next scoring event. They took 2852 first-and-10 plays, divided the field into 10-yard strips, and averaged the value of the next scoring event over all plays. For example, if the next scoring event was a field goal, the value would be +3. If the next scoring event was an opposing touchdown, the value would be −7. The 10 resulting values, representing the expected value on 1st-and-10 from that yard lines, can be seen in Table 8.1 (Carter and Machol 1971).

A 1st-and-10 from the opponent's 35-yard line was worth roughly +3.167 points, while backed up on the team's 5-yard line had an expectation of −1.245. This is the fundamental origin of expected points—the tool we will use to compare everything on a situational level.

It took almost 20 years for the next advancement, this time from Bob Carroll, Pete Palmer, and John Thorn. Together, they wrote *The Hidden Game of Football*, which is the first statistical deep dive into a variety of NFL topics—the biggest of which is why traditional football statistics such as yards do not work for evaluation. Using Carter's work as a baseline, they develop their own linear model that spans from −2 on the team's own goal line to +6 on the opposing goal line. This means that for every 25 yards, a team is expected to score two more points (Carroll et al. 1988). While this is an oversimplification, it provides an initial framework for situational analysis.

The idea is to give an expected point value to every situation, categorized by down, distance-to-go, and yard line. From there, we can look at that expected point value at the start of the play and at the end of the play to see how successful—or unsuccessful—the play was. By applying this to every play throughout a game or season, we can compare everything on a situational level.

8.2 Expected Points

When we think of football, we think of yards. Everything revolves around the 100-yard playing surface. Each new set of downs is a 10-yard increment. We evaluate all different positions based on their yards gained. To this day, offensive and defensive ranks are still based on yards gained or surrendered. There is one giant problem with this: *Not all yards are created equal.*

The best way to look at this is by taking two examples. Take running back A, Alfred. His offense lines up on 3rd-and-15, runs a draw play, and Alfred gains 10 yards. Depending on the situation, his team will most likely punt the ball or kick a field goal given that it is now fourth down. While 10 yards per carry is a great output, in this particular situation, it was not extremely successful. Now, take running back B, Bernard. His team faces a 3rd-and-1 short-yardage situation. Bernard falls through the line for a 2-yard gain, converting a first down and increasing his team's chances of scoring in the process.

Without further reflection, Alfred would look five times better than Bernard, having gained 10 yards versus 2 yards. In reality, Bernard's play was actually much more helpful to his team's cause and should be seen as a more successful result. This is why focusing only on yardage gained or points scored does not work. Everything in the NFL is situational. A field goal is not helpful if your team is down seven with 30 seconds to play, yet it is extremely helpful in a tie game in sudden death. As such, plays, players, teams, and games need to be evaluated with different metrics than more naïve methods.

We have already mentioned two expected points models and now there are several others available—all with the goal of making the most accurate tool for evaluation. As the Chief Analyst for numberFire, we developed our own internal expected points model based on a Markov chain (more on this shortly). Today, we see much more efficient offenses than we did 10 years ago, particularly due to advances in the passing game. Expectation has changed over time—a situation that was worth two points 10 years ago could be worth three points now. Thus, expected points models must always be evolving. Any of the aforementioned models can be used in analysis, but it is important to understand the foundation of the model and how it was created.

Let's discuss three methods for creating an expected points model.

8.2.1 Empirical

This is the method Virgil Carter used. Divide football into a set of states—typically based on down, distance-to-go, and yard line. Average historical data to see how many points were scored across each situation. This is the most intuitive way to create the model—and definitely the easiest way—but it has two big shortcomings. First, this does not account for any changes in the game of football. Using empirical data from 30 years ago does not apply to today's game. This can create a large out-of-sample bias. Second, there will be abnormalities in situations that occur very infrequently. For example, it will be hard to get an accurate expected point value for a 2nd-and-33 from the 34-yard line since that happens so infrequently. One way to fix this is by grouping states together. We could group all distances-to-go over 20 to ensure higher frequencies. Another way to fix this is to use regression and smoothing techniques where applicable in low-frequency areas, which leads us to the next method.

8.2.2 Regression and Smoothing

There are a variety of different statistical techniques that have been used to deal with low-frequency states. The most common is a simple multivariate weighted least-squares regression on down, distance-to-go, and yard line. A linear approximation works reasonably well across the field, but the values near either goal line tend to diverge. As a result, it makes more sense developing a series of parametric equations through regression with

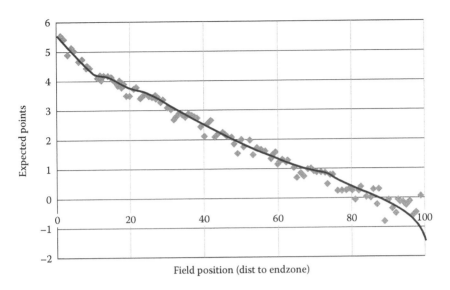

FIGURE 8.1
Brian Burke's expected points model on first down by field position.

different cut points based on yard line. Popular smoothing techniques include moving averages and kernel-weighted local polynomial smoothing.

On the heels of *The Hidden Game of Football* came several improvements in the football analytics community. The two biggest contributors have been Aaron Schatz at Football Outsiders and Brian Burke of Advanced Football Analytics—formerly Advanced NFL Stats. Burke uses a combination of regression and smoothing techniques on top of empirical data to create his own expected points model—which can be seen in Figure 8.1 (Burke 2014a). While Schatz and company at Football Outsiders do not use expected points, they use an idea called "success points," which conveys the same concept. Success points are scaled based on the down, distance-to-go, and yard line, with extra points and deductions awarded as described in Schatz (2003). While not ideal, the results are extremely similar mathematically and they do a great job approximating team and player production based on the situation.

Notice that Burke's model is not entirely linear as it acts differently near each of the goal lines.

8.3 A Markov Model of Football

Each possession consists of a series of discrete plays in an effort to move up the field and eventually score. A set of plays may start with a 1st-and-10 on the 20-yard line after a touchback. Following an incompletion on first down, we then have a 2nd-and-10 from the 20-yard line. After back-to-back 3-yard runs, we are left with a 4th-and-4 from the 26-yard line and decide to punt the ball to the opposing team. Each of these plays has defining initial characteristics: down, number of yards until achieving a new set of downs, and yard line.

We will refer to this as the current "state" of the game. From a statistical modeling perspective, we can estimate the probability of achieving certain states, recognizing that certain states can transition into others. A model that can be used for describing the probabilistic transition of football states is a stochastic process.

A stochastic process is "any process in nature whose evolution we can analyze successfully in terms of probability" (Doob 1996). Essentially, a stochastic process is any process for which we can use probability to estimate the outcome over time. One example of a stochastic process is the Markov chain. The defining factor of a Markov chain is the Markov property: the probability of transitioning into a next state depends only on the current state and not on any earlier states. Suppose $X_1, X_2, \ldots, X_k, X_{k+1}$ are a sequence of states in a team's possession. The Markov assumption can be stated formally in Equation 8.1 as

$$P(X_{k+1} = x | X_1 = x_1, \ldots, X_k = x_k) = P(X_{k+1} = x | X_k = x_k) \qquad (8.1)$$

Equation 8.1 means that the probability of being in state x on the $(k+1)$ play of a drive, given the entire history of the drive, is the same as just knowing only the current state of the drive. More generally, the Markov property of a stochastic process is an assumption about conditional independence. In a Markov chain (of events), the next event is conditionally independent of all past events given the current state.

A football drive can be argued to be a discrete-time Markov chain with a finite number of states. In football, if a team is in a certain situation, previous events have arguably no effect on future events. For example, if we have a 2nd-and-10 from our own 30, it does not matter if the previous play was an incompletion on 1st-and-10 or a false start penalty on 2nd-and-5. In both cases, we have a new situation that only directly affects the next play. One qualification: previous play calls can affect future decision-making based on the tenets of game theory. In other words, teams will run the ball to set up the pass or pass the ball to run draw plays, etc. By disregarding the specific play call, only current down, distance-to-go, and yard line determine the future probabilities of a drive. This fulfills our Markov property.

A drive is a particular type of Markov chain, namely an *absorbing Markov chain*. That is, there are a distinct set of states known as absorbing states that are special. Once the chain moves into an absorbing state, it will never transition out of that state. Some examples of absorbing states of a drive are turnovers, scoring events, and the end of a half or game. Assuming the Markov chain possesses two other properties, namely that the chain is acyclic (no states that return to themselves after a deterministic number of steps) and that the absorbing states can be reached from every nonabsorbing state in a finite number of steps (that the absorbing states are a communicating class to the nonabsorbing states), then the Markov chain will converge to one of the absorbing states. Football drives arguably meet these criteria. In analyzing absorbing Markov chains, it is of interest to determine the probability of ending up in each of the absorbing states, which can help with optimal strategic decision-making. Furthermore, we can characterize the number of steps (plays) to reach a particular absorbing state (e.g., a touchdown) given the current state (e.g., 1st-and-10 in the red zone after the 2-minute warning).

To build a Markov model, we must determine *transition probabilities*. A transition probability is the conditional probability of arriving in a second state in one step given the current state. In a chain with finite states, like a football game, these transition probabilities can be written in the form of a *transition matrix*, as seen in Equation 8.2.

Transition matrix for Markov chain with "k" states:

$$\mathbf{P} = \begin{bmatrix} \mathbf{P}_{1,1} & \mathbf{P}_{1,2} & \cdots & \mathbf{P}_{1,k} \\ \mathbf{P}_{2,1} & \mathbf{P}_{2,2} & \cdots & \mathbf{P}_{2,k} \\ \cdots & \cdots & \cdots & \cdots \\ \mathbf{P}_{k,1} & \mathbf{P}_{k,2} & \cdots & \mathbf{P}_{k,k} \end{bmatrix} \quad (8.2)$$

Each $\mathbf{P}_{x,y}$ represents the conditional probability of arriving in state y in one step given the state is currently x. Using these transition probabilities, we can calculate the probability of being absorbed into any of our special absorbing states.

One simple way to create this transition matrix is to use an empirical approach. That is, using actual historical data, how frequently did we move from state x to state y in one step. For each state, we take the frequency of the transition, divided by total transitions from that state, to obtain transition probabilities. If we moved from state x to state y 25 times in 100 total trips to state x, the entry for $\mathbf{P}_{x,y}$ would be 0.25.

We can also calculate the probability of moving from state x to state y in exactly n steps using this transition matrix. For transition matrix $\mathbf{P}$, our n-step transition matrix is $\mathbf{P}^n$. This matrix operation essentially encodes the so-called "multiplication rule" for conditional probabilities and the law of total probability. For example, suppose in a very simplistic scenario that we want to know the probability a team on its 20-yard line advances 10 yards in two plays. The team first attempts a pass to advance 7 yards; if successful, then the second play only requires 3 yards, and if not successful then the team requires the full 10 yards on the second play. Letting A be the event that the team reaches 10 yards by the end of the second play, B be the event that the team reaches 7 yards after the first play, B' be the complement of B, and C be the current state (on the 20-yard line), the desired probability is given by Equation 8.3:

$$P(A|C) = P(A|B,C) * P(B|C) + P(A|B',C) * P(B'|C)$$
$$= P(A|B) * P(B|C) + P(A|B') * P(B'|C) \quad (8.3)$$

This is precisely the matrix computation for $\mathbf{P}^n$ with $n = 2$ in this simple context. Note that the second part of the equation is due to the Markov property of the sequence of events. For the n-step transition matrix, we denote the probability of going from state x to state y in exactly n steps is equal to $\mathbf{P}^n_{x,y}$. In an absorbing Markov chain, as n goes to infinity (as our number of steps goes to infinity), all transition probabilities will converge to zero except for those entering absorbing states.

For our Markov expected points model, we use these absorption probabilities to calculate the final values. For each state based on down x, distance y, yard line z, all three collectively represented by v, we use Equation 8.4:

$$EP_{x,y,z} = f(x,y,z) = \rho_{v,TD} * 7 + \rho_{v,FG} * 3 + \rho_{v,SAF} * (-2) + \text{Opponent Expected Points} \quad (8.4)$$

where $\rho_{v,a}$ is the absorption probability of drive-ending state a given a team is in state v (Goldner 2012).

Since our absorption probabilities only show the value of a team's current drive—and not the potential value of the opponent's subsequent drive if the team does not score—we need to quantify future expected points if the drive does not end in a score. To include the opponent's drive, we can make state estimates based on every absorbing state—answering questions like, where will the defense likely get the ball if the result of this drive is a punt? Then, by combining the expectation of the two drives, we get a resulting final expected points model.

8.4 Net Expected Points

Net Expected Points, also known as Expected Points Added (EPA), is an effective metric to evaluate plays on a situational basis. By taking the difference between expected points before the play and after the play, we end up with a value for how many points a play added (or subtracted) from that team's scoring potential. The formula (8.5) is

$$\text{NEP} = \text{EP}_{\text{End-of-Play}} - \text{EP}_{\text{Start-of-Play}} \qquad (8.5)$$

Take a real-life situation from Week 3 in 2015, Kansas City Chiefs versus Green Bay Packers on Monday Night Football. Late in the first quarter, the Packers had a 1st-and-10 on the Chiefs 49-yard line. To start the play, they have +2.27 expected points (meaning that if a league-average team played out that drive repeatedly, the offense would increase their score differential by +2.27 points on average). Aaron Rodgers completes a short pass to running back James Starks who runs for a 19-yard gain, resulting in a 1st-and-10 on the Kansas City 30-yard line. The new expected point value jumps to +3.60. We can then attribute that +1.33 points to Aaron Rodgers, James Starks, and the Packers offense. That +1.33 value is our measured NEP on the play.

Since an expected points model represents a league-average team, an NEP of 0 would mean a perfectly league-average play. Any positive NEP means the team performed above expectation and a negative play means the team performed worse than expectation. Level of success can be summarized by a success rate. There are varying degrees of success, but the most basic metric for effectively evaluating success is binary—either a play is successful or not successful. Then, if over 100 plays, a team was successful on 55 of them, we can confidently say their success rate is 55%. Similarly, using NEP, we can find the most successful plays in a game, a week, a season, or of all time! (Table 8.2).

All of this evaluation of a play is after the fact, which is not particularly helpful for decision-making. Decisions should not be evaluated based on the outcome, but rather based on the decision-making process. One major benefit of an expected points model

TABLE 8.2

Top NEP Performances by a Quarterback in Week 5, 2015 from numberFire

Player	Total NEP	Traditional Passing Stats
Eli Manning, NYG	+25.11	41/54, 411 Yards, 3 TD, 1 Int
Josh McCown, CLE	+21.71	36/51, 457 Yards, 2 TD
Brian Hoyer, HOU	+19.17	24/31, 312 Yards, 2 TD, 1 Int
Blake Bortles, JAX	+14.28	23/33, 303 Yards, 4 TD, 1 Int
Colin Kaepernick, SF	+12.76	23/35, 262 Yards, 2 TD

is that it can be used as a tool to effectively gauge optimal decisions—from fourth down decisions, to when to accept a penalty, and much more.

8.5 Optimal Decision-Making: Using Expected Points

Let us take a look at another real-life situation. In the same Week 3 Monday Night game, Aaron Rodgers and company face a 4th-and-5 from the Chiefs 39-yard line with 13:13 left in the second quarter. Since it is so early in the game, we will disregard score and time remaining for this analysis—as the goal early in the game is to increase your point differential as much as possible. From the 39-yard line, the Packers have three choices: "Go for it," attempt a 56- to 57-yard field goal, or punt the ball. Today, most coaches will punt the ball as they are risk averse, but let's perform the relevant calculations. To do this, we will compute the expected value of each situation using expected points and conversion probabilities.

8.5.1 Go For It

If the Packers go for it and convert, for simplicity sake, let's assume they would have a 1st-and-10 on the Chiefs 34-yard line. If they fail, we will assume an incomplete pass, which would lead to Kansas City ball on their own 39-yard line. Expected points on a success (1st-and-10 from the 34) are +3.28 according to our model. On a failure, the Chiefs would be expected to score +1.51, so the expectation for Green Bay is −1.51. Using a simple expected value calculation, we get the following equation, where x is the probability of converting on fourth down (Equation 8.6):

$$\text{EP}_{\text{GO-FOR-IT}} = (3.28) * (x) + (-1.51) * (1 - x) \tag{8.6}$$

We will come back to this equation shortly, but let's look empirically at the expected value. Historically, 4th-and-5 converts around 40%, which means that $\text{EP}_{\text{GO-FOR-IT}}$ is +0.41.

8.5.2 Attempt A Field Goal

If the Packers successfully convert the 56- to 57-yard field goal, their expected points will be +3 (for a made field goal) minus the value of giving the ball back to the Chiefs on a kickoff—approximately +0.40 points. So, a made field goal equates to +2.60. Notice that this is less than a successful conversion on fourth down. If the Packers miss the field goal, the Chiefs get the ball from the spot of the kick, in this case, that would be the Chiefs' 46- or 47-yard line or +2.02 expected points. This gives us Equation where y is the probability of a successful field goal:

$$\text{EP}_{\text{FGA}} = (2.60) * (y) + (-2.02) * (1 - y) \tag{8.7}$$

Historically—after adjusting for trends in field goal attempts, as kickers have become dramatically more accurate—a field goal from the 39-yard line converts at approximately 55%. That means EP_{FGA} equates to +0.32.

8.5.3 Punt

The last possible scenario—and most common—is a punt, trying to pin the Chiefs deep in their own territory. The expected starting position following a punt from the 39-yard line is between the 10- and 11-yard lines. This results from regression-based models to estimate the probability of a touchback (26.7%), fair catch (32.6%), return (2.2%), or some other non-return scenario along with estimated gross punt yardage (32.0 yards) and expected return yards (6.6 yards). The general formula can be seen in Equation 8.8:

$$\text{EP}_{\text{PUNT}} = E[\text{Opponent Field Position}] = P(\text{Touchback}) * (20) + (P(\text{Fair Catch})$$
$$+ P(\text{No Return})) * (\text{Yard line} - E[\text{Gross Punt Distance}]) + P(\text{Return})$$
$$* (\text{Yard line} - E[\text{Gross Punt Distance}] + E[\text{Return Yards}]) \tag{8.8}$$

A 1st-and-10 for the opposing team on the 10- or 11-yard line equates to an EP_{PUNT} of +0.20.

Based purely on expectation, it looks like going for it is the best option at +0.41 compared to the next best option, a field goal at +0.32, followed by punting at +0.20. The most common option—the punt—is the lowest value decision. This is a huge inefficiency in today's NFL. The reason? A punt from the 39-yard line has such a high likelihood of a touchback (over one-in-four) and generally does not have the opportunity to net many yards.

Situational factors, though, can play an enormous factor in this analysis by shifting the probabilities. What if the team is going up against a particularly stout pass defense and does not believe they can convert at 40%? What if it is snowing or raining or windy and the field goal conversion rate is much lower? What if the team is playing in Denver where the altitude is higher and field goal probability is higher? Each situational shift in probability can dramatically affect the outcome of the decision analysis. All that being said, these league-average baselines are the perfect starting point when trying to evaluate a decision.

Given that field goal conversion rates are relatively consistent across the league (not including external factors like game conditions), but fourth down conversion rates can vary dramatically based on team strength, we will take this analysis one step further. By holding the field goal probability constant, we can go back to Equation 8.6 and solve for x, setting the equation equal to EP_{FGA} at +0.32. This will give us the break-even fourth down conversion rate where it makes sense to go for it versus attempting the long field goal. In this case, solving for x yields a break-even point at 38%—pretty close to the estimated 40% conversion rate, making this a tough decision. The break-even point against a punt is roughly 35% conversion probability.

Why not take into account score differential or time remaining, though? If the Packers were down 30 points in the aforementioned scenario, it probably makes more sense for them to go for it, as a field goal will not really help the cause. Adding score and time gives us a new metric: Win Probability.

8.6 Win Probability

Expected points and net expected points are extremely valuable tools. They measure efficiency, regardless of score or time. A 20-yard gain down 30 is just as efficient as a 20-yard gain in a tie game. What differs is the *importance* of those two plays. An expected points model is to efficiency as a win probability model is to importance. Generally, until later in the game, the game state variables of score and time should not have a big impact on decision-making. Teams should act in the most efficient manner possible. Later in the game, or in extreme situations—like blowouts where it makes sense for the losing team to act with a high-variance strategy—these game state variables are imperative for situational analysis. It is therefore more compelling to focus on the probability of winning the game, and how the success and failure of different plays impact the change in win probability. Just like with an expected points model, there have been three primary methods used to develop win probability models to date—empirical models, models built through regression and smoothing, and a Markov model.

The win probability state space is an extension of the expected points state space. In addition to down, distance, and yard line, we now have score differential, time remaining, and even a few other potential variables such as team strength, time-outs remaining for both teams, and pace of play. Each additional variable has a multiplicative effect on the size of the state space. Imagine all the possible down, distance, and yard line combinations. Now imagine all of those combinations with every possible score differential. Now imagine all of those throughout every point in the game. Just like with expected points, we are forced to make grouping decisions to limit the number of states. For example, maybe we group all states in the first quarter together. We could limit the score differential between -30 and +30 since the game is essentially over after that point. Some score differentials could potentially be grouped together—there is not a huge difference between being up four and being up five. Either way, we need to cut down the state space reasonably before developing the model and we will still likely run into sample size issues from an empirical standpoint. How many times has a team been up nine points with a 3rd-and-12 from their own 15 with 3:30 left to play? Certainly not enough to make a realistic win probability estimate from that data point alone.

8.7 Existing Models

One of the first publicized win probability models comes from baseball—a simpler sport to model due to the state-dependent nature of the game and minimal amount of interaction. In *The Book: Playing the Percentages in Baseball*, Tom Tango discusses his win expectancy model as an extension of run expectancy (Tango et al. 2007). Tango develops both of these models using Markov chains, again, due to how the game lends itself to be divided into states: score, inning, outs, runners on base, count, etc. If the Phillies are winning six to five with two outs runners on the corners and a 1-1 count, it does not matter how those runners earned their positions on first and third, or how the first two outs were made. All that matters is the current situation, which can be used to inform estimates of the future of the game.

Similarly, at numberFire, we extended our drive-based Markov chain to a full-game Markov chain, including score differential, time, and a few other state-defining variables, in order to develop an accurate win probability model. In football, in particular, strategy changes dramatically based on the current score and time remaining. A team down 20 points will act drastically different than a team up by three with 2 minutes remaining. Complicating things further, in situations that seem similar, due to the odd increments

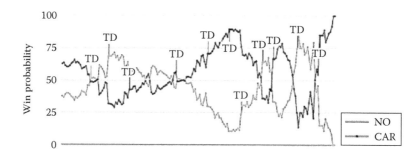

FIGURE 8.2
New Orleans Saints versus Carolina Panthers win probability chart. (From numberFire Live, https://live.numberfire.com/nfl/6302, accessed December 6, 2015.)

in scoring in the NFL, teams will act much differently. A team down four needs to be aggressive to score a touchdown, whereas a team down three is much more likely to play it safe once in field goal range. Other interesting information also falls out of the Markov model like expected scoring events remaining, probability of any final score, and expected drive outcome.

In Figure 8.2, we can see the full win probability graph from the New Orleans Saints versus Carolina Panthers matchup in Week 13 of 2015. Notice, in particular, how the largest shifts in win probability come toward the end of the game. This follows since a team scoring a touchdown early in the game leaves plenty of time for a change in outcome, whereas a go-ahead score with a minute remaining has a dramatic impact on who will ultimately win the game. This shows the volatility of win probability in the NFL throughout the game.

For the majority of the game, the win probabilities for New Orleans and Carolina will sum to 100%. There are, however, late game and overtime scenarios where there is a nonzero probability of tying in football.

A Markov chain is not the only way to build a win probability model. Brian Burke, at Advanced NFL Analytics, built his model by piecing together a variety of regression and smoothing techniques (Burke 2014b). Pro-Football Reference improves upon a methodology outlined by Wayne Winston in his book *Mathletics*. The first step is to fit the score differential to a normal random variable. Using historical data, Winston looked at final score margin in NFL games. He then approximated win probability with the cumulative distribution function of a normal variable with a mean representing team strength—in his case, the Vegas spread—and the standard deviation of the historical scoring margin. Once the game has started, Winston adjusts this mean and standard deviation for the diminishing amount of time remaining in the game and the current score differential. For example, if the spread started with one team favored by six, at halftime, that team strength element would only account for three points. Then, by adding in expected points, we see adjustments for the current down, distance-to-go, and yard line (Paine 2013). This model works extremely well for the first three quarters or so but runs into problems in more complicated late game situations. This is, again, due to the odd increments in scoring in the game of football.

One other method for developing a win probability model that is gaining more and more popularity is the use of machine learning. In particular, there have been several attempts using a random forest (Lock and Nettleton 2014). Trey Causey describes his reasoning for using random forests as a tool on his blog *The Spread:*

> Random forests are popular for a number of reasons. First, they don't overfit as easily as many other methods. Second, they're really robust to non-linear interactions among your features. Third, they're surprisingly accurate in lots of modeling situations. And, fourth, they're easy to run in parallel, which means that you can estimate random forests on really honking big data sets across lots of computers with lots of CPUs.
>
> **Causey (2013)**

A random forest is a generalization of classification and regression trees (Breiman et al. 1984). Rather than constructing one classification tree that predicts whether a team wins given multiple predictor variables at the particular point in the game (e.g., point differential, other relevant factors), multiple classification trees are constructed in the following manner: many (typically 500) bootstrapped versions of the original data set are formed. For each bootstrapped sample, a classification tree is grown, but unlike the usual classification tree the random forest procedure forces each tree to be a "naïve" tree by allowing only a limited number of predictors when constructing the trees. The end result is 500 fitted naïve classification trees. The average fitted values across all the trees are the final prediction from the random forest classifier. After building a random forest win probability model, we found that this was better in theory than in practice. Like all win probability models, it struggled mightily in end-of-game scenarios. One instance was an inability to correctly distinguish between two and three time-outs in a situation where that meant the defense getting the ball back versus the offense kneeling down. A third down with 35 seconds left and a defense with one time-out still gives the defense a chance, whereas no time-outs remaining allow the offense to kneel down and end the game. More work is required to make machine learning approaches, including random forests, usable for end-of-game situations.

8.8 Optimal Decision-Making: Using Win Probability

We saw before that we can use expected points to determine the most efficient play call. As we move later into the game, though, efficiency is not as important as win probability. We want to know what decision gives our team the best chance to win the game?

Let's return to one of the most controversial fourth down calls in recent memory, Bill Belichick's 4th-and-2 versus Peyton Manning's Colts. Here, there are only two possible decisions: Go for it, or punt. Almost all coaches would punt in this situation, but Belichick is not your average coach. The Patriots were up six facing a 4th-and-2 on their own 28-yard line. With 2:08 left to play, the Colts had just one time-out remaining.

8.8.1 Go For It

If the Patriots convert, for simplicity sake, let's assume they have a first down at their own 31-yard line. Since the clock would be moving on a conversion, we will let it hit the 2-minute warning. With a conversion, the Patriots win probability jumps to 94.6%—not quite a guaranteed win as the Colts could still get the ball back with around 18 seconds remaining. If the Colts stop the Patriots, Indianapolis would likely have the ball on the Patriots 28-yard line with one play before the 2-minute warning. That gives the Patriots a 50.9% chance of winning. This is a lot different from public opinion. People believed that giving Peyton Manning the ball so close to the end zone was a guaranteed loss, but it is actually closer to a coin flip. A big reason for this is the Patriots' chance to get the ball back and kick a game-winning field goal in the event that the Colts score a go-ahead touchdown. To calculate the Patriots expected win probability, we can use formula (8.9):

$$\text{WP}_{\text{GO–FOR–IT}} = (94.6\%) * (z) + (50.9\%) * (1 - z) \tag{8.9}$$

Here, z is the probability of converting on 4th-and-2. League-wide, outside of the red zone, this converts an estimated 53% of the time. That gives an expected win probability of 74.1% for the Patriots if they go for it.

8.8.2 Punt

The other option is for the Patriots to punt. A punt from the 28 generally gives the opponent the ball on their own 33-yard line. An average punt would give the Patriots a 73.6% to win the game. So, to find the break-even conversion rate on 4th-and-2, we can set formula (8.9) equal to 73.6%. In this specific case, the break-even conversion rate is 52%, just below the league-average 53%. This suggests that going for it is a better option, but with the numbers so close, there is not a bad decision.

Let's consider the possibility that Peyton Manning would give the Colts a 75% chance to win on a failure to convert and a 50% chance to win on a punt—much closer to how the public perceives these probabilities. With these assumptions, we can compute an expected win probability of 61.9% going for it versus only 50% on the punt. In this case, the break-even conversion rate drops dramatically to 35.9%. In other words, the better the Patriots believe the Colts to be, the more they should want to go for it. Similarly, the Patriots are no average offense. The higher they believe their fourth-down conversion rate to be, the more they should want to go for it.

Unfortunately for Belichick (and as is the case with most coaches who make these "risky" decisions), he was crucified by the media when his team failed to convert and the Colts went on to win. All too often in sports media, decisions are evaluated by the outcome rather than the process. However, our analysis demonstrates that Belichick knew what he was doing using a principled approach to optimal decision-making. When evaluating these decisions, we must look at everything that goes into the decision-making process. Using tools like expected points and win probability is a quality data-driven step toward optimal decision-making.

8.9 What's Next?

While we have seen a lot of recent advancements across football analytics, there is still an expansive unknown for future work in the field. Due to the complexity of football, win probability and expected points models will never be perfect. We have come a long way since Virgil Carter's initial analysis but still have a long way to go. One area that is rarely touched on is the nature of uncertainty in these estimates. Saying that a team has a 75% chance to win the game could mean that there is a 73%–78% chance they win or a 60%–90% chance depending on the uncertainty in the model. These have drastically different implications and would greatly affect the decision-making process.

Another area that is on the verge of analytical revolution is player tracking. With new developments in baseball's FIELDf/x and basketball's SportVU technology, we are collecting massive amounts of valuable unused data. In football, similar technology is possible—particularly at the skill positions (essentially anywhere away from the offensive line where it would be difficult to identify individual players in a pile of bodies). Recently, the NFL announced the use of Zebra Technologies player tracking using motion sensors in players' shoulder pads. The possibilities with this kind of data are endless: automatic recognition of formations, routes, and plays, evaluating defenders based on the distance between their coverage assignments, accurately assess "closing speed." This new information combined with statistics like net expected points and win probability add another layer to decision-making analysis in football. What are the most efficient formations? Which routes perform best against different coverages? How does expected play outcome change based on who is on the field?

Next time your favorite team decides to punt on 4th-and-1 or kick a field goal down 21, think about the process that went into that decision. Usually, if a coach says "it was a gut call," there were not many numbers that went into it. That being said, many coaches are where they are today due to how accurate their gut and the numbers match up. Bill Walsh was very susceptible to Virgil Carter's analysis because it generally lined up with his understanding of the game. There is, however, still a long way to go to close the gap toward optimal decision-making in the NFL. Football analytics remain in a stage of infancy, but, with growing trends in other sports and new data and technology every year, it is only a matter of time before coaches begin to take advantages of these inefficiencies.

References

Barras, M. Former Bengals QB Virgil Carter found a home off the field teaching statistics at Xavier. *Xavier Magazine*, October 2, 2013. http://xtra.xavier.edu/xavier-magazine/off-the-field-former-bengals-qb-virgil-carter-found-a-home-teaching-statistics-at-xavier/.

Breiman, L., J. Friedman, C.J. Stone, and R.A. Olshen. *Classification and Regression Trees*. Pacific Grove, CA: Wadsworth & Brooks/Cole Advanced & Software, 1984.

Burke, B. Expected points and EPA. Advanced Football Analytics, December 7, 2014a.

Burke, B. What I'm working on. Advanced Football Analytics, March 10, 2014b.

Carroll, B., P. Palmer, and J. Thorn. *The Hidden Game of Football*. New York: Warner Books, 1988.

Carter, V. and R.E. Machol. Operations research on football. *Operations Research*. 19(2) (March–April 1971), 541–544.

Causey, T. Building a win probability model part 1. The Spread, December 8, 2013.

Doob, J.L. The development of rigor in mathematical probability (1900–1950). *American Mathematical Monthly*. 103 (1996), 586–595.

Goldner, K. A Markov model of football: Using stochastic processes to model a football drive. *Journal of Quantitative Analysis in Sports*. 8(1) (2012), 1–18.

Levin, J. Why doesn't football have a Bill James? Slate, December 19, 2003.

Lock, D. and D. Nettleton. Using random forests to estimate win probability before each play of an NFL game. *Journal of Quantitative Analysis in Sports*. 10(2) (2014), 197–205.

numberFire Live. https://live.numberfire.com/nfl/6302, accessed December 6, 2015.

Paine, N. The P-F-R Win Probability Model. Pro-Football-Reference.com, 2013.

Schatz, A. Methods to our madness. Football Outsiders, 2003.

Tango, T.M., M.G. Lichtman, and A.E. Dolphin. *The Book: Playing the Percentages in Baseball*. Washington, DC: Potomac, 2007.

9

Probability Models for Streak Shooting

James Lackritz

CONTENTS

In Game 5 of the 2014 National Basketball Association (NBA) Finals, San Antonio Spurs guard Tony Parker missed his first 10 shots and then made his next 7. With a career field goal shooting percentage of just under 50%, Parker's making 7 out of 17 shots would not have been a surprise, but to do so in the sequence of missing 10 in a row and then making 7 in a row immediately raised the question of streak shooting.

Streak shooting in basketball has been a debated topic among statisticians and sports strategists for nearly 30 years, and there has been renewed interest with the growth of sports analytics. The existence or nonexistence of streaks has a direct bearing on sports strategy: namely, should a team focus on more recent events (e.g., a current performance streak) or longer-term trends (e.g., season statistics) when setting strategy.

Streak shooting is said to exist when the likelihood that a player will make his next shot is, in part, a function of whether he made his prior shot(s) rather than strictly a random event. In theory, the existence of streak shooting would suggest that there are times when a player has a higher (or lower) than expected probability of making a prolonged series of shots, which can have immediate implications for game-time strategy. More generally, however, streak shooting would suggest the potential for training interventions to more frequently achieve such performance levels.

This chapter systematically addresses models for not just the hot hand hypothesis but the "cold hand" hypothesis as well over an entire season to assess the potential impact from a supposed hot (or cold) hand. That systematic examination results in mixed conclusions about the hypotheses.

The next section examines some of the research and methodology used to date to evaluate and test hypotheses on streak shooting. Section 9.2 lays out the theory and models that have been used in some of the research and that will be illustrated in this chapter. Section 9.3 then introduces and describes the data used as an example and walks the reader through the methodological details for one player, discussing the results from all of the selected players. The final sections summarize the results and offer a discussion on and suggestions for future research.

9.1 Background

Gilovich et al. (1985) published what many consider to be the initial major study on the hot hand hypothesis, evaluating conditional probabilities that a player makes his next shot after missing or making one to three consecutive shots. They used field goal and free throw data from the NBA as well as a controlled experiment using Cornell varsity basketball players taking repetitive unguarded shots from the same location. They concluded that the likelihood of a successful shot after a made shot did not differ significantly from that after a missed shot. They applied the Wald–Wolfowitz runs test (Siegel, 1956), examining the number of runs (change from a made shot to a missed shot or vice versa) in the Cornell data.

In the runs test, if the null hypothesis (randomness in the shot outcomes) is assumed true, then the distribution of r, the number of runs, is approximately normal with a mean of

$$R = \frac{2N_{made}N_{missed}}{N} + 1$$

and a variance of

$$V = \frac{(\mu - 1)(\mu - 2)}{N - 1},$$

where
   N_{made} is the total number of made shots
   N_{missed} is the total number of missed shots
   $N = N_{made} + N_{missed}$ are the total number of shots

The test statistic is

$$z = \frac{r - R}{\sqrt{V}}.$$

As an example, consider a sequence of *MMmmmMmmMMmmm* of shots, where M indicates a made shot and m a missed shot. There would be a total of six runs, with $N_{made} = 5$ and $N_{missed} = 8$. Therefore, $R = (2(5)(8)/(5+8)) + 1 = 7.154$ and $V = (((7.154 - 1)(7.154 - 2))/(13 - 1)) = 2.643$, so $z = ((6 - 7.154)/\sqrt{2.643}) = -0.710$ and $p = .478$ (two-tail).

Gilovich et al. concluded no significant deviation from chance (actually 1 player out of 26 showed a significant difference at the .05 level, but this would be not be unexpected in random data). Yet their polling of fans confirmed overwhelmingly fan belief in streak shooting for both field goal and free throw attempts.

Koehler and Conley (2003) used data from the NBA three-point shooting contest over 4 years, examining the number of runs in a 25-shot sequence, but found that only 2 of the 23 players examined had a significant number of runs at the .05 level, again within the expected number in a binomial distribution with $n = 23$ and $p = .05$. Bar-Eli et al. (2006) reviewed hot hand research on streaks in different sports, finding limited support for the hot hand hypothesis, and when so, with flawed models. Arkes (2010) used a logit model to examine the outcome of the second free throw of a two-shot foul for an NBA player given the outcome of the first free throw with an estimated 2%–3% increase in the probability of making the second shot given that the first shot was made. He follows up (Arkes, 2013) suggesting that there are times during the season when the hot hand does, in fact, exist.

Stone (2012) addressed streak shooting using an autocorrelation model, where the first-order autocorrelation would be

$$\rho = \frac{\text{cov}(x_t x_{t-1})}{\sqrt{\text{var}(x_t)\text{var}(x_{t-1})}}$$

with x_t as a binary variable with probability p_t if shot t is made, and runs the first-order autoregressive model of

$$p_t = \rho p_{t-1} + (1 - \rho)\mu_p + \varepsilon_t,$$

where μ_p is between 0 and 1.

In the first-order model,

$$x_t = \beta_0 + \beta_1 x_{t-1} + \upsilon_t$$

the ordinary least-squares estimate $\hat{\beta}_1$ provides a consistent estimate of ρ. Stone ran this model on simulated data, concluding that the hot hand exists.

Miller and Sanjurjo (2014) ran a series of controlled experiments in which they identified an area in which the player should make approximately 50% of his or her shots and then observed 300 shots over two different periods 6 months removed. In their analysis, they suggest a bias correction, further discussed by Gelman (2015), resulting from a conditional probability analysis from a standard binomial experiment when one studies the distribution of what happens after a made shot. If one examines sequences of n consecutive shots for $n \geq 3$, $p = .5$ and lists all of the possible sequences, one consistently finds that the probability of the next shot made given an outcome of a previously made shot is less than .5.

For example, consider a series of three coin flips (shots) in which the outcomes are HHH, HHT, HTH, HTT, THH, THT, TTH, TTT. Each sequence occurs with an equally likely probability of $1/8 = .125$. When focusing on the six sequences in which an H could potentially follow another H (eliminating TTH and TTT), for HHH, 100% of the time, H is followed by another H. This also occurs for THH. For HHT, 50% of the time H is followed by another H. For all other sequences, H is followed by T. Averaging out these percentages across these six cases results in 42% of the time an H following an H. Therefore, in a three-flip (shot) sequence, it is more likely that heads (made shot) is expected to be followed by tails (missed shot, approximately 58%) than by heads (approximately 42%). This difference from .5 gets larger as the sequence length becomes greater and when considering longer streaks.

Adjusting for this bias, Miller and Sanjurjo (2014) found from their probability model that "some players systematically get the hot hand, other players appear to always shoot with the same ability, and still others actually underperform immediately following a streak of hits" (p. 30). When they pooled the data over all players, they found a significant hot hand effect.

Much of the early research on streaks focused on the fixed/controlled model, which assumes that all outcomes would be independently and identically distributed (iid). This can only truly exist under conditions such as free throw shooting, three point contest (Koehler and Conley, 2003; Arkes, 2010), or a designed experiment where subjects took all of their attempts from the same location (Miller and Sanjurjo, 2014). For example, if one wanted to test the hot hand hypothesis for basketball players' shooting or golfers' putting, one might set up an experiment where they take a series of shots/putts from the same spot for 3, 5, 10, 15, and 20 foot shots/putts. When these conditions can be achieved, then the approach to determine if the hot hand exists focuses on tests to be discussed in the next section.

Purists argue that while controlled experiments make for a simpler analysis, actual competition does not allow for this and therefore adjustments have to be made. LeBron James rarely takes two shots from the same place (except for foul shots), and Jordan Spieth almost never hits two putts from the same location on the green (and his work comes from 18 greens per round). Therefore, the collection of data on other mitigating factors allows the analyst to take these into account with a more sophisticated multivariate model (regression, logit, or probit), which can better estimate the likelihood of a made shot or other outcome given the conditions at the time of the attempt. This type of model is represented by the more recent work of Bocskocsky et al. (2014), Rao (2009), and Csapo et al. (2015).

Bocskocsky et al. (2014) applied a more complex regression model to NBA data, with the theory that each shot occurs under totally different conditions (shot distance, defensive metrics, player's intrinsic ability, etc.). Using SportVU, an optical tracking system to collect this data on over 83,000 shots from the 2012–2013 NBA season, they modeled the probability of making the next shot based on these conditions, as the situational conditions under which the shot occurs can create a different expected probability for making each individual shot. They also introduced a "Heat" factor, the percentage of shots made over the past n shots, adjusted for the difficulty of the shots. This differed from previous hot hand studies that treated the probability of making the next shot as a constant, under a hypothesis of no streak. Bocskocsky et al. (2014) also concluded that players with high success results on recent shots are more likely to take their next shot from a further distance than previous shots and are also more likely to take their team's next shot.

Rao (2009) studied the 2007–2008 Los Angeles Lakers, using play-by-play data on shot distance, time left in the quarter, shot clock time, double team (if a player was defended by two or more opponents), shot zone, and type of shot (as a 0/1 variable) among other variables. He used a nonlinear probit model to estimate the probability of a made shot from his data, including a cubic effect on distance and testing for interactions. He concluded from a small sample of eight players over 60 games that some players tended to take more difficult shots after a string of made shots, leading to a smaller expected shot value.

Csapo et al. (2015) concluded from a study of the NBA that more consecutive shots made (missed) correlated with more difficult (easier) subsequent shots, on three dimensions.

While this chapter focuses on basketball shooting streaks, McCotter (2010), Horowitz and Lackritz (2013), and Green and Zweibel (2015) suggest that there can be a hot hand for hitting streaks in baseball. There are opportunities for research on streaks in other sports as well, as demonstrated by previous research summarized by Bar-Eli et al. (2006) in golf, tennis, billiards, darts, and bowling.

9.2 An Approach for Fixed/Controlled Data

The fixed/controlled data model assumes that the distribution of shot outcomes is iid with a fixed probability of making each shot. Let M represent a made shot, m represent a missed shot. Define a streak to be a sequence of repeated makes (misses) followed by a miss (make).

Let M_k be a streak of k consecutive made shots (the next shot is missed), and m_k be a streak of k consecutive missed shots (the next shot is made). Then, a sample sequence of $MMmm$ $mMmmMMmM$ represents a discrete alternating renewal process, which could be stated as an alternating set of events of $M_2m_3M_1m_2M_2m_1M_1$ of made and missed shots. Let n_{M_k} and n_{m_k} represent the number of streaks of k made and missed shots. For the aforementioned sample sequence above, we have $n_{M_1} = 2$, $n_{M_2} = 2$, $n_{m_1} = 1$, $n_{m_2} = 1$, and $n_{m_3} = 1$. It is assumed that streaks continue or roll over from one game to the next, so that if a player missed his last three shots to finish a game and missed his first two shots at the beginning of the next game before making a shot, there would be a streak of m_5 bridging the two games. Letting p be the fixed probability of a made shot, if one assumes that individual outcomes of made/missed shots are iid, the probabilities of exactly k consecutive made shots (the $k+1$st shot was missed) and k consecutive missed shots (the $k+1$st shot was made), respectively are given by

$$P(M_k) = (1-p)p^k \qquad (9.1)$$

and

$$P(m_k) = p(1-p)^k \qquad (9.2)$$

for $k = 1, 2, \ldots$.

Let S_k (s_k) represent a sequence of k consecutive made (missed) shots. Consider the probability of the next shot made following a sequence of k consecutive made shots (continuing the made sequence), denoted P_{MS_k}. Analogously, let P_{Ms_k} denote the probability of the next shot made following a sequence of k consecutive missed shots (ending the sequence of missed shots). The probabilities of the next shot missed after a sequence of k consecutive made or missed shots are given by P_{mS_k} and P_{ms_k}, respectively.

Under the iid assumption, $\hat{P}_{MS_k}$, a sample estimate for P_{MS_k}, can be computed by totaling the number of $k+1$st made shots after a sequence of k made shots divided by the total number of $k+1$st shots (made + missed) after a sequence of k made shots. Having earlier defined n_{M_i} as the number of made shot streaks of length i, the numerator of $\hat{P}_{MS_k}$ is the number of streaks of $k+1$ made shots or longer $= n_{M_{k+1}} + n_{M_{k+2}} + n_{M_{k+3}} + \cdots$. The number of missed shots after a sequence of k made shots would be n_{M_k}, since if the streak is exactly k made shots, then the next shot was missed. Therefore, the sample estimate would be

$$\hat{P}_{MS_k} = \frac{\sum_{i=k+1}^{N} n_{M_i}}{\sum_{i=k}^{N} n_{M_i}} \qquad (9.3)$$

where N is the length of the sequence. The probability of a missed shot after a sequence of S_k makes is therefore $\hat{P}_{mS_k} = 1 - \hat{P}_{MS_k}$.

Similarly, a sample estimate for P_{ms_k} can be computed by totaling the number of $k+1$st missed shots after a sequence of k missed shots divided by the total number of $k+1^{st}$ shots (made + missed) after a sequence of k missed shots. From earlier, n_{m_i} represents the number of missed shot streaks of length i, so

$$\hat{P}_{ms_k} = \frac{\sum_{i=k+1}^{N} n_{m_i}}{\sum_{i=k}^{N} n_{m_i}} \qquad (9.4)$$

and $\hat{P}_{Ms_k} = 1 - \hat{P}_{ms_k}$.

TABLE 9.1

Example of a Made/Missed Streak Distribution

Consecutive Shots Made		Consecutive Shots Missed	
k	n_{M_k}	k	n_{m_k}
1	10	1	5
2	5	2	5
3	3	3	6
4	2	4	3
5	0	5	1
Total	20		20

Defining $n_M = \sum_{i=1}^{N} n_{M_i}$ and $n_m = \sum_{i=1}^{N} n_{m_i}$ as the total number of made and missed streaks, we note that the absolute difference $|n_M - n_m| = 0$ or 1, since in the renewal process, we are alternating made and missed streaks.

In the hypothetical example of Table 9.1, the player had a streak of one made shot (and then missed the next shot) on 10 occasions, exactly two consecutive made shots on 5 occasions, 3 consecutive made shots on 3 occasions, and a streak of 4 made shots on 2 occasions. The total number of made streaks is 20 and the total number of missed streaks is 20. From Equation 9.3, $\hat{P}_{MS_1} = 0.5$, $\hat{P}_{MS_2} = 0.5$, and $\hat{P}_{MS_3} = 0.4$.

Similarly, from Equation 9.4, $\hat{P}_{ms_1} = 0.75$, $\hat{P}_{ms_2} = 0.67$, $\hat{P}_{ms_3} = 0.4$, and $\hat{P}_{ms_4} = 0.25$.

Then, under the assumption that shot outcomes are iid, the standard test for streak shooting would be a test of the hypothesis:

$$H_0 : P_{MS_k} = P_{Ms_k} = p \quad \text{for } k = 1, 2, \ldots$$

For the data in Table 9.1, there are a total of 37 made shots and 50 missed shots. To make an individual test, such as $H_0 : P_{MS_1} = p$, once there is a streak of one made shot, there were $10(5 + 3 + 2)$ next shots made and 10 (n_{M_1}) missed. From the remaining data, there were 27 shots made and 40 missed. Creating a two by two contingency table and applying χ^2 test (or a two group percentage test) yields $\chi^2 = 0.593$, $p= .441$.

Examining the *distribution* of streaks collectively over the entire data allows for two possible tests. The first is the Wald–Wolfowitz runs test (Siegel, 1956), as we described previously (change from a made shot to a missed shot or vice versa). For the data in Table 9.1, the number of runs is 40, the totals of the n_{M_k} and n_{m_k} terms. In Table 9.1, $N_{made} = 37$, $N_{missed} = 50$, and $N = 87$. Therefore,

$$R = \frac{2(37)(50)}{87} + 1 = 43.53$$

and

$$V = \frac{(42.53)(41.53)}{86} = 20.54.$$

Therefore, the runs test of $z = (40 - 43.53)/\sqrt{20.54} = -0.78$ ($p = .436$, two-tail) is not statistically significant at the .05 level.

Another distributional test would be a goodness-of-fit test to examine the overall distribution of the number of streaks of each size. The χ^2 test from earlier examines any specific streak length on an individual basis. Does the number of streaks $\{n_{M_i}, n_{m_i}\}$ for $i = 1, 2, \ldots$ fit collectively within the anticipated distributions if we assume H_0 is true? In other words, in the example of Table 9.1, how many made streaks out of the 20 would we expect to be of 1, 2, 3, and more than 3 made shots if the distribution was truly random? To examine the expected number of made and missed streaks of length k, one first needs to find the conditional probability of a specific made (missed) streak length within the group of made (missed) streaks. For example, in Table 9.1, applying Equation 9.1 with $p = 37/87 = .425$ yields the individual probabilities .244, .104, .044, and .033 for made streaks of 1, 2, 3, and >3 consecutive shots, yielding conditional probabilities just within the made streaks of .575, .244, .104, and .077, respectively. Therefore, $E(n_{M_1}) = 20(.575) = 11.50$, $E(n_{M_2}) = 20(.244) = 4.88$, and so on. The first term of the goodness of fit test would be $(10 - 11.50)^2/11.50$, and this would be done for each streak length (both made and missed streaks), collapsing/merging any group(s) that had an expected streak length of less than five occurrences. This yields a χ^2 test based on $g - 1$ degrees of freedom, where g is the number of groups with an expected group size of five or more.

A rejection of the hypothesis does not guarantee that streak shooting exists but suggests that the outcomes are either not completely iid or random (or both). This would lead to one of several possibilities, one of which is that the probability of making a shot is not stable over the duration of the data.

9.3 An Example: NBA Regular Season 2013–2014

In this section, we examine the possibility of streak shooting over an entire NBA season. The data for the example come from all 1312 NBA games played during the 2013–2014 regular season from nbastuffer.com. There were approximately 560,000 lines of play-by-play data in the files, where each line represents an individual outcome (shot, foul, foul shot, rebound, turnover, etc.) for one of the 1312 games. Data were sorted for more than 200,000 field goal (FG) attempts, including three-point field goal (3P) attempts, and almost 60,000 free throw (FT) attempts, recognizing that FTs are taken under a far more controlled environment and come closer to satisfying the requirements of an iid variable. The 10 players that had the greatest number of attempts in each category were identified from the NBA web site to calculate the number of made/missed streaks, for each player in the appropriate categories for the 2013–2014 season. Five additional players were tracked who were not good FT shooters, to see whether they had a streak pattern that could be related to their confidence (or the lack thereof) of making or missing consecutive FTs.

Table 9.2 illustrates, for the 2013–2014 season, the made/missed shot streaks for FG and FT shots for LeBron James, generally acknowledged as being the premier player in the league at that time. For FG attempts, there is no pretense about his shot distribution being random and iid, but the results are illustrated as an example. He had 160 streaks of 1 made shot, 84 streaks of 2 made shots, etc. Similarly, he had 227 streaks of 1 missed shot, 71 streaks of 2 missed shots, and 1 streak where he missed 9 consecutive shots.

Since James' made-shot percentage for the season was 0.567 (rounded to the third decimal), we have under the iid assumption that the probability of a made streak of exactly k shots just within the made streak group, denoted $p_M(k)$, follows a geometric distribution

TABLE 9.2

LeBron James FG and FT Streaks for 2013–2014

	Consecutive Shots Made					Consecutive Shots Missed			
k	n_{M_k}	$p_M(k)$	$E(n_{M_k})$	$\hat{P}_{MS_k}$	k	n_{m_k}	$p_m(k)$	$E(n_{m_k})$	$\hat{P}_{Ms_k}$
Field Goals									
1	160	0.433	155.05	0.553	1	227	0.567	202.95	0.634
2	84	0.246	87.90	0.576	2	71	0.246	87.90	0.542
3	60	0.139	49.83	0.474	3	40	0.106	38.07	0.667
4	28	0.079	28.25	0.481	4	11	0.046	16.49	
5	14	0.045	16.01		5	5	0.020	7.14	
6	8	0.025	9.08		6	2	0.009	3.09	
7	3	0.014	5.15		7	1	0.004	1.34	
8	1	0.008	2.92		8	0	0.002	0.58	
≥ 9	0	0.011	3.82		≥ 9	1	0.001	0.44	
Total	358					358			

Goodness-of-fit test: $\chi^2 = 17.624$, $df = 13$, $p = .172$

Wald–Wolfowitz runs test: $p = .005$

Made   767

Missed   586

Total   1353, Made Percentage: 56.7%

	Consecutive Shots Made					Consecutive Shots Missed			
k	n_{M_k}	$p_M(k)$	$E(n_{M_k})$	$\hat{P}_{MS_k}$	k	n_{m_k}	$p_m(k)$	$E(n_{m_k})$	$\hat{P}_{Ms_k}$
Free Throws									
1	32	0.250	28.70	0.722	1	89	0.750	86.30	0.774
2	19	0.187	21.54	0.771	2	21	0.187	21.54	
3	13	0.141	16.16	0.797	3	5	0.062	7.16	
4	15	0.105	12.13	0.706					
5	11	0.079	9.10	0.694					
6	7	0.059	6.83						
7	1	0.045	5.13						
8	4	0.033	3.85						
9	4	0.025	2.89						
10	4	0.019	2.17						
11	4	0.014	1.63						
12	0	0.011	1.22						
13	0	0.008	0.92						
≥ 14	1	0.024	2.75						
Total	115					115			

Goodness-of-fit test: $\chi^2 = 6.940$, $df = 11$, $p = .804$

Wald–Wolfowitz runs test: $p = .275$

Made   439

Missed   146

Total   585, Made Percentage 75.0%

with $p_M(k) = .433(.567)^{k-1}$ for $k = 1,2,\ldots$ and $p_m(k) = .567(.433)^{k-1}$ for probabilities of missed streaks of exactly k shots within the missed streak group. To estimate the probability of the next shot made after a sequence of k consecutive made shots, we use Equations 9.3 and 9.4 to generate $\hat{P}_{MS_k}$ and $\hat{P}_{Ms_k}$. For James' FG results in Table 9.2, we have $\hat{P}_{MS_1} = 198/358 = 0.553$.

Given that he was on a one-shot made sequence, the estimated probability that James makes his next shot is .553. Given that his overall shot-made percentage for the entire season was .567, running the first individual test from Section 9.2 results in no significant change in the probability of a made shot given S_1 to the probability of a made shot for a non-S_1 outcome ($\chi^2 = 0.38$, $p = .558$). Therefore, the occurrence of a sequence of one made shot does not significantly alter the chances of his making the next shot.

The estimates for the distributions of p_{MS_k} and p_{Ms_k} are given in Table 9.2. If these were truly random and iid outcomes (no streak shooting), the probability of the next made shot should not differ significantly from the probability of the made shot from any other situation. The individual test was run for each sequence of k that occurred *cumulatively* more than 30 times for k or more. Note the total number of streaks of three (or more) includes streaks of four or longer. The results were statistically significant ($\alpha = 0.05$) for sequences of S_3 (probability of the next made shot is significantly lower) and s_1 (higher).

However, the Wald–Wolfowitz test resulted in James having more total runs (716 streaks) than would have been expected. While the total number of streaks is higher than expected, the distribution of the streak lengths (goodness-of-fit test from Section 9.2) is not significant ($p = .172$). Yet, on an individual level, two of the seven streak lengths that had sufficient occurrences had the chance of him making the next shot being significantly different from what would have been expected. The final section of the chapter discusses this mixture of results and offers further commentary.

For James' FT distribution for the 2013–2014 season, there is no value of k for which individual test on the chance of extending the streak is significantly different from the remaining FT-made percentage for the season. And, the runs test ($p = .275$) and goodness-of-fit test for the overall distribution of streaks ($p = .804$) were both statistically insignificant.

Tables 9.3 through 9.5 give the results for the players who, in the 2013–2014 NBA season, had 10 highest number of attempted FG, FT, and three-point field goal (3P) attempts. Intuitively, one would expect the most reliable results to come from the FT data, but, anecdotally, three-point shooters believe they are going to get hot. Therefore, even though shots vary in length and position, the FG and 3P results are used as an illustration of the techniques using a fixed/controlled analysis. In the fourth column, the p-values for the goodness-of-fit test are given first, with the results of the Wald–Wolfowitz Test (WW) in parentheses. Those that are significant at the .05 level are shown in boldface.

Only two players listed in Table 9.3 had a goodness-of-fit test result close to being statistically significant. In the final column, $S_1(D)$ indicates that when Durant had a sequence of one made shot, his chances of making the next shot decreased (D) significantly compared to his chances otherwise of making the next shot. Analogously, $S_4(I)$ indicates that when he had a sequence of four made shots, his chances of making the next shot increased (I) significantly. Therefore, for at least some players on their FG attempts, at some point during their made/missed shot sequences the probability of making the next shot did change significantly. Note there is no consistency for the individual sequences listed in the last column in which the chances of making the next shot is higher or lower when significant.

Using a Bonferroni correction, the significance level for each player was adjusted to .05/s, where s is the number of shot sequences being tested. With this adjustment,

TABLE 9.3

Top 10 NBA Players on Field Goal Attempts, 2013–2014

Player	FGA	%Made	Goodness-of-Fit Test (WW) p-Value	Longest Streak Made	Longest Streak Missed	Sequence Lengths with Significant Individual Tests at .05 Level
Kevin Durant	1688	50.3	.050(.435)	9	14	$S_1(D)$ $S_4(I)$
Carmelo Anthony	1643	45.2	.168(.881)	9	13	$s_4(D)$
LaMarcus Aldridge	1423	45.8	.931(.472)	9	10	
Kevin Love	1422	45.7	.501(.499)	8	11	
DeMar DeRozan	1407	42.9	.681(.263)	7	12	$s_2(I)$
Stephen Curry	1383	47.1	.987(.715)	9	8	
Al Jefferson	1376	50.9	.074(**.039**)	9	8	$S_1(D)$ $S_4(I)$ $s_2(I)$
Paul George	1362	42.4	.251(.622)	8	13	$S_3(D)$
Blake Griffin	1359	52.8	.690(**.020**)	9	7	
LeBron James	1353	56.7	.172(**.005**)	8	9	$S_3(D)$ $s_1(I)$

Note: The last column gives the sequence for which the individual (first test in Section 9.2) chances of making the next shot are significant at the .05 level. If (I), the chances increased. If (D), the chances decreased.

Abbreviation: FGA, field goal attempts.

TABLE 9.4

Top 10 NBA Players on Free Throw Attempts plus Five Additional Players, 2013–2014

Player	FTA	%Made	Goodness-of-Fit Test (WW) p-Value	Longest Streak Made	Longest Streak Missed	Sequence Lengths with Significant Individual Tests at .05 Level
Kevin Durant	805	87.3	.288(.508)	45	3	$S_1(D)$
Blake Griffin	674	71.5	.711(.879)	14	5	
James Harden	665	86.6	.748(.385)	27	3	
Dwight Howard	638	54.7	.247(.146)	10	9	$S_1(D)$
Kevin Love	633	82.1	.783(.443)	25	3	
DeMar DeRozan	630	82.4	.183(.102)	26	3	
DeMarcus Cousins	595	73.3	.972(.403)	17	3	
LeBron James	585	75.0	.804(.275)	14	3	
Carmelo Anthony	541	84.8	.532(.886)	30	3	$S_2(D)$
Paul George	464	86.4	.708(.676)	40	3	
Deandre Jordan	374	42.8	.851(.285)	7	10	
JJ Hickson	290	51.7	.803(.798)	10	10	
Josh Smith	301	53.2	.805(.991)	8	10	
Tony Wroten	216	64.1	.870(.808)	16	5	
Andre Drummond	328	41.8	.995(.613)	6	11	

Note: The last column gives the streak lengths for which the individual (first test in Section 9.2) chances of making the next shot are significant at the .05 level. If (I), the chances increased. If (D), the chances decreased.

Abbreviation: FTA, free throw attempts.

TABLE 9.5

Top 10 Players on Three-Point Field Goal Attempts, 2013–2014

Player	3PA	%Made	Goodness-of-Fit Test (WW) p-Value	Longest Streak Made	Longest Streak Missed	Sequence Lengths with Significant Individual Tests at .05 Level
Stephen Curry	615	42.4	.152(.384)	6	9	
Damian Lillard	554	39.4	.991(.960)	6	12	
Klay Thompson	535	41.7	.816(.852)	6	10	
Wesley Matthews	511	39.3	.287(.261)	5	8	
Gerald Green	510	40.0	.286(.592)	6	12	
Kevin Love	505	37.6	.762(.107)	7	11	
Kyle Lowry	500	38.0	.816(.595)	5	12	
Paul George	500	36.4	.921(.542)	7	10	
Randy Foye	498	38.0	.772(.368)	6	12	
Kevin Durant	491	39.1	.507(.435)	6	13	$S_1(D)$

Note: The last column gives the sequence lengths for which the individual (first test in Section 9.2) chances of making the next shot are significant at the .05 level. If (I), the chances increased. If (D), the chances decreased.

there were still four significant sequence lengths (both of Durant's, Jefferson's $s_2(I)$, and James' $s_1(I)$) for which the chance of making the next shot changed significantly. Note the 10 players in Table 9.3 each had six to eight shot sequences that were tested, for a total of 73 sequences. Having four sequences for which the chances of continuing the streak were significantly different at the .05 level (using the Bonferroni adjustment) from the chances of making a shot in other situations would be expected ($n = 73$, $p = .05$).

For FT attempts (Table 9.4), when considering the top 10 players in terms of attempts, no one player has an overall streak distribution that differs from chance, no one has a significant runs (Wald–Wolfowitz) p-value, and there are only three shot sequence lengths overall for which the chances of the next shot being made change significantly, none of which would be significant with a Bonferroni adjustment. As this is the most controlled environment and FTs come closest to satisfying the requirements of an iid variable, this would indicate that the distribution for FTs does not differ significantly from chance given their season percentage.

We hypothesized that poor foul shooters, such as Dwight Howard (54.7% made percentage for the season), would be more prone to missed streaks due to confidence issues. The five worst foul shooters with close to 300 attempts for the season were identified and listed at the bottom of the table. None of the streak results for these "bad" foul shooters were statistically significant.

The 3P shooters in Table 9.5 are considered to be the best (and streakiest) shooters, yet none of the top 10 players that had the greatest number of attempts showed anything to indicate a "nonrandom" distribution in a significant runs test. Furthermore, for only one sequence was there a significant change in the chances of making the next three-point shot.

To run a multivariate model on LeBron James' data, we ran heat factors for 2–10 shots, allowing for a continuation in a sequence of made/missed shots from one game to the next. This continuation of shots from the previous game has not been done in previous research, even though we are used to hearing announcers on TV mention streaks of consecutive shots made over multiple games or a multiple game series where the player has made "X" percent of his FG attempts. Regressing (binary logit) the shot outcome on the individual heat factors individually, the prediction based on $heat_3$ (number of shots made in the last three shots) had the lowest p-value for the heat factor. We ran a multiple binary logit model to predict the likelihood of the shot made based on $heat_3$, streak length (negative if a missed streak), and additional variables that were available from the original data: quarter (indicator variables for the second, third, and fourth quarters), absolute value of the score differential at the time of the shot, time left in the quarter, time left on the 24-second, shot clock, and distance of the shot, all as linear variables. The second quarter indicator variable, score differential, time left on the 24-second clock, and shot length were all significant at the .05 level (two-tail), but neither the streak length nor the heat variable was close to being significant.

9.4 Comments on Existing Methods

An ongoing concern in the literature has been that it is difficult to assess the true likelihood of streak shooting, due to the different conditions under which a shot is attempted. There is a difference between a contested shot versus an uncontested shot, and a layup/dunk as opposed to a long jump shot. Shots taken at the end of the shot clock or at the end of a period occur under conditions that differ from those taken during the normal 24-second time range of the shot clock. And, the game score can dictate a difference of conditions, in that a shot taken during a close game in the fourth quarter can be taken when there is more pressure and meaning and the player is more focused (and possibly more closely guarded) than a shot taken during a blowout (when one team is ahead by so many points that there is not enough time left for the outcome of the game to have a chance to change). Furthermore, when a player is fouled on a shot, the shot is counted as an attempt if he makes the shot and it is not counted as an attempt if he misses the shot. Thus, for FG shooting, we would expect some bias in favor of made streaks.

Yet, streaks arguably do exist. In Tables 9.3 through 9.5, all players had some made and missed streaks with large numbers. All 10 players listed in Table 9.3 had streaks of at least seven made shots in a row. So when LeBron made nine shots in a row, was he taking better shots? Was he not guarded as closely as at other times? At no point would one expect or predict that LeBron or Durant would make their next nine shots. So, within the normal framework of a game, one would not be prone to predicting streaks of such length. But given that there was such a streak, did the player have a higher than normal value for the probability of making the next shot, or more likely, is it just something that occurs in the normal course of events within the statistical framework of taking over 1300 shots in a season? And when a player misses nine or more consecutive shots, is he "cold"? Taking bad shots? Is their probability of making the next shot lower than normal?

The data for the FG and 3P analysis in the previous section were not under controlled conditions. The FT data are as close to a controlled experiment as one can replicate in live game conditions. And, that data produced results that were not statistically significant from random.

Using the techniques of Section 9.2 to estimate the chance of making the next shot amid a sequence of made/missed shots is best evaluated under a controlled experiment, such as those run by Miller and Sanjurjo (2014) and Koehler and Conley (2003). Koehler and Conley's study that combines the best of a contest and control, but with only five shots from each spot on the floor, does not generate the sample size that one might want to make for a more complete study. Miller and Sanjurjo's study is the best experimental design but will never come close to simulating game conditions. How many times has a player in an NCAA or NBA game just made a shot after shot in warm-ups and then not be able to replicate those results in a live game situation?

For fixed/controlled studies, options for analysis include the Wald–Wolfowitz test on the total number of streaks, conditional probability analysis on the chance of making the next shot given a sequence of k made/missed shots for $k = 1, 2, \ldots$, and the goodness-of-fit test on the overall distribution of the frequencies of the streak lengths. Each is designed to test a different dimension of the streak. The Wald–Wolfowitz test addresses the initial question of the total number of streaks being appropriate for the number of shots. The goodness-of-fit test evaluates the frequencies of streak lengths collectively, rather than individually.

One must be careful in performing a conditional probability analysis on individual streak lengths, in that there can be individually long streaks of made/missed shots within the full data set, as observed in Tables 9.3 through 9.5. Even in a controlled environment, one should be cautious about conclusions that could be reached performing this type of analysis. With random data, if one examines 100 situations of any type, one would expect up to 9 of these to have statistically significant outcomes at the .05 level, fewer if Bonferroni adjustments are made.

Our opinion is that the best option for evaluating a streak under game conditions is FT shooting, as the conditions are the same for each attempt, save for time/score pressure. Yet, not one of the 15 players evaluated in the NBA 2013–2014 regular season data was close to having a distribution of made/missed streaks that would differ significantly from chance, given his season percentages of made free throw shots. And, for only three FT sequences did the chance of making the next free throw change significantly; for all three, the chances were significantly *lower*, even though the sequence was for a made shot. Even by chance, one would expect collectively for the group fewer than seven sequences for the FT shooters for whom the next shot would have a significantly higher (or lower) chance of being made, so the three sequences for which this happened fit easily within this range.

Table 9.6 shows the results for 10 run simulated players with $n = 1500$ shots and $p = 0.5$. No streak distribution was significantly different from chance alone, as would be expected. And, of the 82 individual sequences with more than 30 opportunities to extend the streak,

TABLE 9.6

Results from 10 Simulated Seasons with 1500 Shots and $p = 0.5$

Player	FGA	%Made	Goodness-of-Fit Test p-Value	Longest Streak Made	Missed
1	1500	49.0	0.335	8	9
2	1500	49.1	0.931	10	10
3	1500	47.2	0.205	9	12
4	1500	48.6	0.844	12	11
5	1500	47.5	0.321	12	10
6	1500	48.8	0.166	9	11
7	1500	52.0	0.610	12	10
8	1500	49.3	0.624	9	11
9	1500	51.1	0.412	11	8
10	1500	50.1	0.529	12	9

Abbreviation: FGA, field goal attempts.

not one displayed a significant change in the chances of making the next shot. Furthermore, none of the 10 simulations had a total number of streaks that was significant in the Wald–Wolfowitz test at the .05 level.

When one examines these results, would we conclude that at times the computer had a "hot" or "cold" hand, or attribute the distributions to chance? Is this really any different from what appears in Tables 9.2 through 9.4? So, when a player has a good (or bad) run of made (or missed) shots, does this merit special attention, that is, is it a human nature reaction to a perception of randomness as concluded by Sun and Wang (2010)? Or as Berry (2006) concludes, does it give the fans and talking heads (TV commentators and writers) justification to try to explain something that is not really a statistical phenomenon?

Alternatively, a better way to examine streak shooting under actual game conditions is to use the approach of Rao (2009) or Bocskocsky et al. (2014). Collecting data on all factors relevant to the shot allows for the chance of making the next shot to be forecast from game conditions. The simple heat factor accounts for the current streak of made/missed shots, and the complex heat factor adjusts for defensive metrics such as the proximity and height/length of the nearest defender, presence of a second defender, and type/difficulty of the shot. This adjustment on the heat factor, not done prior to these two studies, differentiates between the contested and the uncontested shot. Logit and probit models can isolate the importance of the current streak versus the other conditions that can impact the chances of a made shot. Now, available online sites track all of these potentially relevant statistics, which with sufficient time and manpower can create the database that allows for the more complex modeling option to estimate the importance of each factor, and to determine if, in fact, the hot (or cold) hand exists.

9.5 Concluding Remarks

Further discussion is still warranted to address issues that have not yet been resolved in the literature. Is the game location (home/away) an issue, given that players may be more comfortable in their home arena, which also can negate any travel/jet lag impact? What about the time between games (back-to-back versus a day or two from the previous game), and travel or no travel prior to the game? Should last-second shots/heaves at the end of the quarter/shot clock in which the player gets the ball with a second or two left be further discounted in the data? Should dunks be totally eliminated for streak consideration? When a player is fouled on a shot attempt, and does not make the shot, it does not count as a shot attempt. Yet, if the player was not fouled, would they have been more likely to make the shot? Or, do the defensive metrics suggested by Bocskocsky et al. (2014) suggest that this probability would be lower than an unguarded shot?

Their study combines players in the data. Yet, is it fair to compare the shot distribution data of Tim Duncan to LeBron James to Stephen Curry? Duncan takes most of his shots inside of eight feet from the basket, while Curry leads the league in three-point attempts, and James has a shot distribution chart that is somewhere in between those of Duncan and Curry. Bocskocsky et al. (2014) include factors to identify players as good, mediocre, and bad shooters, but is that enough?

Should data be limited to just the game being played, or should data be rolled into previous games? Nonscientifically, one would hypothesize that players coming off a good game (or series of good games) bring more confidence to the next game than players who are in a perceived slump. The initial examination of the data from the NBA 2013–2014 regular season might suggest that the heat factor is still limited to a very small series of previous shots, but more research is needed in this area.

From a modeling approach, a nonlinear impact of shot distance needs to be evaluated and tested. The potential for multiple interactions also exists, which only appear to have been previously included by Rao (2009) but not in most of the other papers.

The research, debate, and discussion on the hot hand hypothesis are far from over. There are two parts of the discussion: the outcome and the methodology used to evaluate the data and test for the outcome. There are three possible outcomes from the analysis: (1) randomness, (2) hot/cold streak, or (3) change of conditions. The answers are yet to be finalized. Yet, the psychology of randomness (Sun and Wang, 2010) will always allow for a lively discussion.

References

Arkes, J. 2010. Revisiting the hot hand theory with free throw data in a multivariate framework. *Journal of Quantitative Analysis in Sports* 6:1–12.

Arkes, J. 2013. Misses in 'hot hand' research. *Journal of Sports Economics* 14:401–410.

Bar-Eli, M., Avugos, S., and Raab, M. 2006. Twenty years of "hot hand" research: Review and critique. *Psychology of Sport and Exercise* 7:525–553.

Berry, S. 2006. A statistician reads the sports pages. *Chance* 19:50–54.

Bocskocsky, A., Ezekowitz, J., and Stein, C. 2014. The hot hand: A new approach to an old "Fallacy." *MIT Sloan Sports Analytic Conference*, Cambridge, MA, 10pp.

Csapo, P., Avugos, S., Raab, M., and Bar-Eli, M. 2015. The effect of perceived streakiness on the shot-taking behavior of basketball players. *European Journal of Sport Science* 15:647–654.

Gelman, A. 2015. Hey-guess what? There really is a hot hand! http://andrewgelman.com/2015/07/09/hey-guess-what-there-really-is-a-hot-hand/, accessed January 30, 2016.

Gilovich, T., Vallone, R., and Tversky, A. 1985. The hot hand in basketball: On the misperception of random sequences. *Cognitive Psychology* 17:295–314.

Green, B. and Zweibel, J. 2015. The hot hand fallacy: Cognitive mistakes or equilibrium adjustments? Evidence from major league baseball. http://papers.ssrn.com/sol3/papers.cfm?abstract_id=2358747, accessed August 25, 2016.

Horowitz, I. and Lackritz, J. 2013. The value of *s* at which a major league player with multiple hitting streaks of at least *s* games might be termed streak prone. *Chance* 26:26–33.

Koehler, J. and Conley, C. 2003. The 'hot hand' myth in professional basketball. *Journal of Sport & Exercise Psychology* 25:253–259.

McCotter, T. 2010. Hitting streaks don't obey your rules: Evidence that hitting streaks aren't just by-products of random variation. *Chance* 23:52–57.

Miller, J.B. and Sanjurjo, A. 2014. A cold shower for the hot hand fallacy, 58pp. http://papers.ssrn.com/sol3/papers.cfm?abstract_id=2450479, accessed August 25, 2016.

Rao, J. 2009. Experts' perception of autocorrelation: The hot hand fallacy among professional basketball players, 25pp. http://www.justinmrao.com/playersbeliefs.pdf, accessed January 30, 2016.

Siegel, S. 1956. *Nonparametric Statistics*. New York: McGraw-Hill.

Stone, D.F. 2012. Measurement theory and the hot hand. *The American Statistician* 66:61–66.

Sun, Y. and Wang, H. 2010. Perception of randomness: On the time of streaks. *Cognitive Psychology* 61:333–342.

10

Possession-Based Player Performance Analysis in Basketball (Adjusted +/− and Related Concepts)

Jeremias Engelmann

CONTENTS

For general managers, gamblers, and fans alike, it is important to know a player's impact on the outcome of an NBA basketball game. Managers of basketball teams, often bound by a salary cap or restrictions by team owners, want to sign players that have large positive impact on game outcomes, preferably to cheap contracts. Furthermore, they want to avoid paying maximum salary to players who may be perceived as superstars but aren't significantly helping the team win. Traditionally, front offices of basketball teams have relied on scouts for information on player ability, talent, and impact. In the last 20 years, advanced player metrics have found their way into the decision-making process. Advanced player metrics have some advantages over scouts, including teams being able to synthesize information from all games played in a season in a matter of minutes, or even seconds. Furthermore, advanced player metrics cannot be deceived by highlight plays or the opinion of others and thusly provide a more objective assessment of player impact.

Gamblers want to know each player's impact as precisely as possible so that they can have an edge on the bookmakers or other gamblers. Knowing a player's per-possession impact is crucial, as team strength (and thus the money line and spread) is significantly influenced by the availability of its players, and significant changes of betting lines can be observed the moment a player injury is announced.

For casual fans, single number player metrics can provide useful information on players that can rarely be seen on TV and can provide some objective measurement in arguments. Estimating a player's true impact can be a difficult task in any sport but can be even more difficult in a game like basketball where straight up head-to-head situations rarely exist: A team's performance with a certain player on the court is heavily influenced by the players with whom he plays and the players on the opposing team. Furthermore, many player actions that benefit the team, such as boxing out or setting screens, are generally not recorded in the standard NBA box score.

When people first started building player metrics for basketball, they built linear weights metrics using a player's box score statistics, such as steals, rebounds, points scored, etc. One simple example of such a metric would be the NBA's "efficiency metric" [1], which gives one point to positive-impact box score statistics (points scored, rebounds, assists, steals, blocks) and subtracts a point for turnovers and missed shots.

The birth of play-by-play data in basketball gave way for computing several kinds of possession-based player performance analyses, such as "+/−" (PM), "NET Rating" (NET), and "Adjusted +/−" (APM). These metrics, instead of using individual player statistics, try to assess player impact by evaluating how the team plays with a certain player on the court, how they perform without him, or using regression techniques with information on who is on the court at what time to arrive at player impact estimates. PM-based metrics have the benefit over purely box score based player metrics of potentially (intrinsically) catching all player actions that have an influence on the outcome of a game. This can especially be helpful on the defensive end, where there is a shortage of valuable box score statistics, and those that exist can sometimes be misleading indicators for player performance. For example, gambling for steals gets rewarded by pure box score based metrics, as the additional steals the player records have a positive impact on his rating. PM-based metrics instead punish this player if he is often out of position because of his frequent gambles, assuming the opposing offense has a higher average efficiency once a defender is out of position. But the PM-based metrics are not free of problems: Due to noisy data and that some players are often being substituted in and out of the game together or for one another, it can be difficult for these techniques to identify the player(s) that are helping or hurting a team's performance. Over time, better methods were created to adjust for strength of teammates and opponents and to better combat the noise and multicollinearity in the data, leading to advanced metrics such as "Regularized Adjusted +/−" (RAPM). Additional methods, called "4-factor RAPM," were developed to estimate a player's influence on a team's effective field goal percentage, turnovers, rebounds, and free throw rate.

This chapter lays out the history of development of possession-based player metrics, then shows the steps required to compute player metrics such as RAPM and 4-factor-RAPM, including guidelines for parsing play-by-play data, setup of regression matrices, and finally the computation of player estimates.

10.1 Plus Minus

The PM statistic was originally developed for hockey [2]. PM has been in use since the 1950s and has found its way more into basketball since early 2000. A player's PM in basketball is computed as the difference between the points his team scores while he is on the court and the points scored by the opposing team during his time on the court. The measure is an indication of the number points a player's team outscores the opponent's team while the player is on the court. It can take on a negative number if a player's team was outscored during his time on the court. PM can be computed for quarters, halves, full games, full seasons, and even over multiple seasons. Information on the identity of players scoring, or any other box score information, is ignored in a PM analysis. PM is inherently descriptive. Several concerns exist in the use of PM, including

- Mediocre or bad players that frequently share court time with good players can potentially have a strong positive PM. The good player being on the court concurrently will make it more likely that their team outscores the opponent's team during their time on the court.
- Likewise, good players on bad teams are likely to have negative PM because of the poor quality of their teammates.

The other players on the court have a large influence on a player's PM, making PM a suboptimal player metric.

10.2 NET Rating

NET Rating is another PM-based metric that is often used in conjunction with PM. The team's PM with a player off the court gets subtracted from that player's PM. As such, it measures how good a team performs with a certain player on the court compared to being off the court. This can be helpful, for example, to identify weak players that have a good PM due to playing with good players; their NET rating has a high probability of being negative, assuming that their replacements perform even better in similar lineups.

Generally, great players will have both positive PM and NET rating. Their team is good overall but suffers when the great player sits on the bench. The worst players of the league will tend to have negative PM and NET; their team is weak and is performing even worse with these players on the court. One problem with NET is that great bench players on great teams can have a negative NET simply because the players they replace are top players, for example, when backing up Steph Curry or LeBron James. For this reason, NET should rarely be used on its own, but the player's PM and information on who he generally replaces in lineups should be taken into account as well.

10.3 APM

In 2004, Rosenbaum [3] described a new player metric, "Adjusted +/−," often abbreviated as "APM." Winston and Sagarin [4] had implemented the same idea earlier (dubbed "WinVal") but didn't make results or techniques public, as they were consulting for the Dallas Mavericks. The idea was to estimate a player's impact on the game outcome, taking into account his PM but also controlling for the quality of the player's teammates currently on the court. This is accomplished by performing a regression with each player as a binary indicator variable for being on the court, and possession outcomes making up the response vector. NBA play-by-play data, which record all substitutions and many player actions such as turnovers, steals, made baskets, rebounds, etc., provided the necessary information to derive which 10 players were on the court during any time of the game. Furthermore, the amount of points scored in every possession are included in the play-by-play data. This allowed them to construct the following regression model:

- A game has T "shifts." A shift is a period in the game without substitutions.
- Each 10-man unit shift $t = 1, \ldots, T$ represents a single observation.
- The point differentials during shift t, from the view of the hometeam, make up the vector of our dependent variable: $y = (y_1, \ldots, y_T)$.
- The players, coded as dummy variables, are the independent variables in design matrix X. For player i during shift t, we let X_{it} be 1 if player i is present and playing for the home team, -1 if player i is present and playing for the away team, and 0 otherwise.

When computing APM, we seek to find M-vector β that minimizes the expression

$$\sum_{t=1}^{T} p_t (\beta X_t - y_t)^2$$

where
  p_t is the number of possessions during shift t
  $X_t = (X_{1t}, \ldots, X_{Mt})$ is the row vector of dummy player variables
  M is the number of players in the analysis

The results of the optimization, which is the weighted least-squares solution to the regression model, provide an estimate on the influence of a player on the expected score differential between two 5-man units.

The original version of APM used ordinary least-squares regression (OLS) to compute player estimates. The least-squares estimate is given by

$$\beta = (X'X)^{-1}X'y. \tag{10.1}$$

Because the columns of X are linearly dependent, it is common to set one of the elements of β to 0, though other linear constraints are possible. There were several difficulties with this approach.

- The degree of multicollinearity in the data is large, as some players are often substituted in and out simultaneously, or for one another. Without any adjustment, OLS is generally ill-suited for problems in which a large degree of multicollinearity exists.
- Because of this collinearity and that some players compete in very few shifts, coefficient estimates can be quite unstable and unreliable. It is not uncommon to see extraordinary large estimates of elements of β derived from APM.

To combat the latter problem, one all-encompassing independent variable for players who played less than a certain amount of minutes has been used instead of having one variable per player. If several players on court have played less than a certain amount of minutes, the dummy variable can simply be turned into a counter. Such a method to address this group of players essentially corresponds to assuming that all players who played less than a certain amount of minutes have equal influence on the outcome of a possession. Such an approach represents a less-than-perfect solution. With OLS, it was nevertheless necessary to use this type of solution to obtain more stable results.

10.3.1 Computing Separate Estimates for Offense and Defense

It may be desirable to compute player estimates for both offensive and defensive impact. To do this, let each possession represent a single shift and have $X_{it}^{(O)}$ set to 1 if player i is on offense during shift t, $X_{it}^{(D)}$ set to 1 if player i is on defense during shift t, and horizontally concatenate $X^{(O)}$ and $X^{(D)}$ to create X. The response variable y_t contains the points scored by the team in possession t of the ball. The vector of coefficients can thus be estimated using least-squares regression as in Equation (10.1).

10.3.2 RAPM

Regularized Adjusted +/− (RAPM) was first introduced by Joe Sill at the Sloan Sports Analytics Conference in 2010 [5]. While using the exact same data as APM, he used ridge regression instead of OLS to compute the coefficients. Ridge regression, also known as "Tikhonov regularization," is a regression technique that, through the use of a penalization term, often leads to more stable solutions that are less susceptible to noise in the data than OLS. In the presence of multicollinearity, it is almost always preferable to use ridge regression instead of OLS. Computing RAPM coefficients, then, amounts to minimizing

$$\sum_{t=1}^{T}(\beta X_t - y_t)^2 + \lambda \sum_{m=1}^{M} \beta_m^2$$

and the estimated coefficients β can be found by solving

$$\beta = (X'X + \lambda I)^{-1}X'y$$

where
  I represents the identity matrix
  λ represents the penalization parameter, found through cross-validation

RAPM arguably has significant advantages over APM.

- It is a more accurate player metric. It is significantly better at predicting out-of-sample data.

- There is no need for using a single dummy variable for all players with less than a certain amount of minutes. The regularization will ensure that collinearity is no longer a difficulty, and that these players will not have unreliable estimates.

- With APM, it was necessary to use multiple years of data in one regression, as results with (less than) a single season of data would lead to unstable results. With RAPM, it is helpful to use multiple seasons of data in one regression, but single-year results lead to stable and useful results, as well.

10.3.3 Potential Additional Variables

Many additional variables can potentially be added to an APM model. One could try to control for schedule-related factors, such as days rest, or control for the influence of coaching, as some coaches influence player performance more positively than others. Controlling for current score difference in the game is another possibility, as it has been shown that players play worse the further they are ahead and/or play better the further they are behind. This effect was first described by Goldman and Rao [6], then further investigated by Engelmann [7].

As multiyear RAPM can often serve to build other metrics (see Section 10.3.4.2), running RAPM with as much data as possible is sometimes desired. For example, 14-year-RAPM provides the basis for basketball-reference's BPM [10]. As players age and thus have a different impact on the game at different phases in their careers, it is important to control for aging factors when using multiple years of data in one regression. A quantitative age variable, or dummy variables for each age, could be added to the model, and a smoothed aging curve could be derived from the results [8].

10.3.4 Other Applications of the APM Framework

10.3.4.1 4-Factor RAPM

A similar framework as APM can be used to estimate each player's influence on a team's "four factors" [9], as denoted by Dean Oliver:

- Effective field goal percentage (eFG%)
- Rebounding
- Turnovers
- Free throw rate

These metrics help in determining the different facets of a player's influence on a team's performance. For example, a player might help a team's defensive rebounding but hurt their eFG% (an example is a rebounding specialist like Reggie Evans). Star players generally excel in more than one area. These additional metrics also offer an alternative view of player influence on game aspects that have been considered to be well explained by standard box score metrics. For example, several players exist that have decent individual defensive rebounding numbers but actually seem to hurt team defensive rebounding, most likely due to their not properly boxing out. On the other side of the spectrum, there are some players that help team defensive rebounding significantly by doing a good job of boxing out the opponents' bigs but rarely actually grab the rebound themselves, often leaving it for teammates to grab. As such, 4-factor-RAPM can be a useful tool to realign our senses in certain areas where our eyes have been deceived by individual statistics.

Transforming the original APM framework to compute 4-factor RAPM is quite simple. In case of "Turnover-RAPM," the vector of responses, originally consisting of points scored in each possession, needs to be replaced by 1s and 0s, indicating whether that possession ended in a turnover. In this analysis, the response variable is thus binary, and logistic or probit regression can be used to estimate the coefficients. For eFG%-RAPM, instead of possessions making up observations, each field goal attempt (FGA) becomes an observation, and the points scored with each FGA make up the response vector.

10.3.4.2 Statistical Plus Minus

Any form of APM can be used to create box score based metrics. Metrics created this way are generally called "Statistical Plus Minus." This concept was also originally developed by Rosenbaum [3]. The most current metric of this type that is publicly available is currently basketball-reference.com's "Box Plus Minus" [10]. Statistical Plus Minus metrics weight player box score or advanced statistics (such as true shooting percentage, offensive rebound percentage, etc.) by performing a regression with each player representing an observation, the player's statistics making up the matrix of predictor variables, and using (preferably multiseason, regularized) APM as the dependent variable. This method of creating a box score metric has the benefit, for example, over John Hollinger's player efficiency rating (PER), that the weights for box score statistics are not arbitrarily chosen, leading to better prediction accuracy. The drawback of these metrics is common to all pure box score metrics; not all actions that have an influence on game outcomes are recorded in the box score. Furthermore, this approach does not adjust for strength of teammates and the opposing team's players which does have an influence on a player's statistics. To try to combat the former problem, additional individual statistics not found in the standard box score, such as goaltends or fouls drawn, can be extracted from play-by-play data or information obtained from video recording.

10.3.4.3 Influence on Win Probability

Some work has been done to estimate win probability of the team with the ball, given the current score difference and time left in the game. Stern [11] used a Brownian motion model and applied the model to basketball and baseball. Lock and Nettleton [12] used random forests to model the win probability at any point in the game and applied the concept in the NFL. In basketball, one could potentially, instead of having "points scored in possession" as the dependent variable, use changes in the estimated win probability as the dependent variable. This has the potential advantage over the standard APM model to not overvalue good performance when the game is in "garbage time."

10.4 Implementation

The following section describes the data that are necessary to compute PM, APM, RAPM, and 4-factor-RAPM, how to transform/parse the data, and how to finally compute estimates using linear regression techniques.

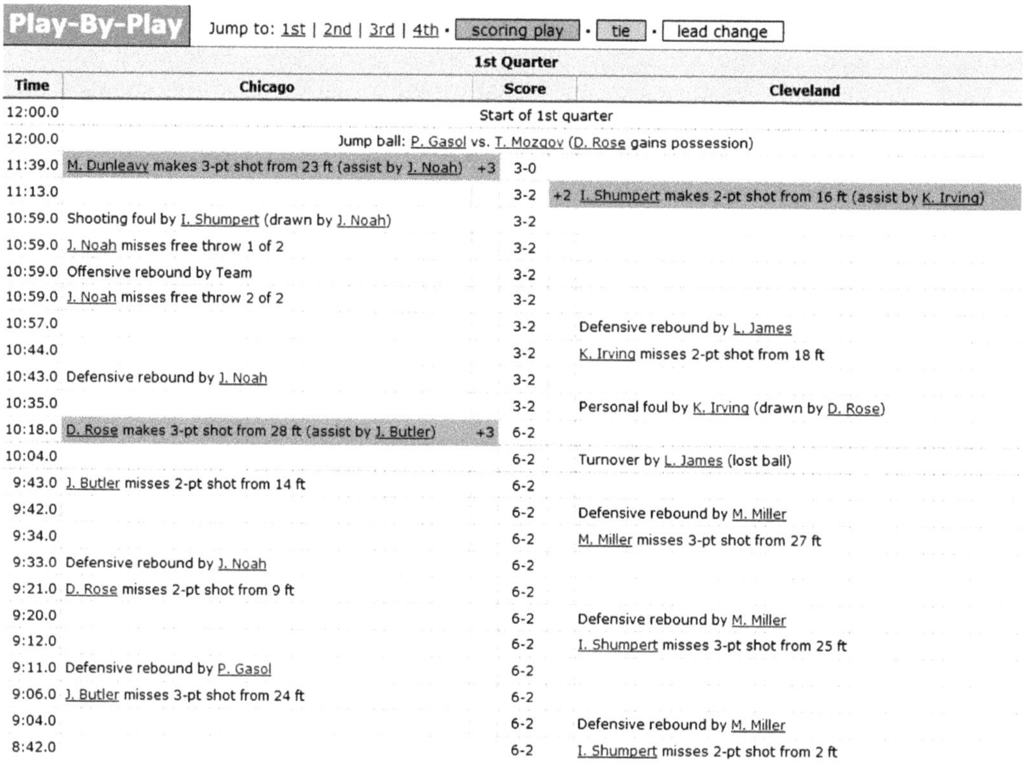

Play-By-Play	Jump to: 1st	2nd	3rd	4th · [scoring play] · [tie] · [lead change]		

		1st Quarter	
Time	**Chicago**	**Score**	**Cleveland**
12:00.0		Start of 1st quarter	
12:00.0		Jump ball: P. Gasol vs. T. Mozgov (D. Rose gains possession)	
11:39.0	M. Dunleavy makes 3-pt shot from 23 ft (assist by J. Noah) +3	3-0	
11:13.0		3-2	+2 I. Shumpert makes 2-pt shot from 16 ft (assist by K. Irving)
10:59.0	Shooting foul by I. Shumpert (drawn by J. Noah)	3-2	
10:59.0	J. Noah misses free throw 1 of 2	3-2	
10:59.0	Offensive rebound by Team	3-2	
10:59.0	J. Noah misses free throw 2 of 2	3-2	
10:57.0		3-2	Defensive rebound by L. James
10:44.0		3-2	K. Irving misses 2-pt shot from 18 ft
10:43.0	Defensive rebound by J. Noah	3-2	
10:35.0		3-2	Personal foul by K. Irving (drawn by D. Rose)
10:18.0	D. Rose makes 3-pt shot from 28 ft (assist by J. Butler) +3	6-2	
10:04.0		6-2	Turnover by L. James (lost ball)
9:43.0	J. Butler misses 2-pt shot from 14 ft	6-2	
9:42.0		6-2	Defensive rebound by M. Miller
9:34.0		6-2	M. Miller misses 3-pt shot from 27 ft
9:33.0	Defensive rebound by J. Noah	6-2	
9:21.0	D. Rose misses 2-pt shot from 9 ft	6-2	
9:20.0		6-2	Defensive rebound by M. Miller
9:12.0		6-2	I. Shumpert misses 3-pt shot from 25 ft
9:11.0	Defensive rebound by P. Gasol	6-2	
9:06.0	J. Butler misses 3-pt shot from 24 ft	6-2	
9:04.0		6-2	Defensive rebound by M. Miller
8:42.0		6-2	I. Shumpert misses 2-pt shot from 2 ft

FIGURE 10.1

An example of NBA play-by-play data, of the CHI@CLE game on May 4, 2015. (From Oliver, D., Basketball on paper, http://www.basketball-reference.com/boxscores/pbp/201505040CLE.html, 2002.)

10.4.1 Data

To be able to compute any kind of PM metrics, it is necessary to have play-by-play data. Play-by-play data contain information on the following actions in a basketball game (Figure 10.1):

- Rebounds (offensive, defensive; individual or team rebounds)
- Turnovers (including type of turnover, whether it was a steal, etc.)
- Shots (type of shot: field goal, free throw, or three-pointer; whether the shot was made or missed; distance)
- Player substitutions

For every one of these actions (unless the action is team-specific), the play-by-play data provide the name(s) of the player(s) involved, and a time stamp of the action. For computing a player's PM, we need to know when this player was on the court. For (R)APM, we need to know which exact 10 players were on the court at any given time in the game. As these metrics try to estimate player influence *per possession*, we need to parse the play-by-play for "possession-ending" actions. These include

gameID	poss. end time	H1	H2	H3	H4	H5	A1	A2	A3	A4	A5	Home_points	Away_points	Home_poss	Away_poss
201505040CLE	47 39.0	I. Shumpert	M. Miller	K. Irving	T. Mozgov	L. James	P. Gasol	M. Dunleavy	J. Butler	J. Noah	D. Rose	0	3	0	1
201505040CLE	47 13.0	I. Shumpert	M. Miller	K. Irving	T. Mozgov	L. James	P. Gasol	M. Dunleavy	J. Butler	J. Noah	D. Rose	2	0	1	0
201505040CLE	46 57.0	I. Shumpert	M. Miller	K. Irving	T. Mozgov	L. James	P. Gasol	M. Dunleavy	J. Butler	J. Noah	D. Rose	0	0	0	1
201505040CLE	46 43.0	I. Shumpert	M. Miller	K. Irving	T. Mozgov	L. James	P. Gasol	M. Dunleavy	J. Butler	J. Noah	D. Rose	0	0	0	1
201505040CLE	46 18.0	I. Shumpert	M. Miller	K. Irving	T. Mozgov	L. James	P. Gasol	M. Dunleavy	J. Butler	J. Noah	D. Rose	0	3	0	1
201505040CLE	46 04.0	I. Shumpert	M. Miller	K. Irving	T. Mozgov	L. James	P. Gasol	M. Dunleavy	J. Butler	J. Noah	D. Rose	0	0	1	0
201505040CLE	45 42.0	I. Shumpert	M. Miller	K. Irving	T. Mozgov	L. James	P. Gasol	M. Dunleavy	J. Butler	J. Noah	D. Rose	0	0	0	1
201505040CLE	45 33.0	I. Shumpert	M. Miller	K. Irving	T. Mozgov	L. James	P. Gasol	M. Dunleavy	J. Butler	J. Noah	D. Rose	0	0	1	0
201505040CLE	45 20.0	I. Shumpert	M. Miller	K. Irving	T. Mozgov	L. James	P. Gasol	M. Dunleavy	J. Butler	J. Noah	D. Rose	0	0	0	1
201505040CLE	45 11.0	I. Shumpert	M. Miller	K. Irving	T. Mozgov	L. James	P. Gasol	M. Dunleavy	J. Butler	J. Noah	D. Rose	0	0	1	0
201505040CLE	45 04.0	I. Shumpert	M. Miller	K. Irving	T. Mozgov	L. James	P. Gasol	M. Dunleavy	J. Butler	J. Noah	D. Rose	0	0	0	1

FIGURE 10.2
An example of NBA matchup data, of the CHI@CLE game on May 4, 2015. (After parsing Play-by-Play data from http://www.basketball-reference.com/boxscores/pbp/201505040CLE.html.)

- Turnovers
- Made field goals with no shooting foul call
- The last of a possible series of free throws, if made (i.e., "2 of 2," but not "1 or 2")
- Defensive rebounds

10.4.2 Determining the Players Are on the Court

For (R)APM analysis, it is necessary to determine which 10 players are on the court at any given time in a basketball game. To determine which players are on the court, XY-coordinate data from services such as SportVU or play-by-play data can be used. The play-by-play data will have to be parsed for substitutions and player names to determine who is currently on the court. At the end of each possession, the sum of the points scored in this possession and which team was in possession of the ball needs to be recorded. This should ultimately lead to matchup data. An example of matchup data is shown in Figure 10.2.

10.4.3 Computing the Different Metrics

10.4.3.1 PM

A player's PM can be determined the following way: For every possession that a player is on the court, sum all points scored by his team and subtract points scored by the opponent team. PM is often presented on a per-possession basis. Thus, the number of possessions a player has played needs to be recorded as well. A player's total PM will have to be divided by the his total number of possessions played. The results from January of the 2015/2016 NBA season can be seen in Table 10.1. When examining the results, it is evident that players of the same team can often have very similar PM, as some of them spend a lot of time together on the court. Good teams tend to have many players at the top of the PM leaderboard, while poor teams tend to have many players among the worst in PM. Critics might argue that Patty Mills' PM, for example, could simply be the product of his sharing a lot of his time on the court with other good Spurs players and an argument could be made that Mills' PM would suffer significantly if he wasn't playing on the 2015/2016 Spurs. In similar vein, one could make the argument that Noel's PM is dragged down by sharing lots of court time with Jahlil Okafor. It is also important to note that only players with more than 400 min played are shown in the table. Without any cutoff, different players, including many lesser-known ones, would appear at the top of PM rankings, but a large part of their PM is a result of randomness.

TABLE 10.1

PM and NET, for the 2015/2016 NBA Season,
January 12

Player	PM	NET
Steph Curry	+21.8	+28.5
Manu Ginobili	+20.0	+ 7.4
Draymond Green	+19.9	+27.3
Tim Duncan	+18.4	+ 5.7
Patrick Mills	+18.4	+ 5.6
...		
Kobe Bryant	−13.8	−8.2
Roy Hibbert	−14.4	−9.9
Nerlens Noel	−15.5	−6.7
Nik Stauskas	−16.0	−4.9
Jahlil Okafor	−18.1	−12.1

Source: http://www.basketball-reference.com.

10.4.3.2 NET Rating

For computing NET ratings, we need a player's raw PM and we need to know the team(s) on which he has played during the season(s). For these teams, we accumulate the PM of the lineups that did not include him (which we call the OFF Rating). Then we simply subtract this player's OFF Rating from his PM to get his NET rating. NET ratings for selected players are shown in Table 10.1. One will notice that Steph Curry and Draymond Green, both playing for the Warriors, have a vastly higher NET rating as those players on the Spurs. This is very likely a function of the talent distribution arguably being narrower on the Spurs roster.

10.4.3.3 APM

The calculation of APM requires regression analysis. Given that the matrices in APM are generally rather large, having software that is able to work with sparse matrices is almost a necessity. Sparse matrices are often used when the majority of entries in a matrix is 0. The matrix is then stored in a specialized format that takes up a fraction of the space that a standard ("dense") matrix would take. The programming languages MATLAB®, R, and Python all possess capabilities to work with and store sparse matrices.

The design matrix X consists of rows indexed by observations and columns that correspond to individual player dummy variables. As indicated earlier, when computing (R)APM, and computing both offensive and defensive players' impact, every possession in a basketball results in an observation. There are roughly 190 possessions in a basketball game, on average. In the regular season, 1230 games are played, with an additional approximately 90 play-off games. In order to run (R)APM with 1 year of NBA data, we thus need to work with a design matrix with about $1320 \times 190 = 250{,}800$ rows. Furthermore, the matrix needs to have as many columns as predictor variables. In the analysis for computing both offensive and defensive ratings for every player, the number of columns in the design matrix is twice the number of players in the league (one offensive and one defensive dummy variable per player). In most NBA seasons, there are approximately 500

TABLE 10.2

Single- and Multiseason APM and RAPM

APM (15/16)		APM (13/14–15/16)		RAPM (15/16)		RAPM (13/14–15/16)	
Name	Rating	Name	Rating	Name	Rating	Name	Rating
Channing Frye	+14.3	Steph Curry	+13.7	Steph Curry	+6.1	Steph Curry	+8.0
LeBron James	+13.3	Chris Paul	+12	Draymond Green	+5.5	LeBron James	+7.3
Steph Curry	+12.7	Ricky Rubio	+12	LeBron James	+4.3	Draymond Green	+6.8
Kyle Lowry	+12.3	LeBron James	+11.5	Kawhi Leonard	+3.8	Chris Paul	+6.5
Omri Casspi	+12	Kevin Durant	+10.2	Manu Ginobili	+3.8	Kawhi Leonard	+6
..							
Reggie Bullock	−11.6	Montrezl Harrell	−11	Gary Neal	−2.9	Nik Stauskas	−4
Caron Butler	−12.2	Jabari Brown	−11.2	Terrence Jones	−2.9	John Henson	−4.1
Jason Richardson	−12.5	Jason Richardson	−11.5	Nerlens Noel	−3.2	Anthony Bennett	−4.3
Richard Jefferson	−12.9	Tyus Jones	−12.5	Nik Stauskas	−3.7	Tony Wroten	−4.4
Tyus Jones	−13	Tyshawn Taylor	−12.7	Jahlil Okafor	−4.9	Andrea Bargnani	−4.5

different players that play at least one possession, so our design matrix would need about 1,000 columns, pushing its final size to roughly 250,000 × 1,000 for one NBA season. In each row of our X-matrix, exactly 10 entries have a nonzero value. This leaves us with a design matrix in which approximately $10/500 = 2\%$ entries are nonzero. The left-hand columns in X will represent the variables for offensive impact, and the right-hand columns represent the variables for defensive impact.

With this design matrix and response vector, we are now able to compute our regression coefficients. In Python, using the package "scikit-learn," the code to compute the coefficients is as follows:

```
clf = linear_model.LinearRegression()
clf.fit(X, y)
```

As can be seen in the results in Table 10.2, player ratings have a wide deviation from 0 in APM, and players that are not regarded as superstars and players with a relatively low amount of minutes played can find themselves at the top (e.g., Channing Frye, Omri Casspi). Multiyear APM shows more of the well-known stars at the top with Curry, Paul, James, and Durant, but also, somewhat surprisingly, has Rubio in the top five. While the player names might seem familiar at the very top, APM has the tendency to overstate the impact of very good players.

For computing RAPM estimates, we can use the exact same design matrix and response vector we used for APM, but we need to call a different regression function. In Python, one could run

```
clf = linear_model.Ridge(alpha = 3000)
clf.fit(X, y)
```

with 3000 representing the optimal penalization value found by Joe Sill and others, using cross validation. This value can mildly vary with the amount of data used.

In Ridge Regression, one needs to determine the optimal penalization parameter alpha. For standard RAPM, this parameter was found to be 3000. Different techniques for determining alpha have been suggested, but the author of this chapter strongly prefers cross validation over other methods. Using cross-validation tends to lead to the best prediction accuracy. Cross validation, as described here, involves splitting the data in two complementary subsets, a "training set" and a "test set." With different values of the ridge parameter alpha, linear coefficients for the player variables are estimated on the training set. Finally, a prediction error is measured on the test set for the given alpha and the set of estimated player coefficients. Some software packages, such as python's `scikit-learn`, provide built-in cross-validation methods to determine the ridge parameter. A call to this function might look like the following:

```
clf = linear_model.RidgeCV(alpha = [1, 10, 100, 1000, 10000])
clf.fit(X, y)
print clf.alpha_
```

Once the best alpha in a certain range is determined, furthermore, more fine-grained searches for the optimal alpha can be made.

10.5 Outlook

Real Plus Minus, an SPM/RAPM hybrid metric, has found its way into a wide array of articles on the web, essentially replacing PER as the go-to metric at ESPN.com. NBA players, such as Khris Middleton who would probably not have received big contracts in the past, are now getting (near) maximum deals due to their strong performance based on PM-based metrics. Several of the creators of both APM and its derivatives are currently working for NBA teams. This list includes Joe Sill who works for the Washington Wizards and Dan Rosenbaum who works for the Atlanta Hawks. These analysts are influencing decisions when it comes to trades and free-agent signings. PM-based metrics (RAPM, ESPN's Real Plus Minus, SPMs such as BPM) currently represent the most accurate publicly available player metrics, with those metrics that consist of, or are informed by, an SPM prior, performing slightly better than their pure PM counterparts. With multiyear RAPM building the basis of these SPMs, work should be focused on creating more refined SPMs through the addition of play-by-play and camera data, which allow extracting more individual player information. The APM framework itself can further be modified to find the impact players have on their teammates' box score statistics [13], which can lead to more accurate estimates of what a player's box score statistics would be on an "average team." Player impact on changes in win probability needs to be further studied, and its prediction accuracy needs to be compared to that of RAPM.

References

1. https://en.wikipedia.org/wiki/Efficiency_(basketball), last modified September 2, 2016.
2. https://en.wikipedia.org/wiki/Plus-minus, last modified August 28, 2016.

3. Rosenbaum, D. Measuring how NBA players help their teams win. http://www.82games.com/comm30.htm, accessed January, 2015.
4. https://en.wikipedia.org/wiki/Jeff_Sagarin#Winval, last modified September 8, 2016.
5. Sill, J. Improved NBA adjusted +/− using regularization and out-of-sample testing. http://www.sloansportsconference.com/?p=2798, accessed January, 2015.
6. Goldman, M., Rao, J.M. Live by the three, die by the three? The price of risk in the NBA. http://www.sloansportsconference.com/?p=10181, accessed January, 2015.
7. Engelmann, J. The effect of leading. http://apbr.org/metrics/viewtopic.php?f=2&t=8501, accessed January, 2015.
8. Engelmann, J. The effect of aging. http://apbr.org/metrics/viewtopic.php?f=2&t=8308, accessed January, 2015.
9. Oliver, D. (2002) Basketball on paper. http://www.basketball-reference.com/boxscores/pbp/201505040CLE.html.
10. Myers, D. About box plus/minus. http://www.basketball-reference.com/about/bpm.html, accessed January, 2015.
11. Stern, H.S. (1994). A Brownian motion model for the progress of sports scores. *Journal of the American Statistical Association*, 89(427), 1128–1134.
12. Lock, D., Nettleton, D. (2014). Using random forests to estimate win probability before each play of an NFL game. *Journal of Quantitative Analysis in Sports*, 10(2), 197–205.
13. Engelmann, J. Estimating a player's influence on his teammates' box score statistics using a modified adjusted +/− framework. https://www.youtube.com/watch?v=OuC0YZTADcE.

11

Optimal Strategy in Basketball

Brian Skinner and Matthew Goldman

CONTENTS

11.1 Introduction

The game of basketball, like just about any other game, can be described as a sequence of random events, with the probabilities of different outcomes determined by the skill levels of the players involved. More specifically, during each offensive possession of a basketball game, the team with the ball attempts to score, and this attempt results in some number of points ranging between 0 and 3 (or, rarely, 4). At the end of the game, what matters is not the absolute score of a given team but the differential score, defined as $\Delta =$ (total points scored by team A) − (total points scored by team B). The goal of team A is for the game to end with $\Delta > 0$. As such, this differential score is the primary random variable for describing a basketball game in the statistical sense. Understanding its expected value and variance, and how basketball teams can influence them through strategic decisions, is the main focus of this chapter.

Despite some important exceptions (which will be discussed in detail in Section 11.5, and which do not fundamentally change the strategic considerations of this chapter), empirical studies of scoring in basketball suggest that scoring events can be modeled with a fairly high degree of accuracy as independent and identically distributed (IID). This statistical independence implies that, as the game progresses, the differential score undergoes a biased random walk. Its random fluctuations are therefore nearly identical to those of Brownian motion (Gabel and Redner, 2012; Clauset et al., 2015), while the net drift of the score is determined by the skill levels of the two teams and by the scoring strategies they employ.

In the limit where there are many possessions remaining in the game, one can invoke the central limit theorem, which says that regardless of the process used by each team to score during a given possession, the probability density $P(\Delta)$ for many possessions is described by a Gaussian distribution:

$$P(\Delta) = \frac{1}{\sqrt{2\pi\sigma^2}}e^{-(\Delta-\mu)^2/2\sigma^2}. \tag{11.1}$$

where

　μ is the expected value of the differential score
　σ^2 is its variance

The intent of strategic decisions in basketball, at the level of a single game, is to maximize the probability of victory, $\mathcal{P} = \int_0^\infty P(\Delta)d\Delta$, that is, the probability of ending the game with $\Delta > 0$. In most cases, maximization of $\mathcal{P}$ is as straightforward as trying to maximize the expected number of points scored by the team, and to minimize the expected number of points scored by the opponent. As we will discuss later, however, this is not always the case, particularly in late-game situations.

　The remainder of this chapter is dedicated to discussing the principles associated with optimal strategy in basketball—that is, with the maximization of $\mathcal{P}$. Of course, it should go without saying that trying to maximize the variable $\mathcal{P}$ is not as straightforward as your typical math problem. Generally speaking, it is as tricky and nuanced as basketball itself and requires one to think about the principles of teamwork, leadership, and performance under pressure. Nonetheless, our goal in this chapter is to review how these all-important concepts can be described quantitatively, and how they ultimately relate back to a mathematical maximization of $\mathcal{P}$.

　We focus our discussion in this chapter around three general principles associated with optimal strategy in basketball. We refer to these principles as allocative efficiency (Section 11.2), dynamic efficiency (Section 11.3), and the risk/reward trade-off (Section 11.4). For each of these three principles, we provide an illustrative example and present an overview of general analytical results. We then provide some preliminary discussion on how these principles can be combined with data from real basketball games to evaluate and improve the decision-making of basketball teams.

11.2 Allocative Efficiency

When planning its offensive strategy, the primary decision faced by a basketball team is this: which shots should the team take and which players should take them? In more general language, this question becomes: how does the team optimally allocate its shot attempts between different offensive options? Or, from the perspective of players in the flow of the game, which shooting opportunities are *good enough* to be taken, and which should be passed up?

　The most naïve approach to this optimization problem would be to simply determine which play (or player) provides the highest expected return in terms of points scored, and then to run that play exclusively. For example, in the 2014–2015 season of the National Basketball Association (NBA), Kyle Korver led the league with a true shooting percentage of 0.699.* This statistic essentially means (Kubatko et al., 2007) that for every shot Kyle Korver took, his team, the Atlanta Hawks, scored $2 \times 0.699 = 1.398$ points,[†] as compared to only 1.089 points on their average possession. Despite his high efficiency, Korver only attempted 8.0 shots per game in that season (less than 10% of his team's total). Clearly, the Hawks are failing to adhere to such a simple model of optimality. But should Kyle Korver really be taking many more shots?

To answer this question, consider what it would mean for Korver to take a significantly larger proportion of his team's shots. A substantial increase in Korver's shooting rate could be accomplished in two different ways. First, the Hawks could run many more plays that are designed to end with a Korver shot. In doing so, however, the Hawks would risk becoming highly predictable offensively—that is, the defense would learn to focus their attention on Korver, and his effectiveness as an offensive option would plummet. A second option for the Hawks would be to run their usual offense, but to have Korver simply increase his willingness to shoot, taking contested shots that he might usually have passed on. By definition, these *marginal* shots are not ones that Korver would have taken previously, and they will (presumably) result in fewer points on average.

In either scenario, it is reasonable to expect that the average return of a Korver shot decreases as Korver's offensive load is increased. In more general terms, this example highlights the crucial idea that the effectiveness of a particular play (or player) is not adequately described by a single number. Instead, a play should be characterized by a *function* that describes how the expected return of the play declines with increased use.

In the seminal work by Oliver (2004), this relationship between efficiency and usage is called a "skill curve." Oliver proposed that this relation could be described by a function f that represents the average number of points scored by a given player (per possession used) as a function of the proportion p of possessions used by that player while he/she is on the court. Under this definition, the relation $f(p)$ is descriptive of the player's skill level: highly skilled players have a skill curve $f(p)$ that declines very little with increased offensive load p, while less skilled players have a curve $f(p)$ with a more strongly negative slope.[‡] The first attempts to measure skill curves involved tallying the number of points scored by a given player per possession, f, as well as the fraction of possessions used by that player, p, for all games within a particular time period. Estimates for $f(p)$ were then produced by sorting games into bins by their value of p and making a simple average of f for each bin.

In this chapter, on the other hand, we adopt a more generalized definition of the relation $f(p)$. In particular, we take p to be a fraction that represents the frequency with which a particular play is used (i.e., the number of times that that play is used, normalized by the total number of possessions), and we take f to be the average number of points scored per possession when that play is run. We generally refer to $f(p)$ as a "usage curve," which can refer to a specific play rather than to the aggregate effectiveness of all shots taken by a particular player.

While a quantitative determination of the function $f(p)$ for a given play or player can be a difficult statistical problem, the idea that the effectiveness of a play declines with increased

---

* Data from http://www.basketball-reference.com/.
† This number does not include points the Hawks might score after rebounding Kyle Korver's missed shots. It is likely that the Hawks averaged more than 1.5 points in total on possessions where he took a shot.
‡ Indeed, this pattern was documented in Goldman and Rao (2011), who found that guards tend to have flatter usage curves than forwards and centers, and that higher scoring players have flatter usage curves more generally.

usage is a fairly intuitive one. Among players and coaches, this idea is often expressed by saying that the team needs to "keep the defense honest," or to "take what the defense gives them." But using usage curve relations to reach quantitative decisions about optimal strategy requires more careful consideration.

To get a sense of how this optimization process works, imagine a very simplified hypothetical example of a strategic decision in basketball, in which a team is attempting to optimize the frequency p with which it takes corner three-point shots (i.e., p is the fraction of all shots that are corner threes). Suppose it is known that the average number of points scored per attempted corner three, f, follows the relation $f(p) = 2(1-p)$. In other words, in this example the corner three produces as much as two points on average when attempted very rarely (i.e., the shot is made 67% of the time), and produces close to zero points if it is attempted every time down the court. All other offensive options, taken together, are assumed to produce a constant 1.0 points per possession. What is the optimal use of the corner three by this offense?

The answer to this question can be found in a simple way by writing down a function F that describes the expected total number of points scored as a function of p:

$$F = p \times f(p) + (1-p) \times 1.0. \tag{11.2}$$

In this equation, the first term corresponds to the number of points scored from corner threes, and the second term is the number of points scored by all other shots. The optimal value of p is the one that maximizes F, and it can be found by taking the derivative of F with respect to p and equating it to zero:

$$\frac{dF}{dp} = \frac{d}{dp}\left(pf(p)\right) - 1.0 = 0. \tag{11.3}$$

Solving this equation gives $p = p_{\text{opt}} = 0.25$. That is, in this example the optimal proportion of corner threes is 25%. This is illustrated in Figure 11.1.

It is important to note that in this optimal strategy, the corner three pointer still produces significantly more points on average than other, nonthree point options (1.5 points as compared to 1.0 points). Noticing this disparity, one would be tempted to suggest that more corner threes should be taken, since they produce a larger average return. Such a conclusion, however, is a mistake, and it arises from focusing on the *average* shot quality $f(p)$ rather than the *marginal* shot quality $d(pf(p))/dp$. In the correct solution, it is the marginal values of the two options that become equal, and not the average values.

In fact, one can generalize this result into a principle that we refer to as the *allocative efficiency criterion*:

> In an optimal strategy, all offensive options have the same marginal efficiency.

Formally, the allocative efficiency criterion can be written as follows. Suppose that different plays $1, 2, 3, \ldots$ are characterized by functions $f_1(p_1)$, $f_2(p_2)$, $f_3(p_3) \ldots$, which describe the average number of points scored per possession by each play as a function of the frequencies $p_1, p_2, p_3 \ldots$ with which the plays are run. The play frequencies are defined such that $p_1 + p_2 + p_3 + \cdots = 1$. The choice of different play frequencies defines an offensive strategy. The expected number of points scored per possession for a given strategy is given by

$$F = p_1 f_1(p_1) + p_2 f_2(p_2) + p_3 f_3(p_3) + \cdots \tag{11.4}$$

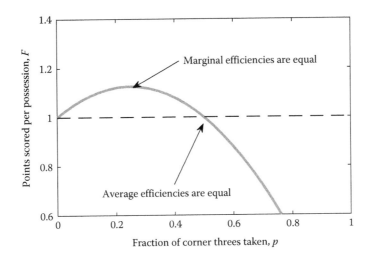

FIGURE 11.1
Optimal usage of the corner three-point shot in the simplified example of this section. The solid line shows the average number of points scored per shot when a fraction p of the team's shots are corner threes. The dashed line shows the effectiveness of all non-corner-three options. Choosing $p = 0.5$ equates the average effectiveness of all offensive options, but $p = 0.25$ equates their marginal effectiveness and provides the optimal solution.

This function is maximized when

$$\frac{d(p_1 f_1(p_1))}{dp_1} = \frac{d(p_2 f_2(p_2))}{dp_2} = \frac{d(p_3 f_3(p_3))}{dp_3} = \cdots \tag{11.5}$$

Since the quantity $d(p_i f_i(p_i))/dp_i$ represents the marginal efficiency of play i, Equation 11.5 is the quantitative form of the allocative efficiency criterion.

If the usage curves $f_i(p_i)$ are all known, then in principle the optimal strategy can be found by solving the set of equations in (11.5). In the case where all functions $f(p)$ are linear, so that $f_i(p_i) = \alpha_i - \beta_i p_i$, there is a relatively simple analytical solution. (Here α_i and β_i are constants and the index i runs from 1 to N, where N is the total number of plays being considered by the offense.) In this case, the optimal frequency of play i is given by (Skinner, 2010)

$$p_{i,\text{opt}} = \frac{\alpha_i + \lambda}{2\beta_i}, \tag{11.6}$$

where λ is a constant given by*

$$\lambda = \frac{2 - \sum_{i=1}^{N} \alpha_i/\beta_i}{\sum_{i=1}^{N} 1/\beta_i}. \tag{11.7}$$

In cases where the functions $f_i(p_i)$ are not linear, one can find the optimum strategy using a numerical search for the maximum of Equation 11.4.

---

* It is possible that, under certain, somewhat unrealistic, conditions, Equation 11.6 can give negative solutions for the frequency p_j of some play j. In this case, one should set $p_j = 0$ and resolve for the remaining p's.

The topic of shot allocation in basketball has been the subject of a significant number of studies during the past two decades, with most studies focusing on the choice between two-point and three-point shots. For example, in Vollmer and Bourret (2000); Alferink et al. (2009); Neiman and Loewenstein (2014), the proportion of three-point shots was compared to predictions from the so-called "matching law" from psychology. The matching law, generally speaking, holds that there is a direct relation between the rate of incidence of a behavior and its rate of reinforcement. In the context of two-point and three-point shots, the matching law predicts that the proportion of attempted threes should match the proportion of points scored by such shots, and this law seems to closely match the shooting habits of NBA players. Of course, adherence to this matching law does not necessarily indicate optimal shot selection. Other studies have examined the two-point/three-point choice through the lens of reinforcement learning and have shown that NBA players tend to overgeneralize from previous successes (Neiman and Loewenstein, 2011) and to use only the two-point/three-point distinction as the primary characteristic for generalization in learning (Neiman and Loewenstein, 2014). Thus, it may be the psychology of reinforcement, rather than a detailed understanding of optimal behavior, that drives shot selection in basketball players.

Some previous studies have examined the question of optimal play selection in soccer, football, and baseball through the paradigm of the "minimax" rule in game theory (Palacios-Huerta, 2003; Kovash and Levitt, 2009). The minimax rule generally dictates that an optimal game-theoretical strategy (say, for the defense) is the one that minimizes the worst-case scenario (the number of points given up). These previous studies, however, evaluated adherence to the minimax strategy only by examining the *average* return from different offensive options, rather than the marginal return. Directly analogous studies in basketball are therefore unlikely to be able to gauge whether basketball teams are exhibiting optimal allocation between plays.

Finally, a number of studies have sought to quantify or refine the definition of "shot value" by either examining the spatial structure behind patterns of shot effectiveness (Reich et al., 2006; Cervone et al., 2014; Chang et al., 2014; Miller et al., 2014; Shortridge et al., 2014) or by attempting to describe the offense as a branching network of options (Fewell et al., 2012; Skinner and Guy, 2015). In Goldman and Rao (2011, 2014b), a method is presented for approximating the usage curve of a given player (or a given unit) by examining the shooting efficacy and the shooting rate as a function of the shot clock time. In particular, these authors show that one can estimate how an offensive option performs under the pressure of increased usage by observing its performance under the pressure of a dwindling shot clock. However, constructing a more robust method of measuring usage curves for specific plays/players remains a prominent challenge.

11.3 Dynamic Efficiency

In addition to the strategic considerations associated with allocation, which are outlined in the previous section, basketball strategy also has an important time element. Specifically, in most competitive basketball games, the teams' decision-making is constrained by a shot clock. As the clock winds down, the expected return of a given possession declines, and as a team becomes increasingly desperate, it must reconsider its previous unwillingness to take certain lower-quality shots. In this way, optimal decision-making in basketball is by nature *dynamic,* and depends on the amount of time remaining on the shot clock.

In this sense, the problem of shot selection in basketball falls into the class of "optimal stopping problems," wherein one is concerned with trying to choose the optimal moment to perform a specific action in the face of finite time and uncertainty about future events.* Similarly, in basketball the player with the ball must decide at the instant of a shot opportunity whether the available shot is good enough to take or whether it is smarter to pass up the shot and wait for a better opportunity. This decision necessarily involves weighing the expected return of the remainder of the possession. In mathematics and economics, such decisions are commonly described analytically by a Bellmann equation. As we show in the following, writing down an analog of the Bellmann equation for the shot selection process brings us to a general criterion for dynamic efficiency.

As a way of introducing the central ideas associated with dynamic efficiency, consider the following hypothetical game. Imagine that you are challenged to make a basketball shot and that the difficulty of the shot is chosen by spinning a roulette-style wheel. For example, the wheel could dictate the distance from which the shot has to be taken. After each spin of the wheel, you are given two choices: you can either attempt the shot or you can choose to spin the wheel again. You are given a finite number N of wheel spins (which we'll call "shot opportunities"), and the game ends when you have either taken a shot or run out of shot opportunities. This leads to a crucial strategic question: how do you decide whether a shot is good enough to take, or whether you should spin again? What will be the winning percentage associated with the optimal strategy?

When $N = 1$, meaning that you are only given one shot opportunity, the answers to both of these questions are simple. Since you only get one shot, you should take it. Your winning percentage will then be equal to the expectation value of your shooting percentage over all possible distances written on the wheel. In other words, if one defines the "shot quality" η for each possible distance as the probability that you will make the shot from that distance, then the chance that you will win the game is $V_1 = E[\eta]$ when $N = 1$.

When $N = 2$, on the other hand, you can afford to be more selective about your first shot. In particular, if the first available shot has a quality η_2 that is smaller than V_1, then it makes sense to pass up that first shot and spin the wheel again. This strategy will produce an expected winning probability V_2 that is strictly larger than V_1. Similarly, when $N = 3$, the first shot should be taken only when its quality exceeds the combined value V_2 of the next two shots, taken together.

This line of inductive reasoning leads to a recurrence equation for the winning probability V_N, based on the principle that when N shots are remaining, the first shot opportunity should be taken only when its quality η exceeds V_{N-1}. This recurrence relation was studied analytically by Skinner (2012). The numeric values of the sequence V_N depend on the probability distribution for η at each opportunity (i.e., which distances are written on the wheel, and how good you are at shooting from those distances).

Of course, this hypothetical game is significantly different from the real shot-selection problem faced by basketball teams. For example, in basketball the shot clock runs continuously rather than in discrete turns, and there is no reason to think that the distribution of shot quality $P(\eta)$ will be independent of the shot clock time. Nonetheless, the line of

---

* The most famous optimal stopping problem is the so-called "secretary problem," which imagines the process of choosing a secretary from a long line of applicants under the constraint that a candidate must be hired or dismissed immediately following the interview.

inductive reasoning used in our hypothesis can be extended to derive a general criterion for dynamic efficiency in basketball.

Specifically, let us first divide up the full length of the shot clock (usually, either 24 or 30 s) into 1-second intervals. Let η_1 be a random variable that describes the quality of a shot opportunity available to a team when there is only 1 second remaining on the shot clock. (η_1 can be defined as the probability that the shot will be made multiplied by the point value of the shot.) When only 1 second is remaining on the clock, the team with the ball must shoot regardless of the value of η_1, and as a consequence the expected number of points scored by the shot is just $V_1 = E[\eta_1]$.

When there are $t = 2$ seconds remaining, on the other hand, the team can afford to be slightly more selective. In particular, a shot should be taken immediately only if its quality η_2 exceeds V_1. The average return from this strategy, in terms of the number of points scored, is $V_2 = P(\eta_2 \leq V_1) \cdot V_1 + P(\eta_2 > V_1) \cdot E[\eta_2|\eta_2 > V_1]$, which is the weighted average of the value the team gets when it waits until $t = 1$ (the first term) and the average quality of the shot the team takes in instances when it shoots at $t = 2$ (the second term).

This line of thinking can be generalized for arbitrary time t into the following set of Bellmann-like equations:

$$V_t = P(\eta_t \leq c_t) \cdot V_{t-1} + P(\eta_t > c_t) \cdot E[\eta_t|\eta_t > c_t]$$

$$c_t = V_{t-1}. \tag{11.8}$$

The quantity V_t can be called the *value* of the possession when t seconds are remaining, and c_t is the *reservation threshold* for shooting with t seconds left. The optimal strategy is defined by shooting if and only if the quality η_t of a given shot opportunity exceeds c_t, with c_t given by Equation 11.8.

In conceptual terms, one can state the conclusion of Equation 11.8 as the following *dynamic efficiency criterion*:

> Shots should be taken only when their expected return exceeds the average value of continuing the possession.

Empirically testing whether teams adhere to dynamic efficiency is a complicated statistical problem. While the value V_t of continuing a possession can be readily measured from data, the reservation threshold c_t is not directly observed. Instead, we can only see average efficiencies from *all* shots taken with t seconds remaining. That is, we see $e_t = E(\eta_t|\eta_t > c_t)$, the average number of points scored by shots taken at time t, which is a number strictly greater than c_t. Another way to say this is that we are interested in the very worst shot a team is willing to take with t seconds remaining on the shot clock, but can only observe the outcomes of the average of all shots taken with t seconds remaining. This average will include some high-value, wide open shots that were easy decisions for the offensive team, and as such is not representative of the team's reservation threshold. One can, of course, apply a simple check for dynamic inefficiency by measuring whether $e_t > V_{t-1}$ for all values of the shot clock time t. If this condition is violated, then the team is overly willing to settle for low-quality shots at time t, since $c_t = V_{t-1}$ represents the lowest quality shot that a team should take and therefore it should never exceed e_t. However, $e_t > V_{t-1}$ does not by itself guarantee optimal dynamic efficiency.

Goldman and Rao (2014b) studied the question of dynamic efficiency by estimating a structural model in which the time remaining on the shot clock is used as an instrumental variable for shot selection. In this way, they were able to model the marginal (or worst)

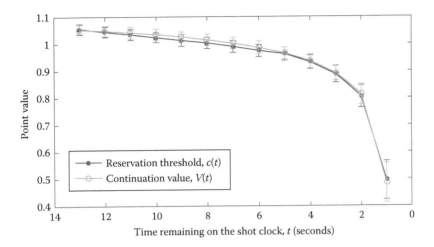

FIGURE 11.2
Average continuation values (open circles) and reservation thresholds (filled points) as a function of the shot clock time for the most common lineups in NBA between 2006 and 2012. (Data taken from Goldman, M. and Rao, J.M., Optimal stopping in the NBA: An empirical model of the Miami heat. SSRN Scholarly Paper ID 2363709, Social Science Research Network, Rochester, NY.)

shot a team is willing to take and compare its quality to the average number of points returned from continuing the possession. Figure 11.2 shows how both the average value of a team's possession and the reservation threshold decline in perfect unison as the shot clock ticks toward zero (from left to right), thus demonstrating impressive adherence to dynamic efficiency.

11.4 Risk and Reward

In the preceding two sections, we have essentially approached the question of optimal strategy as equivalent to the question of "how does a team maximize the expected differential score?" In other words, we have only discussed the process of trying to maximize the expected number of points scored by the offense per possession. But, in fact, this is not the same question as "how does a team maximize its chance of winning?" A true optimal strategy considers not just the expected outcome, but the full distribution of possible outcomes.

To see the difference, consider the following simple example. Imagine that a friend challenges you to a shooting contest. The rules of the game are that you each get to take 10 shots, either from the free throw line or from the three-point line. The free throw shots are worth 2 points each, and the three-point shots are worth 3. Suppose, further, that you know yourself to be a 60% free throw shooter and a 40% three-point shooter. Which shots should you choose to take?

As it turns out, the answer to this question depends entirely on whether your friend is a better shooter than you are.

In this example, the shooting percentages are such that both strategies (taking either all twos or all threes) have the same expected outcome: 12 points scored. But the two strategies differ in the variance of the outcome. In particular, the three-point-shooting strategy offers both a higher "best case scenario," and a lower "worst-case scenario." So the decision of whether to pursue that strategy depends on whether you expect that beating your friend is a likely or an unlikely event. If winning is likely, then you should go with the more conservative two-point shooting strategy. If it is unlikely, then you should choose the more risky three-point shooting strategy.

More formally, the "riskiness" of a strategy can be quantified by the variance σ^2 in the total number of points scored. Under the assumption that individual possessions are IID, the variance is simply the sum of the variances σ_i^2 of each possession i. This single-possession variance is given by

$$\sigma_i^2 = \sum_j P_j \cdot (f_j - \mu_i)^2, \tag{11.9}$$

where
  j indexes the set of possible outcomes of the possession i
  P_j is the probability of each outcome
  f_j is the number of points scored in outcome j

$$\mu_i = \sum_j P_j \cdot f_j \tag{11.10}$$

is the expectation value of the number of points scored in possession i. So, for example, in the shooting contest above, the expected number of points scored for a single two-point shot, $\mu_2 = 1.2$, is the same as the expected number of points scored for a three-point shot. However, the variance for a single two-point shot, $\sigma_2^2 = $ (probability of making the shot) $\times$ $(2 - \mu_2)^2 + $ (probability of missing the shot) $\times (0 - \mu_2)^2 = 0.6 \times (2 - 1.2)^2 + 0.4 \times$ $(0 - 1.2)^2 = 0.96 \, \mathrm{pts}^2$, is substantially smaller than the variance for a single three-point shot, $\sigma_3^2 = 0.4 \times (3 - 1.2)^2 + 0.6 \times (0 - 1.2)^2 = 2.16 \, \mathrm{pts}^2$.

In this example, a strategy of taking 10 two-pointers has a variance $\sigma^2 = 10 \times 0.96 = 9.6 \, \mathrm{pts}^2$ while a strategy of taking 10 three-pointers has a variance $\sigma^2 = 21.6 \, \mathrm{pts}^2$. Thus, one can say that the two-point strategy has a range of outcomes $\mu \pm \sigma = 12 \pm 3.1$ pts, while the three-point strategy has a range 12 ± 4.65 pts. So, for example, if you expect that your friend is going to score 15 points, then it makes more sense to shoot threes. If you only expect your friend to score 8 points, on the other hand, it's smarter to go with the more conservative two-point shots.

This example is relatively straightforward, since it involves choosing between two strategies with the same expected return. Choosing the correct strategy is therefore as simple as deciding whether large variance is a benefit (since it increases the probability of an unlikely upset, if you are the underdog) or a drawback (for the same reason, if you are the favorite). In most realistic scenarios, however, one must choose between strategies that do not have the same expected return. In such cases the ideal strategy may involve sacrificing from the expected number of points scored in order to increase the variance (if you are the underdog) or reduce it (if you are the favorite).

Graphically, one can say that the strategy adopted by each team is characterized by a distribution of possible final scores. For an underdog team, it is in the team's best interest to choose a strategy that maximizes the overlap between the two teams' distributions.

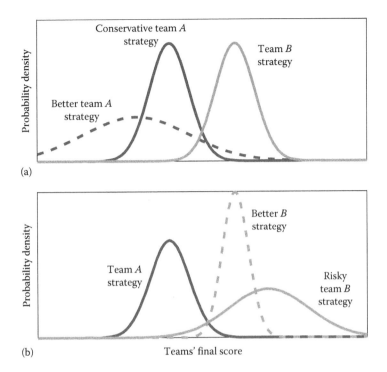

(a)

(b)

Teams' final score

FIGURE 11.3
Schematic picture of the distribution of final scores for two competing teams. (a) The underdog team (labeled A) improves its chance of winning by pursuing the risky strategy corresponding to the dashed line rather than the conservative strategy corresponding to the solid line, even though this lowers the expected final score. (b) Conversely, a favored team (labeled B) can improve its chance of winning using a conservative strategy (dashed line) rather than a risky strategy (solid line), even if this lowers its expected final score.

The favored team, on the other hand, should choose the strategy that minimizes the overlap between the two distributions. This is shown schematically in Figure 11.3.

At the most general level, one can describe the problem of optimal strategy in the following way. Let the final number of points s_A scored by team A be described by a probability distribution $P_A(s_A)$. Let $P_B(s_B)$ similarly be the probability distribution for the score s_B of team B. The probability of a victory by team A is given by*

$$\text{Prob}(s_A > s_B) = \int_0^\infty \int_{s_B+1}^\infty P_B(s_B)P_A(s_A)ds_A ds_B. \tag{11.11}$$

One can think of all strategizing by team A as an attempt to choose its probability distribution $P_A(s_A)$ in such a way that Equation 11.11 is maximized.

One can notice, of course, that the optimal strategy for A requires one to know the strategy for B, which is reflected in the probability distribution $P_B(s_B)$. But team B is similarly engaged in strategic decision-making that requires knowledge about the strategy of A.

---

* Technically, this equation yields only the probability of an outright victory, $s_A > s_B$, without considering ties and overtime situations.

Thus, in general, all strategic decisions are game-theoretical in nature and can in principle require complicated analytical descriptions. In practice, however, the game-theoretical equilibria are often fairly obvious: each team can use its opponent's previous record as an indicator of their intended strategy and ability and adopt their own strategy accordingly. The degree to which this approach can be expected to succeed, or can be exploited by a clever opponent, has not been explored in a rigorous way, but may prove a fascinating topic for future studies.

In general, Equation 11.11 is the quantity that is to be optimized by team A. However, as mentioned in the Introduction, if one assumes that all possessions are independent of each other, then this description can be simplified by writing only a probability density function for the differential score Δ. The description becomes particularly simple in the limit of many remaining possessions, for which the distribution becomes Gaussian:

$$P(\Delta) = \frac{1}{\sqrt{2\pi\sigma_\Delta^2}}e^{-(\Delta-\mu_\Delta)^2/2\sigma_\Delta^2}. \tag{11.12}$$

Here, the symbols μ_Δ and σ_Δ^2 denote the expected value and variance of the differential score, respectively. These are related to the values for each team by

$$\mu_\Delta = \mu_A - \mu_B, \tag{11.13}$$

$$\sigma_\Delta^2 = \sigma_A^2 + \sigma_B^2. \tag{11.14}$$

In this simplified description, the probability of a victory by team A is $\mathcal{P} = \int_0^\infty P(\Delta)d\Delta$. This function increases monotonically with the parameter

$$Z = \frac{\mu_\Delta}{\sigma_\Delta} = \frac{\mu_A - \mu_B}{\sqrt{\sigma_A^2 + \sigma_B^2}}. \tag{11.15}$$

Therefore, in the limit where there are many possessions remaining in the game, one can express the problem of optimal strategy in the following simple way (Skinner, 2011):

> A team's optimal strategy is the one that maximizes Z.

In general, Z can be increased either by increasing the expected number of points μ_A of the team's strategy or by altering the variance σ_A^2 of the team's strategy. In particular, when $\mu_A - \mu_B$ is negative (team A is the underdog), then Z is increased by increasing the variance σ_A^2. This is the mathematical formulation of the intuitive idea that underdogs must be willing to accept larger than average risk. Similarly, when $\mu_A - \mu_B$ is positive, Z is increased by *reducing* the variance—favored teams should play conservatively. Thus, the example from the beginning of this section can be understood in terms of Equation 11.15.

Most of the discussion surrounding the performance of actual basketball teams concerns efforts to increase a team's expected scoring (μ_A) or to decrease their opponents' expected scoring (μ_B). Nonetheless, there are a variety of strategies designed primarily to alter the variance of the outcome (the denominator of Equation 11.15 rather than the numerator). The most common technique is the choice of shot selection—in particular, the choice to either increase (when trailing) or decrease (when leading) the proportion of three-point shots taken. Another way to alter the variance of the outcome is by judicious manipulation

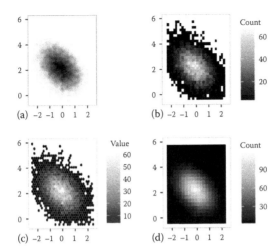

FIGURE 2.3
Four plots of Yu Darvish's pitch locations from 2012 to 2014. These plots take the perspective of the umpire at the pitch crosses the front of home plate. (a) All pitches with alpha transparency, (b) 2D histogram with rectangular binning, (c) 2D histogram with hexagonal binning, and (d) bivariate normal kernel density estimate.

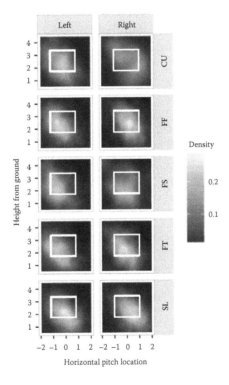

FIGURE 2.4
Bivariate normal kernel density estimates of Yu Darvish's pitches for each combination of batter handedness and pitch type.

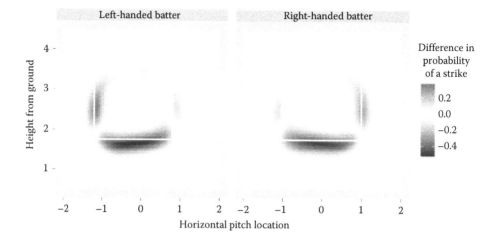

FIGURE 2.5
The difference in the probability of a called strike in 2008 versus 2014 for both left- and right-handed batters. A pitch over the plate at the knees was about 40%–50% more likely to be called a strike in 2014. A pitch on the inner/outer portion of the plate (above the knees) was about 20%–30% more likely to be called a strike in 2008.

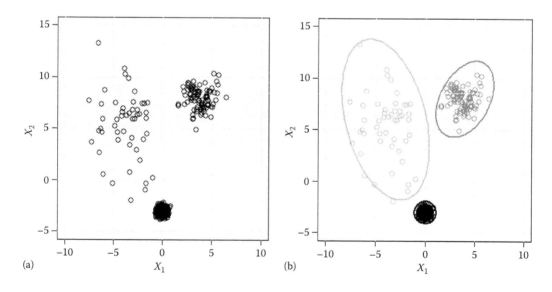

FIGURE 2.6
Exhibition of unclassified data (a) and cluster assignment (b) of generic, randomly generated data.

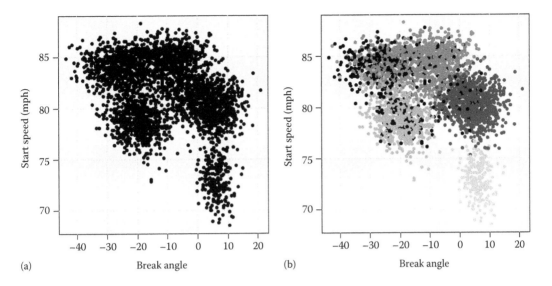

FIGURE 2.7
Exhibition of unidentified pitches from Mark Buehrle's 2013 regular season (a) and pitch cluster assignment (b) using *k*-means clustering with five clusters.

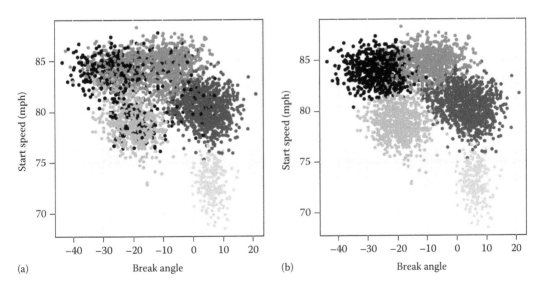

FIGURE 2.8
Exhibition of Mark Buehrle's 2013 pitch cluster assignment using *k*-means clustering (a) and cluster assignment using model-based clustering (b). The latter shows more clear cluster assignment, particularly for the black-colored cluster.

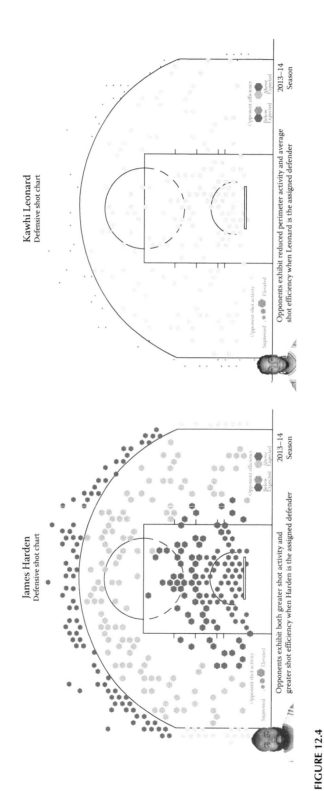

FIGURE 12.4

Kawhi Leonard tends to suppress shots on the perimeter (small dots). James Harden gives up more shots than average almost everywhere on the court (larger dots) and opponents tend to make more perimeter shots than average (red).

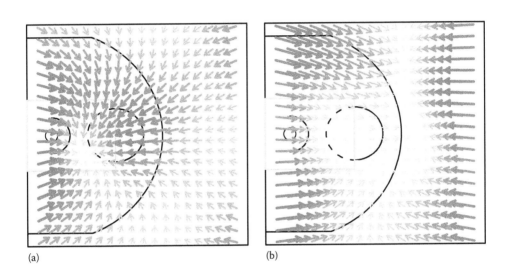

(a)   (b)

FIGURE 12.5
Acceleration fields $(\mu_x(\mathbf{z}(t)), \mu_y(\mathbf{z}(t)))$ for Chris Paul (a) and Tim Duncan (b) with ball possession. The arrows point in the direction of the acceleration at each point on the court's surface, and the size and color of the arrows are proportional to the magnitude of the acceleration. Comparing (a) and (b) for instance, we see that when both players possess the ball, Paul more frequently attacks the basket from outside the perimeter. Duncan does not accelerate to the basket from anywhere within the three point arc.

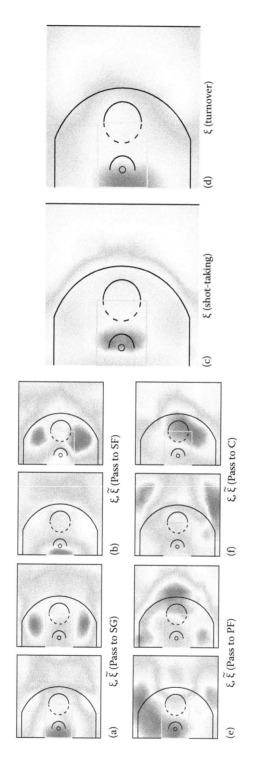

FIGURE 12.6

Plots of estimated spatial effects ξ for Stephen Curry as the ballcarrier. For instance, plot (c) reveals the largest effect on Curry's shot-taking probability occurs near the basket, with noticeable positive effects also around the three-point line. Plot (a) shows that he is more likely to pass to the center when in the wing areas—more so when the center is positioned near the free throw line. Plots (b)–(f) display other spatial effects as labeled.

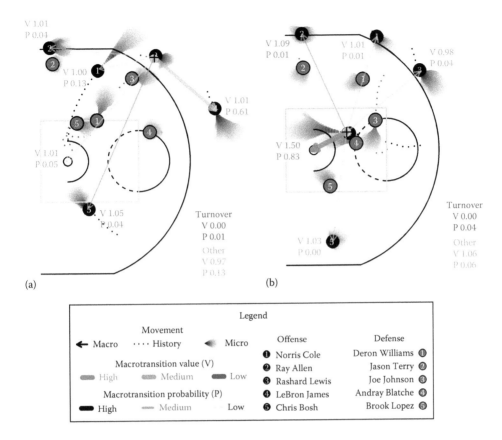

(a)

(b)

Legend

Movement
← Macro ···· History ◄ Micro

Macrotransition value (V)
High   Medium   Low

Macrotransition probability (P)
High   Medium   Low

Offense
❶ Norris Cole
❷ Ray Allen
❸ Rashard Lewis
❹ LeBron James
❺ Chris Bosh

Defense
Deron Williams ①
Jason Terry ②
Joe Johnson ③
Andray Blatche ④
Brook Lopez ⑤

FIGURE 12.8
Detailed diagram of EPV in terms of parameters of the underlying statistical models for player movement and decision-making. (a) The beginning of Rashard Lewis's possession in Figure 12.7 and (b) a fraction of a second before LeBron James attempts the layup at the end of the possession. Two seconds of players' predicted motion are shaded (with forecasted positions for short time horizons darker) while actions are represented by arrows, using color and line thickness to encode the value and probability of such actions. The value and probability of the "other" category represents the case that no pass/shot/turnover occurs during the next 2 seconds.

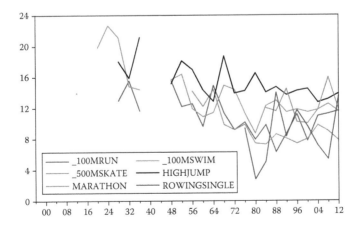

FIGURE 24.2
Relative difference (in percentages) between Gold medal performances of women and men in the 100 m running, the 100 m freestyle swimming, the 500 m skating (data for 1994–2010 are plotted as 1996–2012), the high jump, the marathon, and the single sculls rowing.

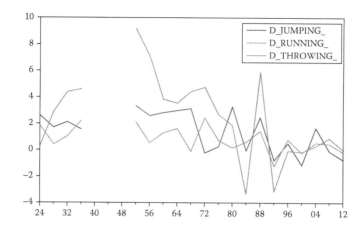

FIGURE 24.4
Average rate of progress of the Olympic Gold performances in Jumping, Running, and Throwing (1924–2012, both men and women).

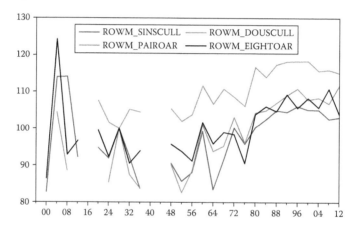

FIGURE 24.6
Performance Index (1928 = 100) of the Olympic Gold performances of men's rowing events: Single Sculls, Double Sculls, Pair-Oared Shell without Coxswain, Eight-Oared Shell with Coxswain (1896–2012).

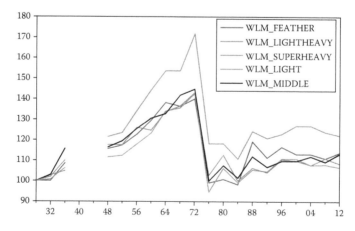

FIGURE 24.9
Performance Index (1928 = 100) of the Olympic Gold performances of men's weightlifting events: Featherweight, Lightweight, Light heavyweight, Middleweight, and Super heavyweight (1928–2012).

of the game clock—either by calling timeouts or by intentionally using the full length of a possession—in order to manipulate the number of possessions remaining in the game. Intentional fouling ("Hack a Shaq") can also be a useful way to alter the variance of the outcome, either by intentionally shortening possessions or by replacing potentially high-variance three-point opportunities with low-variance free throws.

Some elements of the reasoning outlined in this section have been appreciated in sports for a long time. For example, statistician Bill James developed as early as 40 years ago a heuristic rule for when a lead in college basketball is "safe" (James, 2008). His rule, taken in the limit when there is time remaining for $N \gg 1$ possessions, is that a lead is safe when the lead is larger than the square root of the number of seconds remaining in the game. The origin of this rule can be understood by examining Equation 11.15. Since the probability of a victory by the leading team depends only on Z, then the probability that a lead is "safe" is equivalent to the condition $Z > Z_c$, where Z_c is a constant number. For example, if you define "safe" to mean a 98% chance of victory, then $Z_c = 2.1$. If one takes μ_Δ to be equal to the current value of the lead Δ, then the condition for a safe lead becomes $\Delta > \text{const.} \times \sqrt{\sigma_\Delta^2}$. As discussed earlier, the variance σ_Δ^2 of the outcome is directly proportional to the number of possessions remaining, and therefore it is proportional to the amount of time τ left in the game. So the condition $Z > Z_c$ is equivalent to Bill James's heuristic condition for a safe lead: $\Delta > \text{const.} \times \sqrt{\tau}$. This rule has since been studied carefully and refined in Clauset et al. (2015).

Other authors have made empirical studies of optimal strategy in specific scenarios that fall within the risk/reward paradigm. For example, Annis (2006) studied situations where the defensive team is protecting a three-point lead in the final seconds of a basketball game and found that the optimal strategy is generally to foul intentionally (thereby giving up two free throws) rather than allow for the possibility of a three-point shot by their opponent. In the language of this section, one can say that this optimal strategy involves sacrificing points from the mean differential score in order to reduce variance in the score. In Wiens et al. (2013), the authors studied the question of how much effort an offensive team should put into offensive rebounding. Sending multiple players to "crash the boards" for an offensive rebound has a potentially large reward, but comes with the risk of a poorly defended fast break by the opponent. The study quantified the costs and benefits of different offensive rebounding strategies and reached the tentative conclusion that "risky" offensive rebounding is usually more valuable than the more conservative strategy of hedging against fast breaks.

In Goldman and Rao (2014a), the authors studied the risk/reward choices of NBA players, focusing on the frequency of three-point shots as an indicator of risk preference. They found that players correctly respond to trailing situations with increased preference for risk but also (incorrectly) exhibit increased preference for risk when their team has a large lead. In this sense, the general concepts of risk/reward trade-off are not strongly ingrained in the habits of professional basketball players.

11.5 Conclusion

To summarize, this chapter has presented an analytical discussion of the different facets associated with optimal strategy in basketball. Our primary message is that an optimal strategy must satisfy three conditions:

1. Allocative efficiency: All offensive options must have the same marginal efficiency.
2. Dynamic efficiency: The value of a shot taken at any instant in time must exceed the continuation value of the possession.
3. Risk/reward trade-off: An optimum strategy maximizes the probability of victory and not just the expectation value of the differential score.

Methods for evaluating these criteria, and the degree to which they are met by actual basketball teams, have been discussed in Sections 11.2, 11.3, and 11.4, respectively.

It is worth emphasizing, of course, that our results have been largely predicated on the assumption that individual possessions in a basketball game are IID. The validity of this assumption has been a topic of some controversy in the literature for more than 30 years. Most studies have focused on the existence of the "hot hand" in basketball (Gilovich et al., 1985; Wardrop, 1995; Arkes, 2010; Yaari and Eisenmann, 2011; Bocskocsky et al., 2014; Csapo and Raab, 2014; Miller and Sanjurjo, 2015), a purported positive correlation between successive shots taken by a given player. In fact there is some evidence for such a correlation (Arkes, 2010; Yaari and Eisenmann, 2011), and other studies have found additional violations of the IID assumption, such as a decline in shooting effectiveness in high-pressure situations (Goldman and Rao, 2014c) and an increase immediately following timeouts (Saavedra et al., 2012). There is also some evidence for a "momentum effect," in which the probability of a win is influenced by the outcome of recent previous games (Arkes and Martinez, 2011). Nonetheless, these correlations are generally very small in magnitude, so that any attempts to account for them in strategic decision-making would likely be swamped by statistical uncertainty in other variables.

At a practical level, the biggest hindrance to quantitative basketball strategy is usually the difficulty of accurately estimating the efficiency of different offensive options. The usage curves $f(p)$ are particularly difficult to estimate from easily measurable statistics and are necessary for a quantitative determination of optimal allocation. What's more, the usage curves are really only robustly defined relative to a particular defense, and can vary strongly depending on the quality of the team's opponent. A major advance in their determination may therefore provide the most important step toward enabling quantitative optimization of basketball strategy.

It is also worth noting that, while the majority of this chapter has focused on offensive choices, another major challenge is to create a succinct analytical description of basketball *defense*. The defense also faces choices about how to allocate its effort and resources, and in this sense the considerations outlined in this chapter can be recast in terms of the defense as well. Unfortunately, to our knowledge the necessary concept of "defensive usage curves" has not been explored with sufficient depth to enable a real analysis.

Despite these challenges, the level of sophistication associated with describing and optimizing the performance of basketball teams has increased enormously in the past decade. This growth has been driven largely by a revolution in the way that descriptive data are recorded and analyzed (Lowe, 2013), which has enabled hundreds of quantitative studies of basketball teams and players that would have been impossible only a decade ago.* We fully expect that within the coming decades many more important ideas will be developed that can be added to the ones presented here.

---

* To say nothing of the many more studies that are done privately by analysts working for professional basketball teams.

References

Alferink, L. A., Critchfield, T. S., Hitt, J. L., and Higgins, W. J. (2009). Generality of the matching law as a descriptor of shot selection in basketball. *Journal of Applied Behavior Analysis*, 42(3):595–608.

Annis, D. H. (2006). Optimal end-game strategy in basketball. *Journal of Quantitative Analysis in Sports*, 2(2), Article 1.

Arkes, J. (2010). Revisiting the hot hand theory with free throw data in a multivariate Framework. *Journal of Quantitative Analysis in Sports*, 6(1):2.

Arkes, J. and Martinez, J. (2011). Finally, evidence for a momentum effect in the NBA. *Journal of Quantitative Analysis in Sports*, 7(3):13.

Bocskocsky, A., Ezekowitz, J., and Stein, C. (2014). Heat check: New evidence on the hot hand in basketball. SSRN Scholarly Paper ID 2481494, Social Science Research Network, Rochester, NY.

Cervone, D., D'Amour, A., Bornn, L., and Goldsberry, K. (2014). A multiresolution stochastic process model for predicting basketball possession outcomes. *arXiv:1408.0777 [stat]*. arXiv: 1408.0777.

Chang, Y.-H., Maheswaran, R., Su, J., Kwok, S., Levy, T., Wexler, A., and Squire, K. (2014). Quantifying shot quality in the NBA. In *MIT Sloan Sports Analytics Conference*, Boston, MA. Available http://www.sloansportsconference.com/wp-content/uploads/2014/02/2014-SSAC-Quantifying-Shot-Quality-in-the-NBA.pdf. Accessed: August 30, 2016.

Clauset, A., Kogan, M., and Redner, S. (2015). Safe leads and lead changes in competitive team sports. *Physical Review E*, 91(6):062815.

Csapo, P. and Raab, M. (2014). Hand down, Man down. Analysis of defensive adjustments in response to the hot hand in basketball using novel defense metrics. *PLoS ONE*, 9(12):e114184.

Fewell, J. H., Armbruster, D., Ingraham, J., Petersen, A., and Waters, J. S. (2012). Basketball teams as strategic networks. *PLoS ONE*, 7(11):e47445.

Gabel, A. and Redner, S. (2012). Random walk picture of basketball scoring. *Journal of Quantitative Analysis in Sports*, 8(1), Manuscript 1416.

Gilovich, T., Vallone, R., and Tversky, A. (1985). The hot hand in basketball: On the misperception of random sequences. *Cognitive Psychology*, 17:295.

Goldman, M. and Rao, J. M. (2014a). Misperception of risk and incentives by experienced agents. SSRN Scholarly Paper ID 2435551, Social Science Research Network, Rochester, NY.

Goldman, M. and Rao, J. M. (2014b). Optimal stopping in the NBA: An empirical model of the miami heat. SSRN Scholarly Paper ID 2363709, Social Science Research Network, Rochester, NY.

Goldman, M. and Rao, J. M. (2014c). Loss aversion around a fixed reference point in highly experienced agents. Presented at the *2014 North American meeting of the Economics Science Association*, Fort Lauderdale, FL.

Goldman, M. R. and Rao, J. M. (2011). Allocative and dynamic efficiency in NBA decision making. In *MIT Sloan Sports Analytics Conference*, Boston, MA. Available http://www.sloansportsconference.com/wp-content/uploads/2011/08/Allocation-and-Dynamic-Efficiency-in-NBA-Decision-Making1.pdf. Accessed August 30, 2016.

James, B. (2008). The lead is safe. *Slate*. Available http://www.slate.com/articles/sports/sports_nut/2008/03/the_lead_is_safe.html.

Kovash, K. and Levitt, S. D. (2009). Professionals do not play minimax: Evidence from major league baseball and the national football league. Working Paper 15347, National Bureau of Economic Research, Cambridge, MA. Available http://www.nber.org/papers/w15347.

Kubatko, J., Oliver, D., Pelton, K., and Rosenbaum, D. T. (2007). A starting point for analyzing basketball statistics. *Journal of Quantitative Analysis in Sports*, 3(3):Article 1.

Lowe, Z. (2013). Seven ways the NBA's new camera system can change the future of basketball. Available http://grantland.com/the-triangle/seven-ways-the-nbas-new-camera-system-can-change-the-future-of-basketball/.

Miller, A., Bornn, L., Adams, R., and Goldsberry, K. (2014). Factorized point process intensities: A spatial analysis of professional basketball. *arXiv:1401.0942 [stat]*. arXiv: 1401.0942.

Miller, J. B. and Sanjurjo, A. (2015). Surprised by the Gambler's and hot hand fallacies? A truth in the law of small numbers. SSRN Scholarly Paper ID 2627354, Social Science Research Network, Rochester, NY.

Neiman, T. and Loewenstein, Y. (2011). Reinforcement learning in professional basketball players. *Nature Communications*, 2:569.

Neiman, T. and Loewenstein, Y. (2014). Spatial generalization in operant learning: Lessons from professional basketball. *PLoS Comput Biol*, 10(5):e1003623.

Oliver, D. (2004). *Basketball on Paper: Rules and Tools for Performance Analysis*. Potomac Books, Dulles, VA.

Palacios-Huerta, I. (2003). Professionals play minimax. *The Review of Economic Studies*, 70(2):395–415.

Reich, B. J., Hodges, J. S., Carlin, B. P., and Reich, A. M. (2006). A spatial analysis of basketball shot chart data. *The American Statistician*, 60(1):3–12.

Saavedra, S., Mukherjee, S., and Bagrow, J. P. (2012). Is coaching experience associated with effective use of timeouts in basketball? *Scientific Reports*, 2, Article 676.

Shortridge, A., Goldsberry, K., and Adams, M. (2014). Creating space to shoot: Quantifying spatial relative field goal efficiency in basketball. *Journal of Quantitative Analysis in Sports*, 10(3):303–313.

Skinner, B. (2010). The price of anarchy in basketball. *Journal of Quantitative Analysis in Sports*, 6(1):Article 3.

Skinner, B. (2011). Scoring strategies for the underdog: Using risk as an ally in determining optimal sports strategies. *Journal of Quantitative Analysis in Sports*, 7(4):Article 11.

Skinner, B. (2012). The problem of shot selection in basketball. *PLoS ONE*, 7(1):e30776.

Skinner, B. and Guy, S. J. (2015). A method for using player tracking data in basketball to learn player skills and predict team performance. *PLoS ONE*, 10(9):e0136393.

Vollmer, T. R. and Bourret, J. (2000). An application of the matching law to evaluate the allocation of two- and three-point shots by college basketball players. *Journal of Applied Behavior Analysis*, 33(2):137–150.

Wardrop, R. L. (1995). Simpson's paradox and the hot hand in basketball. *The American Statistician*, 49(1):24–28.

Wiens, J., Balakrishnan, G., Brooks, J., and Guttag, J. (2013). To crash or not to crash: A quantitative look at the relationship between offensive rebounding and transition defense in the NBA. In *MIT Sloan Sports Analytics Conference*, Boston, MA. Available http://www.sloansportsconference.com/?p=10196. Accessed August 30, 2016.

Yaari, G. and Eisenmann, S. (2011). The hot (invisible?) hand: Can time sequence patterns of success/failure in sports be modeled as repeated random independent trials? *PLoS ONE*, 6(10):e24532.

12

Studying Basketball through the Lens of Player Tracking Data

Luke Bornn, Daniel Cervone, Alexander Franks, and Andrew Miller

CONTENTS

12.1 Introduction

Until recently, the field of sports analytics was constrained by the data collected by officials and fans. In basketball, for example, play-by-play data has existed for decades, capturing who is on the court, as well as outcomes for each play. Even with a coarsening of the action to the level of possession or play outcomes, significant analysis of the game has been possible. Examples include basic statistical summaries such as field goal percentage and usage rate, as well as more advanced metrics such as regression-adjusted plus-minus variants and win shares (Kubatko et al., 2007). See Oliver (2004) and Shea and Baker (2013) for book-length treatments of the types of analyses possible with traditional, hand-collected basketball data. Despite the relative success of employing play-by-play and box score data to understand basketball, such data inherently lacks information on actions leading to the outcome of the play and fails to capture contributions of the other nine players on the court for each play-by-play event.

By only examining possession outcomes, much of the important strategic aspects of basketball are ignored. For example, when a catch-and-shoot player scores a basket, the success of the play often stems from the efforts of his teammates in running a play that leads to an open shot. Similarly, looking at possession outcomes might expose a given player as that team's primary intended shooter, when in fact the defense is working to direct shots toward that player. From these examples, it is clear that summarizing a basketball game using a handful of statistics captured at the conclusion of a play misses many of the salient characteristics of the game.

In the last decade a technical solution has arisen to this problem in the form of player tracking systems. Specifically, these systems employ image processing or GPS-like tools to measure the locations on the court of all players and the ball multiple times per second. So while a traditional play-by-play log would capture a handful of numbers for a given 10-second play (who the shooter was, the outcome, any assisters, etc.), we now additionally have access to thousands of numbers summarizing a given play in the form of player and ball coordinates at a resolution as high as 25 frames per second.

With player tracking technologies becoming ubiquitous across sports, there has been a tremendous effort in using this vast trove of information to better understand the game at hand. To give a small sample, in soccer Lucey et al. (2013) and Bialkowski et al. (2014) explore team and individual play strategy using tracking data, while in rugby Cunniffe et al. (2009) use GPS to understand an athlete's workload and physiological demands. In basketball, the focus of this chapter, tracking data has been used to characterize rebounds (Maheswaran et al., 2012), understand shooting quality (Weil, 2011; Franks et al., 2015a), typify playing styles (Miller et al., 2014), assign defensive credit (Franks et al., 2015b), and value decisions (Cervone et al., 2014). The latter four references will be the focus of the work herein.

We begin by describing the tracking data, highlighting the specific data available for the NBA. We subsequently use this data to characterize the offensive and defensive playing styles of various players in Section 12.3. These styles can then be used to inform and stabilize inference about a player's defensive skill—we explore this idea in Section 12.4. Finally, we look at the notion of decision-making in Section 12.5. It is important to note that all of these problems would be nearly impossible without tracking data. For example, understanding a player's style of play relies on his court occupancy patterns, which are not captured by simple box-score summaries. Further, defensive play is as much about preventing shots as it is about reducing their efficiency, thus features of defensive skill are by nature absent from the box score. Finally, assessing a player's decision-making requires looking at the full context of a play in order to learn the available alternatives to the possession outcomes ultimately realized. From these studies, it is clear that the advent of player tracking technologies alongside advanced statistical modeling has created a tremendous opportunity in the field of sports analytics to better understand the game at a much finer level of detail than was previously possible.

12.2 Tracking Data

Though different technologies are employed across different sports, there is a common aim to capture the trajectories of all players and the ball throughout the play. The various approaches to obtaining this data fit into two camps, those based on image processing and those based on sensors. The former approach works by filming the play with multiple traditional cameras, then postprocessing this video data to produce the player and ball trajectories. The image-processing approach has the merits of being unobtrusive; specifically, there is no impact on the players or the game, as the only change from an operational standpoint is the camera installation, typically in a stadium's rafters. However, the image-processing approach is limited in sports with scrums and flurries of players, where the post-processing is unable to differentiate the distinct players within the scrum. In contrast, sensor-based technologies rely on installing a sensor on the players and the ball. While this approach has the potential benefit of improved accuracy, the obvious downside is the impact on the game as uniforms and jerseys are altered to accommodate the sensors.

Since the 2013–2014 season, the NBA has been collecting player-tracking data in all of its stadiums using SportVU, a technique based on camera installations and image processing. The data, collected at 25 frames per second, captures all 10 players, the referees, and the ball. However, the referee data are not released. When thinking of the data, it is helpful to think in terms of the underlying video. As such, one may view one "row" of the data as a 1/25 of a second snapshot capturing the locations of the players and the ball, and the collection of all rows as a video containing all of the trajectories.

It is critical to understand not just the source of the data and the resultant opportunities but also the size and subsequent limitations. The 82-game regular season of the NBA consists of 1230 games each at least 48 minutes long. At 25 frames per second, this yields 88,560,000 rows of data, with each row containing 23 columns—the two-dimensional coordinates of each player and the three-dimensional coordinates of the ball. In addition to this, the data also consists of annotations for dribbles, passes, shots, turnovers, and other actions on the court.

Because of the sheer size of the data, one of the foremost challenges in working with it—beyond setting up the basic infrastructure to access and explore it—is building models and techniques that exploit the richness of the full data, while being computable on subsets. Totaling in the hundreds of gigabytes, working with the entire data in memory is typically not feasible, so for computational tractability we explore modeling strategies where the data are naturally partitioned. The partitioning strategies we use include dividing the data across players, across possessions, across spatial regions, and across teams. Each different segmentation has its own use. For example, in Section 12.3, we first segment by player to obtain an image for each player representing where they spend their time on the court. In Section 12.4, we divide the data by possessions to build a hidden Markov model capturing the defensive matchups across the play. Finally, in Section 12.5, we divide the data according to a combination of spatial locations and ballhandler identity in order to capture transactions between different combinations thereof.

12.3 Typifying Players

The conventional player position labels—point guard, shooting guard, small forward, power forward, and center—are a rough simplification of a player's role. While these labels do convey some information, they overlook the variability of player style and skill within the same position, and obfuscate performance comparisons between positions. Despite this shortcoming, position labels are still useful categorizations for coaches and fans to assess the performance of a particular player with respect to their group; we tend to compare point guards to point guards and centers to centers.

This section describes methods that use player-tracking data to expand our quantification of player roles to better promote apples-to-apples comparisons. We define player roles for players on both offense and defense based on spatial patterns in their play. This organization of players offers a more data-driven grouping of players for performance comparison and provides us structured priors for hierarchical pooling of information in models of defensive and offensive performance down the line. The rest of this section will describe our methods for defining player roles.

12.3.1 Offensive Player Roles

A useful source of information to describe an offensive player's role is the location of his shot attempts. These locations offer a sense of how a particular player's skills are used by an offensive scheme. They also paint a picture of the variability of an offensive player's skills—both Tony Parker and Steph Curry are nominally point guards, but their shooting habits and skill sets are quite different. To characterize this difference, we propose to describe an offensive player in terms of their shot selection. This characterization is also crucial for models that describe how defensive performance modulates shot selection (Section 12.4).

The locations of shot attempts throughout the season can naturally be viewed as a point process. In this section, we develop a method to construct a low-dimensional representation of this point process such that "similar" shooters will be nearby in some low-dimensional space. Like other player roles, shot selection in professional basketball is a structured phenomenon, and we can use this structure to find a concise description of a player's attempts in terms of spatially intuitive *shot types*. A shot type is a cluster of similar shots characterized by a spatially smooth intensity surface over the court. This surface indicates where shots from that cluster tend to come from (and where they do not come from). Each player's shooting habits are then represented by a positive linear combination of shot types. Intuitively, shot types correspond to a soft partitioning of regions of the court, and hence players shooting habits are summarized by a weighting over these regions.

12.3.2 Point Process Decomposition

An intuitive characterization of a player's shot attempt point process is the spatial rate function—the function whose integral over some area of the court, S, describes the rate at which a player shoots from S. Alternatively, conditioned on a single shot being taken, the spatial rate function can be viewed as an unnormalized probability density over the court that describes the likelihood of a shot coming from a particular location. Our goal, therefore, may be reframed as building a model for an individual player's spatial rate function. Our modeling approach describes each player's rate function as a positively scaled sum of canonical basis functions that will define our shot types.

To compute these canonical basis functions, we follow a two-step procedure. We first find a nonparametric estimate of each player-specific intensity surface, modeled as log-Gaussian Cox process (LGCP) (Møller et al., 1998). We then combine surfaces from all players and, through nonnegative matrix factorization (NMF), find a low rank representation that best reconstructs the player-specific surfaces. This second step combines information from all players, yielding rate surfaces that generalize better.

We denote each player's set of shot attempts on the offensive half court (47 ft × 50 ft) as $x_k = \{x_{k,1}, \ldots, x_{k,N_k}\}$, where N_k is the number of shots taken by player k, and $x_{k,m} \in [0, 47] \times [0, 50]$. We discretize the court into $V = 47 \times 50 = 2350$ one-square-foot tiles for computational tractability of both the LGCP surfaces and the NMF optimization. Our procedure is as follows: given shot locations for each of K players, $x_1, \ldots, x_K$

1. Construct the count matrix X_{kv} = number of shots by player k in tile v on a discretized court.
2. Fit an intensity surface $\lambda_k = (\lambda_{k1}, \ldots, \lambda_{kV})^T$ for each player k over the discretized court (LGCP).
3. Construct the data matrix $\Lambda = (\lambda_1, \ldots, \lambda_K)^T$, where λ_k has been normalized to have unit volume.
4. Find low-rank matrices L, W such that $WL \approx \Lambda$, constraining all matrices to have nonnegative entries (NMF).

The result of this procedure is a spatial basis that describes canonical shot types, L, and basis loadings for each individual player, $\hat{w}_k$. As will be seen later, the nonnegative assumption helps create disjoint, interpretable spatial bases. Given such an interpretable spatial basis, the weights for each player concisely summarize shooting habits in terms of shot types.

12.3.3 Fitting the LGCPs

The generative model for player k's shot selection is

$$\lambda_k(\cdot) \sim \mathcal{LGCP}(0, \mathcal{C}) \tag{12.1}$$

$$x_k \sim \mathcal{PP}(\lambda_k), \tag{12.2}$$

where $\mathcal{LGCP}$ and $\mathcal{PP}$ denote distributions over log-Gaussian Cox processes and Poisson processes, respectively. Due to the completely spatially random nature of Poisson processes, this model can be discretely approximated as

$$P(\boldsymbol{x}_k|\lambda_k(\cdot)) \approx \prod_{v=1}^{V} p_{pois}(\boldsymbol{X}_{kv}|\Delta A \lambda_{kv}),$$

where
  $\lambda_k(\cdot)$ is the exact intensity function
  λ_k is the discretized intensity function (vector)
  ΔA is the area of each tile (set to one)
  $p_{pois}(\cdot|\lambda)$ is the Poisson probability mass function with mean λ

We assume a spatial dependence on the λ_k by letting $\lambda_k = \exp(z_k + z_0)$ where z_k is a Gaussian random field. The subsequent posterior has the following form:

$$P(z_k|\boldsymbol{x}_k) \propto P(\boldsymbol{x}_k|z_k)P(z_k)$$

$$= \prod_{v=1}^{V} e^{-\lambda_{kv}} \frac{\lambda_{kv}^{X_{kv}}}{\boldsymbol{X}_{kv}!} \mathcal{N}(z_k|0, \boldsymbol{C}),$$

where the prior over z_k is a mean zero normal with covariance matrix defined as

$$C_{vu} \equiv c(\boldsymbol{x}_v, \boldsymbol{x}_u) = \sigma^2 \exp\left(-\frac{1}{2} \sum_{d=1}^{2} \frac{(x_{vd} - x_{ud})^2}{v_{k,d}^2}\right),$$

and z_0 is an intercept term that parameterizes the mean rate of the Poisson process. This kernel encodes the prior belief of spatial smoothness. We place a gamma prior over the length scale, $v_k = \{v_{k,1}, v_{k,2}\}$, for each individual player. This gamma prior places mass dispersed around 8 ft, indicating the reasonable a priori belief that shooting variation is locally smooth. We obtain posterior samples of λ_k and v_k by iteratively sampling $\lambda_k|\boldsymbol{x}_k, v_k$ and $v_k|\lambda_k, \boldsymbol{x}_k$ using Markov Chain Monte Carlo methods, specifically elliptical slice sampling (Murray et al., 2010).

12.3.4 NMF Optimization

Given the collection of player intensity surfaces, we can pool information to find global structure and a low-dimensional space that describes player shooting habits. Nonnegative matrix factorization (NMF) models a matrix $\boldsymbol{\Lambda}$, in our case, the matrix of player-specific intensity functions, as the product of two low-rank matrices

$$\boldsymbol{\Lambda} \approx \boldsymbol{WL},$$

where
  $\boldsymbol{\Lambda} \in \mathbb{R}_+^{N \times V}$
  $\boldsymbol{W} \in \mathbb{R}_+^{N \times \mathcal{B}}$
  $\boldsymbol{L} \in \mathbb{R}_+^{\mathcal{B} \times V}$ are loading and basis matrices, respectively, and we assume $\mathcal{B} \ll V$

The matrices of best approximation, $\boldsymbol{W}^*$ and $\boldsymbol{L}^*$, are determined by an optimization procedure that minimizes a loss function, $\ell(\cdot, \cdot)$, that measures the divergence between the approximation WL and Λ with the constraint that all elements remain nonnegative

$$W^*, L^* = \underset{W_{ij}, L_{ij} \geq 0}{\arg\min} \ell(\Lambda, WL).$$

Different choices of ℓ will result in different matrix factorizations. A natural choice is the matrix divergence metric

$$\ell_{KL}(A, B) = \sum_{i,j} A_{ij} \log \frac{A_{ij}}{B_{ij}} - A_{ij} + A_{ij},$$

which corresponds to the Kullback–Leibler (KL) divergence if A and B are discrete distributions, that is, $\sum_{ij} A_{ij} = \sum_{ij} B_{ij} = 1$ (Lee and Seung, 2001). Although there are several other possible divergence metrics (i.e., Frobenius), we use this KL-based divergence measure for reasons outlined in Miller et al. (2014). We solve the optimization problem using techniques from Lee and Seung (2001) and Brunet et al. (2004).

Due to the positivity constraint, the basis L^* tends to be disjoint, exhibiting a more "parts-based" decomposition than other, less constrained matrix factorization methods, such as PCA. This is due to the restrictive property of the NMF decomposition that disallows negative bases to cancel out positive bases. In practice, this restriction eliminates a large swath of "optimal" factorizations with negative basis/weight pairs, leaving a sparser and often more interpretable basis (Lee and Seung, 1999).

12.3.5 Basis and Player Summaries

The inferred spatial bases are depicted in Figure 12.1. Each player's shot-selection rate function is modeled as a positive linear combination of these surfaces (for $\mathcal{B} = 6$). There is wide variation in shot selection among NBA players—some shooters specialize in certain types of shots, whereas others will shoot from many locations on the court. We visualize the basis loadings, w_k (the kth row of W), that correspond to the landscape of shooting roles in Figure 12.2a. In Section 12.4.2, these weights are used to specify an informative prior over offensive skill parameters in a possession outcome model.

12.3.6 Defensive Player Roles

We can define defensive roles in a similar way by examining the spatial patterns of the offensive player that each defender guards (in Section 12.4.1, we introduce a model to infer defensive matchups). One approach is as follows: using the existing basis depicted in Figure 12.1, count the number of frames a defender's matchup spends in each basis component (thresholded to be disjoint) and normalize into a vector d. This will categorize defensive players according to the types of offensive players they tend to guard. We visualize the loadings for each defensive player, d_k, in Figure 12.2b. The players tend to naturally cluster into three groups, corresponding roughly to "bigs," "forwards," and "guards."

With these spatially informed characterizations of offensive and defensive types, we now move on to valuing offensive and defensive skill.

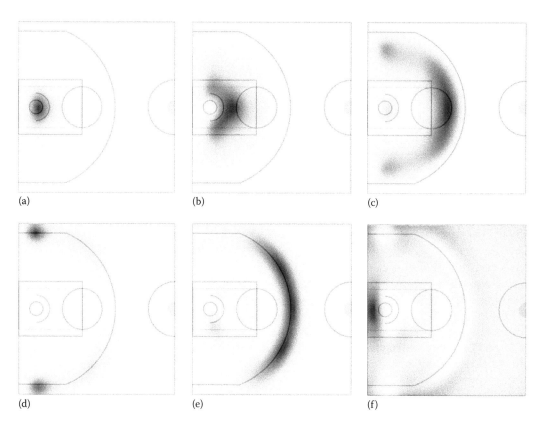

FIGURE 12.1
Basis vectors (surfaces) identified by LGCP-NMF for $\mathcal{B} = 6$. Each basis surface is the normalized intensity function of a particular shot type, and players' shooting habits are a weighted combination of these shot types. Conditioned on a certain shot type (e.g., corner three), the intensity function acts as a density over shot locations, where shading indicates likely locations. (a) Basis 1. (b) Basis 2. (c) Basis 3. (d) Basis 4. (e) Basis 5. (f) Residual.

12.4 Understanding Defense

There is a striking lack of qualitative metrics for individual defensive skill in the NBA. Steals, blocks, a small handful of aggregate metrics based on box score statistics, and plus/minus are the only conventional defensive statistics (Kubatko et al., 2007), but these metrics integrate over the details of play and thus ignore *why* and *how* a player contributes defensively. With player-tracking data, we can go beyond the box score. By leveraging the spatiotemporal information in these data, we can drastically improve our characterization of defensive abilities. However, although the data are incredibly rich, it does not include one of the most vital pieces of information for assessing player ability: *intent*.

Our approach to measuring defensive impact is to quantify a defensive player's impact on where and who is the shooter, as well as the subsequent probability of the basket being made. Before we can measure how well a player performs, though, we must first understand what he is trying to do. This is of particular importance for evaluating

(a)

(b)

FIGURE 12.2
Landscape of offensive shooting roles and defensive guarding roles. Left, a two-dimensional depiction of a $\mathcal{B}$ space of player loadings onto spatial bases. Centers who shoot mostly in the restricted area are in the upper right, whereas three-point specialists are located on the left-most part of the spectrum. These loadings can be used to determine the "most similar" player in terms of offensive shot selection. Right, defensive player groupings based on the type of offensive player they tend to guard. (a) Offensive similarity. (b) Defensive Clustering.

defensive ability, where individual player intent is often less evident. We know broadly that defenders are trying to prevent the opponent from scoring, but we do not necessarily know how they are coached to do so, the details of their defensive strategy, or even who they are guarding at any moment of a possession. As a starting point, our goal for defensive analysis is to estimate how individual defensive matchups evolve over the course of a possession. Identifying who's guarding whom is not a comprehensive reflection of defensive intent but is still quite useful. Given inferred matchup information, we then construct models that assess how well defenders suppress shots (e.g., the probability a shot is taken) and disrupt shots (the probability the shot is made, given that it is taken).

12.4.1 Modeling the Evolution of Defensive Matchups

There are many ways in which one might try to infer defensive intent from player-tracking data. One possible approach is supervised learning, which would require expert annotation of defensive strategy for a large training set of possessions. The annotated data would then be used to train classifiers to "predict" defensive intent on unlabeled possessions. Although this can be a powerful way to model more complex defensive schemes, annotation is labor intensive and requires access to domain experts. As an alternative, we focus on simply estimating man-to-man defensive matchups over the course of a possession using an unsupervised algorithm that does not require labeled examples. We are able to model the evolution of defensive assignments by incorporating only basic basketball knowledge without ground truth matchup information.

To accomplish this, we make several simple assumptions about the basics of one-on-one defense. Specifically, we assume that for all players, the defender's average location over time lies somewhere in the triangle between the offensive player he is guarding, the ball, and the basket (Figure 12.3). A defender may deviate from this average position due to player- or team-specific tendencies, game situations, and unobserved covariates. Let μ_{tk} be the average defender location for a particular offensive player (indexed by k) at time t. Then,

$$\mu_{tk} = \gamma_o \text{Offense}_{tk} + \gamma_b \text{Ball}_t + \gamma_h \text{Hoop}$$
$$\gamma_o + \gamma_b + \gamma_h = 1,$$

where $\gamma = \{\gamma_o, \gamma_b, \gamma_h\}$ represents the weights associated with the offensive player location, ball location, and hoop location. Let I_{tjk} be an indicator for whether a defender (indexed by k) is guarding offensive player j at time t and D_{tk} be the location of the defender k and time t. Under this model, each defender can only be attending to one offensive player at any moment, but multiple defenders can guard the same offensive player.

We further assume the observed location of a defender j guarding offensive player k is normally distributed about the average location

$$D_{tk}|(I_{tjk} = 1) \sim \mathcal{N}(\mu_{tj}, \sigma_D^2 \mathbf{I}_2).$$

The parameter σ_D^2 represents how much the defensive players typically deviate from the average location over the course of the possession. Finally, we model the evolution of the defensive assignments, I_{tjk}, using a hidden Markov model (HMM). The hidden states correspond to the offensive player that a defender is guarding. The HMM enforces smoothness in the evolution of these indicators over time.

To complete the specification, we model the prior transition probabilities, $P(I_{tk} \mid I_{(t-1)k})$, corresponding to the probability of a defender switching his assignment from offensive player j to player j' in any instant. We assume the prior probability of maintaining the same matchup each iteration is

$$P(I_{tjk} = 1 | I_{(t-1)jk} = 1) = \rho.$$

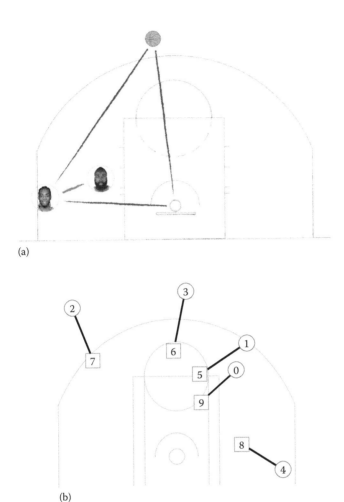

(a)

(b)

FIGURE 12.3
(a) The average defending location is constrained to be a convex combination of the offensive player, ball, and hoop locations. (b) A snapshot of inferred defensive assignments. Estimates incorporate spatial and temporal information and thus are clearly not just based on defender proximity (e.g., 5 is closer to 0 but we still infer that 5 is guarding 1 based on the spatiotemporal information available from the full possession).

We also assume the prior probability of a switch to any other offensive player occurs with equal probability:

$$P(I_{tj'k} = 1 | I_{(t-1)jk} = 1) = \frac{1 - \rho}{4}, \quad j' \neq j.$$

We use a special variant of the expectation-maximization (EM) algorithm known as the Baum-Welch algorithm (Bishop et al., 2006) to fit the HMM to the data. Using this algorithm, we estimate all model parameters; most importantly, we can use this algorithm to estimate the probabilities associated with all defensive matchups at each moment t (e.g., the expected value of I_{tjk}). We found that on average, defenders position themselves slightly more than two-thirds of the way between the hoop and the offensive player, hedging slightly toward the ball (see Figure 12.3). Specifically,

$$(\gamma_o, \gamma_b, \gamma_h) = (0.62, 0.11, 0.27).$$

This matches basic basketball intuition that a defender should be close to the offensive player they are guarding, but also that each defender needs to attend to the location of the ball. We used this model to infer defensive matchups for every moment of every game in the 2013–2014 NBA season. Defensive matchup information can be used in a plethora of ways to define new metrics for understanding the game. This includes summaries like defensive attention drawn or number of defensive switches per possession by team (see Franks et al., 2015b). Next, we illustrate how defensive assignments can be used to construct models for individual defensive ability.

12.4.2 Characterizing Defensive Ability

With reasonable estimates of "defensive intent," it is now possible to construct models for how well defenders affect individual offensive play. We differentiate between a defender's ability to *suppress* shots (e.g., change shot selection) and *disrupt* shots (affect the odds that a shot is made). Importantly, in order to accurately model defensive skill, we must also account for offensive skill. Consequently, we construct joint models for the skills and preferences of offensive players as well as how individual defenders affect those preferences and skills.

Parameter regularization is an important part of these models as well. Despite the large amount of data overall, many individuals only face certain situations a handful of times, so sample sizes can be quite small. We deal with this by exploiting similarities between players and pooling information in different ways. This reflects the idea that players with similar roles should have similar skill sets and thus comparable parameter values. Next, we briefly introduce suppression and disruption models that exploit the player characterizations presented in Section 12.3.

12.4.2.1 Shot Suppression

We model shot selection among the five offensive players in five discretized court regions (Figure 12.1) using a multinomial distribution with logit link function. There are $\mathcal{R} \times 5 + 1$ discrete outcomes for this model, reflecting the $\mathcal{R} = 5$ court regions and five possible shooters, plus one for "no shot" (e.g., turnover). We model shot selection as a function of all the shot propensities of all offensive players on the court and the fraction of time within a possession that each offensive player is guarded by each defender. Letting $S_n(j, r)$ be an indicator for whether shooter j shot in location r in possession n,

$$P(S_n(j, r) = 1 | \alpha, \beta, Z_n) = \frac{\exp\left(\alpha_{jr} + \sum_{k=1}^{5} Z_n(j, k) \beta_{kr}\right)}{1 + \sum_{j,r} \exp\left(\alpha_{jr} + \sum_{k=1}^{5} Z_n(j, k) \beta_{kr}\right)},$$

where

    α_{jr} represents the propensity for offensive player j to take a shot from region r

    β_{kr} corresponds to the suppression ability of defender k in court region r, relative to the average defender

The covariate $Z_n(j, k)$ is the fraction of time defender k is guarding offensive player j in possession n, with $\sum_{j=1}^{5} Z_n(j, k) = 1$. We calculate $Z_n(j, k)$ for each possession using the defensive matchup model outlined in Section 12.4.1. We assume normal distributions for the random effects α and β, but use slightly different specifications for each. For the offensive shot selection parameters, α_{jr}, we use a conditional autoregressive (CAR) model based on the offensive player similarity graph illustrated in Figure 12.2a. Under this prior, the coefficients for each offensive player shrink toward a unique point based on the player's position in the graph. For defense, we use a simple hierarchical prior based on the clustering of defensive players (Figure 12.2b). Under this prior, the coefficients for all defenders in the same group shrink to the same value.

We fit this multinomial logistic regression over 150,000 possessions from the 2013–2014 season. There are several computational challenges in fitting a model of several hundred parameters to so much data. We found the variational inference algorithm introduced by Murphy (2012) to be the most tractable way to estimate the parameters. The estimates of β reflect individual defensive ability to suppress shots in different regions and are discussed in Section 12.4.3.

12.4.2.2 Shot Disruption

We model shot outcomes using Bayesian logistic regression. The make/miss probabilities are a function of each shooter's scoring ability, the disruption ability of the defender, and the defender's distance to the shooter. For a possession n, let Y_n be an indicator for the outcome of the shot (make/miss), then

$$P(Y_n = 1 | \mathcal{S}_n(j, k, r), \mathcal{D}_n, \theta, \phi, \xi) = \frac{\exp(\theta_{jr} + \phi_{kr} + \xi_r \mathcal{D}_n)}{1 + \exp(\theta_{jr} + \phi_{kr} + \xi_r \mathcal{D}_n)},$$

where now $\mathcal{S}_n(j, k, r)$ represents the occurrence of a shot by j in region r, which was defended by k. $\mathcal{D}_n$ is the distance in feet between the shooter and defender at the moment of the shot. The parameter θ_{jr} describes the scoring ability of offensive player j shooting in region r, whereas ϕ_{kr} describes the disruption skill of defender k in that region. The log-odds of a made shot also decreases linearly in the distance of the defender, $\mathcal{D}_n$, with rate ξ_r.

We again employ hierarchical models to pool information between players, using the defensive clusters for the disruption parameters and the CAR prior for the offensive efficiency parameters, as in the shot suppression model in Section 12.4.2.1. We fit the regression to over 100,000 possessions ending in a shot attempt in the 2013–2014 season. We used the probabilistic programming language STAN (Stan Development Team, 2014) to conduct full Bayesian inference on all model parameters. Estimates of ϕ_{kr}, the shot disruption ability for defenders, are briefly presented next.

12.4.3 Results

The coefficients from the suppression and disruption models (e.g., β and ϕ) provide interpretable quantifications of defensive skill. These results are informative about how opposing shooters perform against any defender in any region of the court, even when a defender rarely defends shots in a particular region. For instance, we found that Chris Paul defends relatively few shots in the region closest to the basket, yet the players he guards take fewer shots in this area relative to similar defenders. This suggests that he is good at

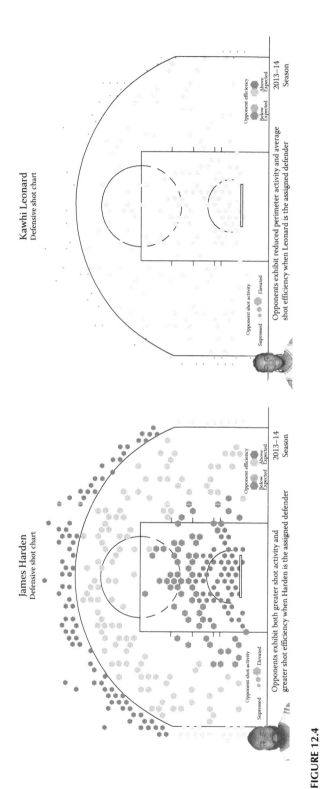

FIGURE 12.4
(See color insert.) Kawhi Leonard tends to suppress shots on the perimeter (small dots). James Harden gives up more shots than average almost everywhere on the court (larger dots) and opponents tend to make more perimeter shots than average (red).

keeping players from driving toward the rim, even though as a defender he rarely occupies the space under the basket.

These parameters can be used to rank a player's ability to both suppress and disrupt shots. We can also visualize the results of our model by extending the original offensive shot chart (Goldsberry, 2012). In the "defensive shot chart," the circles represent all shots faced by a defender. The size of the shots represents the relative magnitude of the suppression score of the defender, and the color of the circles represents the relative magnitude of the disruption score. Figure 12.4 shows two example defensive charts for defenders with contrasting defensive reputations in the NBA.

By constructing models that incorporate player-tracking data, we believe that we have taken a crucial first step toward better quantitative characterizations of defensive ability. Our methods allow understanding individual defensive matchups and how defenders affect opponents' offensive performance. Using optical tracking data and a model to infer defensive matchups at every moment throughout, we are able to provide novel summaries of defensive performance.

There are still many significant challenges in truly understanding defensive competence. "Who's guarding whom" is only one way to understand intent; without ground-truth team strategy, it is very difficult to know whom a defender is supposed to be guarding or when he is supposed to rotate or switch on defense. Future work on quantifying defensive ability ought to incorporate defensive schemes into models for both the team and individual skills. This can be accomplished by combining knowledge from basketball experts with more complex (yet accurate) models of player interactions.

In this section, we presented an approach to evaluate defensive ability by solely modeling shot selection and shot efficiency. A major opportunity for discovering and evaluating both offensive and defensive skill is to use outcomes at a finer level, quantifying the effect of every decision a player makes to understand their impact on how the play ultimately unfolds. We now turn to this idea, building a stochastic process model to capture the value of transitions throughout the play, with a specific focus on valuing the decisions of offensive players.

12.5 Valuing Decision

Player tracking data reveals the spatiotemporal context of the usual play-by-play actions, namely, passes, shots, turnovers, fouls, and other stoppages. Not only does this improve our ability to measure skill associated with such actions (as seen in Section 12.4.2.2), it also allows us to quantify the value in *selecting* these actions in the first place. Decision-making is a key component of a player's impact in basketball, distinct from his usage and skill in traditional metrics. For instance, while a player who can make 40% of three-pointers while double teamed is impressive, he is more impressive if when doubled teamed beyond the arc, he will pass to an open teammate with a 75% chance of scoring a layup.

One way to approach the value of decision-making is to consider the expected success of a given possession. In our own work, we quantify decision-making through a quantity called *expected possession value* (EPV), introduced originally in Cervone et al. (2014). EPV is the expected number of points the offense will score in a particular possession, given the full spatiotemporal history of that possession up to time t. By providing an instantaneous summary of the possession's future point value given all available information, EPV acts like a basketball stock ticker, rising and falling in line with the opportunities created by players' movement and actions.

EPV is calculated using a stochastic process model for basketball possession outcomes. This naturally provides us with counterfactuals for basketball events of interest, which can be used to quantify the value created or destroyed by players' decisions. As a simple illustration, if the EPV is 1.0 after a shot attempt but was 1.1 immediately before the shot attempt, we can surmise that the decision to attempt a shot cost the offense 0.1 points, regardless of the shot outcome. In this section, we introduce the modeling framework that allows us to estimate EPV and provide examples of insights EPV provides on decision-making in the NBA.

12.5.1 Expected Possession Value

The EPV at time t during a possession provides the offense's expected point total, given the spatiotemporal evolution of that possession up to time t. More formally, let $X \in \{0, 2, 3\}$ be the number of points the offense scores on a particular possession*, and let Z_s represent the full set of player locations and actions at time s. Z_s thus includes locations of all 10 players (and the ball), as well as event annotations, such as a pass or shot occurring at time s. We can then write EPV as

$$\nu_t = \mathbb{E}[X|Z_0, \ldots, Z_t]. \tag{12.3}$$

Calculating EPV requires implicitly averaging over all possible future possession paths. To do so in a way that is computationally feasible, while also leveraging the precise spatial detail of our data, we require representing the possession at separate levels of resolution. The raw spatial data Z_t is the *micro* level, while the *macro* level is a coarsening of the raw data into a collection of discrete states defining the ordered triple

$$C_t = \{\text{ballcarrier ID at } t, \text{court region at } t, \text{defended at } t\}.$$

We assume the model for micro transitions $P(Z_{t+\epsilon}|Z_0, \ldots, Z_t)$ is quite complicated, but that the model for macro transitions $P(C_{t+\epsilon}|C_0, \ldots, C_t)$ is semi-Markov—that is, the sequence of new states visited, $C^{(0)}, C^{(1)}, \ldots, C^{(k)}$, is a homogeneous Markov chain. The Markov assumption for C_t is similar to models for game evolution in other sports using discrete, categorical data, such as baseball (Bukiet et al., 1997) and football (Goldner, 2012).

To link these multiresolution models in computing EPV, we use the micro model to predict a *future* coarse state C_{δ_t}. Here, δ_t represents the ending time of the next pass, shot attempt, or turnover after time t, thus C_{δ_t} is the state at which the next pass/shot/turnover ultimately arrives. Passes, shots, and turnovers take some time to transpire, and "reset" the dynamics of the possession, as players are reacting to a new ballcarrier. Thus, we assume that for such a future state C_{δ_t}, states even further in the future do not depend on the *present* full-resolution data:

$$\text{For } s > \delta_t, P(C_s|C_{\delta_t}, Z_0, \ldots, Z_t) = P(C_s|C_{\delta_t}).$$

---

* The offense may also score exactly one or four points due to fouls, but for simplicity and due to limitations with our data, we remove possessions that end in free throws.

This allows us to rewrite EPV (12.3) as

$$\nu_t = \sum_c \mathbb{E}[X|C_{\delta_t} = c]P(C_{\delta_t} = c|Z_0, \ldots, Z_t). \qquad (12.4)$$

Basic properties of homogeneous Markov chains make $\mathbb{E}[X|C_{\delta_t}]$ an easy calculation, so the main modeling task is to predict the endpoint of the next pass, shot, or turnover: $P(C_{\delta_t}|Z_0, \ldots, Z_t)$.

This prediction problem encompasses two tasks, as we need to predict both when a pass, shot, or turnover event will occur after time t as well as the outcome of this event, C_{δ_t}. We thus break $P(C_{\delta_t}|Z_0, \ldots, Z_t)$ into two submodels. First, taking ϵ as the temporal sampling frequency of our data (1/25th of a second), let $M(t)$ be the event {pass/shot/turnover action begins in $(t, t + \epsilon]$} and $M(t)^c$ is its complement. Now, we have $P(Z_{t+\epsilon}|M(t)^c, Z_0, \ldots, Z_t)$, which models player movement conditional on the ballcarrier retaining ball control, and $P(C_{\delta_t}, M(t)|Z_0, \ldots, Z_t)$, which models the outcome of a pass/shot/turnover beginning in the next ϵ time. These submodels correspond to player movement and player decision-making, respectively. Together, they can be used to sample C_{δ_t} given the possession history $Z_0, \ldots, Z_t$, which we need for EPV (12.4):

1. Simulate a possible pass/shot/turnover $M(t)$ or ball control retention $M(t)^c$ from $P(C_{\delta_t}, M(t)|Z_0, \ldots, Z_t)$.
2. If $M(t)^c$, update player positions using $P(Z_{t+\epsilon}|M(t)^c, Z_0, \ldots, Z_t)$.
3. Update $t \leftarrow t + \epsilon$ and repeat (1)–(2) until we draw $M(t)$ and corresponding C_{δ_t}.
4. Calculate EPV using $\mathbb{E}[X|C_{\delta_t}]$.

12.5.2 Parametric Modeling and Inference

The movement and decision-making (sub)models we introduced are both parameterized using familiar statistical models. For player movement, we model $P(Z_{t+\epsilon}|M(t)^c, Z_0, \ldots, Z_t)$ as an integrated autoregressive process with spatial innovations. Specifically, let the location of offensive player ℓ at time t be $\mathbf{z}^\ell(t) = (x^\ell(t), y^\ell(t))$. We then model movement in each of the x and y coordinates separately using

$$x^\ell(t + \epsilon) = x^\ell(t) + \alpha_x^\ell[x^\ell(t) - x^\ell(t - \epsilon)] + \eta_x^\ell(t) \qquad (12.5)$$

(and analogously for $y^\ell(t)$). The innovations, $\eta_x^\ell(t)$, capture nonstationary spatial behavior that describes players' movement tendencies; for instance, when possessing the ball, players typically accelerate toward the basket when beyond the arc, and decelerate when very close to the basket in order to attempt a shot. We assume $\eta_x^\ell(t) \sim \mathcal{N}(\mu_x^\ell(\mathbf{z}^\ell(t)), (\sigma_x^\ell)^2)$, where μ_x^ℓ maps player ℓ's location on the court to an additive effect in (12.5), encoding a spatial pattern of acceleration. These are illustrated in Figure 12.5. Note that we estimate parameters of this offensive movement model separately for every player in the NBA, both as ballcarrier and not. Furthermore, we use a different model for defensive players, adapting (12.5) to include information on who is guarding whom (see Section 12.4.1). This ensures that the locations of the offense and defense evolve cohesively, respecting the spacing inherent in their adversarial roles.

The other model necessary in calculating EPV is $P(C_{\delta_t}, M(t)|Z_0, \ldots, Z_t)$, which gives the joint probability of a pass/shot/turnover occurring in the next ϵ time, along with the

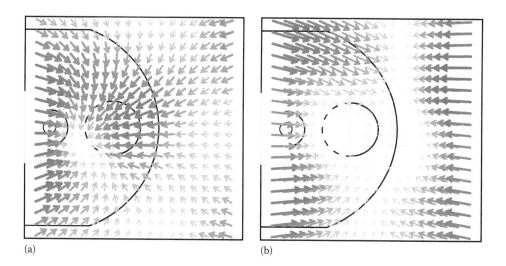

(a)   (b)

FIGURE 12.5
(See color insert.) Acceleration fields ($\mu_x(\mathbf{z}(t)), \mu_y(\mathbf{z}(t))$) for Chris Paul (a) and Tim Duncan (b) with ball possession. The arrows point in the direction of the acceleration at each point on the court's surface, and the size and color of the arrows are proportional to the magnitude of the acceleration. Comparing (a) and (b) for instance, we see that when both players possess the ball, Paul more frequently attacks the basket from outside the perimeter. Duncan does not accelerate to the basket from anywhere within the three point arc.

outcome of that pass/shot/turnover. This model represents players' decision-making tendencies, as we infer spatiotemporal features that influence passing and shooting behavior. At any time t, there are seven possible choices for C_{δ_t} given $M(t)$—four corresponding to each passing option, one for a made shot attempt, one for a missed shot attempt, and one for a turnover. Thus, it is convenient to define events $M_j(t) = \{M(t)$ and $C_{\delta_t} = c_j\}$, where j indexes the seven possible shot/pass/turnover outcomes at time δ_t, so that $P(C_{\delta_t}, M(t)|Z_0, \ldots, Z_t) = P(M_j(t)|Z_0, \ldots, Z_t)$.*

We parameterize these using log-linear models. For small ϵ, this is equivalent to modeling actions $M_j(t)$ using competing risks (Prentice et al., 1978). Assuming player ℓ possesses the ball at time t,

$$P(M_j(t)|Z_0, \ldots, Z_t) = \lambda_j^\ell(t)$$

$$\log(\lambda_j^\ell(t)) = [\mathbf{W}_j^\ell(t)]'\boldsymbol{\beta}_j^\ell + \xi_j^\ell\left(\mathbf{z}^\ell(t)\right) + \left(\tilde{\xi}_j^\ell\left(\mathbf{z}_j(t)\right)\mathbf{1}[j \leq 4]\right), \qquad (12.6)$$

where
  $\mathbf{W}_j^\ell(t)$ is a $p_j \times 1$ vector of time-varying covariates
  $\boldsymbol{\beta}_j^\ell$ a $p_j \times 1$ vector of coefficients
  $\mathbf{z}^\ell(t)$ is the ballcarrier's location on the court at time t
  ξ_j^ℓ is a function over the court space providing a spatial effect

---

* In Cervone et al. (2014), instead of treating shot makes and shot misses as distinct events to model, a single model is fit for shot attempts, and then a separate shot probability model is estimated.

The last term in (12.6) only appears for pass events (without loss of generality, $j \leq 4$) to incorporate the location of the receiving player for the corresponding pass: $\mathbf{z}_j(t)$ provides his location on the court at time t, and $\tilde{\xi}_j^{\ell}$, analogously to ξ_j^{ℓ}, provides the corresponding spatial effect. The four different passing options are identified by the (basketball) position of the potential pass recipient: point guard (PG), shooting guard (SG), small forward (SF), power forward (PF), and center (C).

This decision-making model (12.6) represents the ballcarrier's actions as interpretable functions of the unique basketball predicaments he faces. For example, for shots, the time-varying covariates ($\mathbf{W}_j^{\ell}(t)$) used are the distance between the ballcarrier and his nearest defender, an indicator for whether the ballcarrier has dribbled, and a constant representing a baseline shooting rate. Moreover, the spatial effects ξ_j^{ℓ} reveal locations where player ℓ is more/less likely to attempt a shot, holding fixed the time-varying covariates $\mathbf{W}_j^{\ell}(t)$. As we have seen in Section 12.3.5, these spatial effects (illustrated in Figure 12.6) are nonlinear in distance from the basket and asymmetric about the angle to the basket.

All model components in (12.6) differ across action types j for the same ballcarrier ℓ, as well as across all ballcarriers in the league during the 2013–2014 season. This reflects the fact that players' decision-making tendencies and skills are unique; a player such as Dwight Howard will very rarely attempt a three-point shot even if he is completely undefended, while someone like Stephen Curry will attempt a three-point shot even if closely defended. As such, all unknown parameters are modeled hierarchically: a priori, parameters are correlated for "similar" players (following ideas from Section 12.3). This requires using a basis representation for the spatial effects in our models, such that we have a low-dimensional representation for these effects; otherwise, hierarchical models would be computationally infeasible.

12.5.3 Analyzing Possessions Using EPV

Given estimates of all unknown parameters, EPV can be calculated using (12.4) in order to produce EPV curves that trace the progress of a possession. An example is shown in Figure 12.7. Because EPV derives from an interpretable statistical model, at any point in time we can understand how an EPV estimate is realized as a function of the possession history.

For example, Figure 12.8 shows EPV, at two time points during the possession in Figure 12.7, as the weighted average over future player actions represented by (12.4). This reveals how players' movement and decision-making tendencies impact the expected point value of the offensive possession. For instance, in Figure 12.8a, we see that EPV derives primarily from Lewis' passing options, of which LeBron James is the most likely target. And in Figure 12.8b, we see EPV is driven by an 83% chance of a close-range shot attempt from James, which given the lack of rim protection by the defense, has an expected value of 1.50 points.

Such diagrams can be recreated on any of hundreds of thousands of possessions available in a season of optical tracking data. As such, EPV provides powerful insight as to how players' movements and decisions contribute value to the offense. Using this, coaches and analysts can formulate strategies and offensive schemes that make optimal use of their players' ability—or, conversely, defensive strategies to suppress the situations that generate value for the opposing offense.

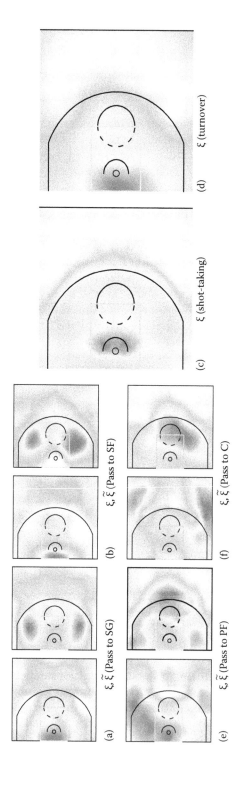

FIGURE 12.6

(See color insert.) Plots of estimated spatial effects ξ for Stephen Curry as the ballcarrier. For instance, plot (c) reveals the largest effect on Curry's shot-taking probability occurs near the basket, with noticeable positive effects also around the three-point line. Plot (a) shows that he is more likely to pass to the center when in the wing areas—more so when the center is positioned near the free throw line. Plots (b)–(f) display other spatial effects as labeled.

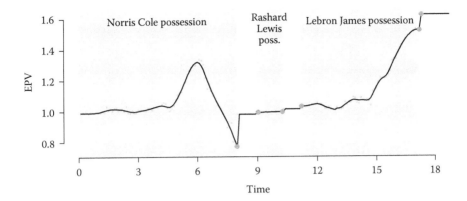

FIGURE 12.7
EPV curve during a Miami Heat possession against the Brooklyn Nets from December 2013. Changes in EPV are induced by changes in players' locations and dynamics of motion; actions such as passes and shot attempts produce immediate, sometimes rapid changes in EPV.

12.5.4 Derived Metrics

While detailed possession case studies provide insight on players' tendencies and decision-making profiles, we can also aggregate player events over the course of a season to discover players who consistently create value for the offense—through passing, shooting, or simply positioning themselves effectively. This is not a trivial procedure, as by construction (12.3) EPV is a martingale, meaning changes in EPV are always mean 0, regardless of the ballcarrier.

Rather than naively aggregating EPV changes to players, we can consider EPV added (EPVA) by a player's actions and decisions relative to the value of a league average player handling the ball in the same situations. If player ℓ has ball possession starting at time t_s and ending at t_e, then $\nu_{t_e} - \nu_{t_s}^{r(\ell)}$ estimates the value contributed player by ℓ relative to the hypothetical league-average player during his period of ball possession (represented by $\nu_{t_s}^{r(\ell)}$). We calculate EPVA for player ℓ (EPVA(ℓ)) by summing such differences over all moments where the player possesses the ball (dividing by the number of games played to provide standardization):

$$\text{EPVA}(\ell) = \frac{1}{\# \text{ games for } \ell} \sum_{\{t_s, t_e\} \in \mathcal{T}^\ell} (\nu_{t_e} - \nu_{t_s}^{r(\ell)}), \tag{12.7}$$

where $\mathcal{T}^\ell$ contains all intervals of form $[t_s, t_e]$ that bookend player ℓ's ball possession. Cervone et al. (2014) provide specific details on calculating $\nu_t^{r(\ell)}$.

Table 12.1 provides a list of the top and bottom 10 ranked players by EPVA during the 2013–2014 NBA season. Generally, high EPVA players adapt their decision-making process to the spatiotemporal circumstances confronting them upon gaining possession. They receive the ball in situations suited to their abilities, so that on average they outperform the rest of the league in these circumstances. Low EPVA players may be skilled; however, their actions simply tend to lead to fewer points than other players given the same options. Of course, EPVA provides a limited view of a player's overall contributions since it does not

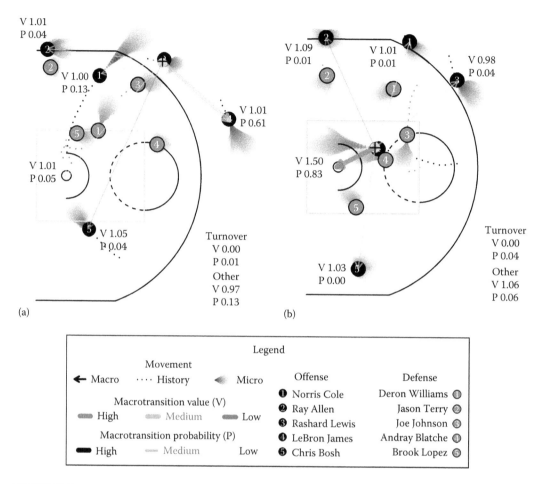

FIGURE 12.8

(See color insert.) Detailed diagram of EPV in terms of parameters of the underlying statistical models for player movement and decision-making. (a) The beginning of Rashard Lewis's possession in Figure 12.7 and (b) a fraction of a second before LeBron James attempts the layup at the end of the possession. Two seconds of players' predicted motion are shaded (with forecasted positions for short time horizons darker) while actions are represented by arrows, using color and line thickness to encode the value and probability of such actions. The value and probability of the "other" category represents the case that no pass/shot/turnover occurs during the next 2 seconds.

quantify players' actions on defense, or other ways that a player may impact EPV while not possessing the ball (though EPVA could be extended to include these aspects).

EPVA quantifies value players provide through all of their actions and decisions, extending the scope of basketball analytics beyond the box score statistics. However, EPV can also be used to construct metrics related to traditional basketball events, such as shots, passes, rebounds, etc. For instance, by implicitly providing counterfactual possession paths, we can assess the (un)selfishness of shooters across the league.

To create a *shot satisfaction* metric, for each shot attempt a player takes, we wonder how satisfied the team is with the player's decision to shoot—what was the expected value of the shot relative to his most reasonable passing option at the time of the shot? If for a particular player, the EPV measured at his shot attempts is higher than the EPV conditioned on his

TABLE 12.1

Top 10 and Bottom 10 Players by EPV-Added (EPVA) per Game in 2013–2014, Minimum 500 Touches During Season

Rank	Player	EPVA	Rank	Player	EPVA
1	Kevin Durant	3.26	277	Zaza Pachulia	−1.55
2	LeBron James	2.96	278	DeMarcus Cousins	−1.59
3	Jose Calderon	2.79	279	Gordon Hayward	−1.61
4	Dirk Nowitzki	2.69	280	Jimmy Butler	−1.61
5	Stephen Curry	2.50	281	Rodney Stuckey	−1.63
6	Kyle Korver	2.01	282	Ersan Ilyasova	−1.89
7	Serge Ibaka	1.70	283	DeMar DeRozan	−2.03
8	Channing Frye	1.65	284	Rajon Rondo	−2.27
9	Al Horford	1.55	285	Ricky Rubio	−2.36
10	Goran Dragic	1.54	286	Rudy Gay	−2.59

TABLE 12.2

Top 10 and Bottom 10 Players by Shot Satisfaction in 2013–2014, Minimum 500 Touches During Season

Rank	Player	Shot Satis.	Rank	Player	Shot Satis.
1	Mason Plumlee	0.35	277	Garrett Temple	−0.02
2	Pablo Prigioni	0.31	278	Kevin Garnett	−0.02
3	Mike Miller	0.27	279	Shane Larkin	−0.02
4	Andre Drummond	0.26	280	Tayshaun Prince	−0.03
5	Brandan Wright	0.24	281	Dennis Schroder	−0.04
6	DeAndre Jordan	0.24	282	LaMarcus Aldridge	−0.04
7	Kyle Korver	0.24	283	Ricky Rubio	−0.04
8	Jose Calderon	0.22	284	Roy Hibbert	−0.05
9	Jodie Meeks	0.22	285	Will Bynum	−0.05
10	Anthony Tolliver	0.22	286	Darrell Arthur	−0.05

possible passes at the same time points, then the team is satisfied with that player's shooting behavior, as it generally represents the best decision for the team. On the other hand, players with pass options at least as valuable as shots could be labeled "selfish shooters." Specifically, we calculate

$$\text{SATIS}(\ell) = \frac{1}{|\mathcal{T}_{\text{shot}}^{\ell}|} \sum_{t \in \mathcal{T}_{\text{shot}}^{\ell}} (\nu_t - \mathbb{E}\left[X \mid \text{pass}, Z_0, \ldots, Z_t\right]), \tag{12.8}$$

where $\mathcal{T}_{\text{shot}}^{\ell}$ indexes times a shot attempt occurs for player ℓ. Unlike EPVA, shot satisfaction SATIS(ℓ) is expressed as a per-shot average, which favors players such as three-point specialists, who take fewer—but more valuable—shots than their teammates. Table 12.2 provides the top and bottom 10 players in shot satisfaction for our 2013–2014 data. While players who mainly attempt three-pointers (e.g., Miller, Korver) and/or close range shots accrue the most satisfaction, players who primarily take mid-range or long-range two-pointers (e.g., Aldridge, Garnett) or poor shooters (e.g., Rubio, Prince) have the least. Note, however, that shot satisfaction numbers are almost entirely positive league-wide, meaning players still shoot relatively efficiently.

12.6 Conclusion

In this chapter, we explored the basketball insights that become possible through the use of player-tracking data. While the initial focus of player-tracking data was on fitness, endurance, and injury prevention, the data also allow for improved understanding of team and individual skill, strategy, and performance. We have shown how tracking data allow us to understand player types by studying the space where players tend to spend their time. Further, they allow us to identify defensive matchups and subsequently characterize defensive performance. Finally, we have seen how the full-resolution player-tracking data allow us to characterize opportunity and decision-making.

While the hardware employed across sports varies, the potential for improving insights is universal. In particular, we hope that this chapter not only highlights the potential for better understanding the sport of basketball, but also the opportunity that exists across other sports with similar data. In particular, we feel that with careful consideration, much of the lessons learned here can be employed in soccer, football, and hockey, where similar data are now being collected.

References

Bialkowski, A., P. Lucey, P. Carr, Y. Yue, S. Sridharan, and I. Matthews (2014). Large-scale analysis of soccer matches using spatiotemporal tracking data. In *IEEE International Conference on Data Mining (ICDM), 2014*, pp. 725–730. IEEE.

Bishop, C. M. et al. (2006). *Pattern Recognition and Machine Learning*, Vol. 1. Springer, New York.

Brunet, J.-P., P. Tamayo, T. R. Golub, and J. P. Mesirov (2004). Metagenes and molecular pattern discovery using matrix factorization. *Proceedings of the National Academy of Sciences of the United States of America 101.12*, 4164–4169.

Bukiet, B., E. R. Harold, and J. L. Palacios (1997). A Markov chain approach to baseball. *Operations Research 45*(1), 14–23.

Cervone, D., A. D'Amour, L. Bornn, and K. Goldsberry (2014). A multiresolution stochastic process model for predicting basketball possession outcomes. 111, 585–599.

Cunniffe, B., W. Proctor, J. S. Baker, and B. Davies (2009). An evaluation of the physiological demands of elite rugby union using global positioning system tracking software. *The Journal of Strength and Conditioning Research 23*(4), 1195–1203.

Franks, A., A. Miller, L. Bornn, and K. Goldsberry (2015a). Counterpoints: Advanced defensive metrics for nba basketball. *MIT Sloan Sports Analytics Conference*. Boston, MA.

Franks, A. et al. (2015b). Characterizing the spatial structure of defensive skill in professional basketball. *The Annals of Applied Statistics 9*(1), 94–121.

Goldner, K. (2012). A Markov model of football: Using stochastic processes to model a football drive. *Journal of Quantitative Analysis in Sports [online] 8*(1). DOI: 10.1515/1559-0410.1400.

Goldsberry, K. (2012). Courtvision: New visual and spatial analytics for the NBA. *MIT Sloan Sports Analytics Conference.*

Kubatko, J., D. Oliver, K. Pelton, and D. T. Rosenbaum (2007). A starting point for analyzing basketball statistics. *Journal of Quantitative Analysis in Sports 3*(3), 1–22.

Lee, D. D. and H. S. Seung (1999). Learning the parts of objects by non-negative matrix factorization. *Nature 401*(6755), 788–791.

Lee, D. D. and H. S. Seung (2001). Algorithms for non-negative matrix factorization. *Advances in Neural Information Processing Systems (NIPS) 13*, 556–562.

Lucey, P., D. Oliver, P. Carr, J. Roth, and I. Matthews (2013). Assessing team strategy using spatiotemporal data. In *Proceedings of the 19th ACM SIGKDD International Conference on Knowledge Discovery and Data Mining*, pp. 1366–1374. ACM, New York, NY.

Maheswaran, R., Y.-H. Chang, A. Henehan, and S. Danesis (2012). Deconstructing the rebound with optical tracking data. In *The MIT Sloan Sports Analytics Conference.* Boston, MA.

Miller, A., L. Bornn, R. Adams, and K. Goldsberry (2014). Factorized point process intensities: A spatial analysis of professional basketball. In *Proceedings of The 31st International Conference on Machine Learning*, Beijing, China, 2014. JMLR: W&CP, vol. 32, pp. 235–243.

Møller, J., A. R. Syversveen, and R. P. Waagepetersen (1998). Log Gaussian Cox processes. *Scandinavian Journal of Statistics 25*(3), 451–482.

Murphy, K. (2012). *Machine Learning: A Probabilistic Perspective.* The MIT Press, Cambridge, MA.

Murray, I., R. P. Adams, and D. J. MacKay (2010). Elliptical slice sampling. *Journal of Machine Learning Research: Workshop and Conference Proceedings (AISTATS) 9*, 541–548.

Oliver, D. (2004). *Basketball on Paper: Rules and Tools for Performance Analysis.* Potomac Books, Inc., Washington DC.

Prentice, R., J. Kalbfleisch, A. Peterson Jr., and N. Flournoy (1978). The analysis of failure times in the presence of competing risks. *Biometrics 34*(4), 541–554.

Shea, S. M. and C. E. Baker (2013). *Basketball Analytics: Objective and Efficient Strategies for Understanding How Teams Win.* CreateSpace Independent Pub., Charleston, SC.

Stan Development Team (2014). Stan: A C++ library for probability and sampling, version 2.2. https://cran.r-project.org/web/packages/rstan/citation.html.

Weil, S. (2011). The importance of being open: What optical tracking data can say about nba field goal shooting. In *2011 MIT Sloan Sports Analytics Conference*, Boston, MA.

13

Poisson/Exponential Models for Scoring in Ice Hockey

Andrew C. Thomas

CONTENTS

When a game takes on a flow, rather than a series of sequential events, it's often described as "too complicated for statistics" by the very same people who then try to describe a game by statistical count measures: goals scored, saves made, and fights won or lost.

Beyond simply collecting information about the game, or the people who play it, there is an interest in describing the flow of a game in terms of the events that *might* happen, and there's no better tool for accomplishing that than a stochastic model based on arrival rates. That goal scoring is relatively rare in hockey on a per-game basis makes it a perfect candidate for models that use the Poisson distribution as their base.

Of course, there are two processes at work in any game: each of the two teams is trying to score goals and prevent the other team from doing so. Our models for scoring are best composed as pairs: one model for each of the teams playing in the game, traditionally designated as home or away. Since each team can be expected to have differing skills in offense and defense from the rest of their league, the two scoring processes will each contain some information about the two teams in the game.

This chapter is thus dedicated to the exploration of Poisson-type stochastic models for ice hockey, borrowing liberally from soccer and other sports that have similar patterns, while having its own distinct features that make it unlike many other sports one might study. I use these models to address a series of questions, starting from simpler models to describe game flow. Once this is established, more complexities are introduced, including a move from single-variable to bivariate models for each of the two teams' counts of useful outcomes during given time intervals. Finally, these processes are modulated by adding information on specific teams and players.

I'll be using data from the National Hockey League from 10 full seasons (2005–2006 until 2014–2015) throughout this chapter; the raw data are available online as data files formatted for the R programming language.

13.1 Simple Poisson Models and Assumptions

The Poisson distribution itself is an imperfect way of describing count events in sports when any fan can see the violations on a short time scale. The nature of the Poisson assumption for any sport is that it first depends on an underlying rate at which events occur, and that events occurring in one time interval are independent of events occurring in another nonoverlapping time interval.

These assumptions are demonstrably false in two ways with opposite consequences. First, following any goal, play resumes with a face-off at center ice, after which the likelihood of a goal is noticeably smaller. In fact, no goal in the NHL has ever been scored less than 4 seconds after a face-off at center ice, and the scoring rates at the beginning of each period are noticeably less than that in the rest of the game, as seen in Figure 13.1.

Second, following a shot on goal that does not result in a goal, the game is more likely to produce another shot on goal by the same team, because the shooting team can quickly recover possession and make another attempt before the defending team can take possession and get to the other end of the rink.

Ultimately, these are not reasons to dismiss an imperfect model for the game, especially since what data the National Hockey League makes available are at such a rough scale for goals but at a finer scale for shots on goal. There are representations of hockey that cast it as a semi-Markov process on a set of discrete game states (Thomas 2006) that when extended

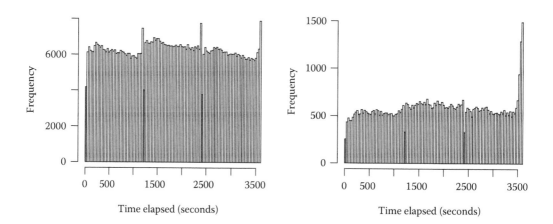

FIGURE 13.1
Shooting and scoring rates for NHL teams by 30-second periods. Period openings are distinctly lower scoring mainly due to the time after opening face-offs resulting in fewer shots. The scoring peak at the end of regulation time is due to "empty-net" goals (a) Shots on goal during regulation time. (b) Goals scored during regulation time.

to the time dimension is aperiodic, recurrent, and irreducible; goals scored for the home and away teams are two of the states in this system. When factoring in other considerations, like the natural effect of game score or player fatigue within states, the transition matrix will certainly be different at various points during the game, but the underlying state model remains adequate to describe the game at a rough scale.

And yet there is still room to doubt it on first glance. The goal spike at the end of the game is for empty net goals, scored by the leading team when the trailing team sacrifices their last line of defense for a chance at tying the game before the end of regulation. This is clearly a state-dependent condition that wouldn't be accounted for in an ordinary model; special circumstances would need to be brought into play to make this work, and even if they were themselves Poisson processes, care would be needed to stitch them together.

13.1.1 A Simple Test Case: Overtime Rules

If the question is more simply posed, however, the assumptions may be more clearly justified. One such example occurred in the summer of 2015, when the NHL discussed and eventually decided to replace the format of the overtime period from a four-on-four format to a new format, using three-on-three play instead. The stated goal was to reduce the number of games that would require a shootout, so knowing how often any goal would be scored in the overtime period would be ideal in making the decision.

Pettigrew (2015) did this using the Poisson model for all scoring at 4-on-4 and 3-on-3 observed in the NHL since 2005. For simple one-parameter models, the Poisson rate parameter is simply the number of goals scored divided by the total time, and the simplest models consider both teams scoring as one process. When combining goal scoring rates, teams scored 5.46 goals per 60 minutes at 4-on-4, a rate only marginally greater than the rate at 5-on-5, but a whopping 10.08 goals per 60 minutes at 3-on-3. When comparing different formats and combinations of overtime, the quantity of interest—how many games will go to overtime?—can be determined directly from the respective cumulative distribution functions or by simulation.

The estimated rate for the chosen option, 5 minutes of 3-on-3, has been shown to reduce the number of shootouts in 2015–2016 from 60% of overtime results to 40%, just as the distribution suggested given prior data.

13.2 Poisson Models for Goals Scored and Allowed

The simplest tests for goals scored in a game compare the Empirical Cumulative Distribution Function (ECDF) to the CDF of the corresponding Poisson distribution. In this case, one can fit two separate simple Poisson models for the number of goals scored in regulation time in a particular game by the home team and for the away team separately:

$$Y_{\text{home}} \sim \text{Poisson}(\lambda_{\text{home}})$$
$$Y_{\text{away}} \sim \text{Poisson}(\lambda_{\text{away}})$$

Table 13.1 shows the empirical and Poisson model CDF for all home and away teams through the entire sample of 10 seasons of data, and not only is there an excellent match

TABLE 13.1

Empirical CDF and Poisson CDF for Scoring by Home and Away Teams

Goals	ECDF (Home)	Poisson (Home)	ECDF (Away)	Poisson (Away)
0	0.06	0.06	0.08	0.08
1	0.22	0.22	0.27	0.27
2	0.45	0.46	0.52	0.53
3	0.68	0.68	0.74	0.74
4	0.84	0.84	0.88	0.88
5	0.93	0.93	0.96	0.95
6	0.98	0.97	0.99	0.98

Note: General agreement suggests that the simple Poisson model is adequate for predicting a single team's total score.

TABLE 13.2

Differences in Percentage Points of Theoretical Independent Poisson Models from 10 Years of NHL Game Scores in Regulation; Home Score Is on the Rows, Away Score Is on the Columns

	Away 0	Away 1	Away 2	Away 3	Away 4	Away 5
Home 0	0.24	−0.17	−0.31	0.31	0.09	−0.01
Home 1	−0.23	1.11	−1.13	−0.32	0.78	−0.22
Home 2	−0.28	−1.37	2.10	−1.30	−0.02	0.66
Home 3	0.25	−0.59	−1.07	1.88	−0.92	0.15
Home 4	0.02	0.99	−0.21	−0.87	0.67	−0.47
Home 5	−0.06	−0.01	0.97	0.03	−0.34	0.19

Note: The strong positive diagonal indicates that more NHL games ended in regulation with a tied score than one could expect from the model; for example, 8.03% of all NHL games ended regulation in a 2-2 tie compared to 5.92% of theoretical games.

between the empirical and modeled CDFs for both home and away rates, it is clear that there is a distinct home ice advantage in scoring.

This suggests that the univariate Poisson distributions will be sufficient for predicting individual results. What about the independent bivariate case? Table 13.2 gives the difference in observed and theoretical game scores using two independent Poisson models for home and away scoring and compares it to all scores in the 10 NHL seasons used for the model. The differences are stark:

Note the strong and positive diagonal: there are more ties at the end of regulation than should be expected from a pair of independent Poisson models. Is this because of the pulling the goalie at the end to tie the game? Removing the last 2 minutes of regulation, each game (Table 13.3) gives an improved result, but one that is qualitatively similar.

The diagonal in Table 13.3 is improved—closer to zero than before—but still consistently positive. This suggests that there's an extra factor in game play that causes a general team to tie a game when down by one than would have been present otherwise, and that for analyzing specific game strategies, a model of pure goal count is not sufficient.

TABLE 13.3

Differences in Percentage Points of Theoretical Independent Poisson Models from 10 Years of NHL Game Scores in the First 58 minutes of Regulation; Home Score Is on the Rows, Away Score Is on the Columns

	Away 0	Away 1	Away 2	Away 3	Away 4	Away 5
Home 0	−0.67	0.14	−0.20	−0.10	−0.03	−0.03
Home 1	0.08	0.52	0.01	−0.40	−0.31	−0.18
Home 2	−0.06	0.14	1.02	0.47	−0.09	−0.11
Home 3	−0.47	−0.20	0.26	0.70	0.31	−0.09
Home 4	−0.07	−0.05	−0.04	0.22	0.24	0.00
Home 5	−0.03	−0.02	−0.07	−0.28	0.09	0.04

Note: There are still more tied scores near the end of regulation.

Would a bivariate Poisson model, or one that specifically inflates the diagonal, be sufficient for modeling purposes? That depends on the use case; if the only goal is to predict final game scores, a parameter can be introduced that would accomplish this and give a better fit; Karlis and Ntzoufras (2003) cover the application in soccer and water polo of the Bivariate Poisson distribution, which adds a parameter for a positive correlation in scores between the two teams by adding and integrating out an auxiliary Poisson contribution to both teams. This has the advantage of maintaining a marginal Poisson scoring distribution for both teams, which matches historical data; it also allows for an inflationary parameter to be added to the diagonal just as a pair of simple Poisson distributions would do. In that way, it is certainly useful and sufficient for many problems involving the distribution of the final score.

But this in itself does not answer a key question of considerable interest: what should be used to predict the final score of a game, given the current score and time remaining? If an extended Poisson-like model has Poisson characteristics like independent intervals, it suggests that the two segments of the game should be additive as well. This would hold up well for the bivariate Poisson model but not a diagonally inflated model (at least, one where the inflationary component was not Poisson itself). Moreover, if there is dependence between intervals, building a model on this case alone would explain away the reasons for the dependence rather than allowing for their explanation.

Knowing, in particular, if this is an issue of coaching decisions—defensive strategies that may do more harm than good, or player selections designed to maintain rather than increase a lead—or of simple psychology that a team playing from behind will try harder to catch up—makes a more in-depth modeling approach more compelling in the long run, particularly one that allows for the simulation of a game as a sequence of events in time rather than a final distribution.

13.3 Exponential Distributions and Score Effects

The Poisson distribution is directly connected to the Exponential distribution by its definition and its construction: if a sequence of event timings occur with an i.i.d. exponential distribution with rate λ, then the number of events that occur in a window of time t is a Poisson distribution with mean λt. This reflects both the memoryless property of both distributions—future events do not depend explicitly on previous ones—and that a series of events with Exponential waiting times given a fixed time interval will yield a uniform distribution of events within that interval.

From the plots of goals over time, one can see this isn't exactly the case—the beginnings of periods being the biggest violation—but the heavier diagonal and incidence of near-tie games suggest that there might be an "even-up" effect in play at the very least. And if prior events influence future ones, this would break the fundamental Poisson assumptions made to use this model.

The immediate solution is to replace the Poisson process with a semi-Markov process, like that proposed by Thomas (2006), in that the time until the next event can be modeled with a new distribution, exponential, or otherwise. Assuming that the process is locally memoryless—that is, once an event is marked, future events do not depend explicitly on events before our indicated time—the intervals i will be Poisson-based, and one can estimate the parameters of the model using Poisson-distributed intervals of length T_i, with constant scoring rate within:

$$Y_i \sim \text{Poisson}(T_i \lambda_{\text{rate}[i]})$$

A comprehensive model can be specified in which the rates depend on the score in the game for each team; for example, different parameters for when the score is 0–0, 2–2, and 4–1, respectively. A model based on the difference in score between the home and away teams has been found to be sufficient and is conventionally adopted by most public data sources.[*]

Once these scoring rates are estimated, the simulation model for exponential waiting times can be done in two ways:

1. Simulate the times until the next goal for each team from two exponential distributions, one for each team's scoring rate, and select the smaller of the two to represent the time until the next goal by said team

2. Simulate the time to the next goal by either team from a single exponential distribution whose rate is the sum of the two individual rates λ_h and λ_a; assign the goal to the home team with probability $\frac{\lambda_h}{\lambda_h + \lambda_a}$

Comparing the rates at which goals are scored given the differential in score, there is a consistent difference, as seen in Figure 13.2; notably, teams that are trailing by 1 goal are (in general) more likely to score the next goal than the team that's leading.

A simple simulation engine can be built by generating exponential random variables for the next time the home and away teams would score, where the longer observation is censored away. Once these intervals are longer than the time remaining in the game, the simulation stops and the score is final. Doing this with the empirical scoring rates in each situation, the diagonal bias present in Tables 13.2 and 13.3 has disappeared in Table 13.4. This model for generating game outcomes can be used up to near the end of regulation time; scenarios in which a trailing team replaces their goaltender for an extra attacker can be appended when necessary.

---

[*] See for example stats.hockeyanalysis.com and behindthenet.ca.

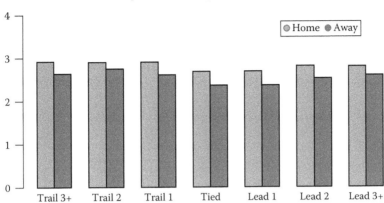

FIGURE 13.2
Goal-Scoring rates per 60 minutes, by home and away, conditional on the score difference. There is a consistent home scoring advantage conditional on that team's lead (or deficit); teams that are tied or lead their opponents by 1 score at a lower rate than in all other states.

TABLE 13.4

The Share of Game Outcomes for the Home Score (Rows) and Away Score (Columns), Expressed as a Difference between Actual Game Outcomes and Those Simulated under the Score-Dependent Model

	Away 0	Away 1	Away 2	Away 3	Away 4	Away 5
Home 0	−0.89	0.02	−0.13	0.00	0.05	0.03
Home 1	0.04	−0.17	0.02	−0.12	−0.02	−0.08
Home 2	0.15	−0.09	0.33	0.36	0.09	−0.03
Home 3	−0.14	0.10	−0.02	0.27	0.40	−0.05
Home 4	0.04	0.24	0.16	0.08	0.03	0.01
Home 5	0.02	0.03	0.03	−0.31	0.08	0.01

Note: The diagonal, formerly higher in real games in all cases, is now neither uniformly positive or negative.

13.3.1 Application: Models for Pulling the Goalie

A team trailing in the waning minutes of a game will often choose to replace their goaltender, a player with exclusively defensive responsibilities, with an additional forward, knowing that the risk in letting up another goal is greatly outweighed by the benefit of tying the game up. In both cases, the Poisson assumption has been the starting point for modeling the goal-scoring process, an approach taken by both Morrison (1977) and Beaudoin and Swartz (2010).

The decision to pull the goalie will ultimately come down to a "trigger" time: at what point the coach would feel comfortable making the substitution, typically with a particular game scenario, like possessing the puck in the offensive zone. There are several goal-scoring rates needed for the simulation problem: for the trailing team, at what rate they score at full strength and with an extra attacker; for the leading team, at the rate they score at full strength and when the opposing net is empty. Also needed is the rate at which teams will be assessed any penalties.

The problem can be analyzed through simulation

- Establish a point in the game where the coach would first consider making the substitution; with a one-goal deficit, 10 minutes remaining in the period is likely sufficient.

- Choose a series of times at which the goalie pull might happen (3:00 remaining, 2:55 remaining, etc.) These are the possible times that a coach might decide to make the substitution.

- Simulate the time to the next goal for each team, as well as to the time until the next penalty; the next event is the one with the shortest time. If all of these times are longer than the "pull time" from the previous step, continue; if a goal was scored by either team, mark it. If a penalty was sustained, continue to simulate under penalty conditions (though strategy would change if the penalty overlapped the pull time).

- Change the simulation rates to what the trailing and leading teams would score under the "goalie pulled" scenario and resimulate starting from the pull time. If the trailing team scores first, mark it as a win (i.e., the game is tied and the strategy worked); if the leading team scores, continue simulating from the model down an additional goal; if both goal times are longer than the time remaining, the leading team has won.

- Repeat this simulation for as many times as iterations are desired. Count how many times the trailing team tied the game under each pull time. The optimal time to pull under this model is then the pull time that led to the greatest fraction of games that ended up tied.

There's far more detail that can be explored in a simulation of this sort; noting the special cases and exceptions that come with penalties, face-off locations, line combinations, and other factors makes this problem worth studying in more detail. In most cases, these models have suggested a pull time closer to 3 minutes remaining when historically this was done with 1 minute left, though the observed time has been increasing in recent seasons. The change seems to have come about primarily by Colorado coach Patrick Roy embracing the strategy for his own reasons; other coaches may have felt more comfortable adapting to the strategy once someone else had executed it with apparent success.

13.3.2 Play-off Overtime

In Stanley Cup play-off games, overtime is not only "sudden death," in that the first goal scored ends the game, but there is no preset limit on the length of overtime, aside from its division into 20-minute periods. One question of interest, if only to broadcasters, is to estimate how long overtime is expected to last.

This situation is a simplified scenario from the previous cases: there is only one score state ("tied"). The 205 play-off games that went to overtime between 2005 and 2015 yielded 105 wins for the home team and an average goal-scoring rate for both teams of 4.87 goals/60 minutes, considerably less than the regular time average in all situations but roughly equal to the rate that goals are scored at standard 5-on-5.

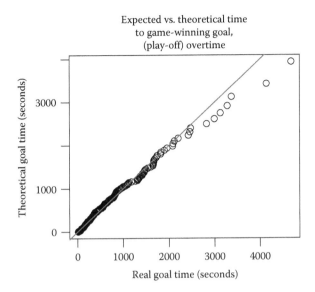

FIGURE 13.3
Observed and theoretical distribution for the time until an overtime game-winning goal. Note that the group of games with the greatest duration lasted longer than the expected time.

We compare the empirical distribution of goal times to the corresponding exponential CDF with the empirical rate in Figure 13.3; the results are nearly consistent. The main difference is in the tail: the longest two overtime games went 10 minutes longer than expected, and the longest 11 were all over the expected time. While the distribution fits considerably well for the bulk of the distribution, there is incentive to suggest that the increased length of the most extreme overtime games is not anomalous. There are two likely explanations: first, players are so fatigued as a game enters its third hour that they are unable to make as many aggressive offensive moves; second, that those teams that last later into overtime have higher defensive and lower offensive skills than those that have games settled earlier.

13.3.3 Team Effects on Scoring Rates

Just as in-game information can be included to these models, so can information on players and teams by adding them directly to the model. This was the premise of linear regression-adjusted plus–minus models such as Macdonald (2011) that was adapted to the Poisson competing process method by Thomas et al. (2013); a purely binary approach was also considered by Gramacy et al. (2016).

There are two main obstacles. One is technical: in fitting a model where a team can appear in the home or away category, we have to be clever about creating a design matrix that allows for an entity to exist in multiple places. The other is more involved—what does it mean to have two players on the ice and in the model at the same time? A direct Poisson model suggests that an indicator for a player being on the ice would yield a multiplicative effect on scoring, either for or against, compared to an average player in that position.

Further information on player ratings of this type can be found in Thomas et al. (2013) and other subsequent works. For this chapter, we'll feature the augmented results from the previous model for teams at Even Strength 5v5 for the 2014–2015 season, by modeling

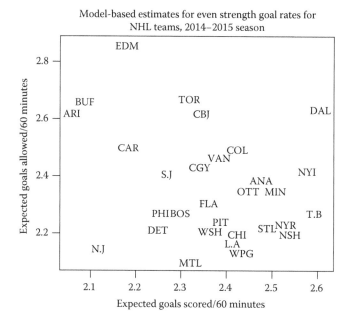

FIGURE 13.4
Model estimates for the rate at which NHL teams would score and allow goals at full strength 5v5 play, accounting for score differential, home/away bias, opponent matchups, and face-off prowess. Rates displayed are for when the team is playing on the road.

the goal-scoring process for the home and away teams. If the NHL schedule were perfectly balanced, we could extract this from the raw statistics themselves, but since the league schedule is unbalanced to favor divisional rivalry, a Poisson model is ideal for adjusting for strength of schedule. The home team's scoring rate during an interval i is a factor of the home offense and the away defense, represented as ω and δ, respectively. The number of events in an interval of length T_i is modeled as

$$Y_{\text{home},i} \sim \text{Poisson}(T_i \lambda_{\text{rate}[i]})$$
$$\log(\lambda_{\text{rate}[i]}) = \mu_{\text{home},i} + \omega_{\text{home}} + \delta_{\text{away}}$$

Similarly, the away team's scoring rate is a function of home defense and away offense. Score differential and home-ice advantage are modeled simultaneously with team effects.

The estimates for each team at full strength (five skaters and one goaltender on each team) in 2014–2015 are seen in Figure 13.4. Not surprisingly, the three teams that performed the worst overall (Edmonton, Arizona, and Buffalo) all appear in the extreme upper-left as teams that scored few and allowed many full-strength goals.

13.4 Drawing and Taking Penalties

The temporary main advantage, in which one team has more players on the ice than their opponent, is a sporting quirk largely unique to hockey. The majority of penalties called

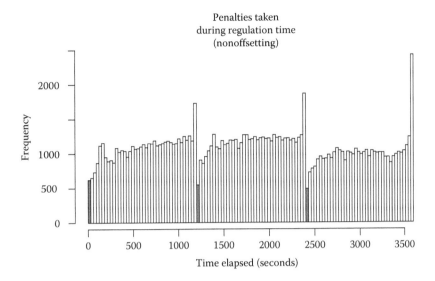

FIGURE 13.5
Penalties called as a function of game time. The spikes at the end of each period correspond to the truncation of penalty calls to when time expires.

are of the "minor" variety, where one player is sent off the ice for 2 minutes, unless the nonoffending team scores first; a minority are of the "major" variety, when the time penalized is 5 minutes with no reprieve in case of a goal.

As can be seen in Figure 13.5, the time in game when penalties are registered is not uniform; when a penalty infraction is committed and when it is registered are different because they are not registered until play stops, typically when the offending team regains possession of the puck. Even if penalty infractions occurred uniformly at random, there would still be a shift in the distribution away from the beginning of periods and a truncation of penalty calls at the 20-minute mark (when regulation periods end). Note that the rate for penalties in the third period is diminished overall (either due to player caution or referee bias) but spikes at the end of the game (when penalties are largely irrelevant.)

Studies of penalties often look not at the total number but at the proportion for one team during a game, such as the "even-up" bias known to all sports fans and identified distinctly in Abrevaya and McCulloch (2014), Davis and Lopez (2015), and Beaudoin et al. (2016).

Simple Poisson models can be used to determine whether certain players are more or less likely than others to take penalties, as well as being more or less likely to *draw* penalties—that is, being the player on whom the infraction is committed. More importantly, the same factors that influence shots or goals being taken can also affect the rate at which a penalty is taken or drawn: game score, the home/away factor, and the presence in the offensive/defensive zone. Two Poisson models can be constructed, just as for goals, but here for penalties taken and drawn, where the intervals in question are for a single player's time on the ice, and a player effect can be included to estimate a player's penalty propensity; this can be as a random effects model (such as in the R package lme4) or a penalized regression model (as in the R package glmnet) so that players with minimal game time do not have outsized estimates for their abilities.

When calculated for the 2014–2015 NHL season, the 10th and 90th percentiles of the rates at which nongoaltenders take penalties are estimated at 0.43 and 0.89 or penalties taken per 60 minutes. For a player who sees 20 min of ice time in a game, that's a range of 14–29 penalties taken over a season. Thomas (2015) estimates the value of a minor penalty in the NHL at 0.17 goals (given the tendency of the other team to score on the power play), meaning that such a shift in discipline when it comes to taking penalties would be worth three goals a season, a surprisingly significant value.

The estimated 10%/90% range is roughly the same for penalties drawn by individual players per 60 minutes, though a little narrower: 0.54–0.91, so that a player who is expert at getting other players to commit offenses, compared to an unaccomplished player, can be worth several goals more than a player who does not incite others.

13.5 Perturbations in the Model

One feature of the Poisson-based model is that intermediate observations can censor the process and allow us to include new information. The face-off, for example, is known to occur at the beginning of the period and after every goal, and depresses the goal-scoring rate at the beginning. Similarly, once a shot has been taken that does not result in a goal, the goal-scoring rate for the team that took the shot will go up in the very short term, because there is new information about the game that negates the stationarity assumption. Measuring goal-scoring rates subsequent to a marked on-ice event was a key contribution of Schuckers and Lock (2009).

To isolate these differences directly, and integrate them into the model as a whole, a more extensive Poisson regression can be fit by adding new covariates beyond just the game score. To start, let's include information on a face-off in the event that follows it: whether it was won or lost by the home team, and in which of the three zones it took place: the neutral zone (between the two blue lines) or either of the team's defensive zones (on either end of the rink). The notion of an "interval" starts with a face-off or scoring event for a team and ends either with a scoring event or something else that censors the time until the next event. With knowledge of the state of the system at this time point, a new interval begins, with a possibly different rate than the previous one given this new knowledge. Both Thomas (2006) and Schuckers and Lock (2009) suggest that if an event hasn't taken place within a period of time—say, 15 or 20 seconds—a marker for "no event recorded" can also be added, since this suggests that the state of the system is different enough that it may signal a return to equilibrium.

Putting these predictors into a Poisson Generalized Linear Model, it is straightforward to estimate both the "score effects" and "game event" components for home and away goals, as well as home and away shots on goal, maintaining a multiplier offset for the total time on ice T_i during each interval i:

$$Y_i^{\text{home}} \sim \text{Poisson}(T_i \lambda_i^{\text{home}})$$

$$log(\lambda_i^{\text{home}}) = \mu_{\text{home}} + \alpha_{\text{last}[i]}$$

The exponentiated coefficients from the model give us multipliers for the relative change in rate, in Table 13.5.

There are six main types of face-offs with respect to a team: in which of the three zones it took place, and whether or not the team won the draw, taking possession of the puck for

TABLE 13.5

Rate Multipliers for Shots and Goals in the Poisson Interval Model, Accounting for Face-off Results and Shot Rebounds That Begin the Next Observable Interval

	Home Goals	Away Goals	Home Shots	Away Shots
Offensive zone win	1.15	1.10	2.29	2.40
Offensive zone loss	0.34	0.36	0.57	0.59
Neutral zone win	0.37	0.35	0.61	0.58
Neutral zone loss	0.29	0.26	0.37	0.36
Defensive zone win	0.35	0.32	0.47	0.46
Defensive zone loss	0.24	0.23	0.31	0.30
Away blocked shot	0.35	1.26	0.31	1.17
Away missed shot	0.25	1.14	0.24	1.15
Away shot on goal	0.24	3.01	0.24	1.43
Home blocked shot	1.21	0.32	1.17	0.27
Home missed shot	1.11	0.20	1.17	0.21
Home shot on goal	2.99	0.22	1.49	0.21

Note: When a face-off was taken, only a team winning in the offensive zone produced a shot rate above what would be seen in open play.

the play that follows. Of those six scenarios, only one yielded rates for shots and goals that exceeded the rate one would see during open play: when a team won in their offensive zone; for all other face-off scenarios the shot and goal rates were considerably depressed.

This model also gives us the relative value of shooting the puck beyond the mere chance that it will result in a goal. It is evident that the puck was in the possession of a team in their offensive zone, and that subsequent shots should be expected depending on the outcome. When a shot was on goal, forcing the goaltender to make a play on the puck, the likelihood of a follow-up shot increases substantially in the next interval compared to open play. This rate was also greater than the baseline when compared to when the shot was missed or blocked by a skater on the other team but to a lesser extent.

The value of a face-off win can be pulled from the data as well: what's the change in scoring during the next interval? From the model, we can estimate this: for an offensive zone win, this is an increased scoring rate of 0.8 times the average (at even strength, roughly 2.5 goals/60 minutes per team) times the length of the average censored shift post face-off (16 seconds) and a decreased scored-against rate of 0.1 times the same factors; this gives an expected return of 0.0094 goals, or 106 face-off wins per marginal goal. This number is far smaller for a neutral zone face-off win, where the difference is 0.0021 goals per face-off win, or 476 face-off wins per marginal goal.

13.6 Conclusions and Extensions

The biggest advantage to modeling hockey with Poisson processes is that these models are extendable. Once the data become available to add into the game flow, we merely need to drop in the modification to see how it affects future play. These extensions come at the usual price: more effort has to be expended into collecting and interpreting the data, including accounting for measurement error and systemic bias.

The previous section shows the other price we must pay: for every new feature we wish to explain, simulating a game with those features becomes all the more complex. Once we examine the impact of a face-off, there is a need to include a simulation for the next face-off, which will occur more frequently than goals but roughly as frequently as shots on goal, and their timing is certainly related.

Similarly, examining a player's impact on team scoring rates, both for and against, seems simple enough: include players in a model just as we would for teams. Indeed, this is the basis of Thomas et al. (2013), which sidesteps the simulation issue: it does not attempt to model when players make substitutions or how they are selected by their respective coaches.

Beyond this, there are known fine-grained violations of Poisson assumptions that are better approached as semi-Markov in nature. Thomas (2006) uses a state space model that has the time until a subsequent event that is nonexponentially distributed; Thomas (2007) shows that there is a correction to be made whenever the deficit to be overcome is two goals or more: namely, that a face-off that follows a goal will effectively take time off the clock by slowing the scoring rate after the goal. Expressing this rate as starting from zero and increasing toward an asymptotic value, the resulting hazard function for events is nonlinear and clearly violates any assumption of independent increments within a predefined interval. There are ways to avoid this—such as inserting a "neutral" event marker 10 seconds after a face-off to signify that nothing has happened, and modeling each increment with a constant event rate.

Finally, the nature of the data itself is changing. More detailed event analyses are being made by recording events at a far greater temporal resolution than previous sources of data. The use of high-resolution data in NBA basketball, through the SportVU system, has been standard for multiple seasons and is expected to expand to other worldwide leagues. Systems like this are in active development in hockey but do not yet have widespread usage in professional leagues. Extracting full value from these systems will demand the use of models sophisticated enough to handle the complexity of the data, but clear enough to show where and how strategic changes and actions can be taken by coaches, managers, and analysts alike.

References

Abrevaya, J. and McCulloch, R. (2014) Reversal of fortune: A statistical analysis of penalty calls in the national hockey league, *Journal of Quantitative Analysis in Sports*, 10(2), 207–224.

Beaudoin, D. and Swartz, T. B. (2010) Strategies for pulling the goalie in hockey, *The American Statistician*, 64(3), 197–204.

Beaudoin, D., Schulte, O., and Swartz, T. B. (2016) Biased penalty calls in the national hockey league, *Statistical Analysis and Data Mining*, 9(5).

Davis, N. and Lopez, M. (2015) Hockey refs are out to get you (if they already got the other guy), http://fivethirtyeight.com/features/hockey-refs-are-out-to-get-you-if-they-already-got-the-other-guy/, accessed September 15, 2016.

Gramacy, R. B., Taddy, M., and Tian, S. (2016) Hockey player performance via regularized logistic regression, http://arxiv.org/pdf/1510.02172v2.pdf, accessed September 15, 2016.

Karlis, D. and Ntzoufras, I. (2003) Analysis of sports data by using bivariate Poisson models, *Journal of the Royal Statistical Society: Series D (The Statistician)*, 52(3), 381–393.

Macdonald, B. (2011) A regression-based adjusted plus-minus statistic for NHL players, *Journal of Quantitative Analysis in Sports*, 7(3).

Morrison, D. G. (1976) On the optimal time to pull the Goalie: A Poisson model applied to a common strategy used in ice hockey, *TIMS Studies in Management Sciences*, 4, 137–144, North Holland, New York.

Pettigrew, S. (2015) How those 3-on-3 overtime rules would cut down NHL shootouts, http://regressing.deadspin.com/how-those-3-on-3-overtime-rules-would-cut-down-nhl-shoo-1692120143, accessed September 15, 2016.

Schuckers, M. and Lock, D. F. (2009) Beyond +/−: A rating system to compare NHL players. Presented at the *Joint Statistical Meetings 2009*, Washington, DC. statsportsconsulting.com/main/wp-content/uploads/LockJSM2009.pptx, accessed September 15, 2016.

Thomas, A. C. (2006) The impact of puck possession and location on ice hockey strategy, *Journal of Quantitative Analysis in Sports*, 2(1).

Thomas, A. C. (2007) Inter-arrival times of goals in ice hockey, *Journal of Quantitative Analysis in Sports*, 3(3).

Thomas, A. C. (2015) The road to wAR, Part 8: Penalties taken and drawn, http://blog.war-on-ice.com/the-road-to-war-part-8-penalties-taken-and-drawn/, accessed September 15, 2016.

Thomas, A. C., Ventura, S. L., Jensen, S. T., and Ma, S. (2013) Competing process hazard function models for player ratings in ice hockey, *Annals of Applied Statistics*, 7(3), 1497–1524.

14

Hockey Player Performance via Regularized Logistic Regression

Robert B. Gramacy, Matt Taddy, and Sen Tian

CONTENTS

A hockey player's plus-minus measures the difference between goals scored by and against that player's team while the player was on the ice. This measures only a marginal effect, failing to account for the influence of the others he is playing with and against. A better approach would be to jointly model the effects of all players, and any other confounding information, in order to infer a partial effect for this individual: his influence on the box score regardless of who else is on the ice.

This chapter describes and illustrates a simple algorithm for recovering such partial effects. There are two main ingredients. First, we provide a logistic regression model that can predict which team has scored a given goal as a function of who was on the ice, what teams were playing, and details of the game situation (e.g., full-strength or power-play). Since the resulting model is so high dimensional that standard maximum likelihood estimation techniques fail, our second ingredient is a scheme for regularized estimation. This adds a penalty to the objective that favors parsimonious models and stabilizes estimation. Such techniques have proven useful in fields from genetics to finance over the past two decades, and have demonstrated an impressive ability to gracefully handle large and highly imbalanced data sets. The latest software packages accompanying this new methodology—which exploit parallel computing environments, sparse matrices, and other features of modern data structures—are widely available and make it straightforward for interested analysts to explore their own models of player contribution.

This framework allows us to quickly obtain high-quality estimates for the full system of competing player contributions. After introducing the measurement problem in Section 14.1, we detail our regression model in Section 14.2 and the regularization scheme in Section 14.3. The remainder of the chapter analyzes more than a decade of data from the NHL. We fit and interpret our main model, based on prediction of goal scoring, in Section 14.4. This is compared to shot-based analysis, and metrics analogous to Corsi or Fenwick scores, in Section 14.5. Finally, Section 14.6 considers the relationship between our estimated performance scores and player salaries. Overall, we are able to estimate a partial plus-minus metric that occurs on the same scale as plus-minus but controls for other players and confounding variables. This metric is shown to be more highly correlated with salary than the standard (marginal) plus minus. Moreover, we find that the goals-based metric is more correlated with salary than those based upon shots and blocked shots. We conclude in Section 14.7 with thoughts on further extensions, in particular by breaking out of the linear framework to use classification models popular in the machine learning literature.

The code for all empirical work in this chapter is provided to the public via a GitHub repository [36] and utilizes open source libraries for R [25], particularly the gamlr [28] package from [30].

14.1 Introduction: Marginal and Partial Effects

Hockey is played on ice, but that's not all that sets it apart from seemingly related sports like soccer, basketball, or even field hockey, at least not from an analytics perspective. The unique thing about hockey is the rapid substitutions transpiring continuously during play, as well as at stoppages in play. In the data sets we have compiled, which we discuss in more detail shortly, the median amount of time observed for a particular on-ice player configuration (determined by unique players on the ice for both teams) is a mere 8 s. Although many "shifts" are much longer than that, a trickle of piecemeal substitutions on both sides, transpiring as play develops, makes it difficult to attribute credit or blame to players for significant events, such as goals or shots.

Plus-minus (PM) is a traditional metric for evaluating player contributions in hockey. It is calculated as the difference, for a given player, between the number of goals scored against the player's team and those scored by the player's team while that player was on the ice. For example, during the 2012–2013 season Stanley Cup Finals, between Boston and Chicago, Duncan Keith of the Chicago Blackhawks was on the ice for eight goals by Chicago and four by Boston, giving him a +4 PM for the series.

The PM score represents what statisticians call a *marginal effect:* the aggregate change in some response (goals for vs. against) with change in some covariate (a player being on the ice) *without accounting for whatever else changes at the same time.* It is an aggregate measure that averages over the contributions of other factors, such as teammates and opponents. For example, suppose that the three authors of this chapter are added to the Blackhawks roster and that Joel Quenville (the coach of the Blackhawks) makes sure that Duncan Keith is with us on the ice whenever we are playing. Since none of us are anything close to as good at hockey as Keith is, and surely our poor play would allow the other team to score, this will cause his PM to drop. At the same time, our PMs will be much higher than they would be if we didn't get to play next to Keith.

Due to its simplicity and minimal data requirements, plus-minus has been a preferred metric for the last 50-odd years. But since it measures a marginal effect, the plus-minus is impacted by many factors beyond player ability, which is the actual quantity of interest. The ability of a player's teammates, or the quality of opponents, are not taken into account. The amount of playing time is also not factored in, meaning plus-minus stats are much noisier for some players than others. Finally, goalies, teams, coaches, salaries, situations, special teams, etc.—which all clearly contribute to the nature of play, and thus to goals— are neither accounted for when determining player ability by plus-minus, and nor are they used to explain variation in the goals scored against a particular player or team.

Instead of marginal effects, statisticians are more often interested in *partial effects*: change in the expected response that can be accounted for by change in your variable of interest *after removing the change due to other influential variables*. In our example, a partial effect for Duncan Keith would be unchanged if he plays with the authors of this chapter or with the current members of the Blackhawks. In each case, the partial effect will attempt to measure how Keith can influence the box-score regardless of with whom he skates. Because such partial effects help us predict how Keith would perform on a different team or with a different combination of linemates, this information is more useful than knowing a marginal effect.

One way that statisticians can isolate partial effects is by running *experiments*. Suppose that now, instead of playing for the Blackhawks, we are coaching them. In order to figure out the value of Keith, we could randomly select different players to join him whenever he is on the ice and send completely random sets of players onto the ice whenever he is not playing. Then, due to the setup of this randomized experiment, Keith's resulting PM score will represent a partial effect—his influence regardless of who he plays with. Of course, no real hockey coach would ever manage their team in this way. Instead, we hockey analysts must make sense of *observational data* that is collected as the games are naturally played, with consistent linemates and offensive-defensive pairings and where Keith tends to play both with and against the best players available.

Partial effects are measured from observational data through *regression*: you model the response (e.g., goals) as a function of many influential variables (*covariates*; e.g., all of the players on the ice). With rich enough data, we can simultaneously estimate the full set of competing partial effects corresponding to all of our influential variables. This is straightforward when there are only a small number of covariates. However, the standard regression algorithms will fail when the number of covariates is large. This "high dimensional regression" setting occurs in hockey analysis, where we would like to regress "goals" onto the set of variables corresponding to whether each NHL player is on the ice (a set of 2500 players in our dataset) while also including effects of team, season, play-offs, and special teams scenarios (e.g., power plays). Moreover, the covariate design is highly *imbalanced*: over the span of several seasons, there may be tens of thousands of goals, but players play with and against only a small fraction of other players and the number of unique player configurations is relatively small. Due to the use of player lines and consistent line match-ups with opponents, where groups of two or three players are consistently on ice together at the same time, the data contain many clusters of individuals who are seldom observed apart. Standard regression algorithms, such as maximum likelihood inference via Fisher scoring, will either massively *over-fit* (e.g., assign large effects to players who rarely play) or simply fail to converge.

However, there has been a tremendous improvement over the past two decades in the techniques available for high-dimensional regression analysis. These advancements are driven by the demands of researchers in genetics and finance, for example, for whom resolving partial effects among large sets of variables is the key to their science. The most successful approaches introduce some amount of *regularization* to the estimation problem—an additional penalty term that rewards simplicity (e.g., [13]). In our context, regularization shrinks toward a model where individual players don't make a huge difference while still allowing for large estimated player effects when the data warrant it. This conforms to what most analysts already believe: many players have a neutral, or "zero," effect (relative to the NHL average), whereas some are stars and others are liabilities. The amount of regularization is chosen to make the model perform as well as possible in out-of-sample prediction, and, again, contemporary statistical learning tools are designed to do exactly this—reliably predict the future. To take advantage of these tools, we need only to phrase partial player effect estimation as a regression problem.

14.2 Regression Model

The goal of our regression analysis will be to estimate a model that relates individual presence on the ice to observable outcomes of interest. We describe the model here for a goals-based analysis but extend it to shots and other metrics in the analysis sections.

Previous attempts at partial player effect estimation range from standard linear regression (usually on aggregate data)—the adjusted plus-minus scores of [1,19,24]—to the complex hazard model of [32], which proposes a proportional hazards process for game events, allowing partial player effects to be backed out from high-resolution game data. Adjusted plus-minus is built from similar ideas for basketball analysis (see [18,23]). Its linear model analysis implies an underlying normality assumption for the error structure; this may be a good approximation for basketball, where scoring is frequent and variability in player configurations is small, but it is inappropriate for disaggregated data with a binary response (e.g., whether an individual goal is for vs. against the home team). Such misspecification becomes especially problematic when combined with the modern regularization techniques necessary for reliable estimation of high dimensional models. On the other hand, more complex stochastic process modeling requires many additional assumptions on the data generating process and can be difficult to validate in practice; moreover, models such as that of [32] take far longer to run than we wish for our analysis. Some other important contributions to estimating player ability and attributing that to team success include [20,22,27].

The goal of our modeling is to provide a correct treatment of the binary "goals" data without introducing significant additional modeling complexity. In particular, we advocate the simple *logistic regression* framework suggested by [12]. In logistic regression, the average log odds of a goal being scored "for" a particular team is modeled as a linear function of predictor variables that may be comprised of an indicator of player configuration and other quantities, which is otherwise identical to the familiar ordinary (least squares) regression setup. We provide a detailed description of the model here, but refer the reader to [26] or similar texts covering *Generalized Linear Models (GLMs)*, of which logistic regression is a special case. The setup is rather straightforward, easy to extend, and highly interpretable. Estimated coefficients describe contributions to the log odds of goals, and we show that these can be converted back onto the scale of goals, resulting in an adjusted plus-minus statistic, but this time one which is a true partial effect.

Given n goals throughout the NHL over some specified time period, define y_i is $+1$ for a goal by the *home* team and -1 for a goal by the *away* team.* Say that $q_i = p$ $(y_i = 1) = p$(**home team scored goal i**). The logistic regression model of player contribution for goal i in season s with away team a and home team h is

$$\log\left[\frac{q_i}{1 - q_i}\right] = \alpha + \mathbf{u}_i'\boldsymbol{\gamma} + \mathbf{v}_i'\boldsymbol{\varphi} + \mathbf{x}_i'\boldsymbol{\beta}_0 + (\mathbf{x}_i \circ \mathbf{s}_i)'(\boldsymbol{\beta}_s + p_i\boldsymbol{\beta}_p), \qquad (14.1)$$

where

- Vector $\mathbf{u}_i$ holds indicators for each team-season (e.g., the Blackhawks in 2013–2014 would correspond to a coordinate of $\mathbf{u}_i$), set $u_{it} = +1$ if team-season t was the home team for goal i, $u_{it} = -1$ for the away team, and $u_{it} = 0$ if team-season t was not on the ice for goal i. This information is included to control for factors beyond the player's control, such as quality of coaching and fan support.

- Vector $\mathbf{v}_i$ holds indicators for various special-teams scenarios (e.g., being short-handed on a penalty kill), again set $v_{ik} = +1$ if the home team is in special-teams scenario k when goal i was scored, $v_{ik} = -1$ if the away team is in scenario k, and $v_{ik} = 0$ if neither team was in scenario k when goal i was scored. We consider 6 non-six-on-six settings (6v5, 6v4, 6v3, 5v4, 5v3, 4v3) and an additional "pulled goalie" indicator; note that more than 35% of the goals occur on some type of special teams scenario.

- Vector $\mathbf{x}_i$ contains player-presence indicators, set $x_{ij} = 1$ if player j was on the home team and on ice for goal i, $x_{ij} = -1$ for away player j on ice for goal i, and $x_{ij} = 0$ for everyone not on the ice. With $\circ$ denoting the Hadamard (element-wise) product, this player vector is also interacted with

  - Season (e.g., 2012–2013) vector $\mathbf{s}_i$, with $s_{ti} = 1$ if goal i was scored in season t, and
  - The post-season indicator p_i for whether or not the goal was scored in the play-offs, with $p_i = 1$ for the play-offs and zero for the regular season.

  By interacting players with seasons and with play-offs in this way, we have the potential to differentiate player ability over time both within (regular vs. post-) season(s) and across seasons. There is potential for confounding with team–season effects $\mathbf{u}_i$ however, as very few players change teams during a season.

In this full specification, the number of parameters, that is, $K = 1 + |\boldsymbol{\gamma}| + |\boldsymbol{\varphi}| + |\boldsymbol{\beta}_0| + |\boldsymbol{\beta}_s| + |\boldsymbol{\beta}_p|$ is on the order of the number of players, p, in the league spanning the seasons/games of interest. The exact number depends on modeling aspects, like the number of special teams scenarios (i.e., constant in p), and on quantities like the number of team-seasons that grow more slowly than p.

---

* *Home* and *away* are merely organizational devices, creating a consistent binary bifurcation for goals that can be applied across games, seasons, etc. Due to the symmetry in the logit transformation, player effects are unchanged when framing away team probabilities as q_i rather than $1 - q_i$, so we lose no generality by privileging home team goals in this way.

To explain the coefficients and their interpretation, $\beta_{0j} + \beta_{sj}$ is the regular-season-s effect of player j on the log odds that, given a goal has been scored, the goal was scored by their team. Coefficient $\beta_{0j} + \beta_{sj} + \beta_{pj}$ is the corresponding effect for post-season-s (note that under the regularization scheme in the next section, β_{pj} will be fixed at zero unless player j reaches the play-offs). These effects are "partial" in that they control for who else was on the ice, special teams scenarios, and team-season effects—a player's β_{0j} or β_{sj} only need be nonzero if that player effects play above or below the team average for a given season. A test of understanding: what does the intercept α represent in (14.2)?*

For intuition, consider a simple "player-only" version of our model that has only who-is-on-the-ice as a time-invariant influence on goal scoring. This is the version of the model that was applied in [12]. Then there are no team-season-specific intercepts ($\alpha_{sh} = \alpha_{sa} = 0$), no special teams effects ($\varphi = \mathbf{0}$), and no season-specific player-effect changes ($\beta_s = \mathbf{0}$ and $\beta_p = \mathbf{0}$) so that $\beta_j = \beta_{0j}$ is the constant effect of player j. The log odds that the home team has scored a given goal become

$$\log\left[\frac{q_i}{1-q_i}\right] = \alpha + \beta_{h_{i_1}} + \cdots + \beta_{h_{i_6}} - \beta_{a_{i_1}} - \cdots - \beta_{a_{i_6}}, \tag{14.2}$$

where the subscripts on the coefficients β are as follows: $h_{i_1}, \ldots, h_{i_6}$ are the six players on the ice for the home team and $a_{i_1}, \ldots, a_{i_6}$ indicate the players for the away time.[†] This is just a rewriting of $\mathbf{x}_i'\beta$ from (14.1), where the vector x_i (of length equal to the number of players) contains the "+1" and "−1" indicators depending on whether that player was on the home or away team, and where all other x_{ij} are zero so that $\sum_j |x_{ij}| = 12$ for full-strength play. See Figure 14.1 for illustration.

The model in 14.1 is simple and transparent; if you wish to control for new variables or situations you just need to add covariates to the logistic regression. In theory, one could fit the model easily in R by typing

```
R> fit <- glm(y~X, family="binomial")
```

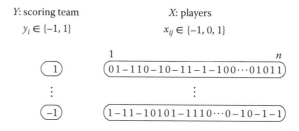

FIGURE 14.1
Diagram of a simple design matrix for a "players only model" and two example goals (rows). The two goals are shown in the same season under the same configuration of teams except that the first goal was scored by the home team while the second goal was by the away team. The configurations of players are only differed by the first player since the home team was on a 6v5 power play for the first goal.

---

* It is the home ice advantage: If you do not know anything about who is playing or on the ice, the odds are e^α higher that the home team has scored any goal.
† In this setup, the goalies *are* included in the calculations, unlike with *plus-minus*.

But unfortunately, that is problematic on both statistical and computational grounds. Hockey data involving player indicators is too high dimensional (too many covariates) for ordinary logistic regression software, and the design matrices are too highly imbalanced to obtain meaningful (low variance) estimates of player effects. In almost all regressions one is susceptible to the temptations of rich modeling when the data set is large, and our hockey setup is no different. One must be careful not to *over-fit*, wherein parameters are optimized to statistical noise rather than fit to the relationship of interest. And one must be aware of *multicollinearity*, where groups of covariates are correlated with each other making it difficult to identify individual effects, as happens when players are grouped into lines.

A first approach to finding a remedy might be to entertain stepwise regression, that is, via step in R with a stopping rule based upon information criteria like Akaike Information Criterion (AIC) and Bayesian Information Criterion (BIC). But that also doesn't work on this data: the calculations take days, and turn up very few nonzero predictors (i.e., players whose presence have any effect on goals). The trouble here is that players can't be judged on their own, since they almost always play with and against 11 others. Therefore the one-at-a-time judgments made by step fail to discover many relevant players despite making a combinatorially huge number of such comparisons. Moreover, stepwise regression results are well known to be highly variable: tiny jitter to the data can lead to massive changes in the estimated model. The combined effect is an unstable algorithm that yields overly simple results and takes a very long time to run.

Instead, a crucial contribution of [12] is to suggest the use of modern penalized and Bayesian logistic regression models, which biases the estimates of player effects toward zero. In the next section, we consider one fast and successful version of these methods: L_1 regularization.

14.3 Regularized Estimation

Our solution is to take a modern regularized approach to regression. If $\eta_i = \log[q_i/(1-q_i)]$ is our linear equation for log odds from (14.1), then the usual maximum likelihood estimation routine (e.g., via glm in R) minimizes the negative log likelihood objective for n goal events

$$l\left(\boldsymbol{\eta};\mathbf{y}\right) = \sum_{i=1}^{n} \log\left(1 + \exp[-y_i\eta_i]\right). \tag{14.3}$$

Instead, a *regularized* regression algorithm will minimize a penalized objective, say, for example, $l\left(\boldsymbol{\eta};\mathbf{y}\right) + n\lambda \sum_{j=1}^{p}\left[c\left(|\beta_{0j}|\right) + c\left(|\beta_{sj}|\right)\right]$, where $\lambda > 0$ controls overall penalty magnitude and $c(\cdot)$ is a coefficient cost function, while n is the total number of goals and p is the number of players. The penalty λ is discussed in some detail shortly. A few common cost functions are shown in Figure 14.2. Those that have a nondifferentiable spike at zero (all but ridge) lead to sparse estimators, meaning that many coefficients are set to exactly zero. The curvature of the penalty away from zero dictates the weight of shrinkage imposed on the nonzero coefficients: L_2 costs increase with coefficient size, lasso's L_1 penalty has zero curvature and imposes constant shrinkage, and as curvature goes toward $-\infty$ one approaches the L_0 penalty of subset selection.

We will focus on the L_1 penalty for its balance between shrinkage of large signals (players tend not to have huge effects) and a preference for sparsity (we can only measure

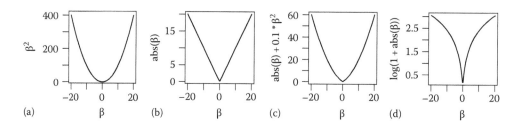

FIGURE 14.2
(a) L_2 "ridge" costs [14], (b) L_1 "lasso" [33], (c) the "elastic net" mixture of L_1 and L_2 [35], and (d) the log penalty [3].

the nonzero effects of a subset of players). For these and many appealing theoretical reasons (and for computational tractability), the L_1 penalty is by far the most commonly used in contemporary regularized regression; see [13] for a broad-audience overview and [30] for details on the algorithms used in this chapter. Under lasso L_1 penalization, estimation for the unknown parameters in our particular hockey model (14.1) proceeds through optimization of

$$l\left(\mathbf{\eta};\mathbf{y}\right) + n\lambda \sum_{j=1}^{p} \left(|\beta_{0j}| + |\beta_{sj}| + |\beta_{pj}|\right). \tag{14.4}$$

It is important to note there that we are *penalizing only the player effects*. The team-season effects (γ) are unpenalized. This strategy of combining penalized and unpenalized estimation is advocated in, for example, [7] and [29]. It works nicely whenever you have a subset of covariates for which there is strong data signal (many repeated observations, which we have for team-season and special teams effects) and whose effect you'd like to completely remove from estimation for other coefficients. In this way, we ensure that the player effect estimates are not *polluted* by confounding effects in $\mathbf{u}$ and $\mathbf{v}$.

Moreover, consider the covariates on β_0, β_s, and β_p in (14.1): for the latter two, x_i interacts with additional binary indicators, such that β_{0j} acts on more nonzero terms than β_{sj}, which itself acts on more nonzero terms than β_{pj}. Thus there is less signal associated with the season and play-off player effect innovations than with the player baseline effects, so that these will tend to be estimated at zero unless there is significant evidence that a player has become better or worse across seasons or in a given post-season.

Penalty size, λ, acts as a *squelch*: canceling noise to focus on the true input signal. Large λ lead to very simple model estimates, while as $\lambda \to 0$ we approach maximum likelihood estimation. Since you don't know optimal λ, practical application of penalized estimation requires a *regularization path*: a $p \times T$ field of $\hat{\beta}$ estimates obtained while moving from high to low penalization along $\lambda^1 > \lambda^2 \dots > \lambda^T$. These paths begin at λ^1 set to infimum λ such that (14.4) is minimized at $\hat{\beta} = \mathbf{0}$, and proceed down to some pre-set λ^T (e.g., $\lambda^T = 0.01\lambda^1$).

A common tool for choosing the optimal λ—that for which we report estimated player effects—is cross validation (CV). In CV, the path of coefficients is repeatedly fit to data subsamples and used to predict the response on the left-out data. The λ leading to minimum error is then selected as optimal. In this chapter, we instead use an analytic alternative to CV that yields models that perform as well or better out-of-sample. The corrected Akaike Information Criterion (AICc), proposed in [17], is defined as

$$AICc = 2l(\hat{\boldsymbol{\eta}}_\lambda; \mathbf{y}) + \frac{2kn}{n - k - 1},$$

where

$\hat{\boldsymbol{\eta}}_\lambda$ are the estimated log odds under penalty λ, via estimates of parameters $(\hat{\alpha}, \hat{\varphi}, \hat{\beta}_{0,s,p})$ in Equation 14.1

$k \leq K$ is the number of nonzero estimated coefficients (likely far fewer than the total number of parameters K, as many players cannot be distinguished from the league average) at this penalty

See [6] and [30] for details on AICc selection in this context; we find AICc preferable to CV because it is computationally efficient (you only need to optimize once) and because there is no random Monte Carlo variation—it always gives the same answer on the same data. However, all of our ideas here apply if you wish to use CV selection instead.

The `gamlr` package [28] can be called via R to implement this procedure:

```
R> fit <- gamlr(X, Y, standardize=FALSE, family="binomial")
```

For CV, just replace `gamlr` with `cv.gamlr`. The `standardize=FALSE` flag tells `gamlr` to *not* weight the coefficient penalties by the standard deviation of the corresponding covariate (i.e., to use penalty $\lambda|\beta_j|$ instead of $\lambda \, \mathrm{sd}(\mathbf{x}_j)|\beta_j|$); this is appropriate here because such standardization would *up-weight* the influence of players who rarely play (and have low $\mathrm{sd}(\mathbf{x}_j)$) relative to those who have a lot of ice time (and thus high $\mathrm{sd}(\mathbf{x}_j)$). The software exploits sparsity in our player effects (X) via the `Matrix` library for R, and is extremely fast to run: no examples in this article require more than a few seconds of computation. Estimated coefficients at optimal λ are available as `coef(fit)`.

One natural way to understand regularized regression is through the lens of Bayesian posterior inference. Judiciously chosen prior distributions lend stability to the fitted model, which is crucial in contexts where the number of quantities being estimated is large. In our setting where larger β-values indicate large positive or negative contributions to player ability, it makes sense to choose a prior that encourages coefficients to center around zero, a so-called *shrinkage* prior. Our *a priori* belief is that most players are members of the rank-and-file: their contribution to goals is *neutral* (e.g., zero on the log-odds scale), and that only a handful of stars (and liabilities) have a strong contribution to the chances of scoring (or letting in) goals. From the perspective of point estimation, adding a prior on β_j centered at zero is equivalent to adding a penalty term for $\beta_k \neq 0$ in our objective function. Different choices of priors correspond to different penalty functions on $\beta_j \neq 0$; a Laplace prior distribution on each β_j corresponds to our L_1 penalty in (14.4). The posterior density is the product of likelihood and prior, and on the log scale that product becomes a sum. So maximizing the posterior to obtain posterior modes is equivalent to maximizing the log likelihood plus a penalty term, which is the log of the prior. Conversely, minimizing (14.4) may be interpreted as Bayesian posterior maximization.

14.4 Analysis: Goal-Based Effects

This section attempts to quantify the performance of hockey players using data from the NHL. It extends the analysis in [12], which used a smaller dataset, assumed constant player effects, and did not control for team-season effects. The data, downloaded from http://www.nhl.com, comprise of play-by-play NHL game data for regular and play-off games during 11 seasons of 2002–2003 through 2013–2014*. The data capture all signifigant events in every single game, such as change, goal, shot, blocked shot, miss shot, penalty, etc. There were $p = 2439$ players involved in $n = 69449$ goals.

The analysis proceeds through estimation of the model from (14.1),

$$\log\left[\frac{q_i}{1 - q_i}\right] = \alpha + \mathbf{u}_i'\boldsymbol{\gamma} + \mathbf{v}_i'\boldsymbol{\varphi} + \mathbf{x}_i'\boldsymbol{\beta}_0 + (\mathbf{x}_i \circ \mathbf{s}_i)'(\boldsymbol{\beta}_s + p_i'\boldsymbol{\beta}_p),$$

by minimization of the implied penalized deviance in (14.4). The estimated player coefficients for each season s are then available as $\beta_{0j} + \beta_{sj}$, the combination of a baseline effect plus a season-specific innovation. These effects represent the estimated change in the log odds that, given a goal was scored, the goal was scored by player j's team.

The estimated effects—our β_{0j} and β_{sj}—might be tough for nonstatisticians to interpret. One option is to translate from the scale of log-odds to that of probabilities. In particular, we define the "partial for-%" functional as

$$\text{PFP}_{sj} = \left(1 + \exp[-\beta_{0j} - \beta_{sj}]\right)^{-1}. \tag{14.5}$$

This feeds the player effects through a logit link to obtain the probability that, given a goal was scored in season s, it was scored by player j's team, *if we know nothing else other than that player j was on the ice*. It lives on the same scale as the commonly used *for-%* (FP) statistic: the total number of events by a given player's team *divided* by the total number of events by either team, while that player was on the ice. Like PM, FP is a marginal effect that does not account for who else was on the ice and other confounding factors. Hence, PFP is the partial effect version of FP.

An important feature of the standard PM statistic that differs from both *for-%* and β or PFP is that it—in a limited sense—accounts not only for player ability but also the amount that they play. For example, a player with a very high PM must both perform well *and* maintain this level of performance over an extended period of time (assuming that you need to be on the ice a long time to be on the ice for many goals). Conversely, similarly estimated β values for two players might hide the fact that one of these two logs much more ice time and is thus more valuable to the team.[†] It is therefore also important to be able to translate our partial player effects back to same scale as PM, and we do this in the *partial plus-minus* (PPM). Suppose player j was on the ice for g_{sj} total goals (for or against) during season s; then the PPM is defined as

$$\text{PPM}_{sj} = g_{sj}\text{PFP}_{sj} - g_{sj}(1 - \text{PFP}_{sj}) = g_{sj}(2\text{PFP}_{sj} - 1). \tag{14.6}$$

Just as PFP is the partial effect version of FP, PPM is the partial effect analog to PM's marginal effect.

---

[*] Season 2004–2005 was a lockout that resulted in a cancellation.

[†] Note that, due to the role of the penalty in our regularized estimation scheme, players with little ice time tend to have their effect estimated at zero; thus, the difference between β or PFP and the PPM statistics should be less dramatic than the difference between FP and PM statistics. In a fully Bayesian analysis, such as the `reglogit` approach discussed in our conclusion, one would be able to separate posterior uncertainty about β from the issue of the number of goals for which a player is on ice; for example, the `reglogit` approach yields posterior uncertainty over a player's PPM.

Table 14.1 lists top and bottom players by their PPM, along with the corresponding β effects and their standard PM. Since these PPMs and effects are calculated for each season, players will occur repeatedly in the table; for example, Sidney Crosby has 4 of the top 10 best player-seasons since 2002; he has been consistently the best, or near best, player in the league. The number one player-season since 2002 by PPM is Peter Forsberg in 2002–2003, with a PPM of 55.5. This is around 25% better than the second best PPM: Crosby's 43.5 in 2009–2010. These tabulated effects are all calculated based upon regular season performance alone. However, for all 10,000 player seasons, using goals data, we never see enough signal to conclude that a given player was significantly better or worse in the post-season than in the regular season. That is, the $\hat{\beta}_{pj}$ are all zero and we have no evidence of "clutch" players who improve their play in the post-season. At the same time, many of the β_{sj} *are* estimated at nonzero values: There is measurable signal indicating that player performance changes across seasons.

The ranking in Table 14.1 differs dramatically from that in [12]. This occurs because we're now *controlling* for additional nonplayer confounding factors (e.g., coaching through team-season effects) and we allow player performance to change over time rather than be fixed at a single "career" value. To help institution on why such control and model flexibility is useful, we note that Sidney Crosby's β effects drop significantly (he falls out of the top five for any season) if you do not control for the special teams effects. This occurs because he spends a lot of time on the penalty kill (shorthanded), which makes it easier for him to get scored upon through no fault of his own. As another example, many of the goalies have large PPM if you do not control for team-season effects; since the goalie is almost always on the ice, they act as a surrogate for aggregate team performance unless you explicitly control for it (unfortunately for Patrick Lalime, there is still enough variation at goal to measure the effect and PPM for some goalies).

Another change from [12] is that we are ranking players here by PPM rather than by β, as described earlier; this rewards those with more ice time. For comparison, Table 14.2 ranks players by their PFP (which is equivalent to ranking by β); the table includes both the goal-based metrics from this section and the shot-based metrics from our next section. While PFP and PPM are clearly related quantities, we do see some major differences. For example, Tyler Toffoli (ranks 9 and 10 by goal-based PFP) was a breakout star in 2013–2014 with the Los Angeles Kings; this was his first full season, after playing only a portion of 2012–2013 in the NHL. As a rookie, his ice time was relatively limited; however, he clearly has talent and this is reflected in his β and PFP but less in his PPM. On the other hand, players ranked at the bottom by PPM in Table 14.1 are those who have a negative β *and* get a large amount of ice time. There are many players who have lower PFPs than Jack Johnson's 0.45 (e.g., John McCarthy at 0.38 and Thomas Pock at 0.40), but they do not get to play as much and thus don't show up in our bottom 20.

14.5 Analysis: Comparison to Shot-Based Metrics

The analysis in the previous section is built around the event of a "goal"; this is the most reasonable baseline analysis, as it removes any subjectivity about whether or not the statistics are related to team performance—you score more you win. However, it has recently become popular in hockey analysis to consider alternative metrics that are built from shots and other events; see [34] for a review. The most popular of such statistics is Corsi, which

TABLE 14.1

Top 25 and Bottom 20 Player-Seasons When Ranked by Their Regular-Season PPM

			Goal-Based Performance Analysis				
Rank	Player	Season	Team	PFP	FP	PPM	PM
1	Peter forsberg	2002–2003	COL	0.68	0.77	55.52	85
2	Sidney crosby	2009–2010	PIT	0.60	0.64	43.47	60
3	Dominik hasek	2005–2006	OTT	0.59	0.67	42.45	80
4	Sidney crosby	2008–2009	PIT	0.60	0.61	42.26	48
5	Sidney crosby	2005–2006	PIT	0.60	0.62	41.86	52
6	Peter forsberg	2005–2006	PHI	0.68	0.77	40.67	61
7	Pavel datsyuk	2007–2008	DET	0.60	0.72	39.49	87
8	Pavel datsyuk	2008–2009	DET	0.60	0.67	39.49	69
9	Sidney crosby	2006–2007	PIT	0.60	0.72	35.62	79
10	Mark streit	2008–2009	NYI	0.59	0.56	35.08	24
11	Matt moulson	2011–2012	NYI	0.60	0.61	34.92	37
12	Lubomir visnovsky	2010–2011	ANA	0.58	0.66	34.52	70
13	Alex ovechkin	2008–2009	WAS	0.57	0.66	34.46	80
14	Joe thornton	2009–2010	SJS	0.60	0.65	33.91	52
15	Joe thornton	2010–2011	SJS	0.60	0.64	33.91	48
16	Ondrej palat	2013–2014	TAM	0.64	0.66	32.75	37
17	Pavel datsyuk	2006–2007	DET	0.60	0.71	32.61	70
18	Joe thornton	2002–2003	BOS	0.60	0.64	32.17	47
19	Joe thornton	2007–2008	SJS	0.60	0.71	32.17	69
20	Andrei markov	2007–2008	MON	0.57	0.60	31.9	47
21	Peter forsberg	2003–2004	COL	0.68	0.72	31.47	39
22	Joe thornton	2008–2009	SJS	0.60	0.67	31.21	56
23	Peter forsberg	2006–2007	PHI	0.68	0.68	31.12	32
24	Pavel datsyuk	2005–2006	DET	0.60	0.74	30.85	75
25	Robert lang	2003–2004	WAS	0.60	0.66	30.8	50
10184	Patrick lalime	2008–2009	BUF	0.43	0.44	−15.79	−15
10185	Jack johnson	2007–2008	LOS	0.45	0.39	−15.82	−34
10186	Brett clark	2011–2012	TAM	0.44	0.35	−16.93	−47
10187	Niclas havelid	2008–2009	ATL	0.45	0.39	−16.97	−40
10188	Jack johnson	2010–2011	LOS	0.45	0.53	−17.21	9
10189	Jack johnson	2011–2012	LOS	0.45	0.5	−17.21	−1
10190	P. J. Axelsson	2008–2009	BOS	0.41	0.49	−17.35	−1
10191	Bryan allen	2006–2007	FLA	0.45	0.45	−17.9	−17
10192	Jack johnson	2009–2010	LOS	0.45	0.49	−19.46	−4
10193	Patrick lalime	2005–2006	STL	0.43	0.40	−19.77	−29
10194	Alexander edler	2013–2014	VAN	0.37	0.27	−20.49	−35
10195	Patrick lalime	2007–2008	CHI	0.43	0.49	−22.29	−4
10196	Tim thomas	2009–2010	BOS	0.43	0.46	−24.22	−16
10197	Andrej meszaros	2006–2007	OTT	0.42	0.48	−27.32	−6

(Continued)

TABLE 14.1 (*Continued*)

Top 25 and Bottom 20 Player-Seasons When Ranked by Their Regular-Season PPM

	Goal-Based Performance Analysis						
Rank	Player	Season	Team	PFP	FP	PPM	PM
10198	Bryce salvador	2008–2009	NJD	0.35	0.37	−34.4	−31
10199	Patrick lalime	2002–2003	OTT	0.43	0.58	−37.81	47
10200	Patrick lalime	2003–2004	OTT	0.43	0.56	−37.81	37
10201	Niclas havelid	2006–2007	ATL	0.34	0.44	−62.64	−22
10202	Niclas havelid	2005–2006	ATL	0.33	0.40	−65.94	−41
10203	Jay bouwmeester	2005–2006	FLA	0.33	0.42	−69.62	−32

TABLE 14.2

Top 20 Player-Seasons by Goal and Corsi-Based PFP

	PFP Player Rankings							
	Goal-Based				Corsi-Based			
Rank	Player	Season	Team	PFP	Player	Season	Team	PFP
1	Peter forsberg	2002–2003	COL	0.68	David van der gulik	2010–2011	COL	0.64
2	Peter forsberg	2005–2006	PHI	0.68	David booth	2012–2013	VAN	0.63
3	Peter forsberg	2003–2004	COL	0.68	Daniel sedin	2012–2013	VAN	0.62
4	Peter forsberg	2006–2007	PHI	0.68	Alexander semin	2003–2004	WAS	0.61
5	Peter forsberg	2007–2008	COL	0.68	Daniel sedin	2010–2011	VAN	0.60
6	Peter forsberg	2010–2011	COL	0.68	Mikhail grabovski	2010–2011	TOR	0.60
7	Ondrej palat	2013–2014	TAM	0.64	Daniel sedin	2007–2008	VAN	0.60
8	Ondrej palat	2012–2013	TAM	0.64	Daniel sedin	2008–2009	VAN	0.60
9	Tyler toffoli	2013–2014	LOS	0.63	Daniel sedin	2011–2012	VAN	0.60
10	Tyler toffoli	2012–2013	LOS	0.63	Patrik elias	2010–2011	NJD	0.60
11	Vincent lecavalier	2006–2007	TAM	0.61	Sidney crosby	2013–2014	PIT	0.60
12	Vincent lecavalier	2003–2004	TAM	0.61	Daniel sedin	2009–2010	VAN	0.60
13	Sidney crosby	2009–2010	PIT	0.60	Justin williams	2010–2011	LOS	0.60
14	Sidney crosby	2008–2009	PIT	0.60	Daniel sedin	2013–2014	VAN	0.60
15	Sidney crosby	2005–2006	PIT	0.60	Patric hornqvist	2013–2014	NSH	0.60
16	Pavel datsyuk	2007–2008	DET	0.60	Pavel datsyuk	2012–2013	DET	0.60
17	Pavel datsyuk	2008–2009	DET	0.60	Alex steen	2011–2012	STL	0.60
18	Sidney crosby	2006–2007	PIT	0.60	Brad richardson	2011–2012	LOS	0.60
19	Matt moulson	2011–2012	NYI	0.60	Eric fehr	2008–2009	WAS	0.60
20	Joe thornton	2009–2010	SJS	0.60	Tyler toffoli	2013–2014	LOS	0.60

counts the number of events that are goals, shots on goal, missed shots, or blocked shots. Fenwick is another statistic; it is Corsi but without counting blocked shots. Although we have seen no evidence that Corsi or Fenwick events are more useful in predicting team performance than goal-based metrics, they do offer a big advantage to the statistician: they lead to a larger sample size, so that you can hopefully better identify the competing influences of

TABLE 14.3

Top 25 and Bottom 20 Players by Corsi-Based PPM

			Corsi-Based Performance Analysis				
Rank	Player	Season	Team	PFP	FP	PPM	PM
1	Daniel sedin	2010–2011	VAN	0.60	0.65	615.14	876
2	Eric staal	2008–2009	CAR	0.58	0.59	605.41	619
3	Mikhail grabovski	2010–2011	TOR	0.60	0.57	597.05	465
4	Joe thornton	2011–2012	SJS	0.59	0.61	596.37	742
5	Alex ovechkin	2009–2010	WAS	0.59	0.66	575.72	1047
6	Daniel sedin	2007–2008	VAN	0.60	0.63	562.11	685
7	Daniel sedin	2008–2009	VAN	0.60	0.62	547.83	680
8	Ryan kesler	2010–2011	VAN	0.58	0.59	530.05	649
9	Sidney crosby	2009–2010	PIT	0.57	0.62	517.86	815
10	Daniel sedin	2011–2012	VAN	0.60	0.67	510.16	880
11	Henrik zetterberg	2011–2012	DET	0.58	0.60	497.04	596
12	Claude giroux	2010–2011	PHI	0.58	0.56	487.53	347
13	Zach parise	2008–2009	NJD	0.58	0.64	486.45	843
14	Joe thornton	2010–2011	SJS	0.58	0.60	482.72	647
15	Alex steen	2010–2011	STL	0.59	0.61	475.5	561
16	Lubomir visnovsky	2010–2011	ANA	0.56	0.56	474.91	446
17	Eric staal	2010–2011	CAR	0.56	0.56	473.92	415
18	Justin williams	2011–2012	LOS	0.59	0.63	471.53	717
19	Alex ovechkin	2007–2008	WAS	0.56	0.65	463.14	1094
20	Patrik elias	2010–2011	NJD	0.60	0.60	461.75	461
21	Sidney crosby	2013–2014	PIT	0.60	0.61	459.92	480
22	Dustin byfuglien	2010–2011	ATL	0.56	0.60	456.04	705
23	Jaromir jagr	2007–2008	NYR	0.58	0.65	455.78	911
24	Alex ovechkin	2008–2009	WAS	0.56	0.64	455.46	1065
25	Jason blake	2008–2009	TOR	0.58	0.55	454.7	278
10605	Mike commodore	2008–2009	CBS	0.43	0.42	−447.91	−537
10606	Scott hannan	2011–2012	CGY	0.42	0.40	−451.04	−591
10607	Chris phillips	2007–2008	OTT	0.43	0.40	−454.09	−644
10608	Jay bouwmeester	2005–2006	FLA	0.44	0.46	−457.01	−305
10609	Karlis skrastins	2008–2009	COL	0.43	0.38	−457.54	−754
10610	Karlis skrastins	2009–2010	DAL	0.42	0.39	−464.49	−655
10611	Mattias ohlund	2008–2009	VAN	0.42	0.47	−465.36	−212
10612	Mattias ohlund	2006–2007	VAN	0.43	0.48	−470.03	−147
10613	Scott hannan	2008–2009	COL	0.43	0.38	−478.83	−788
10614	Douglas murray	2009–2010	SJS	0.42	0.47	−486.16	−184
10615	Scott hannan	2007–2008	COL	0.42	0.42	−507.7	−504
10616	Filip kuba	2011–2012	OTT	0.42	0.49	−509.74	−77
10617	Niclas havelid	2007–2008	ATL	0.41	0.35	−516.86	−883
10618	Johnny oduya	2008–2009	NJD	0.42	0.51	−522.4	51
10619	Douglas murray	2010–2011	SJS	0.40	0.48	−540.83	−117

(Continued)

TABLE 14.3 (*Continued*)

Top 25 and Bottom 20 Players by Corsi-Based PPM

		Corsi-Based Performance Analysis						
Rank	Player	Season	Team	PFP	FP	PPM	PM
10620	Dion phaneuf	2006–2007	CGY	0.42	0.49	−552.69	−42
10621	Niclas havelid	2008–2009	ATL	0.40	0.40	−562.65	−604
10622	Sergei gonchar	2006–2007	PIT	0.42	0.52	−586.55	174
10623	Paul martin	2008–2009	NJD	0.39	0.55	−695.83	283
10624	Bryce salvador	2008–2009	NJD	0.32	0.42	−912.17	−407

different players and confounding factors. Our data contain $n_c = 1,329,679$ Corsi events and $n_f = 1,034,154$ Fenwick events; this is an order of magnitude more events than the $n = 69,449$ goals.

The standard way to report Corsi and Fenwick for a given player is as the *for-%* (FP) described earlier. Again, since the FP score does not reward players for the amount of time that they spend on the ice, we also consider both Corsi and Fenwick versions of the plus-minus statistic. Of course, all of these statistics (Corsi-FP, Corsi-PM, etc.) measure marginal effects. They are thus subject to the same criticisms as the original PM: they fail to control for the influence of other players and confounding factors, and are thus less useful than a partial effect for predicting and measuring player performance. However, we can apply the exact same regression analysis that we've used earlier for goal events to derive *partial* versions of the Corsi and Fenwick statistics: simply replace y_i with a response calculated from Corsi or Fenwick events. For example, a Corsi regression applies the model as in (14.1) but for response $y_i = +1$ if the event was a Corsi event (shot, goal, blocked shot) by the home team and $y_i = -1$ if it was a Corsi event by the away team. The partial *for-%* and plus-minus formulas of (14.5 and 14.6) can similarly be applied to obtain Corsi-PPF and Corsi-PPM values.

The results for regular season Corsi-based performance analysis are in Table 14.3 and on the right-hand side of Table 14.2. A comparison to the goal-based rankings shows a distinctly different set of players are at both the top and bottom. For example, Daniel Sedin is a prominent player who ranks highly in multiple seasons under Corsi-PPM but does not appear in the top 20 for goals-PPM (his best goals-PPM is a still respectable 19.45 in 2010–2011, which ranks 152nd across all player-seasons). At this point, we are not looking to argue for either the goal- or Corsi-based metrics as "best"; however, the fact that they do differ dramatically should be a bit troubling for those who wish to focus exclusively on Corsi statistics (since only goal differentials dictate who wins the game). In the next section, we consider a comparison between all of our metrics and an outside measure of player quality: salary.

14.6 Analysis: The Relationship Between Salary and Performance

In our final analysis section, we consider how our partial effect statistics relate to the *market* value of a player's worth as represented by their annual salary. The salary numbers are

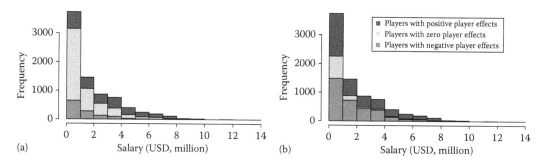

FIGURE 14.3
Distribution of estimated player effects and salaries over all 11 seasons. (a) Goals-based. (b) Corsi-based.

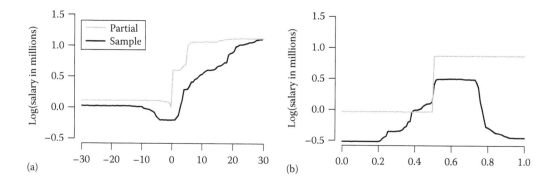

FIGURE 14.4
Nonparametric regression for log average salary (per player) onto their average performance metric: partial (PPM, PFP) and sample/marginal (PFP, FP) analogs. (a) Plus-minus. (b) For-probability.

obtained through a combination of the databases maintained at blackhawkzone.com and hockeyzoneplus.com; we are able to obtain annual salaries for 80% of the player-seasons in our dataset, including almost all player-seasons where the player was on ice for more than a couple of goals. The histograms in Figure 14.3 show, for the salary distribution over all 11 seasons, how the estimated player effects are distributed between negative, neutral (zeros), and positive. Results are shown for both goal- and Corsi-based regressions. In both cases, the ratio of positive to negative effects increases with salary. The main difference between the two plots is that fewer players have zero estimated effects under Corsi—this is a result of its much larger event sample.

Figure 14.4 takes a deeper look at the relationship between salary and performance for two of our goal-based metrics: the goal plus-minuses (PM and PPM) and for-percentages (FP and PFP). For each metric, we use nonparametric Bayesian regression to fit expected log salary as a function of that metric. In particular, we apply the `tgp` package [8,11] to obtain posterior means for the Bayesian regression trees of [4]. The trees are fit to salary and performance data that have first been aggregated by player, such that these surfaces represent the expected log average salary (per player) conditional upon their average performance across the years that they played. This aggregation is done to minimize dependence between observations, and because we assume that a player's salary is not determined by a single season.

The surfaces in 14.4 expose some interesting differences between the partial and marginal performance statistics. In the plus-minus case (PM and PPM), the two curves are similar—they both show little relationship between salary and performance for negative plus-minus, while salaries rise with positive performance. However, there is a larger jump up from zero for PPM than for PM. This occurs because the regression estimation only assigns a positive nonzero effect for statistically significant performances, so that players with small but positive PM values have their PPM shrunk to zero. The difference between FP and PFP is more dramatic. In the case of marginal FP, the salaries are highest for players with FP in the middle of the range (near and above 0.5). This occurs because FPs outside of that range occur only for players with little ice time. In contrast, the PFP relationship with salary is very simple: salaries are low for PFP below 0.5, and high above that number. If you make it more likely than not that a goal-scored is a goal-for, you can expect to make more money.

Finally, we close with a look at the "highest value" players in the 2013–2014 season: those for whom their salary was low relative to their goal-based PPM. Table 14.4 lists the top 20 players by goal-PPM/salary—that is, the top players measured by their goals-added-per-dollar. The cheapest player contribution, by a massive margin, is from Ondrej Palat of Tampa Bay. Palat was an inexpensive seventh round draft pick in 2011, making $500 thousand per year. After spending two seasons in the minors, he moved up to the NHL in 2013–2014 and had a season good enough to be nominated for Rookie of the year. The Lightning re-signed Palat to a new contract in 2014; he now averages around $3.5 million per year.

TABLE 14.4

Top 20 Value Players as Ranked by PPM/Salary

Rank	Player	Team	Goals per million
1	Ondrej palat	TAM	58.27
2	Ryan nugent-hopkins	EDM	19.81
3	Gabriel landeskog	COL	16.74
4	Tyler toffoli	LOS	16.72
5	Gustav nyquist	DET	9.08
6	Jaden schwartz	STL	8.43
7	Eric fehr	WAS	7.51
8	Andrew macdonald	NYI	7.48
9	Benoit pouliot	NYR	6.43
10	Brad boyes	FLA	6.01
11	Tomas tatar	DET	5.83
12	Al montoya	WPG	5.79
13	Brandon saad	CHI	5.5
14	Frans nielsen	NYI	5.5
15	Jaromir jagr	NJD	4.73
16	Logan couture	SJS	4.7
17	Radim vrbata	PHO	4.4
18	David perron	EDM	4.1
19	Henrik lundqvist	NYR	3.76
20	Andrei markov	MON	3.5

14.7 Conclusion

We have provided a sketch for how modern techniques in regularized logistic regression, developed originally to address challenging large-scale problems in genetics, finance, and text mining, can be used to calculate partial player effects in hockey. We have argued that such partial effects are a better measure of player ability compared to the classic plus-minus statistic, and have the benefit of being interpretable on the same scale as plus-minus. We have shown how the framework is flexible, allowing one to control for many aspects of situational play (special teams, overtime, play-offs), and personnel/season (coaches, salaries, season-years). A comparison was provided to the popular Corsi and Fenwick alternatives to plus-minus, and we argued that a recent emphasis in the literature on shots (and blocked shots), does not in general compare favorably to the traditional goals focus in this framework.

Our development has focused on point-estimation via the `gamlr` package, which infers parameters under L_1 penalization. Another software package offering similar features is `glmnet` [35]. The difference between `gamlr` and `glmnet` is in what options they provide on top of the standard L_1 penalty. As detailed in [30], including an example analysis of this same hockey data, `gamlr` provides a "gamma-lasso" algorithm for *diminishing bias* penalization: the penalty on coefficients automatically diminishes for strong signals. If you believe that the current analysis has over-shrunk the influence of, say, total stars like Sidney Crosby or Pavel Datsyuk, then `gamlr` and [30] will offer a preferable analysis framework. On the other hand, if you think that all players should be shrunk *closer* to zero—perhaps you believe that the current results overstate the effect of a few stars—then the elastic net penalization scheme of [35] and `glmnet` will be preferable. In either case, simple L_1 penalization provides a useful reference baseline analysis.

Beyond changes to the penalty specification, we think that there can be considerable value in moving from point-estimation to a fully Bayesian analysis. Reference [12] included exploration of the player effect posterior in their earlier analysis of a related model. They apply the `reglogit` package [9] for R, which implements the Gibbs sampling strategy from [10]. The software takes advantage of sparse matrix libraries (`slam` [16]) and is multi-threaded via `OpenMP` to engage multiple processors simultaneously. It combines two scale mixtures of normals data-augmentation schemes, one for the logit [15] and one for the Laplace prior [21]. Obtaining T samples from the full posterior is straightforward using the following R code

```
bfit <- reglogit(T=T, y=Y, X=X, normalize=FALSE)
```

The full posterior sample for β, residing in `bfit`, is available for calculation of posterior means and covariances of player effects and other posterior functionals relevant to player performance. For example, [12] use the posterior *probability* that one player is better than another as a basis for ranking players, and can even provide posterior credible intervals around these rankings.* This information can be used to construct teams of players under budget constraints and subsequently describe the probability that those teams will score more goals than their opponents.

---

* See, for example, https://github.com/TaddyLab/hockey/blob/master/results/blog/logistic_pranks_betas.csv

Another option is to depart from the restrictions of linear modeling. Anecdotally, some in the sports analytics community (not just in hockey) have embraced a framework built around random forests [2]. An advantage of decision trees, on which random forests are based, is that they naturally explore interactions between predictors, for example, between players and other effects in the hockey analysis. The bagging procedure—averaging across many trees—provides a mechanism for avoiding overfit: structure that is not persistent across trees is eliminated by the averaging. Such work also fits with our advocacy of fully Bayesian analysis: [31] describes random forests as approximating a Bayesian nonparametric posterior over trees, while Bayesian additive regression trees (BART [5]) provide an alternative tree-based scheme that can be extended to logistic regression via the latent variable techniques in [10]. Finally, the proportional hazards model [32], mentioned in Section 14.2, attempts to reproduce more completely the stochastic processes behind scoring in a hockey game. Their fully Bayesian analysis accounts for a wide set of game information, including the time-on-ice information that we are only roughly accounting for in our PPM statistic.

Regardless of these and other possible complex extensions, we argue strongly that our simple L_1 penalized logistic regression has much to recommend it. The model is very simple to interpret and relies upon minimal restrictive assumptions on the process of a hockey game. Our measures are also much faster to compute than any of the alternatives. These qualities make sophisticated real-time analysis of player effects possible as games and seasons progress.

References

1. T. Awad. Numbers on ice: Fixing plus/minus. *Hockey Prospectus*, April 03, 2009, 2009. http://www.hockeyprospectus.com/puck/article.php?articleid=64.
2. L. Breiman. Random forests. *Machine Learning*, 45:5–32, 2001.
3. E.J. Candes, M.B. Wakin, and S.P. Boyd. Enhancing sparsity by reweighted l1 minimization. *Journal of Fourier Analysis and Applications*, 14:877–905, 2008.
4. H.A. Chipman, E.I. George, and R.E. McCulloch. Bayesian treed models. *Machine Learning*, 48:303–324, 2002.
5. H.A. Chipman, E.I. George, and R.E. McCulloch. BART: Bayesian additive regression trees. *The Annals of Applied Statistics*, 4:266–298, 2010.
6. C. Flynn, C. Hurvich, and J. Simonoff. Efficiency for regularization parameter selection in penalized likelihood estimation of misspecified models. *Journal of the American Statistical Association*, 108:1031–1043, 2013.
7. M. Gentzkow, J. Shapiro, and M. Taddy. Measuring polarization in high dimensional data. *Chicago Booth working paper*, 2015.
8. R.B. Gramacy and M. Taddy. Categorical inputs, sensitivity analysis, optimization and importance tempering with tgp version 2. *Journal of Statistical Software*, 33, 2010.
9. R.B. Gramacy. `reglogit`: *Simulation-based Regularized Logistic Regression*, 2012. R package version 1.1.
10. R.B. Gramacy and N.G. Polson. Simulation-based regularized logistic regression. *Bayesian Analysis*, 7:1–24, 2012.
11. R. Gramacy. tgp: An R package for Bayesian nonstationary, semiparametric non-linear regression and design by treed Gaussian process models. *Journal of Statistical Software*, 19, 2007.
12. R.B. Gramacy, S. Jensen, and M. Taddy. Estimating player contribution in hockey with regularized logistic regression. *Journal of Quantitative Analysis in Sports*, 9:97–111, 2013.

13. T. Hastie, R. Tibshirani, and J. Friedman. *The Elements of Statistical Learning: Data Mining, Inference, and Prediction*. Springer, New York, 2001.

14. A. Hoerl and R. Kennard. Ridge regression: Biased estimation for nonorthogonal problems. *Technometrics*, 12:55–67, 1970.

15. C. Holmes and K. Held. Bayesian auxiliary variable models for binary and multinomial regression. *Bayesian Analysis*, 1(1):145–168, 2006.

16. K. Hornik, D. Meyer, and C. Buchta. slam: *Sparse Lightweight Arrays and Matrices*, 2011. R package version 0.1-23.

17. C.M. Hurvich and C.-L. Tsai. Regression and time series model selection in small samples. *Biometrika*, 76(2):297–307, 1989.

18. S. Ilardi and A. Barzilai. Adjusted plus-minus ratings: New and improved for 2007–2008. *82games.com*, 2004.

19. B. Macdonald. A regression-based adjusted plus-minus statistic for nhl players. Technical report, arXiv: 1006.4310, 2010.

20. D. Mason and W. Foster. Putting moneyball on ice? *International Journal of Sport Finance*, 2(4):206–213, 2007.

21. T. Park and G. Casella. The Bayesian lasso. *Journal of the American Statistical Association*, 103(482):681–686, June 2008.

22. S. Pettigrew. Assessing the offensive productivity of nhl players using in-game win probabilities. In *MIT Sloan Sports Analytics Conference*, Cambridge MA, 2015.

23. D.T. Rosenbaum. Measuring How NBA Players Help Their Teams Win. *82games.com*, April 30, 2004, 2004.

24. M.E. Schuckers, D.F. Lock, C. Wells, C.J. Knickerbocker, and R.H. Lock. National hockey league skater ratings based upon all on-ice events: An adjusted minus/plus probability (ampp) approach. Technical report, St. Lawrence University, New York, 2010.

25. R Development Core Team. *R: A Language and Environment for Statistical Computing*. R Foundation for Statistical Computing, Vienna, Austria, 2010. ISBN 3-900051-07-0.

26. S. Sheather. *A Modern Approach to Regression with R*. Springer, 2009.

27. A. Stair, J. Neral, L. Thomas, and D. Mizak. Team performance characteristics which influence wins in the national hockey league. *Journal of International Business Disciplines*, 6(2), 2011.

28. M. Taddy. gamlr: *Gamma Lasso Regression*, 2013. R package version 1.11-2.

29. M. Taddy. Distributed multinomial regression. *The Annals of Applied Statistics*, 9:1394–1414, 2015.

30. M. Taddy. One-step estimator paths for concave regularization. *arXiv:1308.5623*, 2015.

31. M. Taddy, C.-S. Chen, J. Yu, and M. Wyle. Bayesian and empirical Bayesian forests. In *Proceedings of the 32nd International Conference on Machine Learning (ICML-15)*, pp. 967–976. JMLR Workshop and Conference Proceedings, 2015.

32. A.C. Thomas, S.L. Ventura, S. Jensen, and S. Ma. Competing process hazard function models for player ratings in ice hockey. Technical report, ArXiv:1208.0799, 2012.

33. R. Tibshirani. Regression shrinkage and selection via the lasso. *Journal of the Royal Statistical Society, Series B*, 58:267–288, 1996.

34. R. Vollman. Howe and why: Ten ways to measure defensive contributions. *Hockey Prospectus*, March 04, 2010, 2010. http://www.hockeyprospectus.com/puck/article.php?articleid=480.

35. H. Zou and T. Hastie. Regularization and variable selection via the elastic net. *Journal of the Royal Statistical Society: Series B (Statistical Methodology)*, 67(2):301–320, 2005.

36. R.B. Gramacy, M. Taddy, and S. Tian. https://github.com/TaddyLab/hockey. Accessed August 30, 2016.

15

Statistical Evaluation of Ice Hockey Goaltending

Michael E. Schuckers

CONTENTS

In this chapter, we look at some statistical methods for assessing the performance of ice hockey goaltenders. Since the time when ice hockey metrics began being recorded, several measures of goalie performance have been proposed. As we shall see, some of these incorporate factors that are beyond the control of goalies. We will assess metrics on goalie performance on how correlated they are with results from the same season and how correlated they are with results from subsequent seasons. In particular, we look at a goalie's save proportion—their adjusted save percentage based upon shot difficulty and their ability to control rebounds from saved shots.

15.1 Metrics and Notation

Historically, there have been several metrics that have been proposed to assess and evaluate hockey goaltenders or goalies. These roughly are wins, goals against average (GAA), and save proportion (SVP). Each of these metrics has some merits, though SVP and its variants are generally considered the best of these for the evaluation of goalies. In the following text, we discuss each of these in turn.

15.1.1 Notation

Throughout this chapter, we will use a variety of different variables and metrics in order to evaluate goaltender performance. Here, we introduce the basic notation that we will use. We will denote information about a shot using a vector of attributes $\mathbf{z}_i$ for the ith shot from among n total shots with $i = 1, \ldots, n$. We will use an indicator function, Φ_i to denote whether or not a shot is saved using

$$\Phi_i = \begin{cases} 1 & \text{if shot } i \text{ is saved,} \\ 0 & \text{if shot } i \text{ is not saved.} \end{cases} \tag{15.1}$$

Since time on the ice is sometimes important in rate metrics, we will denote the time on the ice for a particular goalie over some interval, for example, a season by T. As needed, we will use a subscript of j for the jth goalie. We will use J to represent the total number of goalies considered in a particular analysis with $j = 1, \ldots, J$. For simplicity, we will drop the goalie subscript if it is clear that we are considering a single goalie. Often we will make use of the additional measurements that are available about the ith shot. The other additional notation that we will use is a result of information about a particular shot, $\mathbf{z}_i$. For such a shot, we will say the strength of the team taking the shot is denoted by s where s takes the values: EV (Even Strength), PP (Power Play), PK (Penalty Kill). As we will see in the following, different types of shots will have different probabilities of becoming goals. The type of shot will be denoted w where the basic shot types are Backhand, Deflection, Slap, Snap, Tip-In, Wrap(around), and Wrist. Additionally, for shots we may have location from which a shot was taken. From that location, we will obtain the coordinates of a shot relative to center ice that will take the place of the origin $(0, 0)$. We will use x and y to denote the vertical and horizontal distances from center ice. The orientation here is that the x-axis goes through the center of both goals. Furthermore, we can also obtain the distance of a given shot from the center of the goal and the angle of the shot relative to the horizontal axis of the shot. We will use d to represent that distance and θ to represent the corresponding angle.

All of the information that we have regarding shots taken from the National Hockey League (NHL) comes from the NHL's Real Time Scoring System (RTSS). A description of the RTSS system can be found in [Kasan, 2008]. Many authors including [Desjardins, 2010a], [Bruce McCurdy, 2010], [Fischer, 2010], [Zona, 2011], and [Awad, 2009], among others have noted that there are issues, specifically measurement error and bias, with the collection of these data. In particular, there seem to be some substantial differences in the recording of shot location from rink to rink in the NHL. We will discuss some of these issues in the following, as well as some of the proposed remedies. Throughout this chapter, we will use data from RTSS for the 2009–2010 through the 2012–2013 regular seasons.

15.1.2 Wins

The evaluation of goaltenders based upon wins is one manner to evaluate the contribution of a player to their team's performance. A win is awarded to a goalie if they are on the ice for the game-winning goal. For many analysts, this suggests that winning is very team dependent. While it is common to have a goalie with a high save proportion win a large number of games, for example, during the 2011–2012 season Jonathan Quick won 35 games with the fifth highest goalie save proportion, it sometimes happens that a goalie wins a very high number of games having had a save proportion that is not well above the league average. The latter can be exemplified by the noting that Marc-André Fleury won 37 games (out of 65 started and 67 total) during the 2009–2010 season while having a 0.905

save proportion. The league average save proportion that season was 0.908. That winning tends to be associated with quality of a team rather than with quality of a goalie suggests that wins is not a strong measure of goalie performance.

15.1.3 Goals against Average

Goals Against Average, or GAA, is the average number of goals that a goalie concedes per 60 min of ice time where 60 min is the regulation length of a hockey game, that is, the length of a hockey game that does not go to overtime. The calculation of GAA is then number of goals conceded divided by time on ice (in minutes) multiplied by 60. Using our notation, this is

$$\text{GAA} = 60 \left[\frac{n - \sum_{i=1}^{n} \Phi_i}{T} \right]. \tag{15.2}$$

We note that the GAA can be written as a function of a goalie's SVP in the following way:

$$\text{GAA} = 60 \left[\frac{n}{T} (1 - SVP) \right]. \tag{15.3}$$

One result of writing GAA in this manner is that we can explicitly see that GAA is a function of time on ice and number of shots faced which are ultimately not under the direct control of a goalie. Consequently, it seems that SVP is a better choice for the evaluation of goalie performance than GAA.

15.1.4 Save Proportion

The most widely accepted traditional measure of goalie performance is the save proportion. One primary reason for this is that it is a rate over which a goalie has some measure of control, unlike wins and GAA which are dependent on the other things that the goalie's team is doing, scoring lots of goals or giving up lots of shots, respectively. A goalie's SVP is calculated as the number of shots saved divided by the number of shots faced. This is often done on a per game or per season basis. We can think of SVP as

$$\text{SVP} = \frac{\sum_{i=1}^{n} \Phi_i}{n}.$$

Figure 15.1 shows the average SVP in the NHL since the 1983–1984 season. Any historical analysis needs to adjust for these changes which are likely due to changes in goalie athleticism and equipment. The increasing pattern here is discussed by Paine (2014). Over this period, the goaltending styles improved and the size of equipment became larger. Both of these effects contributed to this trend.

More generally, we can think of SVP as a function of the quality of goaltending and the quality of shots faced. To that end, we can think of SVP as a weighted average with

$$\text{SVP} = \sum_{u=1}^{u} G(u)S(u). \tag{15.4}$$

where
    $G(u)$ is a probability that a goalie saves a shot with attributes $\mathbf{z}_u$
    $S(u)$ is the probability of facing a shot with attributes $\mathbf{z}_u$

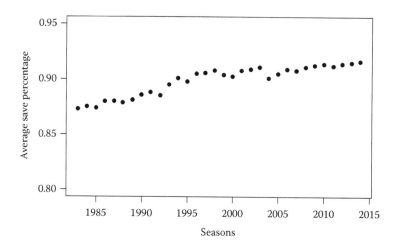

FIGURE 15.1
NHL league average save proportion by season 1983 to present.

For this equation, $\mathbf{z}_u$ is a vector of descriptors for the attributes of a shot and $\mathbf{z}_u = (x, y, d, \theta, w, s, \ldots)^T$ where $u = 1, \ldots, \mathcal{U}$. These attributes might include things like the type of shot, the location of the shot, and others. In this way, we can think of SVP as an expected save proportion taken with respect to some measure of shot distribution. This formulation will be especially useful in the next section where we consider adjustments to SVP. One reason that these adjustments are used is the possibility of differences between the distribution of shot types, that is, different goalies could face different $S(u)$'s.

To understand our notation here, consider the following example using information from Table 15.1. Suppose that we have three goalies, A, B, and C, who play in the same league and who faced four shots with $\mathcal{U} = 4$ different attribute types as listed in Table 15.1. Additionally, n_u is the number of shots with a given set of attributes that each goalie faced where we will let $n = \sum_{u=1}^{\mathcal{U}} n_u$ be the total number of shots faced of all attribute types. The overall SVP of each goalie is given in the last row of the aforementioned table. That is,

$$\text{SVP} = \sum_{u=1}^{\mathcal{U}} [G(u)] \frac{n_u}{n} = \sum_{u=1}^{\mathcal{U}} [G(u)] S(u). \tag{15.5}$$

TABLE 15.1

Example of Notation

Shots Attributes, z_u		Goalie A		Goalie B		Goalie C	
Shot Type	Distance	$G(u)$	n_u	$G(u)$	n_u	$G(u)$	n_u
Slap	Close	0.900	1200	0.880	100	0.880	500
Slap	Far	0.930	200	0.920	700	0.920	500
Wrist	Close	0.920	500	0.900	300	0.900	500
Wrist	Far	0.930	100	0.910	900	0.910	500
Overall		0.9095	2000	0.9105	2000	0.9025	2000

So for goalie A, we have that

$$\text{SVP}_A = (0.900)\frac{1200}{2000} + (0.930)\frac{200}{2000} + (0.920)\frac{500}{2000} + (0.930)\frac{100}{2000} = 0.9095. \qquad (15.6)$$

Similar calculations for goalies B and C yield overall save proportions of 0.9105 and 0.9025, respectively. Moving to our notation for the rest of the chapter, we will use $S(u) = \frac{n_u}{n}$. The more general notation of $G(u)$ and $S(u)$ will be used when we want to make adjustments to the SVP. Two things are also noteworthy from this example. First, note that Goalie A has a lower overall save proportion than Goalie B because of the distribution of shot attributes despite being better than Goalie B at saving each type of shot. These examples are synthetic, but they serve to illustrate that SVP is a function of $S(u)$. Similarly, Goalie C has a lower save proportion than Goalie B despite having the same save proportion for shots with the same attributes.

There has long been a discussion in "hockey analytics" circles about whether or not there are differences between the $S(u)$ faced by different goalies over substantial periods of time, say more than half of an NHL season. This debate revolves around the notion that $S(u)$ for different goalies or different teams converges to a mean $S(u)$, say $\bar{S}(u)$ over time. Consequently, there is not an advantage in the prediction of future outcomes to be gained by adjusting for the differences in shot distributions. This debate is often referred to under the heading of shot quality.* Implicit in these discussions is that what is meant by shot quality is instead shot probability and average shot probability at that. All of these discussions are dependent upon the modeling of shot probability, a topic we take up later in this chapter.

The reliability of SVP has led some commentators and authors to suggest that goaltenders are difficult to assess.[†] Table 15.2 shows the within-season correlation between a goalie's even and odd shots faced in our data. Correlations are taken over goalies and J is the number of goalies involved in each correlation. Here, we are limiting ourselves to looking at even strength five-on-five nonempty net shots as we will do throughout this chapter. We consider even versus odd shots in an attempt to balance the variety of factors that can possibly impact a goalie's SVP such as the players on the ice, the team they are playing against, the rink in which the game is played etc. By using even shots and odd shots

TABLE 15.2

Intraseason Correlation of Save Proportion (SVP) for Even and Odd Shots

Season	NHL Average Save Proportion	More than 500 Shots		More than 750 Shots	
		Correlation	J	Correlation	J
2009/2010	0.9198	0.203	48	0.072	33
2010/2011	0.9221	0.282	46	0.385	32
2011/2012	0.9215	0.180	45	0.165	35
2012/2013	0.9203	0.145	24	0.252	9

---

* It is often said that shot quality in the NHL does not matter or it matters very little. See, for example, this blog post by Eric Tulsky, http://nhlnumbers.com/2012/7/3/shot-quality-matters-but-how-much.

† The phrase that is commonly used in hockey analytics is that *goaltending is voodoo* or *goalies are voodoo*. See, for example, http://grantland.com/the-triangle/2015-nhl-awards-andrew-hammond-carey-price-mark-giordano/

TABLE 15.3

Correlation between a Goalie's SVP in 1 Year and Their SVP in Subsequent Years Given a Number of Shots Faced in Each Season

	More than 500 Shots		More than 750 Shots	
Seasons	Correlation	J	Correlation	J
2009/2010 v 2010/2011	0.186	34	0.120	22
2010/2011 v 2011/2012	0.018	37	−0.038	28
2011/2012 v 2012/2013	0.060	20	0.909	8

we hope to split our data in such a way as to have two groups of data for which the impact of these factors is roughly the same. While the year to year correlations do fluctuate, there is some consistency in these correlations. These results mirror those found in, for example, Macdonald et al. (2012). In both the results for goalies who faced more than 500 shots and for those who faced more than 750 shots, the correlations average around 0.210 which is relatively weak. Note that there are fewer goalies in the results for the 2012/2013 season because it was shortened due to the lockout. In Table 15.3, we look at the correlation in SVP from year to year among goalies facing 500 or more shots in each season. What is clear is that there is not a strong relationship between a goalie's save proportion in a given year and their save proportion in the subsequent year. The correlations range from −0.038 to 0.909 though the latter is anomalous and based upon only 8 goalies from the lockout shortened 2012/2013 season. Again, this uncertainty around the ability to predict future SVP confirms what some authors have suggested: it may be difficult to evaluate goaltenders on less than three years of data, Desjardins (2010b).

While it is the case that save proportion is generally the best traditional metric to evaluate a goaltender, it has some issues. Foremost among these is that there is a good deal of variation in goaltender performance, even for a single goaltender, in the short term and that makes prediction of future performance difficult at best. We have seen this in the correlation results discussed earlier. Note that a goaltender who faces about 1500 shots and saves them at a rate of 0.925 would have a standard error of their save proportion of about 0.007 assuming a binomial distribution. Two standard errors from 0.925 yields an interval of 0.909–0.939. In terms of NHL goalie performance, this is a wide range of values. There are other factors as well including the distribution of shots faced. Next, we will look at some attempts to deal with this inability to predict future save proportion.

15.2 Adjusted Save Proportion

There have been several attempts to create adjusted save proportion metrics. In this section, we describe some of these. The main motivations for these adjustments are to account for variation in shot difficulty, or the distribution of shot difficulty, and to improve upon our ability to predict a goalie's future SVP (or even strength, nonempty net SVP). These approaches fall into two categories and use different information about shots that are part of z_u. Both of these were first formulated by Ryder (2004). The first type of adjusted save proportion essentially takes the league average for goaltending ability for a given set of shots. We will call this $\text{aSVP}_{\overline{G}}$. This can be calculated as

$$\text{aSVP}_{\overline{G}} = \sum_{u=1}^{\mathcal{U}} \overline{G}(u)S(u), \tag{15.7}$$

where $\overline{G}(u)$ is the league average probability that a shot with attributes $\mathbf{z}_u$ is saved. This metric has been traditionally referred to as the adjusted save proportion and is useful in comparison to the SVP. What this adjustment gives is the save proportion relative to what an average goalie's save proportion would have been, given the shots that an individual goalie faced. Comparison of $\text{aSVP}_{\overline{G}}$ for different goalies will have different $S(u)$'s and so are still dependent upon the distribution of shots faced. This metric does allow for a comparison of how a goalie did relative to the league average on the shots they faced but each goalie faces a different distribution of shots which makes comparison across goalies problematic.

The second type of adjusted save proportion which we will refer to as shot quality neutral save proportion. We will denote this by $\text{aSVP}_{\overline{S}}$ and calculate it as the following:

$$\text{aSVP}_{\overline{S}} = \sum_{u=1}^{\mathcal{U}} G(u)\overline{S}(u), \tag{15.8}$$

where $\overline{S}(u)$ is the percentage of all shots in the league that are of attribute type u. This approach is useful because it allows direct comparison of individual goalies since they are compared based upon the same distribution of shot types. It is this general approach that has been adopted by war-on-ice.com for their adjusted save proportion, [War-On-Ice.com, 2014], and by the Defense Independent Goalie Rating of [Schuckers, 2011]. Here, we will look at using this second type of aSVP, $\text{aSVP}_{\overline{S}}$, to predict future aSVP.

As mentioned earlier, all of the various methods that have been proposed for adjusting the save proportion fall into one of these two basic approaches. There is devil in the details of all of the methods that have been proposed by various authors to calculate these aSVPs. In some cases, the authors use statistical smoothing or regression to estimate the goalie's ability functions, the $G()$'s, and the shot distribution functions, the $S()$'s. Further, different authors have used different information about each shot to calculate their adjustments. In the following, we will go through some of the more prominent adjustments and place them into the framework we have established with $\text{aSVP}_{\overline{S}}$ and $\text{aSVP}_{\overline{G}}$.

15.2.1 Shot Probability Models

The goal of any shot probability model is to estimate the probability that a shot results in a goal given information about a particular shot. This is equivalent to the estimation of $G(u)$ in our earlier notation. As mentioned earlier, one of the earliest looks at aSVP was by Ryder (2004), who also developed the notation of shot neutral adjusted save proportion, our $\text{aSVP}_{\overline{S}}$. A logistic, or log odds, model for shot probability was introduced by Krzywicki (2005). The final model that Kryzwicki reports in that paper includes "distance, shot type, rebound, and situation" with each term additively effecting the log odds of the probability of a shot being a goal. Subsequently, Kryzwicki introduced an adjustment for rink biases in the recorded distances in Krzywicki (2009). The adjusted distance which replaced distance in the previous model was a single quantity added (or subtracted) to each shot recorded in a particular rink. Again using a logistic linear model, Kryzwicki added an indicator for whether or not a shot was preceded by a giveaway. Again this model did not include any interaction terms. The following year in Krzywicki (2010), he introduced shot angles as an added addition to his logistic model. All of these models lead to predicted values for a shot. We will call the predicted probability that a given shot results in a goal $\hat{p}(\mathbf{z}_i)$. This can be thought of as one minus the predicted average save probability for this quantity. So that $\overline{G}(\mathbf{z}_i) = 1 - \hat{p}(\mathbf{z}_i)$ where $\overline{G}(\mathbf{z}_i)$ is a smoothed or averaged goalie's ability to save a particular shot.

In many ways, these models and approaches are spatial in nature. They each try to model a probability surface where the x and y axes represent locations on a rink. (Note that in some cases x and y are replaced by polar coordinates, radius, and angle.) Schuckers (2011) introduced the Defense Independent Goalie Rating (DIGR), which is a shot quality neutral adjusted save proportion using a general nonparametric spatial model. That is, their model did not assume a particular parametric form for the relationship between the x and y locations and the probability of a shot being a goal. Most previous models assumed a linear form for the probability surface on either the original or logistic scale. Further this model assumed that the impact of shot type, w, and strength, s, was not consistent across locations. Therefore, DIGR effectively includes interactions for the factors of location, strength, and shot type. Another innovation that Schuckers (2011) introduced was a form of shrinkage analysis by creating shot probability maps for each goalie that were weighted to the league average of all shots of a particular strength and shot type. The DIGR for the jth goalie can be written as

$$DIGR_j = \sum_{u=1}^{\mathcal{U}} \widetilde{G}_j(u)\overline{S}(u)$$

$$= \sum_{u=1}^{\mathcal{U}} \Big[\alpha(u)G_j(u) + (1 - \alpha(u))\overline{G}(u)\Big]\overline{S}(u), \qquad (15.9)$$

where
$\alpha(u) = \frac{n_u}{n_u + n^\dagger}$
n_u is the number of shots faced by goalie $j, j = 1, \dots, J$ of shot type w at strength s
$n^\dagger$ is a "shrinkage" constant

Thus, $\widetilde{G}_j(u)$ shrinks the save proportion of goalie j on shots with attributes u toward the league average, $\overline{G}(u)$, for those types of shots. $\alpha(u)$ determines the amount of shrinkage. If $n^\dagger = 0$ then $\alpha(u) = 1$ and DIGR is just the aSVP$_{\overline{s}}$ and as $n^\dagger$ gets large relative to n_u for each u then $\alpha(u)$ approaches zero and DIGR becomes close to the league average. The larger the value of $n^\dagger$ the more similar we are assuming is the performance of each goalie is to the league average for the characteristics, the $\mathbf{z}_u$'s, of shots. This makes for a broader, more flexible shot probability model than had been previously proposed.

The next improvement to the shot probability model was proposed by Macdonald et al. (2012). Their innovation was to incorporate the change in angle from the

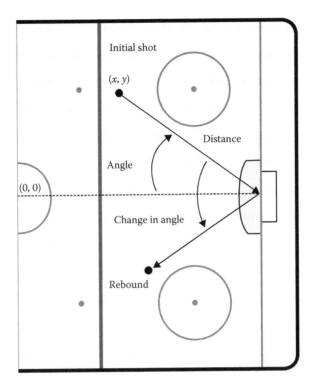

FIGURE 15.2
Visual explanation of distance, shot angle, and change in angle due to a rebound. (From Macdonald, B. et al., Evaluating nhl goalies, skaters, and teams using weighted shots, http://arxiv.org/abs/1205.1746, accessed October 24, 2015, 2012. With permission.)

previous shot of a rebound shot to a logistic regression model of goal probability. This variable, change in angle, which we will denote by δ, is illustrated in Figure 15.2. The model that Macdonald et al. use in that paper also have factors for length of time that the shooter was on the ice (shooter fatigue), length of time that the defense was on the ice (defense fatigue), length of time the offense was on the ice (offense fatigue), whether the shot was by the home team (Byhome), and the score differential (Scorediff). We note here that Schuckers' original DIGR model did not include rebounds in that analysis.

The issues with specific rinks, discussed earlier, are explored further in Macdonald et al. (2012). Specifically, the authors show that the adjusted save proportion is affected by a player's home rink. This is due to the aforementioned measurement errors in shot location by rink. The adjustment that they use, which was previously suggested by others, was to look at only shots faced when a goalie is on the road. The idea behind this approach is that it will average out the impact of any one rink. An alternative which has not been done would be to reweight all shots so that the weights for shots from a particular rink are equal across all rinks. For the analyses that follow, we will use the location from which the rebound shot was taken not the x and y of the initial shot.

Although it was not the focus of their work, Schuckers and Curro (2013) created rink adjustments for shot location using a discrete version of the probability integral transform to convert a shot coordinate, either x or y, from a particular rink to what the equivalent coordinate would have been by matching up the cumulative distributions from all rinks.

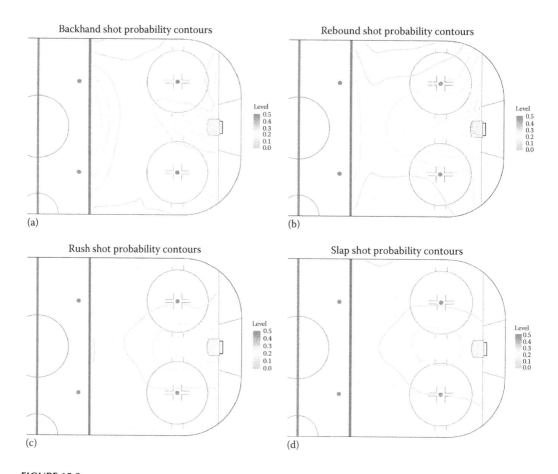

FIGURE 15.3
Using data from the NHL from the 2009 to 2013 seasons, these images are contour plots for shot probabilities of (a) Backhand shots, (b) Rebound shots, (c) Rush shots, and (d) Slap shots.

Their adjustment, which we will use for plotting here, rounds to integer distances and makes the adjustments in a univariate rather than bivariate way. While this method is an improvement over most adjustments, a possible innovation would be to use an inherently bivariate adjustment and to allow for continuous scaling of the x and y (or r and θ) quantities. See, for example, [Pishchulin et al., 2012] for a discussion of how this might be accomplished. Using the adjusted shot locations from [Schuckers and Curro, 2013], we have plotted probability contours for each shot type. These can be found in Figures 15.3 and 15.4. The data we use here for shot locations have been adjusted following this methodology.

In 2014, the popular web site http://war-on-ice.com/ added an adjusted save proportion that was a simpler version of the forementioned shot probability models just based upon shot location [Ventura, 2014]. Their adjusted save proportion is based upon a goalie's save proportion relative to the league average in each of three zones defined by the areas in Figure 15.5. This aSVP is the first type of adjusted save proportion given earlier, that is, $\text{aSVP}_{\overline{G}}$. In the same year, a descendent of Ryder's and Kryzwicki's work on shot probability was introduced by Johnson (2014). Kryzwicki had a term in his model for shots that

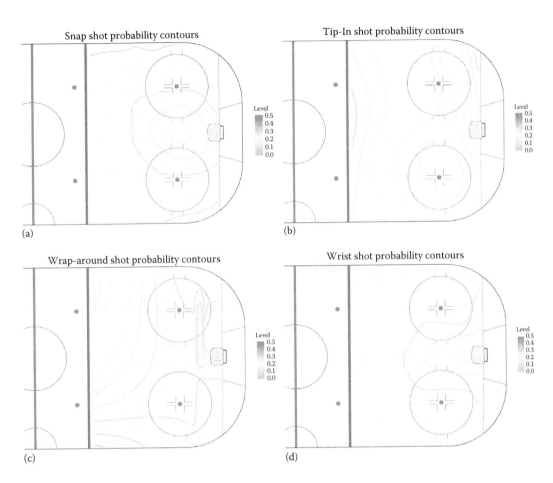

FIGURE 15.4
Using data from the NHL from the 2009 to 2013 seasons, these images are contour plots for shot probabilities of (a) Snap shots, (b) Tip-In shots, (c) Wraparound shots, and (d) Wrist shots.

followed a giveaway. Johnson modified and broadened this to include shots that followed shots by the other team as well as faceoffs, hits, takeaways, or giveaways in the neutral zone or the shooting team's defensive zone. Further, he provided some evidence that these shots have higher probabilities relative to other types of shots. Johnson referred to these types of shots as "rush shots."

Figures 15.3 and 15.4 show the contour plots for the probability of a goal from a given location for each of eight shot types. It is clear from those graphs that shot probability is a function of x and y locations as well as shot type s. In both of those figures, we are plotting the probability that shots from a given location result in goals. This is one minus the save proportion at those locations. The contours are given over the entire offensive zone but for most types of shots there are some locations where few shots of a given type were recorded. Red and yellow contour lines correspond to higher probabilities than green or blue lines. Some of the contour plots for various shot types are clearly a function of few shots taken from certain locations, for example, Wraparound shots and Tip-In shots. Other than averaging the volume of shots at a location are not accounted for that in these graphs.

League-wide success rate

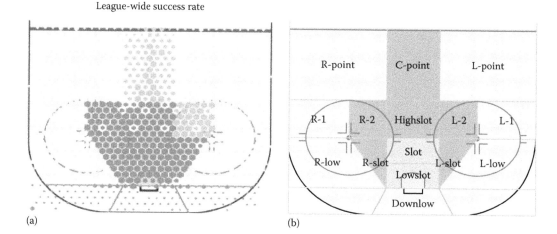

(a)   (b)

FIGURE 15.5
(a) Mapping of shot success rate relative to unblocked shot attempts and (b) represents the definition of the three (colored) regions of shot danger high, medium, and low. (From War-On-Ice.com, Glossary, http://war-on-ice.com/glossary.html, accessed January 4, 2016, 2014. With permission.)

Additionally, there are some combinations of shot type and shot location that are unlikely. For example, it is uncommon to take a backhand shot near the blue line or a slap shot within 5 ft of the goal.

To assess the reliability of adjusted SVP methods, we will look at DIGR as representative of these approaches. The DIGR we use here is slightly different from the version in Schuckers (2011). In this analysis, we have added rebounds and rush shots to the shot types. We define a rebound as a shot that is the next event after another shot by the same team and that occurs within 3 s of the previous shot. For rush shots, we follow [Johnson, 2014] in calling a shot a rush shot if it is a shot taken within 10 s of a "shot attempt by the other team on the other net [or] ... off a face off[sic] at the other end or in the neutral zone [or] ... of a hit, giveaway or takeaway in the other end or the neutral zone." Further, for consistency we will only analyze DIGR for even strength nonempty net shots.

We present results for a modified version of the original DIGR. First, we treated rebounds and rush shots as separate types of shots, s, as in Figures 15.3 and 15.4. Note that the probability contours found in those graphs are based upon the league averages for all shots not just those faced by a particular goalie. Second, we used the adjusted shot location from Schuckers and Curro (2013) as inputs to this version to account for the effect of different rinks. Finally, we altered the values for $n^\dagger$ to be 1000 which is much larger than the value of $n^\dagger$ used in the original DIGR paper. This choice of $n^\dagger$ was driven by a trade-off in the ability to predict future aSVP while trying to minimize shrinkage. That is, we were trying to find the smallest value for $n^\dagger$ that yielded good correlations from one season to the next. We get similar results for those found here with $n^\dagger$ decreasing as n increases. Our results for the intraseason correlation between DIGR calculated on even shots and on odd shots faced by the same goalie can be found in Table 15.4 are stronger than those for just SVP that we saw in Table 15.2. The correlation of DIGR from one year to the next as seen in Table 15.5 is clearly stronger generally than the equivalent correlations for SVP from Table 15.3. However, there is one negative correlation (2011/2012 v 2012/2013 for goalies facing at least 500 shots) that stands out although for the same seasons with a larger number of shots faced,

TABLE 15.4

Intraseason Correlation of DIGR for Even and Odd Shots Using $n^\dagger = 1000$

Season	More than 500 Shots Faced		More than 750 Shots Faced	
	Correlation	J	Correlation	J
2009/2010	0.288	48	0.309	33
2010/2011	0.644	45	0.751	36
2011/2012	0.655	45	0.621	33
2012/2013	0.251	24	0.178	9

TABLE 15.5

Correlation between a Goalie's DIGR in 1 Year and Their DIGR in Subsequent Years

Seasons	More than 500 Shots		More than 750 Shots	
	Correlation	J	Correlation	J
2009/2010 v 2010/2011	0.237	34	0.393	22
2010/2011 v 2011/2012	0.647	37	0.665	28
2011/2012 v 2012/2013	−0.174	15	0.677	6

TABLE 15.6

Top 10 Goalies Ranked by DIGR for 2010–2011 NHL Season Using $n^\dagger = 1000$

Goalie	DIGR(aSVP)	SVP	n
T. Thomas	0.9314	0.9499	1416
S. Clemmensen	0.9308	0.9216	638
N. Backstrom	0.9305	0.9301	1202
J. Reimer	0.9301	0.9330	880
K. Lehtonen	0.9295	0.9290	1522
A. Niemi	0.9295	0.9310	1333
J. Quick	0.9295	0.9243	1255
S. Varlomov	0.9288	0.9370	571
M. Kiprusoff	0.9281	0.9172	1438
P. Rinne	0.9279	0.9364	1462

the correlation is positive and strongly so. These outcomes also mirror results for the aSVP approach in Macdonald et al. (2012). Those authors found substantial improvement in reliability of aSVP (or DIGR) over SVP where reliability can be thought of as the correlation in a metric over time (Tables 15.3 and 15.5). These results are suggestive that adjusted SVP's may be moderately reliable for goalies who face a large number of shots. The top 10 goalies as ranked by DIGR for the 2010–2011 NHL regular season can be found in Table 15.6.

15.3 Rebounds

One area of goaltending that has received surprisingly little attention is a goalie's ability to control rebounds. Myrland (2009) wrote about the relationship between a teams rebound rate and their average shot probability against. Myrland argues that there is a relationship between the rate at which a team gives up rebounds and the percentage of shots that are rebounds. In a similar vein, Pettapiece (2013) looked at the repeatability of a player's ratio of rebounds to unsaved shots. Pettapiece notes that "a goaltender who has the ability to cover up rebounds, or control them enough that a teammate can clear the zone, can significantly increase his value to his team." To assess this repeatability he compared the year 1 rebound rate to the year 2 rebound rate. Here, we will define the rebound rate as the ratio of the number of rebound shots allowed to the number of saved shots. As we did in the previous section, we will use even strength, 5-on-5, nonempty net shot data from the 2009–2010 through the 2012–2013 regular seasons.

We begin by looking at the correlation between the rebound rate of even and odd shots. Table 15.7 shows the correlation between the rebound rate of even and odd shots. The counts in the table are the number of goalies involved in the correlations. As is perhaps not surprising, there are higher correlations between those goalies that had more than 750 shots total in a season. In order for a goalie to be considered for this analysis, they had to have faced more than 500 shots in a season. Among goalies who faced at least 500 shots, the correlations here are moderate ranging from quite low, below 0.1, for 2009–2010 to fairly strong, about 0.5, for the 2011–2012 season. As always, the results of the lockout-shortened 2012–2013 season should be taken with a grain of salt since that season had only 48 regular season games relative to the usual 82. The correlations between even and odd shots in the same season are much stronger, all at least 0.3 for those goalies that faced more than 750 shots. These correlations suggest that the rebound rate is possibly something that would be considered repeatable. We will next look at the rebound rate from season to season.

The correlations for year to year rebound rate found in Table 15.9 are not great. For those that faced at least 500 shots, the correlations have different signs, which do not suggest consistency. For those that faced over 750 shots the correlations uniformly have the same sign, positive, but the values are diverse ranging from just under 0.1 to nearly 0.5. This suggests that past rebound rates are mildly predictive of future rates though they are consistent within a given season.

We next investigated the relationship between rebound rates and SVP. Summaries of these relationships can be found in Table 15.8. It is clear from the correlations in that table

TABLE 15.7

Intraseason Correlation of Rebound Rate for Even and Odd Shots

Season	NHL Average Rebound Rate	More than 500 Shots Faced		More than 750 Shots Faced	
		Correlation	J	Correlation	J
2009/2010	0.0823	0.066	46	0.327	33
2010/2011	0.0825	0.329	48	0.395	32
2011/2012	0.0842	0.512	45	0.642	35
2012/2013	0.0838	0.268	24	0.336	9

TABLE 15.8

Correlation between a Goalie's Rebound Rate and Their Save Proportion

Seasons	More than 500 Shots		More than 750 Shots	
	Correlation	J	Correlation	J
2009/2010	−0.164	48	−0.114	33
2010/2011	−0.440	46	−0.210	32
2011/2012	−0.261	45	−0.252	35
2012/2013	−0.066	24	−0.195	9

Note: The numbers in parentheses correspond to the number of goaltenders included in the correlation calculations.

TABLE 15.9

Correlation between a Goalie's Rebound Rate in 1 Year and Their rebound rate in Subsequent Years

Seasons	More than 500 Shots		More than 750 Shots	
	Correlation	J	Correlation	J
2009/2010 v 2010/2011	−0.086	61	0.217	22
2010/2011 v 2011/2012	0.483	37	0.474	28
2011/2012 v 2012/2013	−0.261	20	0.067	8

TABLE 15.10

Top 10 Goalies Ranked by Rebound Rate for 2010–2011 NHL Season among Those Goalies Facing More than 500 Shots

Goalie	Rebound Rate	SVP	n
J. Reimer	0.0572	0.9330	880
J. Quick	0.0586	0.9243	1255
S. Clemmensen	0.0663	0.9216	638
R. Luongo	0.0699	0.9333	1349
S. Mason	0.0714	0.9115	1152
P. Rinne	0.0723	0.9364	1462
C. Crawford	0.0726	0.9269	1218
O. Pavelec	0.0734	0.9295	1319
T. Vokoun	0.0735	0.9191	1347
B. Elliott	0.0748	0.9003	1144

that an increased rebound rate is associated with a lower current season save proportion. As might be expected a goalie's rebound rate is correlated with their save proportion although the relationship is not particularly strong (Table 15.9). While not dramatic the relationship is consistently negative as would be expected. All of the goalies in the calculations given in Table 15.8 have faced at least 500 shots. Finally, we report the goalies with the lowest rebound rates for the 2010/11 season. This can be found in Table 15.10.

15.4 Discussion and Future Directions

In this chapter, we have looked at some of the metrics that have been used to evaluate goalies. It is clear that prediction of goalie performance in the National Hockey League is difficult. The primary traditional metric is the save proportion (SVP). Here we focused on even strength nonempty net 5 on 5 save proportion. Our analysis confirms what others have found which is that SVP is mildly correlated within a season using split data but not correlated with subsequent season's data. Adjusting for shot difficulty is one way that has been proposed to improve upon the prediction of subsequent season prediction. Methods for adjusting the SVP fall into one of two approaches either adjusting for the performance of the goalie relative to an average goalie's shots faced or adjusting for the quality of shots relative to the performance of a particular goalie. Extending the work of Schuckers (2011), which is a method that falls into the latter category, we confirmed results found by [Macdonald et al., 2012] that adjusting the save proportion can improve upon the season to season reliability of SVP. Additionally, we have added to the small literature on rebound rates among goalies to suggest that rebound rates are somewhat reliable within a season and between seasons. These relationships get stronger the more shots that a goalie has faced.

There are certainly other factors that impact how a goalie will perform in the future and one of these is their age. Certainly, it is well known that athletes' performance decline as they get older. There has also been some work done on looking at player trajectories. This goes along with and could be used to improve the prediction of future goalie performance. Several authors including [Lundsford, 2012; Tulsky, 2013; garik16, 2014] have added to our knowledge in this area. In general, this work shows that goalies' save proportion decreases with age and estimates the rates at which this occurs. Another metric that has received some interest is the Quality Starts metric [Vollman, 2009]. A Quality Start is defined as one in which "the goaltender's save proportion is 0.913 or better, or at least 0.885%, but allowed fewer than 3 goals." A derivative of this is the percentage of a goalie's starts that were quality.

One future direction for analysis would be to try to build a model, perhaps a generalized additive model, that might incorporate a variety of these metrics on a goalie to predict their save proportion or adjusted save proportion in future years. The variables in such a model might include the factors discussed in detail here are past save proportion, past rebound rate, and adjusted save proportion, as well as other variables such as age and performance in minor leagues such as the American Hockey League.

There are some ongoing efforts to improve the data in the NHL. As we (and many others) have noted, there are issues with data from the NHL's system for recording of data. See, for example [Schuckers and Macdonald, 2014]. Some efforts have also been made to record NHL data by hand. In particular, Boyle (2013) has introduced a "Shot Quality Project" that tried to track large numbers of NHL shots by hand to assess the impact of a variety of factors especially things like whether or not a goalie's view of the puck was obstructed. Having an obstructed view is something that likely has an impact on a goalie's ability to save a shot but is not in the current publicly available NHL data. Another project on which Boyle has collaborated is the so-called "Royal Road" of Valiquette (2014). The idea here is that an imaginary line extending from one goal mouth to the other exists—along the *x*-axis of our notation—and passes that cross the royal road before a shot have an increased probability of being a goal. This project is in the early stages and the evidence they present is based upon goals rather than shots but this is an area that has potential for future work. Recently, the NHL has moved to add player tracking data for large amounts of additional automated information collection and for improved accuracy [Masisak, 2015]. The amount and quality of these new data will no doubt provide new and important insights going forward which will impact our evaluation of goaltenders.

15.5 Addendum: Pulling the Goalie

Another area in which statistical analysis can have an impact is the altering of strategies or the development of new strategies. There are many strategies involved in any hockey game such as how to forecheck, how to carry the puck through the neutral zone, etc. The statistical evaluation of the impact of such strategies is difficult. As [Schuckers, 2014] wrote " with any strategic innovation, its success will lead to (being copied and eventually) its demise." The most common strategy advocated by hockey analysts is pulling the goalie early. The classic evaluation of pulling the goalie was done by [Beaudoin and Swartz, 2010]. They proposed an empirical Bayesian hierarchical Poisson model for goal scoring rates that suggests that being more aggressive about pulling the goalie can lead to an additional expected point or two over the course of a season. Recently, the idea of pulling the goalie has received additional attention in part because the aforementioned popular web site, war-on-ice.com, created an interactive webpage that allowed users to compare the probability that the trailing team ties the game (if trailing by one or two goals) and the time at which there is the biggest increase in the probability of tying the game. These quantities are determined by comparing the scoring rates for the two teams at even strength and when one team has an extra attacker and their opponent is facing an extra attacker.

There is evidence that teams are potentially taking some of the information to heart. Davis and Lopez (2015) recently looked at the times at which teams are pulling the goalie and they report that the length of time at the end of games when goalies are pulled for an extra skater has increased particularly for the 2013–2014 and 2014–2015 seasons. It seems that "pulling the goalie" is becoming more acceptable to players, teams, and coaches. As with any sort of strategy like pulling the goalie, the advantage gained will likely dissipate over time.

References

Awad, T. (2009). Numbers on ice: A castle built on sand. http://www.hockeyprospectus.com/puck/article.php?articleid=351. Accessed September 19, 2013.

Beaudoin, D. and Swartz, T. B. (2010). Strategies for pulling the goalie in hockey. *The American Statistician*, 64(3):197–204.

Boyle, C. (2013). Introducing the shot quality project. http://www.sportsnet.ca/hockey/nhl/introducing-the-shot-quality-project/. Accessed October 22, 2015.

Bruce McCurdy (2010). RTSS stats: Giveaways, takeaways, and throwaways. http://www.coppernblue.com/2010/3/30/1396334/rtss-stats-giveaways-takeaways-and. Accessed September 20, 2013.

Davis, N. and Lopez, M. (2015). NHL coaches are pulling goalies earlier than ever. http://fivethirtyeight.com/features/nhl-coaches-are-pulling-goalies-earlier-than-ever/. Accessed November 2, 2015.

Desjardins, G. (2010a). Clean up your act, Madison Square Garden. http://www.arcticicehockey.com/2010/10/18/1756880/clean-up-your-act-madison-square-garden. Accessed September 23, 2013.

Desjardins, G. (2010b). Fooled by randomness: Goaltender save percentage. http://www.arcticicehockey.com/2010/4/20/1429278/fooled-by-randomness-goaltender. Accessed October 30, 2015.

Fischer, J. (2010). Blocked shots, The New Jersey Devils' rink, and scorer bias—a follow up. http://www.inlouwetrust.com/2010/7/8/1559914/blocked-shots-the-new-jersey. Accessed September 20, 2013.

garik16 (2014). How well do goalies age? a look at a goalie aging curve. http://hockey-graphs.com/2014/03/21/how-well-do-goalies-age-a-look-at-a-goalie-aging-curve/. Accessed November 8, 2015.

Johnson, D. (2014). Introducing 'Rush' shots. http://hockeyanalysis.com/2014/07/09/introducing-rush-shots/. Accessed October 24, 2015.

Kasan, S. (2008). Off-ice officials are a fourth team at every game. http://www.nhl.com/ice/news.htm?id=388400. Accessed September 24, 2013.

Krzywicki, K. (2005). Shot quality model: A logistic regression approach to assessing nhl shots on goal. http://hockeyanalytics.com/Research_files/Shot_Quality_Krzywicki.pdf. Accessed October 16, 2015.

Krzywicki, K. (2009). Removing observer bias from shot distance shot quality model. http://hockeyanalytics.com/Research_files/SQ-DistAdj-RS0809-Krzywicki.pdf. Accessed October 17, 2015.

Krzywicki, K. (2010). NHL Shot Quality 2009-10: A look at shot angles and rebounds. http://hockeyanalytics.com/2010/10/nhl-shot-quality-2010/. Accessed October 16, 2015.

Lundsford, J. (2012). Aging goalies: What does the future hold for Luongo, Miller & Bryzgalov? http://www.coppernblue.com/2012/4/30/2987615/impact-age-goaltender-performance-save-percentage. Accessed November 8, 2015.

Macdonald, B., Lennon, C., and Sturdivant, R. (2012). Evaluating nhl goalies, skaters, and teams using weighted shots. http://arxiv.org/abs/1205.1746. Accessed October 24, 2015.

Masisak, C. (2015). NHL, Sportvision test program to track players, puck. http://www.nhl.com/ice/news.htm?id=750201. Accessed October 31, 2013.

Myrland, P. (2009). The value of rebound control. http://brodeurisafraud.blogspot.com/2009/04/value-of-rebound-control.html. Accessed October 25, 2015.

Paine, N. (2014). Why gretzky had it easy: The butterfly effect. http://fivethirtyeight.com/datalab/the-butterfly-effect/. Accessed January 13, 2016.

Pettapiece, R. (2013). Can goalies control the number of rebounds they allow? http://nhlnumbers.com/2013/7/15/can-goalies-control-the-number-of-rebounds-they-allow. Accessed October 25, 2015.

Pishchulin, L., Gass, T., Dreuw, P., and Ney, H. (2012). Image warping for face recognition: From local optimality towards global optimization. *Pattern Recognition*, 45(9):3131–3140.

Ryder, A. (2004). Goal Prevention 2004: A review of goaltending and team defense including a study of the quality of a hockey team's shots allowed. http://hockeyanalytics.com/Research_files/Goal_Prevention_2004.pdf. Accessed October 16, 2015.

Schuckers, M. (2011). DIGR: A defense independent goalie rating based upon spatial probability maps. http://myslu.stlawu.edu/~msch/sports/Schuckers_DIGR_MIT_2011.pdf.

Schuckers, M. (2014). Estimating the lower bound for the return on investment for hockey analytics. http://statsportsconsulting.com/2014/05/06/return-on-investment-for-hockey-analytics-1/. Accessed November 2, 2015.

Schuckers, M. and Curro, J. (2013). Total Hockey Rating (THoR): A comprehensive statistical rating of National Hockey League forwards and defensemen based upon all on-ice events. *Proceedings of the 2013 MIT Sloan Sports Analytics Conference* http://www.statsportsconsulting.com/thor.

Schuckers, M. and Macdonald, B. (2014). Accounting for rink effects in the National Hockey League's real time scoring system. http://arxiv.org/abs/1412.1035.

Tulsky, E. (2013). Henrik Lundqvist's contract and estimating player value. http://www.sbnation.com/nhl/2013/12/10/5175204/henrik-lundqvist-contract-player-value-nhl. Accessed November 8, 2015.

Valiquette, S. (2014). Red shots and green shots. http://www.msgnetworks.com/content/msgsite/en/videos/index.2.html?search=vally%20view. Accessed October 22, 2015.

Ventura, S. (2014). Adjusted save percentage: Taking into account high, medium, and low probability shots. http://blog.war-on-ice.com/adjusted-save-percentage-taking-into-account-high-medium-and-low-probability-shots/. Accessed October 24, 2015.

Vollman, R. (2009). How and why: Quality starts. http://www.hockeyprospectus.com/puck/article.php?articleid=54. Accessed January 18, 2016.

War-On-Ice.com (2014). Glossary. http://war-on-ice.com/glossary.html. Accessed January 4, 2016.

Zona, D. (2011). Dear Oilers: Find new official scorers. http://www.coppernblue.com/2011/10/17/2495447/dear-oilers-find-new-official-scorers. Accessed September 20, 2013.

16

Educated Guesswork: Drafting in the National Hockey League

Peter M. Tingling

CONTENT

> In the factory we make cosmetics, in the store we sell hope.
>
> **Charles Revson**

Other than the season's opening and final championship games, few events are as defining or as filled with hope and optimism as "the draft"—when teams choose prospects eligible to join the ranks of professional athletes. Practiced by all four of the major North American sports leagues, drafting is a highly regulated selection process developed to reduce the dynasty or hegemony of individual franchises and increase equality and competitiveness by establishing selection order as the inverse of performance. The draft is a rebalancing effort that grants a poorer performing team the first selection and the championship winning team the last choice. Teams that are able to identify and select exceptional talent can acquire athletes that will enrich their team's potential and performance and delight their fans. In the National Hockey League (NHL), fortunate teams choosing very early can lock up generational players such as Lemieux (#1-1984), Lindros (#1-1991), and Crosby (#1-2005), or franchise players such as Ovechkin (#1-2004), Toews (#3-2006), Henrik and Daniel Sedin (#2 and #3-1999), and Stamkos (#1-2008). Conversely, teams that choose later or are not able to find "hidden gems" missed by teams choosing earlier, such as Datsyuk (#171-1998), Zetterberg (#210-1999), Robitaille (#171-1984), Benn (#129-2007), or Fleury (#166-1987), forgo opportunities to lock up entry-priced talent and tie their team's performance to player development, trades, and free agency—paths exacerbated by salary caps and the uneven economics of small markets. Worse off still are the teams that are stuck in the middle, not quite good enough to challenge for the championship but neither losing badly enough that they win the right to choose early and reverse their fortunes. With such high stakes, it is little wonder that the draft is stressful for the athletes awaiting selection and general managers making their decisions. Drafts are a mix of luck, decision science, hope, optimism, and intuition and are often characterized by such uplifting but anodyne clichés as "we had him much earlier on our list," "we were surprised that he was still available when it was our turn to pick," and "looking back, I think this will be viewed as one of the day's great steals."

TABLE 16.1

Draft History of the Four Major North American Sports Leagues

Sport	Official Name	Date Introduced	Round Length	Distinguishing Characteristics
NFL	Player Selection Meeting	1936	7	Implemented specifically to increase competitive parity. Regular picks might be traded, compensatory picks might not.
MLB	Rule 4 Draft	1965	40	Evolved from the 1921 minor league draft to prevent wealthy teams from stockpiling talent. 2012 implemented luxury tax and bonus picks that can be sold or traded (unlike regular picks). Longest draft of any major sport but players rarely enter major leagues.
NBA	NBA draft lottery	1950	2	Has varied in length from 21 rounds in the 1960s to 10 in 1985 and is now the shortest draft in major sport. First implemented a lottery (coin flip) in 1966 to prevent intentional losses.
NHL	NHL Amateur Draft	1963	7	Renamed as Entry Draft following expansion. Uses a hard salary cap with narrowly defined
	NHL Entry Draft	1979		entry contracts and a 1995 lottery to reduce motivation to intentionally lose.

Drafting for the four major North American sports has a long history that is summarized in Table 16.1. Introduced into baseball in 1921 as a way to distribute talent to their minor leagues, professional drafting first moved to the 1936 National Football League (NFL) when the Player Selection Meeting was unanimously adopted as a way of reducing the virtuous cycle of affluent or winning teams monopolizing talent and enhancing the competitive parity and financial viability of all teams. Evolving from a casual chalkboard of names to a sophisticated process of testing and a myriad of scouts and advisors, the NFL draft is now a seven-round, 3-day, televised event where teams often trade and negotiate their eligible picks.

Drafting was implemented by the National Basketball Association (NBA) following the 1949 merger of the Basketball Association of America and the National Basketball League. The lottery draft, as it is known, has varied from a high of 21 rounds to its current number of 2. In 1985 the NBA draft, distinguished as having the shortest duration, was the first to introduce a lottery aspect in order to reduce the motivation to intentionally lose and claim the first choice. In stark contrast to the shortness of the NBA draft, Major League Baseball's (MLB) first-year player or Rule 4 draft is distinctive for its sheer length of 40 rounds and its relatively low rate of success. Unlike other sports in which top picks are expected to make an immediate impact (all of the 30 athletes selected in the first round of the 2005 NFL draft had played in at least one professional game within 3 years), the vast majority of the MLB draftees are developed in the minors and very few ever play in a single MLB game. Two other significant characteristics of the MLB draft are that only North American resident athletes are eligible and regular picks cannot be traded. This trading prohibition, however, does not apply to compensatory picks—those that are awarded as a result of failure to sign selected players or from certain conditions of free agency. (It is interesting to note that these restrictions are essentially the opposite of those imposed on the trading of NFL draft picks, where regular selections might be traded but those granted for compensatory reasons might not.)

The 1963 implementation of the NHL Amateur Draft signaled the last of the major North American professional sports' ability to monopolize talent by rich or large market teams. As with the other leagues, the NHL draft has undergone a number of changes. These include its renaming to the Entry Draft following the absorption of the defunct World Hockey Association in 1979 and varying of its format and length from a traditional sequential order to an occasional "zipper" or "snake" style rotation in which the decision order reverses at the conclusion of each round. Currently the NHL draft, like the NFL draft, is seven rounds (although there can be compensatory picks). Since 1995, the NHL draft, like the NBA draft, has incorporated a lottery aspect to reduce the incentive for intentional losing of games to improve the order of draft selection. Although disdained officially and vehemently denied by contemporary general managers as counter to the true spirit of competition and sport, strategic tanking ("fail for Nail" (Nail Yakupov 2012), "dishonour for Connor" (Connor McDavid 2015)) is a theme that seems to frequently resurface when teams are rebuilding or highly touted players enter the NHL draft.

Although this chapter focuses on hockey, draft analysis has been conducted across each of the four main North American sports. In the NFL, Massey and Thaler (2012) demonstrated that early selections were overvalued when compared to later choices, and Moskowitz and Werthheim (2011) and Schuckers (2011) provided an in-depth analysis of the Dallas Cowboys' exploitation of variance in decision-making ability and application of an NFL pick list to build a superior team at lower cost. Similarly, in MLB, Lewis (2003) described and popularized Billy Beane's analytic approach to player selection and encouraged use of the term "Moneyball" to describe the use of additional information and analytics to make player decisions. The effect of sunk costs—the idea that teams would make increased and disproportionate investments of time and development in players that they had selected by allocating increased playing time to the players selected earlier rather than later—has been examined and found to be present by both Staw and Hoang (1995) and Camerer and Weber (1999) in the NBA.

Decision analytics has had relatively slower adoption in the NHL, but the NHL is an ideal sport for the study of decisions and drafting because of four factors: its length, lack of trading restrictions, hard salary cap, and unique provision of league-wide scouting services (the Central Scouting Service or CSS).

The length or number of rounds in a draft has two implications. First, because drafting is inherently a zero-sum game—athletes selected by one team are unable to be selected by another—teams draw from what is a dwindling supply of talent as the number of rounds increases. As such, a short draft decreases the opportunities for teams to choose talent that might be misclassified. Second, given that prediction is imperfect, a larger number of rounds increases the probability that "late bloomers" have an opportunity to develop into career players (Burger and Walters 2009). At 2 rounds, the NBA draft places an inordinate importance on the pair of decisions each manager might make, and at 50 rounds the MLB draft attenuates decision-making and can result in cavalier choices made simply to complete the process. This situation was perhaps best manifested when the Buffalo Sabres' general manager Punch Imlach (exasperated by his belief that the 1974 NHL draft was too long) selected the nonexistent Taro Tsujimoto of the Japanese Hockey League's Tokyo Katanas as his 11th-round (183rd) pick. Shorted to 7 rounds in 2005, the NHL draft is a workable balance between the extremes of the 2 rounds of the NBA and the 50 rounds of MLB.

Although there is a prohibition against the outright sale of draft choices, something that has occasionally been creatively sidestepped by bundling choices with athletes with extremely limited prospects, drafting in the NHL is the least restrictive of the four leagues. This is important because such trading allows managers to better respond to their individual circumstances and market conditions. If, for example, a manager believes that a player who has been widely touted as the most highly desirable first pick is overvalued when compared to the second-ranked player, the manager holding the first pick can make the second choice after extracting the difference in value with a trade. Such a trade allows managers to side-step the pressure of making conforming choices they believe are incorrect or making discordant decisions that might alienate fans and the owners. Furthermore, because the exact order or value of same-year picks might not be known until the end of the season (and future picks not until several years later), trading picks form an additional layer of ambiguity and complexity that increases its applicability, relevance, and generalizability across environments. This has sometimes resulted in exchanges perceived to be surprising or lopsided, such as the Maple Leafs' then general manager Brian Burke's September 2009 trade of two first (Tyler Seguin and Dougie Hamilton) and one second-round pick (Jared Knight) for the Boston Bruins 2006 first round, fifth selection of Phil Kessel. This decision was essentially concluded in 2015 when the Maple Leafs retained some of the salary but traded Kessel to the Pittsburgh Penguins for two players originally drafted 22nd and 54th and a future first-round pick. Although it is a truism that no one knowingly makes a bad trade, what might first appear to be a good decision can later be viewed as suboptimal, and decisions that appear lopsided might later be perceived in an entirely different light.

Unlike the softer or flexible restrictions of the luxury taxes that exist in the NBA and MLB, the NHL imposes a relatively hard salary cap and rigid rookie salaries. Teams operate with both a minimum payroll floor and a maximum payroll ceiling with little flexibility to pay a fine or to bank and move capacity across years. Creative attempts to exploit loopholes are exemplified by the Vancouver Canucks' general manager Mike Gillis and "capologist" Laurence Gilman's 2009 front-loaded, 12-year, $64,000,000 contract with 30-year-old goalie Roberto Luongo, but such contracts were later prohibited. Salary innovation has been stifled, and the salary caps are so stringent that teams in danger of exceeding the maximum have been forced to reduce their payroll through trades or by maintaining a numerically smaller active roster than their opponents. Provisions in the NHL players' agreement that limit the maximum salary paid to athletes and impose restrictions on how much they can earn for the first part of their careers has allowed veteran salaries to increase and earn a disproportionate share. In most teams, 6 athletes, 30% of the total dressed for a game, command 60% of the available compensation and leave the remainder to be allocated among as many as 17 roster athletes.

Unique among North American professional leagues, the NHL operates its own captive scouting department as a service to member teams. Originally designed to reduce the impact of a scouting arms race among richer teams, the Central Scouting Service (CSS) provides baseline information to all teams on the top North American and European prospects. Although teams are able to augment this information through their use of private information and external scouting services, CSS has proven to be a trove of information to teams as well as an excellent training ground for scouts.

Considering the factors of draft length, absence of trading restrictions, salary cap, and league-wide scouting service, it is likely that the salary cap and rookie salary provisions have fuelled most of the focus on drafting. Although choosing great athletes has of course always been important in order to improve one's own team and to displace the talent available to others, the change in total player compensation from $200 million to $1 billion in the 10 years prior to the 2005 salary cap has exacerbated the salary disparity between players. Teams need the contractually limited compensation paid to rookies, a form of indentured servitude, to offset the higher salaries paid to franchise athletes. (The draft is open to all previously unclaimed North American players who have attained the minimum age of 18 years on September 15 and are younger than 20 years on December 31 of their draft year, and to non-North American players who are 21 or older and have played at least one North American season while aged between 18 and 20.) By some estimates each rookie player frees more than $900,000 in cap space. Far from an egalitarian environment, the NHL is increasingly becoming an environment that supports many minnows and few whales.

Based upon the relatively benign but twin assumptions that choosing earlier is preferable to choosing later and that general managers (representing scouts) are able to exploit these benefits and select more than their fair share of talent, the NHL draft is an eagerly awaited event filled with myths and populated with hope. Teams such as the Detroit Red Wings are said to have excellent scouting departments (Tingling et al. 2011) or be strong in particular areas, and scouts themselves have been "stolen" from one team by another. Still other teams such as the Chicago Blackhawks are said to use external advisors and psychological tools such as the Wonderlic test to assist with the interview and selection processes.* Although recognizing that some popular drafting myths can be true, academics have proffered theories that "drafting is no better than guessing" (Brydon and Tingling 2011a), journalists have called drafting a "coffee and caffeine fuelled crap shoot" (Joyce 2007) and "guesswork at best" (Dowbiggin 2006), and Central Scouting—arguably the most authoritative source for player evaluation—frequently characterizes aspects of the overall ranking outcomes as "luck" and "guesswork" (Tingling et al. 2011).

The idea that some teams have drafted well is not necessarily contradictory with the idea that drafting is largely random. Given an appropriate sample, evidence can be found to support both positions. Teams, scouts, and fans point to the impact that players have had on their teams and the extent to which teams jostle and trade to improve the order in which they pick as proof that both scouting and order matters. Every NHL player selected first between 1968 and 2012 has played in at least 160 games, and only three of the first picks (1964, 1965, and 1967) of all the drafts to date never played a single game. Similarly, late-round choices are cited as evidence that teams can knowingly identify talent despite the fact that such selection ignores the earlier bad choices that failed to play a single game. Each draft class consists of almost 60% that never play a single game in the NHL, and on average four athletes (such as Mark Giordano, Adam Oates, Ed Belfour, Dino Ciccarelli) that were passed over by the draft each year go on to sterling careers. (It is technically true that two of the greatest players of all time, Bobby Orr and Wayne Gretzky, were also undrafted, but this was due more to loopholes and quirks of the draft and expansion processes than poor decision-making. Those decision processes could be viewed as highly astute outcomes—at least on the part of the teams that benefitted.) General managers and scouts also complicate the analysis or discussion of draft success by citing idiosyncratic requirements or invoking game theory explanations that describe how their decisions are dependent upon the expected or actual draft decisions made by other teams or the cascading consequences of their own earlier decisions (Roenick and Allen 2012). Even the definition of what makes a successful draft decision has been contested. A player that never plays a single game is

---

* The validity and applicability of the Wonderlic to professional sports has been questioned. High-quality NFL athletes, such as Dan Marino and Terry Bradshaw, scored 15 and 16 out of 50 on the test, while conversely some high scorers have not had stellar careers.

unambiguously a poor choice, but a common refrain (and unarguably true) is that simply counting up the games played, points scored, or salary earned does not reflect a player's contribution or the decision-making ability of a scout. Mario Lemieux played fewer games than Tie Domi, but one was a much easier selection decision than the other. Another example, in what is often considered the worst NHL draft decision of all time (and said to have resulted in the rookie salary cap), is the Ottawa Senators choice to select Alexandre Daigle first in the 1993 entry draft. In perhaps a sign of hubris, Daigle contributed to this notoriety by saying, "I'm glad I got drafted first, because no one remembers number two" (Hall of Fame defenseman Chris Pronger).

Longitudinal research of drafting has failed to provide any evidence that a team has been able to consistently identify and select talent at a level that is better than guessing or random assignment of players across the length of the draft (Brydon and Tingling 2011a). It is clear that the NHL in general (and the Central Scouting Service in specific) are able to identify top talent among the first 45–60 selections of the approximately 210 made at the draft, but the ability to identify and differentiate players attenuates quickly and the rate of success (measured as a function of athletes playing in more than 160 games*) flatlines after the third round. In much the same way that mutual fund managers seem unable to outperform the market (i.e., the efficient market hypothesis (Fama 1970)), hockey teams, for the most part, have no greater sustained ability to identify talent and outperform the Central Scouting Service. Teams are able to benefit from and augment the information provided by the CSS, but no team has been empirically shown to out-decide others; instead, the occasional, remarkable success of a late-round pick is used as indication of superior ability rather than survivor bias. For many teams the axiom seems to be to try to avoid being unlucky when choosing from the riches-laden talent available in the early part of the draft and to hope to be lucky when choosing later.

Reviewing the nearly 6000 draft decisions made between 1981 and 2003 (Table 16.2) shows that less than half (41.7%) played at least one NHL game and slightly less than a half of those (20.1%) went on to a career of playing in at least 160 games.

The data in Table 16.3 tells an even more compelling story. The average yield from the 1981–2003 draft years shows a decline, measured as a percentage of the 30 players chosen in each round who played at least 160 career games, from 64.1% in the first round to 14.9% in the fourth round. The general decline might not be surprising, given that opportunities to select high-performing players are expected to diminish as the draft progresses, but the rate of attrition is surprising. It fiercely declines between the first and second rounds, it is stable for the second and third, and then it essentially flatlines for the fifth through eighth rounds (the sample frames predate the shortening of the draft to seven rounds). From about the 120th decision onward, there is no statistically significant difference in the yield. Decision-makers, it would appear, are able to identify those more likely to enjoy a career in the NHL for the first few rounds but then are unable to further differentiate athletes, and there is no substantive benefit from being able to make the 120th selection or the 210th. This is in stark contrast to the intraround variance (not illustrated) of the first round, in which the overall yield of 64.1% is made up of the statistically significant differences among four groups: the 100% success rate of the players drafted first who played 160 games, the 84.3% success rate for the first 10 choices, the 57.4% success rate for choices 11–20, and the 50.5% success rate for choices 21–30.

---

* Career success can be hard to define and, although the number of games played might seem to be arbitrary (good players can have abbreviated careers due to injuries), 160 games was the threshold used by the NHL to establish pension eligibility.

TABLE 16.2

NHL Draft Summary

NHL Draft 1981–2003	
Total number of players drafted	5981
Percentage of draft picks who never played in the NHL	58.3%
Number of draft picks having played only one NHL game	1.43%
Percentage of draft picks who played less than 10 NHL games	64.9%
Percentage of draft picks who played more than 160 games (full NHL Pension)	20.1%
Average number of games played by all players having played more than one game	290
Average number of games played by all players having played more than 160 games	556
Most games played by one player (Ron Francis)	1731
Average time for a first-round pick to play first NHL game	1.2 years
Average time for a fourth-round pick to play first NHL game	3 years
Average career length of a first-round pick having played at least one game (1978–1992)	11.5 years
Average career length of a first-round pick having played at least 160 games (1978–1992)	13.9 years

TABLE 16.3

Analysis of 1981–2003 NHL Draft Selections Playing 160 or More Games

Round	Draft Success Rates (%)	P-Values for Full Round Comparisons							
		1	2	3	4	5	6	7	8
1	64.1		0.00	0.00	0.00	0.00	0.00	0.00	0.00
2	28.6			0.49	0.00	0.00	0.00	0.00	0.00
3	23.5				0.01	0.00	0.00	0.00	0.00
4	14.9					0.94	0.94	0.04	0.29
5	11.9						1.00	0.61	0.96
6	11.9							0.61	0.96
7	7.2								1.00
8	9.0								

When presented with this finding, which is consistent with similar research conducted on the NFL, the overwhelming response from team executives and scouts falls into two camps. The first is, "These findings might be true of teams in general but are not true of our team, and, even if it is, we have already made (or are now in the process of making) changes to our drafting processes to address it. If you come back in three to four years, I am certain that you will see a marked improvement" (personal communication). Given that the average tenure of a general manager is 5.4 years and it takes, on average 3 years for a fourth-round pick to advance to the NHL, general managers can defer explanation for a large part of their careers before being called to account or terminated. Furthermore, a new general manager may intentionally make decisions using processes that are quite different from those made by his predecessor in order to demonstrate a highly visible break with the presumably poor decision-making that contributed to his predecessor's exit. General managers are rational and want to make changes that will improve decision-making processes and the long-term drafting process, but, as the Chinese parable suggests, "The generation that plants the tree is not the same one that gets the shade." Wanting and making positive change are desirable, but fans can be fickle, and it is hard to make long-tailed decisions when one is held to short-term goals. It might well be that teams are continually making changes to their draft processes, but—though hope springs eternal—there is no evidence that whatever changes general managers refer to making have in fact paid off.

The second response is more nuanced and harder to address. That is, when pressed to explain the reason that the average yield from a seven-round draft (including one from the high-yielding first round) is a paltry 1.7 career players, general managers and scouts frequently respond that this "is an anomaly and the result of idiosyncratic requirements." That is, the explanation continues, that rather than considering athletes as fungible or interchangeable commodities, the team makes draft decisions that consider their specific portfolio of requirements and they choose only players whom they believe fit or will fit their future direction. If true, this would be a perfectly viable explanation but it is subject to two constraints. (1) It neglects the fact that teams could select the best available player regardless of their immediate need and then trade players (as Vancouver famously did with Cam Neely). (2) It strains belief that scouts and managers can project the capability, potential, and personality of players still in their teens onto the complex and dynamic chemistry and requirements of their particular future team.

General managers and team executives have talked extensively about trying to determine other teams' potential draft decisions in order to improve their own decision-making (Roenick and Allen 2012), but this is an extremely complex and dynamic cognitive task further complicated by the short time window of the draft. Furthermore, even if the approach of selecting only athletes likely to fit with a team's future needs and culture were valid, the team's overall order of choice should remain intact. That is, it is tautological that a team following this strategy must consider that all of the athletes they select fit these criteria and that the order in which they select them should reflect their level of confidence in each decision. It follows therefore, as a reasonable assumption, that a team generally prefers the athlete that they have selected earlier or believe that the athlete selected earlier has a greater probability of a successful career than one that they have chosen later because, they would have otherwise arbitraged the opportunity to select at that position in order to maximize the utility of their choices. This would also be the case if the decision-maker is risk seeking and prefers "athlete b" over "athlete a" but expects "athlete b" to still be available should they choose "athlete a" first. A general manager might realistically claim that his choices are idiosyncratic, but it is much harder to claim that he prefers a player selected later over one selected earlier particularly if the decision-maker is risk averse. As noted earlier, decision order is perceived at least to be very important to general managers and they frequently make trades involving relatively small changes. At the 2010 NHL draft, for example, a general manager initiated a trade in order to advance four positions and be certain that his selection would still be available (personal communication). Following this strategy, players selected by teams following a "fit"-based methodology should exhibit ordinal performance. That is, an athlete selected by a team as its first selection should play in as many or more games than an athlete selected second by the same team, and the athlete selected second should in turn play in as many or more games than the athlete that the team selected third, and so on. The facts, however, testify to the contrary. In an analysis involving the 10 selections made by the Chicago Blackhawks at the 2003 draft,* Brydon and Tingling (2011b) showed that 40% failed to exhibit consistent ordinal performance, a process that Brydon and Tingling define as inversion and quantify as Ordinal Ranking Error (ORE).† Rather than the expected monotonic decline in quality, decision quality varied with poor-quality decision outcomes frequently appearing before higher-quality outcomes. For example, at the time of their analysis, Lasse Kukkonen, selected 151st with their fourth pick, had played 139 games while Michal Barinka, selected 59th with their third pick, had played 34 games. As Tingling and Brydon pointed out in "Better Off Guessing," a presentation (available online) at the 2013 Sloan Sports Analytics Conference, if the Blackhawks truly knew (or believed) that Kukkonen was a better decision that Barinka, why did they wait 91 decisions before selecting him in the 2003 draft and run the risk that another team would have the same insight? A similar situation existed with Dustin Byfuglien (selected 245th), although, as Brydon and Tingling acknowledge, selecting him was demonstrably more risky (though equally more profitable, given his level of physical conditioning). In their analysis, Brydon and Tingling noted that random assignment of the drafted players would result in a Bust Ordinal Ranking Error (BORE)—where an athlete playing zero games (a low bar) would be selected earlier than one playing 160 games or more—would occur 28% of the time but actually occurred in the NHL more than 39% of the time.

As Brydon and Tingling note, Ordinal Ranking Error is a promising way to measure team drafting effectiveness because it focuses on the decision actually made by a team, and its internal comparison and consistency eliminate the idiosyncratic decision claims made by NHL teams. But ORE does have three criticisms. The first is that the bust rate of players as a comparative measure (the almost 60% of drafted players who never play in a single NHL game) sets a very low threshold for success. To address this, Brydon and Tingling suggest a more realistic floor of career games (Career Ordinal Ranking Error or CORE), although they note this makes the team's decision-making abilities look even worse.

The second criticism is that Ordinal Ranking Error does not directly address drafting for position rather than best player available. This is almost certainly a factor when selecting goalies—pundits have suggested that a first round pick should never be wasted on a goalie—but drafting by position is a questionable and debatable strategy because projection is so difficult and assets can be easily traded. Despite these facts, however, many teams point to a draft-for-position strategy when explaining their poor rate of success. Either way, ORE can compensate for this strategy by considering the inversion only of positional decisions, a measure Brydon and Tingling referred to as Positional ORE (PORE). This would normally improve the reported success rate of teams but can occasionally require working with only one or two cases and be complicated by players who change positions (such as Byfuglien). Using PORE would not have affected the analysis of Chicago because all three—Barinka, Kukkonen, and Byfuglien—played defence.

The final criticism with the use of ORE as a single measure of a team's drafting ability is that teams whose drafting yields no career players (a situation that occurs at an alarming rate) would report a low inversion rate. To address this, Brydon and Tingling recommend

---

* The Blackhawks did not participate in this research and the researchers do not claim that the Blackhawks are following any particular drafting strategy, but whatever they are doing seems to be working. The Blackhawks had a stellar 2003 draft yield and won Stanley cups in 2010, 2013, and 2015.

† Regardless of the form Brydon and Tingling propose reporting ranking error as a percentage of the incidents or number of draft decisions where performance is inverted compared to the total number of draft decisions.

reporting the absolute number of drafts where none of a team's players completed 160 career games. The resulting calculations would illustrate the level of a team's drafting effectiveness by showing both the number of drafts where teams were unable to identify and select any career players and the relative risk and anomalies present in their decision-making shown by their choosing superior players later than their analysis would suggest. Interested readers might wish to read the paper in its entirety, but it is worth noting here that the Detroit Red Wings, consistently well regarded by fans and media for late-round drafting, were identified in their analysis as having both an ineffective drafting process and a high rate of Ordinal Ranking Error. Between 1995 and 2003, the Detroit Red Wings earned the worst relative draft performance with five drafts that yielded no career players. This contrasts starkly with the New York (Rangers and Islanders), New Jersey, Chicago, San Jose, St. Louis, Tampa Bay, and Toronto teams (which managed only two zero-career draft events for the same period) and Philadelphia, Ottawa, Montreal, Anaheim, Florida, Calgary, and Boston (which had one).

High-profile Detroit choices such as Jonathan Ericsson (selected 291 in 2002; 226 games), Henrik Zetterberg (selected 210 in 1999; 652 games), and Pavel Datsyuk (selected 171 in 1998; 721 games) have indeed made significant contributions to Detroit's success, and it might seem counterintuitive to criticize a team for "brilliant" late-round drafting; however, this is in fact the case if they do not exhibit such success through all rounds of the draft or do not arbitrage picks that they believe are overvalued. This is precisely the allure of the BORE approach; if Detroit was indeed brilliant and knew that Pavel Datsyuk would become a dominant player, why did they give 29 other teams five or six opportunities each to draft him? A simpler explanation is that no team fully appreciated Datsyuk's potential but Detroit was lucky enough to pick him. Accordingly, a large number of BOREs is not a reliable predictor of team underperformance; instead, it is an indication of the significant role chance plays in the drafting process.

That much of the NHL draft is indistinguishable from chance does not diminish the question of why it is the case. It is certainly true that such decision-making is hard; picking future NHL players among teenagers has been compared at times to identifying Nobel Prize winners among junior high school chemistry students. What is less well known is if this is because of flaws endemic to the process or individual scouts. It is clear, for example, and entirely consistent with natural statistical variability, that if talent represents the average of many components that its distribution, as suggested by the Central Limit Theorem, would be nearly normal. As such the first few rounds of the draft would obviously yield higher rates of success than the later rounds but a 5% probability difference in tails corresponds to a bigger change in talent/performance than a 5% probability difference in the middle of the distribution. Why this is the case, however, is debatable. Two possible factors are the complexity of scouting and the virtuous cycle of greater investments made by teams in the athletes that they have selected in the early rounds. Observing hundreds of athletes over a season and compiling reports that assess their capabilities on a number of dimensions is cognitively taxing. Scouts might consciously or subconsciously focus on the top 100, a number that is both consistent with constraints of the neocortex (Dunbar 1992) and reinforced by the showcasing of the top 100 amateurs at the Central Scouting annual combine. Similarly, being drafted in the early rounds might result in a virtuous cycle of training, feedback, and support that reinforces the original selection. This focus on the sunk costs has been observed in the commitment to play athletes who have been chosen early but could also affect how athletes are supported and developed.

Scouting departments vary considerably in size and makeup. Although more affluent teams have a greater number of scouts (such as the Vancouver Canucks with 20 and the

Toronto Maple Leafs with 23) and less well-off teams have fewer (such as the Nashville Predators with 9), there has been little examination of the extent to which adding more personnel and resources have resulted in better decisions (cf. Tingling and Masri (2010) for a preliminary study). Such analysis, however, is complicated by the fact that scouting departments do more than simply scout and are often used as a placeholder for a future executive, former player, or nepotistic hiring. A surprising number of scouts share filial connections.

Group think, the idea that people tend to confirm their own opinions, has long been cited as a reason for scouts to fail in making optimal decisions and to discard disconfirming evidence (Janis 1972). The majority of teams (and the Central Scouting Service) use an open forum for drafting recommendations, making it difficult to disagree with the collective wisdom, especially among a mix of more senior or respected staff. In many cases it may be better to conform and run the risk of being collectively wrong than to make an individual recommendation where the outcome might be many years hence (as Keynes observed "Worldly wisdom teaches that it is better for reputation to fail conventionally than to succeed unconventionally" (Keynes 1936) and memories may fade). Bruce Dowbiggin, describing the decision by the Ottawa Senators to select Alexandre Daigle, noted, "Not everyone on the Sens scouting staff was ga-ga about Daigle; some wanted to draft hulking defenseman Chris Pronger from Peterborough, or a shifty forward from the University of Maine, Paul Kariya." However this is one of the few documented instances that question a consensus pick, and it is not clear from Dowbiggin's account who the scouts with the dissenting opinion were and if their other drafting recommendations exhibited similar accuracy or if the problem is worse in larger departments. In some franchises, notably the Detroit Red Wings, a single scout such as Håkan Andersson has been acknowledged as having a large influence over scouting decisions and individually credited with many of the key decisions. The question then is, "Do the Red Wings have a good scouting department or a good scout?"

Although technology has been used to improve drafting decisions (e.g., video scouting), a few teams have experimented with advanced group decision systems but these approaches have been hampered by the low quality of data. In an interesting but as yet unpublished analysis conducted by the author, substantial variation in individual scouting ability was found in a 1983–1995 sample of more than 14,000 professional scouting reports provided by NHL teams that were matched to the now-known professional player performance after the related draft period. While there was variation in the specific players observed and the number of times that they were seen (better players receiving more observations for confirmatory reasons), among a group of 13 scouts the ability to predict career NHL success (defined as playing 160) games was six times higher for the best scout than the worst. For 3 of the 13 scouts, the ability to predict career success was no better than the overall draft itself, an ordering that has been shown to lack discrimination beyond the first 100 or so athletes (refer to Figure 16.1). It might be the case, therefore (and illustrated by the Detroit Red Wings Håkan Andersson), that individual scouts are able to identify talent but, in a reversal of *The Wisdom of Crowds* (Surowiecki 2004), this information signal becomes lost in the cacophony of decision-making. In general, despite the fact that drafting has grown in importance, research has shown that for the most part teams have been slow to adopt innovative decision-making ideas. Instead, like money managers, most scouts publically state that they beat the index (in this case CSS), point to their one or two high-profile, late-round "steals," and do not objectively measure their historical performance and recommendations. Each year brings a clean slate of optimism and hope.

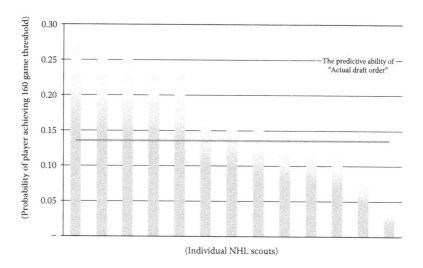

FIGURE 16.1
Predictive ability of scouts to identify career threshold amateur players using composite measures.

As noted in the opening of this chapter, sports is an excellent venue for conducting research because of its relevancy, transparency, and allowances for generalizability and counterfactual testing. Beyond this, however, the NHL in particular offers further opportunity because of the presence of Central Scouting and the seven-round draft. Although drafting in aggregate has been unambiguously shown to be a little better than guessing, especially in later rounds, there are still many unanswered questions. Analytics has come late to the NHL and it is an open question as to whether this will also move to off-ice decision-making and if teams' drafting will improve. Research into how draft decisions are made and how they can be improved is immensely useful, especially with regard to predictive or explanatory variables and how amateur players can be compared across varying junior leagues. As a key feature of the draft, the Central Scouting Service offers significant opportunities for research in that it can be considered as an overall benchmark that operates as a form of information market. Accordingly, the extent to which various teams are able to improve upon or detract from the baseline information provided by CSS is a real-world opportunity to analyze the extent to which private information improves decision-making. The average education of NHL general managers is increasing and this presents a natural experiment to see the positive or negative impact education has on decision-making (measured absolutely by yield rather than by proportional share, which might be expected to remain constant because drafting in aggregate is zero-sum). The depth of each year's draft, measured by the number of athletes that have career impact, has traditionally been described anecdotally and ex ante rather than ex post, and research that is able to assist with prediction of a particular draft year's yield would be highly useful—as would a longitudinal comparison of draft years (though it might quell some journalistic rhetoric). The relationship between trading and performance has also been underexamined in sport, both in drafting and during the regular season. In particular, there has been little empirical analysis of the extent to which trading up or down is a useful strategy, arbitrage is present, and little examination has been undertaken of the relative value of the various picks beyond simple probability of career success. At the level of individual players, few researchers have examined what differentiates one scout from another and the extent to which scouting is an intuitive or analytic process. The lower costs and growing acceptance of wearable technology to track player telemetry, or the unconscious eye movement of scouts watching players, or fMRI machines for use during decision-making offer the opportunity to provide true insight into high-stakes personnel decision-making with far-ranging implications. Given that decision-making is the essence of management and the growing importance of scouting, these are worthy questions.

References

Brydon, M. and P. M. Tingling (2011a). *Aiming for the Mean: Why Isn't Drafting Better than Guessing?* Administrative Sciences Association of Canada, Montreal, Quebec, Canada.

Brydon, M. and P. M. Tingling (2011b). Better off guessing: Measuring the quality of draft decisions. *MIT Sloan Sports Analytic Conference (SSAC)*, Boston, MA.

Burger, J. B. and S. J. K. Walters (2009). Uncertain prospects rates of return in the baseball draft. *Journal of Sports Economics* **10**(5): 485–501.

Camerer, C. F. and R. A. Weber (1999). The econometrics and behavioral economics of escalation of commitment: A re-examination of Staw and Hoang's NBA data. *Journal of Economic Behaviour and Organization* **39**(1): 59–82.

Dowbiggin, B. (2006). *Money Players. The Amazing Rise and Fall of Bob Goodenow and the NHL Players Association*. Key Porter Books, Toronto, Ontario, Canada.

Dunbar, R. (1992). Neocortex size as a constraint on group size in primates. *Journal of Human Evolution* **22**(6): 469–493.

Fama, E. (1970). Efficient capital markets: A review of theory and empirical work. *Journal of Finance* **25**(2): 383–417.

Janis, I. L. (1972). *Victims of Groupthink*. Houghton Mifflin, Boston, MA.

Joyce, G. (2007). *Future Greats and Heartbreaks: A Year Undercover in the Secret World of NHL Scouts*. Doubleday, Toronto, Ontario, Canada.

Keynes, J. M. (1936). *The General Theory of Employment, Interest and Money*. Harcourt Brace, New York.

Lewis, M. (2003). *Moneyball*. WW Norton, New York.

Massey, C. and R. H. Thaler (2012). *The Loser's Curse: Decision Making & Market Efficiency in the National Football League Draft*. Available at SSRN: http://ssrn.com/abstract=697121.

Moskowitz, T. J. and L. J. Werthheim (2011). *Scorecasting: The Hidden Influences Behind How Sports Are Played and Games Are Won*. Crown Archetype, New York.

Roenick, J. and K. Allen (2012). *J.R.: My Life as the Most Outspoken, Fearless, and Hard-Hitting Man in Hockey*. Triumph Books, Chicago, IL.

Schuckers, M. (2011). An alternative to the NFL draft pick value chart based upon player performance. *Journal of Qualitative Analysis in Sports* **7**(2): 1–12.

Staw, B. M. and H. Hoang (1995). Sunk costs in the NBA: Why draft order affects playing time and survival in professional basketball. *Administrative Science Quarterly* **40**(3): 474–494.

Surowiecki, J. (2004). *The Wisdom of Crowds: Why the Many Are Smarter Than the Few and How Collective Wisdom Shapes Business, Economies, Societies and Nations*. Doubleday, New York.

Tingling, P. M. and K. Masri (2010). *Feast or Famine: Does Wealth Help or Hinder Innovation in Sport?* Administrative Sciences Association of Canada (ASAC), Regina, SK.

Tingling, P. M., K. Masri, and M. Martell (2011). Does decision order matter? An empirical analysis of the NHL draft. *Sport, Business & Management: An International Journal* **1**(2): 155–171.

17

Models for Outcomes of Soccer Matches

Phil Scarf and José Sellitti Rangel Jr.

CONTENTS

17.1 Introduction

Soccer (or association football) is the most popular sport in the world, with an estimated 3.5 billion fans and 250 million players worldwide (Sporty Desk, 2015). Betting on outcomes in soccer matches is also very popular, unsurprisingly, and the value of the soccer betting market in 2012 is estimated to be between £500 billion and £700 billion (Keogh and Rose, 2013). Consequently, statistical modelling of outcomes in soccer matches is popular among researchers, both in academia and industry, not only for the potential for financial returns but also for the challenges that such modelling presents. This is not to say that betting drives all research in statistical modelling in soccer and many interesting problems relating to tactical questions (e.g. Wright and Hirotsu, 2003; Hirotsu and Wright, 2006; Brillinger, 2007; Tenga et al., 2010; Titman et al., 2015); team, player and manager rating (e.g. Knorr-Held, 2000; Bruinshoofd and Weel, 2003; Schryver and Eisinga, 2011; Baker and McHale, 2015); competitive balance and outcome uncertainty (e.g. Koning, 2000; Buraimo and Simmons, 2015); match importance (e.g. Scarf and Shi, 2009; Goossens et al., 2012); tournament outcome prediction (e.g. Koning et al., 2002; Groll et al., 2015) and tournament design and scheduling (e.g. Scarf et al., 2009; Goossens and Spieksma, 2012; Scarf and Yusof, 2011; Lenten et al., 2013) have been studied. Nonetheless, modelling results and scores, and other in-match outcomes, both straightforward (e.g. first player to score) and unusual (e.g. number of player cautions), motivated by the search for betting market inefficiency, have been a major motivational factor in the development of the state of the art. In this chapter, our aim is to describe the state of the art in the statistical modelling of match results and scores in particular.

Statistical modelling developments have important implications for both sides of the market. On one hand, bookmakers use them to help set their odds competitively, so that they are both interesting to punters and profitable for the bookmaker. On the other hand, punters may use them to seek opportunities for profit. Given this importance, one could argue that research in modelling soccer match outcomes is to bookmakers and punters as financial mathematics and macroeconometrics is to stock markets.

Following the book (later turned movie) *Moneyball* that discusses the use of statistical analysis by Oakland Athletics MLB for player valuation, which allowed Athletics to be competitive with teams with much greater payrolls, soccer clubs have begun to use statistical modelling, particularly in player performance evaluation, and several top teams around the world now have an analytics team. There are however significant challenges posed by the continuous nature of the 'beautiful game', the significant interactions between players, and the multifaceted and dynamic roles that individual players assume as they take on more or less attacking and defensive responsibilities (McHale et al., 2012). This is an important strand of modelling research in soccer. However, we do not consider it in this chapter.

Match outcome models for soccer can be split into two classes: direct and indirect models. The former directly predict the outcome of a match, as a win, draw or loss, and ordered probit and logit models are generally used (Koning, 2000; Goddard, 2005). The latter indirectly model the bivariate score (goals-for and goals-against), to which a bivariate probability distribution is attributed. Given that the number of goals is a non-negative integer, count data models such as the Poisson or negative binomial distributions are implemented and the seminal papers are Maher (1982) and Dixon and Coles (1997). Other approaches have also been investigated such as models for goal differences through the Skellam distribution (Karlis and Ntzoufras, 2009), models which predict shots for and against (McHale and Scarf, 2007), and ratings model such as ELO (Hvattum and Arntzen, 2010) and pi-rating models (Constantinou and Fenton, 2013). This chapter will describe the direct and indirect models, with emphasis on the latter, especially because these are both richer in the modelling questions that they pose and the applications they underpin.

That indirect models have taken over from direct models appears to be the case, not least because to our knowledge the most recent paper published in the statistical literature that uses the direct approach is Goddard (2005). This shift in interest is probably as a result of the many challenges associated with indirect models that open the way for further research. These challenges include the issue of dependence within the bivariate distribution of goals scored in a match; the modelling of the dynamics of attacking and defensive strengths; the contribution of players' strengths to team strengths; count data model selection. We believe that the main challenges ahead of the literature lie in modelling player and team strength dynamics. The pioneering work here is Knorr-Held (2000) and Rue and Salvesen (2000). The way we structure this chapter reflects this view: the direct models are presented followed by indirect models. Then we describe the literature on team strength dynamics and we conclude with suggestions for future research direction.

17.2 Direct Match Outcome Models

Direct measurement of the result as win, draw or loss (from the point of view of the reference team) leads to the class of generalised linear models suitable for a categorical response variable. To model a dependent variable of this type, it is common to use an ordered response model with an appropriate link function, which links the probabilities for each category to the linear combination of model parameters and covariates. For an ordered probit, the probit link function is used, and for ordered logistic regression the logit link function is used. The idea is that a latent continuous variable exists (but is not observed) which underlies the ordered responses (which are observed) and cut-points divide the real line into a series of intervals corresponding to the categories of the ordered response. For soccer match outcomes, we have three categories (win, draw or loss) and two cut-points. Thus, let y^* be the latent variable that is some linear combination of explanatory variables plus error, denoted

$$y^* = X'\beta + \varepsilon, \tag{17.1}$$

where $\varepsilon \sim N(0,1)$. The latent variable y^* takes values Y as follows

$$Y = \begin{cases} 1(\text{win}) & \text{if } \delta_1 \leq y^* < +\infty \\ 0(\text{draw}) & \text{if } \delta_{-1} \leq y^* < \delta_1 \\ -1(\text{loss}) & \text{if } -\infty \leq y^* < \delta_{-1} \end{cases} \tag{17.2}$$

where δ_{-1} and δ_1 are the cut-points. Thus Y has a trinomial distribution with probabilities p_{-1}, p_0 and p_1 ($p_{-1} + p_0 + p_1 = 1$) obtained by substituting (17.1) into (17.2):

$$\begin{aligned} p_{-1} = \Pr(Y = -1) = \Pr(\text{loss}) &= \Pr\left(-\infty \leq X'\beta + \varepsilon < \delta_{-1}\right) = \Pr\left(X'\beta + \varepsilon < \delta_{-1}\right) \\ &= \Pr\left(\varepsilon < \delta_{-1} - X'\beta\right) = \Phi\left(\delta_{-1} - X'\beta\right) \\ p_0 = \Pr(Y = 0) = \Pr(\text{draw}) &= \Pr\left(\delta_{-1} \leq X'\beta + \varepsilon < \delta_1\right) \\ &= \Pr\left(\delta_{-1} - X'\beta \leq \varepsilon < \delta_1 - X'\beta\right) = \Phi\left(\delta_1 - X'\beta\right) - \Phi\left(\delta_{-1} - X'\beta\right) \\ p_1 = \Pr(Y = 1) = \Pr(\text{win}) &= \Phi\left(+\infty - X'\beta\right) - \Phi\left(\delta_1 - X'\beta\right) = 1 - \Phi\left(\delta_1 - X'\beta\right). \end{aligned}$$

Here $\Phi(\cdot)$ is a cumulative distribution function (CDF) of the standard normal distribution. The parameters of the model are typically estimated by maximising the log-likelihood function. The vast majority of econometric and statistical software packages have in-built functions for this purpose.

In the ordered logistic (or logit) model, often called the proportional odds model, the log odds of the ordered outcomes are assumed to be linearly related to the covariates and parameters:

$$\log\left\{\frac{\Pr(\text{win})}{1 - \Pr(\text{win})}\right\} = \log\left\{\frac{p_1}{1 - p_1}\right\} = \alpha_1 + X'\beta \tag{17.3}$$

$$\log\left\{\frac{\Pr(\text{win or draw})}{1 - \Pr(\text{win or draw})}\right\} = \log\left\{\frac{p_1 + p_0}{1 - p_1 - p_0}\right\} = \alpha_0 + X'\beta. \tag{17.4}$$

Thus

$$\Pr(\text{win}) = \text{logit}^{-1}\left(\alpha_1 + X'\beta\right),$$

$$\Pr(\text{win or draw}) = \text{logit}^{-1}\left(\alpha_0 + X'\beta\right),$$

where $\text{logit}^{-1} : \Re \to [0, 1]$ such that $\text{logit}^{-1}(x) = \exp(x)/(1 + \exp(x))$. Although this model is very similar to the ordered probit model, it is not generally used for modelling match outcomes in soccer. Nonetheless, the model-fit diagnostics are often very similar to the ordered probit.

The motivations for the use of direct models vary. Forrest and Simmons (2000a), for example compare tipster performance in predicting match outcomes, whereas Forrest and Simmons (2000b) compare Pools Panel football match decisions with actual results. Audas et al. (2002) analyse the impact of within-season managerial change on team performance in English football. Forrest and Simmons (2002) study the impact of competitive balance on football attendance. Koning (2000) estimates team strengths and home advantage for the purpose of analysing competitive balance in Dutch football. Dobson and Goddard (2003) study short-term persistence in sequences of match results to conclude that negative persistence effects exist, implying that series of consecutive wins and consecutive losses tend to end sooner than expected if there was no persistence. Goddard and Asimakopoulos (2004) examine the predictability of English league football results and utilise several explanatory variables including match significance for championship, promotion or relegation issues and geographical distance between home and away teams. They test for weak-form inefficiency in the betting market and conclude that their forecasting model contains additional information not included in the bookmakers' odds. Finally, Goddard (2005) compares the forecasting performance of four different direct models using a pseudo-likelihood statistic and concludes that the best performing model is a 'hybrid' model – an ordered probit regression with results-based independent variables and goals-based lagged performance covariates. This model outperforms the other models in 6 out of 10 seasons. Nevertheless, they did not implement the model to produce *ex ante* forecasts or test the percentage return on bookmakers' odds, such as in Goddard and Asimakopoulos (2004).

Regarding the interpretation of the parameter estimates, since the probit and logistic models both apply a non-linear link function, the coefficients in β do not correspond with the linear interpretation as in an ordinary least squares regression. One has to compute marginal effects using

$$\frac{\partial E[y^*|x]}{\partial x} = \left\{ \frac{dF(x'\beta)}{d(x'\beta)} \right\} \beta = f(x'\beta)\beta \tag{17.5}$$

For the probit model, $f(x'\beta)\beta = \phi(x'\beta)\beta$, so that a marginal change in one of the explanatory variables depends on the values that the explanatory variable takes.

Research studies into direct outcome models for forecasting football matches have, since Goddard (2005), not really developed. This is perhaps because indirect match outcomes have presented greater modelling challenges. Additionally, modelling exact scores and goal differences has wider applications in betting, not least to spread betting.

17.3 Indirect Outcome Models

Moroney (1956) observed that football scores were fitted better by a negative binomial than a Poisson distribution. Nonetheless, Maher (1982) proposed a double-Poisson model that is considered the backbone for all indirect outcome models for modelling soccer scores. In this seminal paper, Maher estimates the attack and defence parameters of teams and a home advantage parameter. The model is as follows: consider team i playing at home against team j, and let X_{ij} and Y_{ij} denote the goals scored by the home and away teams, respectively. Then

$$X_{ij} \sim Poisson(\alpha_i \beta_j \gamma) \tag{17.6}$$

$$Y_{ij} \sim Poisson(\alpha_j \beta_i) \tag{17.7}$$

with X_{ij} and Y_{ij} independent. In the first Poisson mean in (17.6), the parameter α_i is interpreted as the attacking strength of the home team i (tendency to score goals), β_j is the defensive weakness of the away team j (tendency to concede goals) and γ is the home advantage effect. These three parameters determine the goal-scoring rate of the home team (X_{ij}). The parameters that determine the goal-scoring rate of the away team (Y_{ij}) in (17.7) are α_j, the attack strength of the away team j and β_i, the defensive weakness of the home team i. In the simplest variation of this model, the home advantage effect is the same for all teams. Maher used the method of maximum likelihood to estimate the parameters and noted that the model underestimated the number of draws. This motivated Dixon and Coles (1997) to enhance the model by adding a dependence parameter:

$$P\left(X_{ij} = x, Y_{ij} = y\right) = \tau_{(\lambda,\mu)}(x,y) \frac{\lambda^x \exp(-\lambda)}{x!} \frac{\mu^y \exp(-\mu)}{y!} \tag{17.8}$$

where

$$\lambda = \alpha_i \beta_j \gamma \tag{17.9}$$

$$\mu = \alpha_j \beta_i \tag{17.10}$$

and

$$\tau_{(\lambda,\mu)}(x,y) = \begin{cases} 1 - \lambda\mu\rho & \text{if } x = y = 0 \\ 1 + \lambda\rho & \text{if } x = 0, y = 1 \\ 1 + \mu\rho & \text{if } x = 1, y = 0 \\ 1 - \rho & \text{if } x = y = 1 \\ 1 & \text{otherwise} \end{cases} \tag{17.11}$$

so that ρ is the dependence parameter that modifies the independence assumption for low-scoring matches ($x \le 1, y \le 1$).

With N teams, there are $2N + 2$ parameters in this model in total. The model is over-parameterised because, under the model specification, strength can only be measured relatively (that is, one can only know for example whether A is stronger than B, and one cannot know how strong is A). Therefore, the parameter space must be constrained, and Dixon and Coles (1997) use the constraint

$$N^{-1} \sum_{i=1}^{N} \alpha_i = 1. \tag{17.12}$$

The authors acknowledged that the strength parameters are only locally constant, developing a pseudo-likelihood function for estimation in which the outcomes of more recent matches are given greater weight. Thus

$$L\left(\alpha_i, \beta_i, \rho, \gamma : \text{all } i\right) = \prod_{k=1}^{J} [\tau_{(\lambda,\mu)}(x_k, y_k) \exp(-\lambda)\lambda^{x_k} \exp(-\mu)\mu^{y_k} S]^{\phi(t-t_k)} \qquad (17.13)$$

where
 (x_k, y_k) is the result of match k played at time $t_k < t$
 J is the number of matches in the sample
 ϕ is a non-increasing function of time

In fact,

$$\phi(t) = \exp(-\xi t),$$

where ξ was chosen to maximise the prediction of match outcomes and $\hat{\xi} = 0.0065$. A large ξ puts large weight on recent matches and $\xi = 0$ is the static model, with all matches given equal weight.

Dixon and Robinson (1998) developed a model in which scoring rates depend on the current match state and proposed a bivariate non-homogeneous Poisson process for goals scored, with the intensities for home and away goals varying linearly with match time t, and strength parameters contingent on the current score (x, y):

$$\begin{aligned}
\lambda_i(x, y, t) &= \alpha_i(x, y)\beta_j(x, y)\gamma + \xi_1 t, \\
\mu_j(x, y, t) &= \alpha_j(x, y)\beta_i(x, y) + \xi_2 t
\end{aligned} \qquad (17.14)$$

They conclude that goal-scoring rates increase for both teams throughout a match and when a goal is scored by the opposition. This model is most useful for in-play betting, but they demonstrated an improvement in match outcome prediction compared to the models of Maher (1982) and Dixon and Coles (1997).

The next interesting development is due to Karlis and Ntzoufras (2003). They describe a bivariate Poisson distribution that captures positive dependence in home and away goals. Consider the (pairwise) independent Poisson random variables $Z_k, k = 1, 2, 3$ with means $\lambda_1, \lambda_2, \lambda_3$, and form the bivariate pairs $X = Z_1 + Z_3$ and $Y = Z_2 + Z_3$. Then

$$P_{X,Y}(x, y) = P\left(X = x, Y = y\right)$$

$$= \exp\left\{-(\lambda_1 + \lambda_2 + \lambda_3)\right\} \sum_{k=0}^{\min(x,y)} \frac{\lambda_1 \lambda_2 \lambda_3}{(x-k)!(y-k)!k!} \qquad (17.15)$$

where

$$E(X) = \lambda_1 + \lambda_3$$
$$E(Y) = \lambda_2 + \lambda_3 \qquad (17.16)$$
$$\lambda_3 = \text{cov}(X, Y).$$

Karlis and Ntzoufras (2003) also 'diagonally inflate' the model to increase the probability of observing draws. This is specified by

$$P_D(x,y) = \begin{cases} (1-p)BP(x,y|\lambda_1,\lambda_2,\lambda_3), & \text{if } x \neq y \\ (1-p)BP(x,y|\lambda_1,\lambda_2,\lambda_3) + pD(x,\theta) & \text{if } x = y \end{cases} \qquad (17.17)$$

where $D(x,\theta)$ is a discrete distribution with parameter vector θ. They recommend to use a simple discrete distributions for D. The Poisson marginal means then become

$$\begin{aligned} E(X) &= (1-p)(\lambda_{1i} + \lambda_{3i}) + p\theta_1 \\ E(Y) &= (1-p)(\lambda_{2i} + \lambda_{3i}) + p\theta_1 \end{aligned} \qquad (17.18)$$

where

$$\begin{aligned} \lambda_{1i} &= \mu + \text{home} + \text{att}_{hi} + \text{def}_{gi} \\ \lambda_{2i} &= \mu + \text{att}_{gi} + \text{def}_{hi} \\ \lambda_{3i} &= \beta^{SS} + \gamma_1 \beta_{hi}^{home} + \beta_{gi}^{away}. \end{aligned} \qquad (17.19)$$

Karlis and Ntzoufras (2003) applied the diagonally inflated bivariate Poisson model to the Italian Series A league to capture the underpredicted number of draws observed in the data set, particularly the 1–1 scores. A possible explanation for the tendency for many draws in this league (Italian Serie A, 1991–1992 season) is that the scoring system of 2–1–0 for win–draw–loss was still in place. They compared the fits of the double-Poisson, the bivariate Poisson and the diagonal inflated models for a variety of diagonal distributions. The best fitting model according to AIC, BIC and LRT was the bivariate Poisson model with an extra parameter for 1–1 draws.

An issue of concern regarding these bivariate models is that they only permit non-negative correlation. Since λ_3 is a Poisson mean, it cannot be negative. A solution was offered by McHale and Scarf (2011) who used particular copula functions that allow negative dependence. In application to international football, these authors found that goals scored in a match (X_{it}, Y_{jt}) are negatively dependent. In general, a copula provides a flexible means of joining marginal distributions to form multivariate distributions with interesting dependence structure, although some care is required with discrete distributions. The copula function itself is a multivariate distribution with all univariate marginal distributions as $U(0,1)$. Hence C is the distribution of a multivariate uniform random vector. For a bivariate distribution F with margins F_1 and F_2, the copula associated with F is a distribution function $C : [0,1]^2 \rightarrow [0,1]$ that satisfies

$$F(x,y) = C\{F_1(x), F_2(y)\}, (x,y) \in \Re^2. \qquad (17.20)$$

McHale and Scarf (2011) makes use of extendable Archimedean copulas that can model both positive and negative dependence. They fit three different copulas and showed that Frank's copula, for which

$$C(u,v) = -\kappa^{-1} \log \left\{ 1 - \frac{(1 - e^{-\kappa u})(1 - e^{-\kappa v})}{1 - e^{\kappa}} \right\}, (\kappa \in \Re), \qquad (17.21)$$

provides the best fit to the international soccer data set. While copulas can capture general dependence structure, often at the expense of estimating only one additional parameter, it is not clear if their use in model specification is preferred to the direct dependence models of Dixon and Coles (1997) and Karlis and Ntzoufras (2003).

What is clear however is that the hot topic in the literature at present is strength dynamics, and although it is some decades since Dixon and Coles (1997) acknowledged strength dynamics but did not really model them (instead they proposed rolling, fitting and forecasting approach in which recent matches were given more weight in strength estimation), the important contributions to this topic have been much more recent. It is this literature that we will discuss next.

17.4 Strength Dynamics

Strength dynamics can be modelled either deterministically or stochastically. Focussing first on the former, Baker and McHale (2015) use a smoothly varying function (barycentric rational interpolation) to specify attack and defence strength for all time, and while their model can be used for short-term forecasting, their motivation is the identification of the historical strongest team. In the aforementioned paper they discuss in particular the strongest club side in England and Wales between 1888 and 2012. These authors have had success with their approach in identifying the all-time best not only in soccer but also in tennis and golf (Baker and McHale, 2014). The advantage of the deterministic approach is that the likelihood function is both easier to specify and to maximise.

Modelling team strength dynamics stochastically has received considerably more attention in the literature and Knorr-Held (2000) and Rue and Salvesen (2000) and later Crowder et al. (2002) pioneered this research. Several different stochastic processes have been used to model dynamics in team strengths, and these models can be estimated using either classical or Bayesian methodologies. A natural place to start when modelling dynamics stochastically is to use some type of autoregressive process, where a team's strength at time t is related to its strength at some time $t - s, s > 0$. Knorr-Held (2000) considered that time-dependent abilities α_i followed a random walk $\alpha_{i,t} \sim N(\alpha_{i,t-1}, \sigma^2)$, where σ^2 was estimated by maximising the in-sample predictability of the model with respect to observed final team rankings. The model was implemented on the 1996–1997 season of the German Bundesliga. As no forecasts were made and no comparison with a benchmark model carried out, there is no indication as to how well the model can forecast results.

In Rue and Salvesen (2000), strength parameters for attack and defence follow a Brownian motion, a stochastic process in continuous time. These authors also include in the Poisson marginal mean, as a new development, a covariate that they call the psychological effect (denoted γ) of the superior team underestimating the strength of the weaker one. Then, letting t' and $t'' > t'$ denote two different points in time, the attack strength dynamics is such that

$$\alpha_i^{t''} - \alpha_i^{t'} = \left\{ B_{\alpha,i} \left(\frac{t''}{\tau} \right) - B_{\alpha,i} \left(\frac{t'}{\tau} \right) \right\} \frac{\sigma_{\alpha,i}}{\sqrt{\{1 - \gamma(1 - \frac{\gamma}{2})\}}} \tag{17.22}$$

where $B_{,,}(t), t \geq 0$ is a standard Brownian motion, the subscript marks denoting attack (subscript β for defence) and team. The time parameter τ is a scaling factor identical for all teams and specifies the inverse loss of memory rate for α_i^t, $\text{var}(\alpha_i^{t''} - \alpha_i^{t'}) \propto \sigma_{\alpha,i}^2/\tau$. The model was estimated through Bayesian Markov chain Monte Carlo. The model used data from the first half of the 1997–1998 Premier League and Division 1 English league and was shown to perform as well as the bookmaker in a simulated betting experiment. In this experiment, they placed bets to maximise a particular utility function: betting on matches which would give a positive expected profit, placing the bets with low variance of profit. They obtained a final return of 39.6%, winning 15 bets out of 48 placed in the English Premier League and 54.0% in Division 1 (27 out of 64). However, the lower bounds of the 95% confidence interval of the betting returns were negative, indicating there was still a risk of losing money. They also showed that combination bets gave lower returns.

There then followed something of a lull in model developments until Owen (2011) presented a dynamic generalised linear modelling framework that allows some or all parameters to be time dependent. He identified the challenge of estimating the 'evolution' variance σ^2 (a volatility parameter in the strength dynamics whereby a higher evolution variance implies more volatile strength dynamics). He estimated this parameter by maximising the one-step ahead predictive probability of the model:

$$P_1 = \exp\left\{\frac{1}{N}\sum_{K=1}^{N}\log_e\left[P(O_k)\right]\right\}, \tag{17.23}$$

which is equivalent to the geometric mean of the one-step ahead match predictive probabilities that were actually observed, $P(O_k)$ being the one-match ahead predictive probability that match k would end with the outcome O_k. However, Owen made use of a cumulative measure of $P_1 : P_1(t)$ which included all matches played up to and including round t, and as a result, allows the parameter σ^2 to be updated as more information becomes available. He also used another measure of predictive performance, P_2, a quadratic loss function or a measurement error, as well as its cumulative counterpart $P_2(t)$, which incidentally was also used in Knorr-Held (2000). Owen used three seasons of the Scottish Premier League (2003–2004 to 2005–2006) to consider the efficacy of his model and concluded that the dynamic model provided a better fit to these results than a non-dynamic one. Owen's approach has significant advantages over the continuous-time model of Rue and Salvesen (2000), particularly in the choices of priors for the attack and defence parameters, which are derived from previous seasons' data. Estimating σ^2 directly in the model also allows greater flexibility in the model, allowing the strength parameters to vary differently at different times in the season – there appeared to be higher volatility in the strength parameters earlier in the season than later, which seems reasonable. However, he does not apply a betting strategy to investigate possible returns of the model.

Koopman and Lit (2015) use a stationary $AR(1)$ process for team strength parameters, including an intercept term, in the form $z_t = \mu + \Phi z_{t-1} + \eta_t$, where z_t is a state vector of the α_{it} and β_{it} elements, μ is a vector of constants, Φ is a square matrix with the autoregressive coefficients in the diagonal and η_t is the error vector which is normally independently distributed with mean 0 and variance matrix H. Nine seasons of the English Premier League were used in the analysis (2003–2004 to 2011–2012), with the first seven seasons used to provide out-of-sample forecasts for the last two seasons. The authors also implement a betting strategy, betting on matches where the expected value of a unit bet exceeds some benchmark, denoted by $\tau > 0$. They use average odds across 40 different bookmakers, giving a total of 760 betting opportunities (all matches in two seasons) when $\tau = 0$. As the value of τ increases, the number of betting opportunities decreases and positive mean returns are obtained when $\tau > 0.12$ (around 270 'value' bets). When $\tau = 0.4$, 1 unit was bet in each of 50 matches and generated a return of 75 units (an expected profit of 50% on the stake).

The autoregressive parameters in Koopman and Lit (2015) were estimated to be $\hat{\phi}_\alpha = 0.9985$ for the attack dynamics, and $\hat{\phi}_\beta = 0.9992$ for the defence, both are very close to one, so that the strength dynamics are nearly non-stationary, although their model constrained the dynamics so that $0 < \phi_K < 1$, $K = \alpha, \beta$. The question then naturally arises as to which assumption is most appropriate for the dynamics: stationary or non-stationary.

17.5 Stationary or Non-Stationary Dynamics, and Structural Breaks

The introduction of dynamics in the modelling of attack and defence parameters has added a layer of depth in soccer match outcome models that has improved forecasts overall. However, authors have made rather strong assumptions as to how attack and defence of teams evolve, and we believe that the questioning of these assumptions is the next step for research in soccer match outcome modelling. Knorr-Held (2000), Rue and Salvesen (2000) and Owen (2011) all use a type of non-stationary processes to capture dynamics, whereas Koopman and Lit (2015) implement a stationary autoregressive model. Certainly, there are advantages to each. However, non-stationary and stationary processes have different properties, which as a result affect forecasting performance if the model is mis-specified. We will briefly describe stationary processes as well as two different non-stationary processes, and subsequently explain how mis-specifying the dynamics could lead to misleading forecasts.

A (covariance) stationary process is one with the following properties:

1. $E(y_t) = \mu$, $\quad -\infty < \mu < \infty \quad \forall t$
2. $V(y_t) = \sigma^2 < \infty \quad \forall t$
3. $C(y_t, y_{t-s}) = \gamma(s) \quad \forall t, s$ (the autocovariance function)

where the mean, variance and covariance do not depend on the time t. If we observe team strengths over time in several papers (e.g. Dixon and Coles, 1997; Knorr-Held, 2000; Koopman and Lit, 2015), it appears that the time series plots of strengths do not exhibit the properties attributed to a stationary process. Two classic examples of non-stationary time series that violate at least one of the aforementioned properties are trend-stationary and difference-stationary processes.

A simple trend-stationary process that could be applied to modelling attack strength dynamics can be formulated as

$$\alpha_t = \alpha_0 + dt + \varepsilon_t \tag{17.24}$$

where dt is the trend component of the time series. The mean is time-dependent, as $E(\alpha_t) = \alpha_0 + dt$. This series can be made stationary by detrending. A simple example of a difference-stationary series is a random walk, (used in Knorr-Held, 2000 and Owen, 2011, where $\alpha_t = \alpha_{t-1} + \varepsilon_t$), for which the second and third criteria are violated: $V(\alpha_t) = t\sigma^2$ and $C(\alpha_t, \alpha_{t-s}) = (t-s)\sigma^2$, so that both the variance and autocovariance functions depend on time t. These types of time series processes are often called $I(1)$ processes (integrated processes of order 1), because one can achieve stationarity by taking first differences once.

Sometimes, a time series can have a trend and also be a difference-stationary process, so that it is said to have a stochastic trend. Despite the fact that trend-stationary and difference-stationary processes have very different properties, realisations and their diagnostics can look rather similar. The implications for forecasting are that the one step ahead forecast $\hat{\alpha}_{t+1}$ is different for all three processes:

1. For a trend-stationary process, $\hat{\alpha}_{T+1|T} = \alpha_0 + d(T+1)$
2. For a random walk, $\hat{\alpha}_{T+1|T} = \alpha_T$
3. For a stochastic trend, $\hat{\alpha}_{T+1|T} = \alpha_0 + \alpha_T + d(T+1)$

Diebold and Kilian (2000) reinforce this point that it is very important to decide which dynamic model to use because different models imply different predictions. They recommend pretesting for a unit root of the autoregressive parameter. There are several tests that can be used to test for the presence of a unit root, most notably the augmented Dickey–Fuller test (Said and Dickey, 1984) and an extension of this test (Elliot et al., 1996). For a comprehensive review of these tests and test alternatives, see Maddala (2001).

A further issue in modelling strength dynamics is the possible presence of a structural break. A structural break occurs when the assumption of parameter 'continuity' fails, and this occurs in the trend-stationary process (17.24) when there is a discontinuity in either α_0 or d or both. Such discontinuities may arise naturally in soccer when teams buy and sell players at transfer deadlines, and when management and team ownership changes. Clements and Hendry (1998) provide Monte Carlo evidence for ex ante forecast failure in the presence of structural breaks in time series. This is quite common in macroeconomic and financial time series, but it has never been considered when modelling team strength dynamics in soccer. However, this adds a further challenge to modelling match outcomes. The presence of structural breaks has important implications when determining the dynamics to use for team strengths, because structural breaks affect unit root tests. If a time series has a structural break, Perron (1989) shows that traditional unit root tests lose power. Furthermore, Pesaran and Timmermann (2004) show that ignoring these breaks leads to inconsistent and biased forecasts.

Several solutions have been offered for forecasting time series in the presence of structural breaks. Koop and Potter (2007) present a model that can forecast a new break partly dependent on the properties of previous 'regimes', taking into account all previous breaks by using a hierarchical prior on regime durations. This could be useful if a trend in team strength dynamics is assumed, where there exist different trend 'regimes' such as growing, steady and declining. Such an approach was used to model U.S. economic output dynamics. Maheu and Gordon (2008) suggest using Bayesian methods of learning and model comparison to generate forecasts. Using two different macroeconomic data sets, they compare the 'break' model to others that do not account for breaks and they conclude that there are substantial gains in terms of out-of-sample forecast error. In a recent paper by Bauwens et al. (2015), several different sophisticated models that address the structural break issue are compared in an extensive out-of-sample forecasting exercise on 60 different macroeconomic series. However, the authors conclude that no one model is consistently the best in the presence of structural breaks. This seems to indicate that there is still some research to be carried out on this topic. Nevertheless, it would be interesting to see these existing ideas being applied to a topic outside of macro or financial economics. A good starting point would be team strength dynamics in soccer.

17.6 Conclusion

In the course of the last 10–15 years, the literature on modelling scores in soccer games has moved into researching dynamics to explain how team strengths change over time. The basic idea is to extrapolate a statistical relationship into the future to forecast where team strengths will be at time $t + 1$ in order to generate predictions for scores at this time. Many questions arise, such as what kind of time series process to choose for the strength dynamics. This has important implications for forecasting accuracy, and it may be detrimental to choose the wrong model.

A further challenge arises when the statistical relationship between the variables breaks down and dynamics change. The use of unit root tests taking into account structural breaks implies that we only observe the changes after they occur. We believe that the future of research in team strength dynamics will be directed towards why these relationships break down and to investigate if it is possible to predict when a break will occur, so that better forecasts can be made. There currently exist ideas in the time series, financial and macroeconometrics literatures that could be used to address the problem of structural breaks in team strength dynamics in soccer. These include time-varying parameterisations and change-point models with regime switching, Markov-switching models and Bayesian learning.

Given the importance of the application of match outcome research to not only the betting industry but also soccer clubs and analytics teams, there exists a strong motivation for researchers to engage in this field. We are sure that there are many statisticians who are also soccer fans who would like to see this '*Moneyball* effect' spill over into the world of soccer. Finally, these ideas could be implemented not only in soccer but also in any sport that predicts scores using strength dynamics. Exciting times lie ahead for research in this area.

References

Audas R, Dobson S and Goddard J (2002) The impact of managerial change on a team performance in professional sports. *Journal of Economics and Business* 54, 633–650.

Baker R and McHale G (2014) A dynamic paired comparisons model: Who is the greatest tennis player? *European Journal of Operational Research* 236, 677–684.

Baker R and McHale I (2015) Time varying ratings in association football: The all-time greatest team is… *Journal of the Royal Statistical Society: Series A (Statistics in Society)* 178, 481–492.

Bauwens L, Koop G, Korobilis D and Rombouts J (2015) The contribution of structural break models to forecasting macroeconomic series. *Journal of Applied Econometrics* 30, 596–620.

Brillinger D (2007) A potential function approach to the flow of play in soccer. *Journal of Quantitative Analysis in Sports* 3(1), article 3.

Bruinshoofd A and Weel B (2003) Manager to go? Performance dips reconsidered with evidence from Dutch football. *European Journal of Operational Research* 148, 233–246.

Buraimo B and Simmons R (2015) Uncertainty of outcome or star quality? Television audience demand for English premier league football. *International Journal of the Economics of Business* 22, 449–469.

Clements M and Hendry D (1998) Forecasting economic processes. *International Journal of Forecasting* 14, 111–131.

Constantinou A and Fenton N (2013) Determining the level of ability of football teams by dynamic ratings based on the relative discrepancies in scores between adversaries. *Journal of Quantitative Analysis in Sports* 9, 37–50.

Crowder M, Dixon MJ, Ledford L and Robinson M (2002). Dynamic modelling and prediction of English Football League matches for betting. *The Statistician* 51, 157–168.

Diebold F and Kilian L (2000) Unit root tests are useful for selecting forecasting models. *Journal of Business and Economic Statistics* 18, 265–273.

Dixon M and Coles S (1997) Modelling association football scores and inefficiencies in the football betting market. *Applied Statistics* 46, 265–280.

Dixon M and Robinson M (1998) A birth process model for association football matches. *Journal of the Royal Statistical Society: Series D (The Statistician)* 47, 523–538.

Dobson S and Goddard J (2003) Persistence in sequences of football match results: A Monte Carlo analysis. *European Journal of Operational Research* 148, 247–256.

Elliot G, Rothenburg T and Stock J (1996) Efficient tests for an autoregressive unit root. *Econometrica* 64, 813–836.

Forrest D and Simmons R (2000a) Forecasting sport: The behaviour and performance of football tipsters. *International Journal of Forecasting* 16, 317–331.

Forrest D and Simmons R (2000b) Making up the results: The work of the football pools panel, 1963–1997. *Journal of the Royal Statistical Society: Series D (The Statistician)* 49, 253–260.

Forrest D and Simmons R (2002) Outcome uncertainty and attendance demand in sport: The case of English soccer. *Journal of the Royal Statistical Society: Series D (The Statistician)* 51, 229–241.

Goddard J (2005) Regression models for forecasting goals and match results in association football. *International Journal of Forecasting* 21, 331–340.

Goddard J and Asimakopoulos I (2004) Forecasting football results and the efficiency of fixed-odds betting. *Journal of Forecasting* 23, 51–66.

Goossens D, Belien J and Spieksma F (2012) Comparing league formats with respect to match importance in Belgian football. *Annals of Operations Research* 194, 223–240.

Goossens D and Spieksma F (2012) Soccer schedules in Europe: An overview. *Journal of Scheduling* 15, 641–651.

Groll A, Schauberger G and Tutz G (2015) Prediction of major international soccer tournaments based on team-specific regularized Poisson regression: An application to the FIFA World Cup 2014. *Journal of Quantitative Analysis in Sports* 11, 97–115.

Hirotsu N and Wright M (2006) Modelling tactical changes of formation in association football as a zero-sum game. *Journal of Quantitative Analysis in Sports* 2(2), 4, 1–22.

Hvattum L and Arntzen H (2010) Using ELO ratings for match result prediction in association football. *International Journal of Forecasting* 26, 460–470.

Karlis D and Ntzoufras I (2003) Analysis of sports data using bivariate Poisson models. *Journal of the Royal Statistical Society: Series D (The Statistician)* 52, 381–393.

Karlis D and Ntzoufras I (2009) Bayesian modelling of football outcomes: Using the Skellam's distribution for the goal difference. *IMA Journal of Management Mathematics* 20, 133–145.

Keogh F and Rose G (2013) Football betting—the global gambling industry worth billions. http://www.bbc.co.uk/sport/football/24354124, accessed April 22, 2016.

Knorr-Held L (2000) Dynamic rating of sports teams. *Journal of the Royal Statistical Society: Series D (The Statistician)* 49, 261–276.

Koning R (2000) Balance in competition in Dutch soccer. *Journal of the Royal Statistical Society: Series D (The Statistician)* 49, 419–431.

Koning R, Koolhaas M, Renes G and Ridder G (2002) A simulation model for football championships. *European Journal of Operational Research* 148, 268–276.

Koop G and Potter S (2007) Estimation and forecasting in models with multiple breaks. *Review of Economic Studies* 74, 763–789.

Koopman S and Lit R (2015) A dynamic bivariate Poisson model for analysing and forecasting match results in the English Premier League. *Journal of the Royal Statistical Society: Series A (Statistics in Society)* 178, 167–186.

Lenten L, Libich J and Stehlı́k P (2013) Policy timing and footballers' incentives: Penalties before or after extra time? *Journal of Sports Economics* 14, 629–655.

Maddala G (2001) *Introduction to Econometrics*, 3rd edn. Wiley, Chichester, West Sussex.

Maher M (1982) Modelling association football scores. *Statistica Neerlandica* 36, 109–118.

Maheu J and Gordon S (2008) Learning, forecasting and structural breaks. *Journal of Applied Econometrics* 23, 553–583.

McHale I and Scarf P (2007) Modelling soccer matches using bivariate discrete distributions with general dependence structure. *Statistica Neerlandica* 61, 432–445.

McHale I and Scarf P (2011) Modelling the dependence of goals scored by opposing teams in international soccer matches. *Statistical Modelling* 11, 219–236.

McHale I, Scarf P and Folker D (2012) On the development of a soccer player performance rating system for the English Premier League. *Interfaces* 42, 339–351.

Moroney M (1956). *Facts from Figures*, 3rd edn. Penguin, London, UK.

Owen A (2011) Dynamic Bayesian forecasting models of football match outcomes with estimation of the evolution variance parameter. *IMA Journal of Management Mathematics* 22, 99–113.

Perron P (1989) The Great Crash, the Oil Price Shock, and the unit root hypothesis. *Econometrica* 57, 1361–1401.

Pesaran M and Timmermann A (2004) How costly is it to ignore breaks when forecasting the direction of a time series? *International Journal of Forecasting* 20, 411–425.

Rue H and Salvesen O (2000) Prediction and retrospective analysis of soccer matches in a league. *Journal of the Royal Statistical Society: Series D (The Statistician)* 49, 399–418.

Said S and Dickey D (1984) Testing for unit roots in autoregressive-moving average models of unknown order. *Biometrika* 71, 599–607.

Scarf P and Shi X (2008) The importance of a match in a tournament. *Computers and Operations Research* 35, 2406–2418.

Scarf P and Yusof M (2011) A numerical study of tournament structure and seeding policy for the soccer World Cup Finals. *Statistica Neerlandica* 65, 43–57.

Scarf P, Yusof M and Bilbao M (2009) A numerical study of designs for sporting contests. *European Journal of Operational Research* 198, 190–198.

Schryver T and Eisinga R (2011) Piecewise linear regression techniques to analyse the timing of head coach dismissals in Dutch soccer clubs. *IMA Journal of Management Mathematics* 22, 129–137.

Sporty Desk (2015) Top 10 most popular sports in the world. http://www.sportyghost.com/top-10-most-popular-sports-in-the-world/, accessed April 22, 2016.

Tenga A, Holme I, Ronglan L and Bahr R (2010) Effect of playing tactics on goal scoring in Norwegian professional soccer. *Journal of Sports Sciences* 28, 237–244.

Titman A, Costain D, Ridall P and Gregory K (2015) Joint modelling of goals and bookings in association football. *Journal of the Royal Statistical Society: Series A (Statistics in Society)* 178, 659–683.

Wright M and Hirotsu N (2003) The professional foul in football: Tactics and deterrents. *Journal of the Operational Research Society* 54, 213–221.

18

Rating of Team Abilities in Soccer

Ruud H. Koning

CONTENTS

18.1 Introduction

Soccer is a team sport, so measurement of team quality is important. Preferably, quality is measured as a single number that can be used to answer questions such as which soccer team is the best or how do two teams compare to each other. Such questions are asked by fans, players, management of teams and other stakeholders alike. In soccer, the standard to rank teams is by having all teams play a double round robin tournament, with each two teams playing each other twice, once at each venue. Ranking is based on the number of points obtained. However, it is not always practical to play a complete round robin tournament. Moreover, intermediate rankings before all matches in a round robin setup are played reflect both abilities of teams and the matches played so far. As long as not all teams have played each other, it is possible that the team that leads the ranking has had the 'easiest' schedule compared with other teams. Intermediate rankings, if available, do not reflect team abilities alone.

Different models have been proposed to measure team quality. Usually, such models are referred to as rating models, a term we use from now on in this chapter. This chapter is about statistical models of team quality, that are applicable both in completed double round robin tournament, but also in a partially observed tournament. First, we review the most important models in Section 18.2, where we distinguish between rule-based models (such as the ELO rating and the FIFA ranking) and purely statistical models (such as the Clarke and Norman (1995) model). In Section 18.3, we discuss the predictive accuracy of some of the different models of Section 18.2 and in Section 18.4, we apply one of the models to measure match significance, a potential important determinant of live attendance and number of television viewers. Throughout this chapter, we use two leading examples: the English Premier League as a competition for club teams which is played according to a double round robin tournament and matches between national teams as an example of matches that are played irregularly.

18.2 Rating Models

One approach to measure team abilities in an objective way is by rating models. In this chapter we discuss different rating models, and we show some applications of such models. A rating model assigns a score to each team based on past results. The ordering of teams is then based on these scores. In the context of soccer, a useful rating model satisfies two properties:

1. The rating score is non-decreasing in performance.
2. Older matches carry less weight than more recent ones in determining the rating score.

Besides these two general criteria, one may impose other requirements on a rating model, for example

- A change in the rating score depends on the quality of the opponent.
- The rating model takes the importance of the match into account.
- A match played at home is treated differently from a match played away.

Stefani (2011) gives a comprehensive review of different rating systems that are used in practice by different international sporting federations. He distinguishes between three categories of rating systems: subjective, accumulative and adaptive systems. In subjective systems, ratings are determined by a panel of experts (e.g. judges in a figure skating contest). Subjective imputation of match results has been used in gambling on soccer results, when a match was not played, for example because of weather conditions. In accumulative systems, the rating depends on all performances in a time window. Performances may be weighed, for example by the type of competition or by the time that has elapsed between the performance and the moment of rating. Finally, in adjustive systems, the *change* in rating is determined by the difference between the actual performance and some measure of the expected performance. The expected performance usually depends on the rating of both contestants and possibly some weighting factor.

We do no intend to provide an exhaustive discussion of rating systems. Instead, we focus on some that are used in soccer. First, we discuss three rule-based rating systems (points, ELO-type ratings and the FIFA World Ranking), and then we proceed to estimate ratings as parameters in statistical models (Clarke–Norman linear regression and the ordered Bradley–Terry model).

18.2.1 Rule-Based Rating Models

The simplest rule-based model to measure team quality in a round robin tournament is the number of points obtained by a team. In all soccer leagues, a win generates three points, a draw one point, and a loss zero points. League rankings are based on the cumulative number of points obtained. This measure is important, since it determines the final ranking of a league, and hence, which team is the winner and which team is relegated. With some positive probability, teams may end up with the same number of points. Ties are then solved using additional rules: first goal difference is used a tie-breaking rule, and if that does not solve the tie, the number of goals scored is the next tie-breaking rule (rules C4–C7 in Premier League 2015). Teams that score identically on all three dimensions are then given the same position in the ranking. Additional tie-breaking rules are used in the group stage of the World Cup: a fourth tie-breaking rule is the result between the tied teams (assuming two teams that are tied), and the FIFA allows for draw of lots to determine ranking if the tie-breaking rules do not provide an ordering (rule 41.5 in FIFA 2014b). In the case of the group stage in a World Cup tournament, it is necessary to have an unambiguous ranking since the best two teams are to proceed in the tournament as there is no time to organise an additional tie-breaking match.

A second type of rule-based rating is the ELO rating. Originally, this rating was developed for chess (Elo, 1978), but since its introduction, it has been used for other sports as well. The basic idea of the ELO rating is very simple and appealing: if a weak team wins against a strong team, it gains points, and the strong team loses points. Unexpected wins are rewarded more than expected wins. In its simplest form, the system can be described as follows (Hvattum and Arntzen, 2010; Lasek et al., 2013). Consider a match between teams i and j, with prematch ratings $P_i^{(t)}$ and $P_j^{(t)}$. The updated number of points after the match is then

$$P_i^{(t+1)} = P_i^{(t)} + K(s_{ij} - \pi_{ij}), \tag{18.1}$$

with s_i the score of team i (0 for a loss, 0.5 for a draw, 1 for a win) and π_{ij} the expected score of team i in a match against team j. The K-factor determines how fast deviations of actual results (s_{ij}) from the expectation (π_{ij}) are transmitted to the actual rating. The expected score is taken to be

$$\pi_{ij} = \frac{1}{1 + c^{\left(P_j^{(t)} - P_i^{(t)}\right)/d}}. \tag{18.2}$$

In this expression, d is a scaling factor and c a constant. This basic ELO model has three parameters c, d and K, that need to be calibrated. In particular, the choice of K is important: if K is too small, improved performance is reflected in the rating very slowly, if it is too large, the rating becomes too volatile.

Different variants of this basic ELO rating methodology have been proposed. First, the method as such does not take home advantage into account. This may bias ratings if some teams play more home games than other teams. One solution to this issue is to add a number of rating points to the rating of the home playing team when calculating the expected result in Equation 18.2. A second extension is to make the K-factor depend on the type of the match (friendlies are less important than league matches). In a third extension, K depends on the actual match outcome, for example $K = K_0(1 + |w_{ij}|)^\lambda$, with $|w_{ij}|$ the absolute value of the goal difference, and K_0 and λ constants to be determined. Other specifications relating K to the actual goal difference have been proposed as well, see for example Lasek et al. (2013).

Two important applications of ELO rating methodology are the FIFA Women's World Ranking and the Euro Club Index. In the FIFA Women's World Ranking, the measure for the actual result s_{ij} depends on the goal difference and the number of goals scored. For example winning a match by 2-1 gives $s_{ij} = 0.84$, and winning by 4-3 gives $s_{ij} = 0.82$. The K-factor depends on the importance of the match, for a FIFA Women's World Cup match $K = 60$, for a continental final $K = 45$, and for a friendly two teams both ranked outside the top 10, $K = 15$. The Euro Club Index (ECI) ranks football teams from different countries in Europe at a given point in time. In that case, the outcome s_{ij} is -1 for a loss, 0 for a draw and 1 for a win. The probability distribution of these outcomes is based on an ordered probit model, and using the estimated probabilities, the expected outcome is calculated as $\pi_{ij} = -1 \times \Pr(\text{loss}) + 0 \times \Pr(\text{draw}) + 1 \times \Pr(\text{win})$. Ratings are then updated according to Equation 18.1 with $K = 35$. The Euro Club Index is interesting because it ranks teams of different national leagues, and therefore can be used to predict international match results.

A third rule-based method for measuring team quality is the FIFA World Ranking. This system is tailored to measure the quality of national teams and is important because of its use in seeding for major international qualification tournaments and final tournaments. Contrary to clubs in national leagues, national teams do not play matches frequently (e.g. in 2014, Spain played 12 matches, in that same year Barcelona played 37 matches in the Spanish league and 10 in the Champions League alone). Moreover, most matches between national teams are played within a confederation (the world is divided in six regional confederations, for example UEFA for Europe and CONMEBOL for South America), which makes it difficult to compare national teams between confederations.

The basic setup of the FIFA World Ranking is simple: the quality of a national team is measured by a weighted average of points obtained during the last four years. The importance of a match, strength of the opposing team and strength of the confederations are all taken into account. The number of points for team i in a match between teams i and j is given by

$$P_{ij} = M_{ij} \times I_{ij} \times T_j \times \frac{1}{2}(C_i + C_j). \tag{18.3}$$

In this equation, M_{ij} is the result of the match, 3 points for a win, 1 for a draw and none for a loss. I_{ij} is the importance of the match, 1.0 for a friendly, 2.5 for a qualifier, 3.0 for a confederation final match and 4.0 for a World Cup final match. The strength of the opposing team is indicated by T_j, measured as $200 - R_j$, with R_j the rank of team j in the FIFA World Ranking. The minimum of T_j is 50, and T_j equals 200 for the team leading the FIFA World Ranking at the moment the match is played. C_i denotes the strength of the confederation team i belongs to, currently (January 2016) 1.0 for CONMEBOL, 0.99 for UEFA and 0.85 for all other confederations. These weights are determined by results during the last three World Cup finals. Finally, matches played during the last 12 months count fully, matches between 1 and 2 year old count 50%, matches 2–3 years old count 30% and matches between 3 and 4 years old count 20%. Matches older than 4 years do not count at all. This weighted average is the number of points used to rank the countries; more details on the procedure can be found in FIFA (2014a).

An example may illustrate this calculation. Consider the match between England and Uruguay, played during the 2014 World Cup in Brazil. At that moment, England was

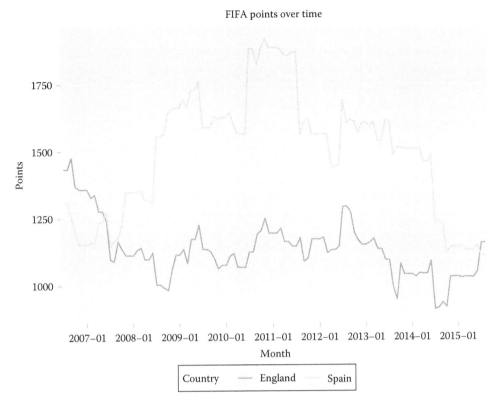

FIGURE 18.1
Development of points in the FIFA ranking over time for England and Spain.

ranked 10th, Uruguay was ranked 7th. The final score was 1-2. This match assigns 0 points to England ($M_{ij} = 0$) and Uruguay gets

$$3 \times 4.0 \times (200 - 10) \times \frac{1}{2}(0.99 + 1.00) = 2268.6$$

points towards their average. Usually, this weighted average number of points is colloquially referred to as the number of points of the FIFA World Ranking.

FIFA publishes the ranking using this methodology once a month, starting on 12 July 2006. In Figure 18.1, we show the development of the number of points over time for both England and Spain. Clearly, Spain has been extremely successful during the 2008–2012 period (winner European Championship 2008, 2012 and the World Cup in 2010), but they did not qualify beyond the group stage in the World Cup of 2014. Then, in August 2014, the 'successful' matches of the 2010 World Cup drop out of the calculation and are replaced by the two losses and one win during the 2014 World Cup. This causes a drop in the average number of points per match as shown in the figure.

18.2.2 Statistical Rating Models

An alternative to the rule-based rating methods discussed in the previous section are statistical rating models. In such models, the quality of teams is captured by one or two parameters that are to be estimated. In this section, I discuss two of such models: one that is estimated by least squares and one that is estimated by maximum likelihood. Statistical rating models have two advantages over rule-based rating models: it is easy to take an incomplete schedule of play into account, and it is simple to assess the effect of interventions by testing whether parameters have changed (see, e.g. Koning, 2003).

18.2.2.1 Rating by Regression

Clarke and Norman (1995) introduced a simple model for soccer outcomes. They separated the outcome of a soccer match in three factors: home advantage, difference in team qualities and random factors. The win margin w_{ij} (i.e. the difference between the number of home goals and away goals) of home team i against away team j is modelled as

$$w_{ij} = h_i + \psi_i - \psi_j + \eta_{ij}, \tag{18.4}$$

with h_i home advantage of the home team, ψ_i is the quality of team i (so $\psi_i - \psi_j$ is the quality difference between the home team and the away team) and η_{ij} is a mean zero random error term, that captures all factors, which determine the outcome, and are not captured by either home advantage or quality difference. Note that the win margin w_{ij} is 0 if the match ends in a draw and is negative if the home team loses. Clarke and Norman assume that the error term follows a normal distribution with constant variance. Obviously, this is an approximation because of the discrete nature of the dependent variable.

The parameters of the model in Equation 18.4 are not identified, identification is obtained by imposing the restriction $\sum_i \psi_i = 0$. This restriction can be interpreted as imposing the constraint that average quality is zero. A team with a positive ψ_i is better than average, a team with a negative ψ_i is worse than average. The home advantage parameter h_i can be interpreted as the expected goal difference of team i if it were to play against a (hypothetical) team of identical quality in a home match. Considering the prevalence of home wins in soccer, one expects h_i to be positive for most teams. It is straightforward to test the hypothesis that home advantage is identical for each team, it would correspond to the (linear) null hypothesis $H_0 : h_1 = h_2 = \cdots = h_K$ with K the number of teams in the league.

Allowing for team-specific home advantage, it is simple to calculate the implied probability distribution of the outcomes:

$$\Pr(y_{ij} = \text{HW}) = 1 - \Phi\left(\frac{1/2 - (h_i + \psi_i - \psi_j)}{\sigma_\eta}\right),$$

$$\Pr(y_{ij} = \text{D}) = \Phi\left(\frac{1/2 - (h_i + \psi_i - \psi_j)}{\sigma_\eta}\right) - \Phi\left(\frac{-1/2 - (h_i + \psi_i - \psi_j)}{\sigma_\eta}\right),$$

$$\Pr(y_{ij} = \text{AW}) = \Phi\left(\frac{-1/2 - (h_i + \psi_i - \psi_j)}{\sigma_\eta}\right).$$

Here, we assume, as in Clarke and Norman (1995), that the error term η_{ij} follows a normal distribution. This expression will be used later to compare home advantage in this model with home advantage in the ordered Bradley–Terry model.

Estimation of the parameters of model (18.4) is simple by least squares, the only covariates are dummy variables corresponding to each particular match. Clearly, it is not necessary to have a complete season of match results to estimate the parameters of the model. If a complete schedule has been played, Clarke and Norman provide a more intuitive estimator as well. The parameters of the model are estimated in three steps:

1. Calculate for each team the total home goal difference HGD_i and away goal difference AGD_i. The total of all individual team's home advantages is $H = \sum_i h_i = \sum_i HGD_i/(K-1)$.
2. For each team, individual home advantage h_i is calculated as $h_i = (HGD_i - AGD_i - H)/(K-2)$. As a consequence, $HGD_i - AGD_i = (K-2)h_i + \sum_i h_i$.
3. Finally, the estimate for team quality ψ_i is $\psi_i = (HGD_i - (K-1)h_i)/K$.

From step 2, it is clear that the difference between home performance and away performance depends both on the home, the individual team's home advantage and the sum of all other team's home advantages. Clarke and Norman show that these estimates coincide with the least squares estimates of the parameters in model (18.4). Obviously, unlike the aforementioned simple approach, model (18.4) can be extended with covariates.

To illustrate the Clarke–Norman model, we use results from the English Premier League from the 15 seasons 2000/2001–2014/2015. The parameters of model (18.4) were estimated for each season separately, so we obtain 15 estimates for home advantage and team quality for each team that is in de English Premier League during that period. In Figure 18.2 we show the development of the estimates of Manchester City and Manchester United over time. As expected, in most seasons home advantage is estimated to be positive, on average home advantage corresponds to half a goal. Team quality varies over time, with Manchester United being the better team until 2009/2010, and since that season both teams are more or

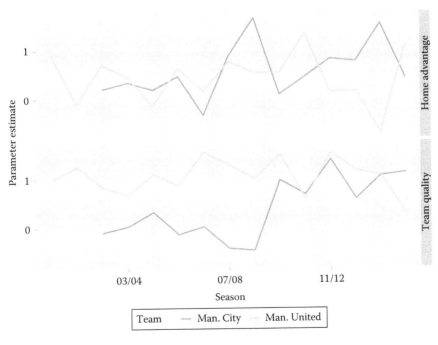

FIGURE 18.2
Estimated home advantage and team from regression model quality for Manchester City and Manchester United, 2000/2001–2014/2015.

less evenly matched. According to the estimates, the quality of Manchester City improved markedly as of the 2009/2010 season. In September 2008, Manchester City was bought by Abu Dhabi United Group, and these new owners have provided funds to improve the quality of the team.

Using the same data and period, we also test the hypothesis that home advantage is constant between all teams in the English Premier League. For each season, we test whether $h_1 = \cdots = h_K$ or not. The smallest p-value we find for these fifteen seasons is 0.052, so we conclude that as a first approximation home advantage may be taken to be constant between teams. The average (over all seasons) home advantage amounts to 0.4 goals for the home playing team. Home advantage is clearly substantial in soccer, but not too large: it can be overcome by one goal. Constant home advantage is imposed by the rating model in the next subsection.

18.2.2.2 Rating by Bradley–Terry

The Bradley–Terry model (Bradley and Terry, 1952) is a general model for paired comparisons, and its simplest specification models the probability that i is better than j as

$$\Pr(i > j) = \frac{\exp(\beta_i)}{\exp(\beta_i) + \exp(\beta_j)}.$$

This specification is easily estimated by maximum likelihood, under an identifying restriction, such as $\beta_1 = 0$, or $\sum_i \beta_i = 0$. This specification is well suited for sports where every contest has a winner and a loser (e.g. tennis), but not for soccer since it does not allow for a draw. In Koning (2000), this model is extended to allow for draws as well, using a latent regression framework. The (unobserved) quality difference between team i (at home) and team j (away) is denoted y_{ij}^*, and is modelled as

$$y_{ij}^* = \theta_i - \theta_j + \epsilon_{ij}, \tag{18.5}$$

where
θ_i and θ_j indicate the team qualities
ϵ_{ij} all random factors that influence the outcome of a match

It is assumed that ϵ_{ij} follows a normal distribution $\mathcal{N}(0, \sigma^2)$. The observed outcome is a home win (HW), draw (D), or away win (AW), according to

$$y_{ij} = \begin{cases} \text{HW} & y_{ij}^* > c_2, \\ \text{D} & c_1 < y_{ij}^* \leq c_2, \\ \text{AW} & y_{ij}^* \leq c_1. \end{cases} \tag{18.6}$$

The probabilities of each outcome are now given by

$$\Pr(y_{ij} = \text{HW}) = 1 - \Phi\left(\frac{c_2 - (\theta_i - \theta_j)}{\sigma}\right),$$

$$\Pr(y_{ij} = \text{D}) = \Phi\left(\frac{c_2 - (\theta_i - \theta_j)}{\sigma}\right) - \Phi\left(\frac{c_1 - (\theta_i - \theta_j)}{\sigma}\right),$$

$$\Pr(y_{ij} = \text{AW}) = \Phi\left(\frac{c_1 - (\theta_i - \theta_j)}{\sigma}\right). \tag{18.7}$$

It is clear that the parameters of this model are not all identified. The scale is not identified, so we take $\sigma = 1$, and the θ parameters are identified if we impose $\sum_i \theta_i = 0$ (other identifying restrictions are possible, at the cost of interpretability), so that average team quality is 0. If team i is better than team j ($\theta_i > \theta_j$), the probability of a home win is large. Home advantage is measured by the two parameters c_1 and c_2. First, note that we have $c_1 < c_2$ if the probability of a draw is to be positive, from now on we assume that this is the case. Suppose two teams of equal quality meet, so that $\theta_i - \theta_j = 0$. The probability that the home team wins is $1 - \Phi(c_2)$, the probability that the away team wins is $\Phi(c_1)$. If there is no home advantage, we have $1 - \Phi(c_2) = \Phi(c_1)$, so no home advantage implies $c_1 = -c_2$. In case $c_1 < -c_2$, we have $1 - \Phi(c_2) > \Phi(c_1)$, so in a match between two equally good teams, the home team is more likely to win. Home advantage is numerically identical for all teams.

The parameters of the model can be easily estimated by maximising the log-likelihood function

$$\ell(\theta_1, \ldots, \theta_{K-1}, c_1, c_2) = \sum_{y_{ij}=\text{HW}} \log\left(1 - \Phi(c_2 - (\theta_i - \theta_j))\right)$$

$$+ \sum_{y_{ij}=\text{D}} \log\left(\Phi(c_2 - (\theta_i - \theta_j)) - \Phi(c_1 - (\theta_i - \theta_j))\right)$$

$$+ \sum_{y_{ij}=\text{AW}} \log\left(\Phi(c_1 - (\theta_i - \theta_j))\right).$$

The maximum likelihood estimate for θ_i is finite if team i does not win all its games (in which case $\hat{\theta}_i \to \infty$) or loses all games ($\hat{\theta}_i \to -\infty$). If the number of matches considered is large enough, this is usually not the case. It is easy to extend the model by having the home advantage parameters c_1 and c_2 depend on covariates. Goddard and Asimakopoulos (2004) extend this model by having the team quality parameters θ_i depend on covariates that vary between teams and over time, while Cattelan et al. (2013) focus on a specification that allows for performance-dependent dynamics of the quality parameters. It is also straightforward to extend the model to allow for change of team quality after a particular intervention (e.g. changing the manager of a team).

We estimate this model using results from the English premier League, 2000/2001–2014/2015. Again, we draw the development of the quality parameters of Manchester City and Manchester United over time. In this model, there is no team-dependent home advantage, so in Figure 18.3 only estimated quality parameters for both teams are drawn. Again, the marked improvement of the quality of Manchester City as of the 2009/2010 season is visible. According to these estimates, Manchester City overtakes Manchester United only in

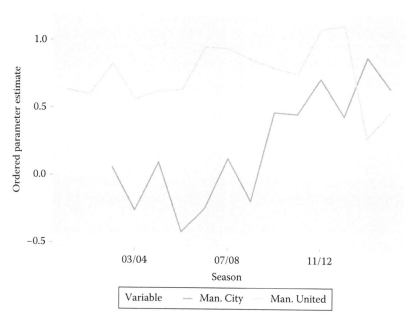

FIGURE 18.3

Estimated team quality from ordered probit model for Manchester City and Manchester United, 2000/2001–2014/2015.

2012/2013, unlike as in Figure 18.2 where the estimated quality of Manchester City exceeds the one of Manchester United already in 2010/2011.

The models can be compared further by looking at the implied probabilities of a home win. In Figure 18.4, we plot the implied probability of a home win when two teams of equal quality meet. That probability is calculated according to Equation 18.5 for the regression model, where we assume that home advantage does not vary by team. In the previous

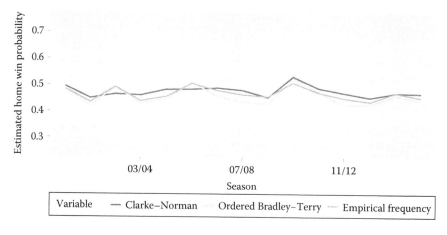

FIGURE 18.4

Estimated home advantage in the regression model, ordered Bradley–Terry model, and actual frequencies in the EPL, 2000/2001–2014/2015.

subsection, we argued that this hypothesis is not rejected by the data. That probability is $1 - \Phi(c_2)$ for the ordered Bradley–Terry model. In Figure 18.4, we also show the empirical frequency of home wins over a complete season. The ordered Bradley–Terry model tracks the empirical frequency very well, the regression model slightly less so. The correlation between the home win probability in the regression model and the empirical frequency is 0.75, while it is 0.95 for the ordered Bradley–Terry model. This difference is partly due to the fact that in essence the regression model is a model for continuous outcomes, while the actual outcome (home win, draw, away win) is discrete in this case.

18.3 An Example of Match Outcome Prediction with a Rating Model

If rating models are useful to measure quality of teams, the numerical measures should be good predictors of the outcome of a soccer match. In this section, we focus on this issue, we assess whether the qualitative outcome of a match (home win/draw/away win) can be predicted well using one of the models from the previous section. This is not the first attempt to predict the qualitative outcome of soccer matches. For example the usefulness of the ELO-type systems is discussed in Hvattum and Arntzen (2010) and Leitner et al. (2010). Predictions based on the FIFA World Ranking have been analysed in, among others, McHale and Davies (2008) and Koning and McHale (2012). Finally, predictions of league matches have been used frequently in the context of assessing efficiency of betting markets. Examples of such studies are Forrest and Simmons (2000) and Goddard and Asimakopoulos (2004). Here, we focus on national team matches.

Prediction of the outcome of national team matches is difficult to the extent relatively few such matches are played in a given year. It is certainly impossible to estimate the statistical rating models of the previous section for national teams: the number of parameters to be estimated would be very large compared to the actual number of matches observed. For that reason, it seems reasonable to impose more structure, by making the quality parameters dependent on a few (easily observable) covariates. For example Koning and McHale (2012) make the quality parameters θ_i in the ordered Bradley–Terry model dependent on World Ranking points and rank, and on GDP per capita and size of the national population. Here, we use only the covariates from the World Ranking, as they are easily observable and updated frequently.

To be more precise, consider a match between team A and team B. World Ranking points and ranks are p_A, p_B, r_A and r_B respectively. Koning and McHale suggest to relate the outcome of a match to these observed variables by using the ordered Bradley–Terry model, with quality parameters dependent on the logarithm of these variables: $\theta_A = \beta_1 \log p_A + \beta_2 \log r_A$. Since only the difference in quality matters according to Equation 18.7, only the difference in covariates matter for the probability distribution of the actual outcome: $\theta_A - \theta_B = \beta_1 \log p_A/p_B + \beta_2 \log r_A/r_B$. The logarithmic transformation ensures that difference between teams ranked, say, 1 and 3 is larger than between teams ranked 51 and 53. In this specification of the relation between quality and World Ranking points and ranks, we need to estimate only two regression parameters β_1 and β_2 instead of one parameter for each team.

Treatment of home advantage requires more attention in matches between national teams: some matches are played during a tournament, where there is no home advantage. For example, neither Germany nor Argentina had home advantage when they played

TABLE 18.1

Estimation Results, Ordered Bradley–Terry Model for
Matches between National Teams

	Estimate	St. Err.	z-Value	p-Value
$\log p_A/p_B$	0.527	0.045	11.746	0.000
$\log r_A/r_B$	−0.226	0.032	−7.041	0.000
c_1	−0.707	0.029	−24.359	0.000
c_2	0.009	0.027	0.355	0.723
d	0.388	0.032	12.216	0.000

the final of the World Cup in Brazil in 2014. For some matches the restriction of no home advantage needs to be imposed. This is easily done by introducing another parameter d, and replacing the parameters c_1 and c_2 in Equation 18.7 by $-d$ and d respectively. The parameters c_1, c_2 and d are separately identified because there are both matches with home advantage and matches without home advantage.

To estimate the relation between outcome of a match, and difference in (logarithmic) points and ranks, we use all official matches between national teams between 1 January 2008 and 31 December 2014. Furthermore, we do not consider friendlies, only qualifiers for major tournaments, and the major tournaments themselves. Home advantage is relevant for qualifiers, 50% of the qualifiers during this period were won by the home team, while 22% ended in a draw. Out of 3048 matches played in that period, 2578 were qualifiers and 470 were played during the final tournaments (such as World Cup, European Cup, etc.).

Estimates of the ordered Bradley–Terry model are given in Table 18.1. The win probability of team A is increasing both in its number of points and rank (the sign of $\log r_A/r_B$ is negative), as a higher rank corresponds to a lower numerical value, keeping characteristics of team B constant. The effects of both covariates are significantly different from 0.

To assess the accuracy of the predictions implied by this rating method and the model that translates the points and ranks into probabilities, we calculate the probability distribution over the three possible outcomes for 275 qualifiers and 116 final tournament games played in 2015, using the point estimates of Table 18.1. In Table 18.2, we give the results of this out-of-sample test. The average probability assigned to a win by the first team (we use 'first team' because the first team listed may not be the home team if the match is played on neutral ground) is 0.59 over all matches won by the first team. Predicting draws is more difficult: 98 out of the 391 observed matches ended in a draw, and the average probability assigned to this event for these matches is only 0.24. Finally, looking only at matches won by team 2, the average probability assigned to that event is 0.44. The model is better at indicating a winning team, than at identifying matches that will end in a draw.

TABLE 18.2

Average Predicted Probabilities for Each Outcome

	Pr(1 wins)	Pr(draw)	Pr(2 wins)
1 wins	0.59	0.21	0.20
Draw	0.46	0.24	0.30
2 wins	0.32	0.24	0.44

18.4 An Application: Prediction and Measurement of Match Importance

In a second illustration of the use of rating models, we measure match significance. Some matches are more significant in determining the winner of a league than others. On the last day of the 2011/2012 season, all matches in the EPL were played simultaneously so that every team started playing with the same information. Before the matches started, Manchester City was leading the table, followed by Manchester United, on goal difference (Manchester City: 86 points, goal difference +63, Manchester United 86 points, goal difference +55). On the last day, Manchester City played a home game against Queens Park Rangers, Manchester United played an away game at Sunderland. Clearly, both matches had very high title significance: If Manchester City would win, they would win the title, if Manchester United would obtain more points in the matches than Manchester City, they would win the title. In the remainder of this section, we discuss a numerical measure of match significance, following Bojke (2007) and Koning (2007). (On that last day of play, Manchester United won their game by 1-0. Manchester City won against Queens Park Rangers by 3-2, with 2 goals in stoppage time. Manchester City won the title).

The final result of a season in a league is a ranking of teams, we denote this final ranking by R. This raking is a function of match results M_1 to M_T. With some slight abuse of notation, t indexes time as well as the match. Since match outcomes are random, so is the final ranking:

$$R = h(M_1, \ldots, M_T).\tag{18.8}$$

The probability distribution of the final ranking R is then obtained by summing over all possible match results:

$$\Pr(R = r) = \sum_{m_1,\ldots,m_T} I_{\{h(m_1,\ldots,m_T)=r\}} \Pr(M_1 = m_1, \ldots, M_T = m_T).\tag{18.9}$$

Clearly, this is not a practical way to estimate the probability distribution of the final ranking: effectively one integrates out a T-dimensional distribution, and T is the total number of matches in a season (380 in the case of EPL). Instead, it is easier to simulate match results, calculate the ranking and average over simulations:

$$\hat{f}(r) = \frac{1}{B} \sum_{b=1}^{B} h(M_1^b, \ldots, M_T^b).\tag{18.10}$$

with $\hat{f}(r)$ the estimated probability distribution of the final ranking and B the number of draws.

Usually, one is not only interested in the probability distribution of the final ranking at the beginning of the season, but also at some intermediate moment, just before the tth match is played. At that moment, we have information (and in particular, observed match results $M_1 = m_1, \ldots, M_{t-1} = m_{t-1}$) up to and including $t - 1$, we denote this information set by $\mathcal{I}_{t-1}$. The conditional distribution of the final ranking is then

$$\Pr(R = r | \mathcal{I}_{t-1}) = \sum_{m_t, \ldots, m_T} I_{\{h(m_1, \ldots, m_T) = r\}} \Pr(M_t = m_t, \ldots, M_T = m_T | \mathcal{I}_{t-1}). \tag{18.11}$$

This distribution can again be estimated by simulation:

$$\hat{f}(r | \mathcal{I}_{t-1}) = \frac{1}{B} \sum_{b=1}^{B} h(m_1, \ldots, m_{t-1}, M_t^b, \ldots, M_T^b), \tag{18.12}$$

with the remaining match results $(M_t^b, \ldots, M_T^b)$ now drawn from $f(m_t, \ldots, m_T | \mathcal{I}_{t-1})$, the probability distribution of outcomes of remaining matches, conditional on information up to and including $t - 1$. If we assume that future matches are conditionally independent, we obtain

$$f(m_t, \ldots, m_T | \mathcal{I}_{t-1}) = f(m_t | \mathcal{I}_{t-1}) \times \cdots \times f(m_T | \mathcal{I}_{t-1}). \tag{18.13}$$

At this moment, rating models as the Clarke–Norman model or the ordered Bradley–Terry model can be used to estimate the relevant conditional probability distributions. $\mathcal{I}_{t-1}$ can be thought of as a dataset with all observed match results up to yesterday. The model is estimated, and the estimated parameters are used to estimate the outcome distribution of the matches that remain to be played. If a match is played and a new result is observed, the parameters are re-estimated, etc. The estimated probability that, say, Manchester United wins the title varies over time because of two reasons. First, results are becoming known over time, the uncertainty of match results is resolved. Second, by re-estimating the model using, say, the 150 most recent observations, parameters are updated and reflect current form. We illustrate this by an example.

   At the start of the 2011/2012 season ($t = 0$), we estimate the parameters of the ordered Bradley–Terry model, and obtain the probability distribution of the outcomes of all 380 matches that remain to be played. The probability distribution of the final ranking is obtained by drawing from this probability distribution. In fact, for each match we draw the outcome-home win/draw/away win-from the probability distribution based on the estimated ordered Bradley–Terry model. Suppose this is a home win. The actual score is then obtained in a second step by drawing from a set of actually observed scores of matches ending in a home win. Repeated draws ($B = 50,000$) allow us to estimate the probability that Manchester United or Manchester City would win the title at the end of the season. After each day one or more matches are played, the parameters are updated using only observations from the previous 200 days. In Figure 18.5 we graph the development of the (smoothed) probability to win the title over time. Manchester United started as the favourite, reflecting the win of the title in the 2010/2011 season. Throughout the season, the odds of Manchester City improved. Early December 2011 Manchester City was leading the table, five points ahead of Manchester United and having a much better goal difference. In the early spring of 2012 fortunes reversed when Manchester City lost one game and drew two in March, and Manchester United won their games. Early April 2012 Manchester United was leading the table by five points over Manchester City, with similar goal differences (Manchester United: +49, Manchester City +50). With four matches left to play, Manchester United was still five points ahead, so the model predicts a very large probability they would retain their title. However, with three matches left in the season, Manchester United lost their away game to Manchester City, and thereby their lead in the table.

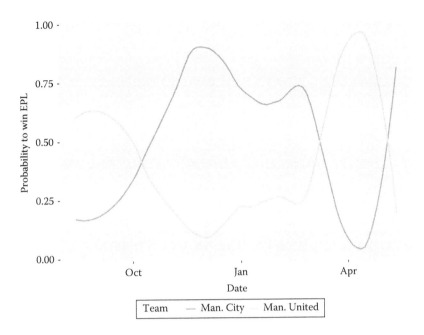

FIGURE 18.5
Estimated probability to win the EPL in 2011/2012.

This approach allows one to measure match significance as well, by means of a very intuitive measure: the significance of next match for a team is the probability of winning the title if the team were to win next match, minus the probability of winning the title if it were to lose next match:

$$\Pr(R_{\text{Man. United}} = 1 | M_1 = m_1, \ldots, M_{t-1} = m_{t-1}, M_t = \text{Man. United wins match } t)$$
$$- \Pr(R_{\text{Man. United}} = 1 | M_1 = m_1, \ldots, M_{t-1} = m_{t-1}, M_t = \text{Man. United loses match } t).$$

Of course, variations to this measure can be calculated, for example the difference in expected rank, or the difference in probability to finish on a position that avoids relegation. We calculated this measure for title significance for the 2011/2012 season, a smoothed version of this measure is graphed in Figure 18.6, both for Manchester City and Manchester United. As expected, matches at the end of the season were very significant for both teams. Early April 2012 matches were not so significant for Manchester City, as the model estimated their probability to win the title to be small.

18.5 Concluding Remarks

In this chapter we have discussed different rating models that can be used to measure team quality in soccer. Also, we have discussed two applications of these models. Clearly, different models are useful in different cases: one would not want to determine a final ranking based on the point estimates of an ordered Bradley–Terry model. Is the team that wins the league significantly better than all others? A rule based method seems preferable,

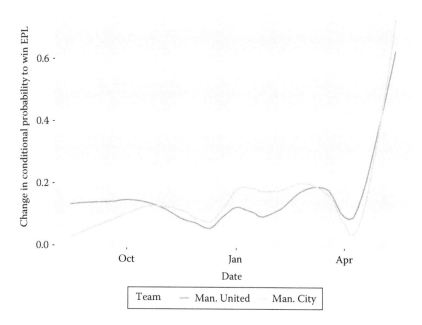

FIGURE 18.6
Change in probability to win EPL in 2011/2012, conditional on the outcome of next match.

also because of its simplicity. Elements of one model may be used in another. For example ELO-type rating models require a measure for the expected score. Usually, it is based on a rule as Equation 18.2, but one could also use the results of one of the statistical rating models of Section 18.2.2.

Arbitrarily, we have applied some models to match prediction. In the first application we related a rule based method to outcome results using the ordered Bradley–Terry model. In the second example, we estimated match significance. Rating models are useful in these cases if one needs a quick estimate of the probability distribution of outcomes of matches yet to be played. Rating models may also be useful in the context of assessing the effect of interventions (change of coach, move to a new stadium).

Even though the models we have discussed are all applied to the rating of soccer teams, they can be applied to other team sports as well. Many sports maintain a world ranking comparable to the FIFA World Ranking, and (variants of) the ordered Bradley–Terry model are easily implemented in both team sports and individual sports.

References

Bojke, C. (2007). The impact of post-season play-off systems on the attendance at regular season games. In J. Albert and R.H. Koning (Eds.), *Statistical Thinking in Sports*, Chapter 11, pp. 179–202. London, U.K.: CRC Press.

Bradley, R.A. and M.E. Terry (1952). Rank analysis if incomplete block designs: I, the method of paired comparisons. *Biometrika* 39(3–4), 325–345.

Cattelan, M., C. Varin, and D. Firth (2013). Dynamic Bradley-Terry modelling of sports tournaments. *Applied Statistics* 62(1), 135–150.

Clarke, S.R. and J.M. Norman (1995). Home ground advantage of individual clubs in English soccer. *The Statistician 44*(4), 509–521.

Elo, A.E. (1978). *The Rating of Chessplayers, Past and Present*. New York: Arco Publishing.

FIFA (2014a). *FACTSheet The FIFA/Coca-Cola World Ranking*. Zurich, Switzerland: FIFA.

FIFA (2014b). *Regulations 2014 FIFA World Cup Brazil*. Zurich, Switzerland: FIFA.

Forrest, D. and R. Simmons (2000). Making up the results: The work of the Football Pools Panel, 1963–1997. *The Statistician 49*(2), 253–260.

Goddard, J. and I. Asimakopoulos (2004). Forecasting football results and the efficiency of fixed-odds betting. *Journal of Forecasting 23*, 51–66.

Hvattum, L.M. and H. Arntzen (2010). Using ELO ratings for match result prediction in association football. *International Journal of Forecasting 26*, 460–470.

Koning, R.H. (2000). Competitive balance in Dutch soccer. *The Statistician 49*(3), 419–431.

Koning, R.H. (2003). An econometric evaluation of the effect of firing a coach on team performance. *Applied Economics 35*(5), 555–564.

Koning, R.H. (2007). Post-season play and league design in Dutch soccer. In P. Rodríguez, S. Késenne, and J. García (Eds.), *Governance and Competition in Profesisonal Sports Leagues*, pp. 191–215. Oviedo, Spain: Ediciones de la Universidad de Oviedo.

Koning, R.H. and I.G. McHale (2012). Estimating match and World Cup winning probabilities. In W. Maennig and A. Zimbalist (Eds.), *International Handbook on the Economics of Sporting Mega Events*, London, U.K.: Edward Elgar.

Lasek, J., Z. Szlávek, and S. Bhulai (2013). The predictive power of ranking systems in association football. *International Journal of Pattern Recognition 1*(1), 27–46.

Leitner, C., A. Zeileis, and K. Hornik (2010). Forecasting sports tournaments by rating of (prob)abiilities: A comparison for the EURO 2008. *International Journal of Forecasting 26*, 471–481.

McHale, I.G. and S. Davies (2008). Statistical analysis of the effectiveness of the FIFA World Rankings. In J. Albert and R.H. Koning (Eds.), *Statistical Thinking in Sports*, pp. 77–90. London, U.K.: Chapman & Hall/CRC.

Premier League (2015). *Premier League Handbook 2015/2016*. London, U.K.: The Football Association Premier League Limited.

Stefani, R. (2011). The methodology of officially recognized international sports rating systems. *Journal of Quantitative Analysis of Sports 7*(4), DOI: 10.2202/1559-0410.1347.

19

Player Ratings in Soccer

Ian G. McHale and Samuel D. Relton

CONTENTS

19.1 Introduction

Ranking is a popular pastime in society. From using league tables of schools and universities to top 10 lists of movies, books and songs, ranking is very much in the main stream of modern-day life. Even the examinations we sit (and that we academics set) are used to rank students so that the top $x\%$ can be identified and accepted onto a course. The same exams are ultimately used as part of the employment selection process.

The interest in rankings related to sport is perhaps even greater. Indeed, the whole purpose of professional sport is to rank the competitors. League competitions take place in soccer, cricket, baseball, basketball, American football and many other sports throughout the world. The team or individual ranked one at the end of the league is declared the champion.

However, when the unit of competitor is a team, the sporting focus is typically not on ranking the individual. Nevertheless, there are many reasons to rank the individuals comprising the team including to entertain and engage fans, to aid in the recruitment of skillful players and to inform strategic decisions on the field of play.

In this chapter, we are going to look at methods for, and issues relating to, ranking individual players in soccer. Of course, there are many ways to do this, and many methodologies exist; we will discuss the merits and drawbacks of some of the alternatives. Before that, in order to make measured judgements and comments on a rating system, and to decide whether it has any merit, it is important to acknowledge and understand the purpose of each system. There are two basic objectives that any rating system aims to fulfil, these are to

1. reward past performance, or to,
2. predict future performance.

Appreciating which of these two purposes a 'yet-to-be-designed' rating system will be primarily used for will influence decisions about the type of methodology to adopt and the final properties that the rating system should have.

In our opinion, the key fundamental difference between rating systems to be used for either of these purposes is that (greater) acknowledgement should be made of the likely influence of *noise* in the performance of competitors when predicting future performance. We will see that in soccer's most important action – scoring goals – there is a considerable influence of random chance in an individual's scoring record. As such, if one is predicting future performance (e.g. for recruitment purposes) rather than rewarding past performance (e.g. for assigning points for a player in a fantasy football game), then one should expect a certain amount of 'regression to the mean' for both over-performing and under-performing players.

We split the remainder of this chapter roughly into two parts. The first part looks at assigning values to player actions with the purpose of rewarding their past performances. The second part looks at how one might approach rating players with a view to predicting future performance. However, before this, we discuss the reliance of rating systems on the available data.

19.2 The Importance of Rich Data

The type of player rating system one can build is enormously dependent on the data available to the analyst. To demonstrate some key concepts, let us consider a simple example of two actions that occur in all soccer matches: passes and tackles. One way to rate players is to assign a value to these actions. Of course, one should assign values to the actions that make sense in relative terms – that is how much is a successful pass worth *relative* to a successful tackle? Of course, there are many variables that would need to be known to estimate this for an actual pass and an actual tackle. For example, is a tackle in the penalty area when the attacking player has just the goalkeeper to beat more or less 'valuable' than a 10-yard pass made on the half-way line? Most people will agree that in this example, the tackle is the most valuable action, and the player performing the tackle should be rewarded more highly than the player performing the pass.

Identifying such subtleties requires 'rich' data: data on the location, timing and success of the action. Until recent years, such data were not available in soccer. Nowadays companies like Opta and Prozone collect a vast amount of data on soccer matches. Prozone even has player tracking data in which the x–y coordinates of all 22 players on the pitch are recorded at a frequency of 10 times per second. However, if one cannot access these data sets (not only are they expensive, but in the case of Prozone tracking data, they are not generally available to people outside of soccer clubs on which the data concern), and only match count data are available (e.g. a player did m passes and n tackles) one needs to assign values to these actions taken as an average. In this case, the question that an analyst can answer with these data is actually: *on average, what is a successful pass worth, relative to a successful tackle?*

We have been given access to Prozone player tracking data, allowing us to extract insightful features about the current state of play at any given time. Later in the chapter we describe many of these features and use the data to develop a ratings algorithm for identifying passing ability in soccer.

19.3 Rewarding Past Performance and Valuing Actions

In soccer, designing a player rating system that rewards past performance is, in many ways, a much simpler task than designing one to predict future performance. As discussed in the previous section, the exercise boils down to assigning values to the actions that a player performs during a match or matches.

The scale of the value is of lower importance than the relative values of the actions. Some rating systems try to give scores to players that make it easy for the reader to use. For example, the EA SPORTS Player Performance Index, described in McHale et al. (2012) was originally designed to give a score to each player on the scale of 'three-points-for-a-win' and the 'one-point-for-a-draw'. As a result, the points for an individual player in a match were typically less than one and the ratings were displayed to three decimal places. However, after the first season of use, feedback from the public meant that the index was multiplied by 100 to give numbers that were more easily processed by the public. Crucially though, the relative worth of different actions remained as it was before.

So then, how might one go about valuing how much is a goal worth relative to a pass or indeed any other action? The approach adopted depends on the data available. For data in which timing and location of events is not given, one way to reward players for the actions they perform on the pitch (and identify the value of these actions) is to use the actions as covariates in a regression model for some dependent variable deemed to be a measure of success in the match.

For example, at the heart of the EA SPORTS PPI is a regression model for the number of shots in a match which uses covariates such as crosses, passes in the opposition half and tackle success ratio. The estimated coefficients represent the 'size' of the relationship of these actions with the number of shots. Shots are then related to goals via a second model, and goals are related to points awarded to the team via a deterministic function (if the team scores more goals than the opposition, it is awarded 3 points; if the team scores the same number of goals as the opposition, it is awarded 1 point; and if it scores less than the opposition, it is awarded 0 points). The end result is a rating system in which players are awarded a fixed number of points for each action they perform on the pitch regardless of where or when the action took place.

Additional factors that may effect the value of an action that one might wish to take account of include the difficulty of the action and the contribution of the action to the final result. For example, a player may successfully pass the ball 70 yards diagonally across the pitch. On the face of it, this is an impressive thing for a player to do. However, the pass may have made no contribution to creating a scoring opportunity, and in fact, a simpler pass, 10 yards down the line, may have been a better choice. This example demonstrates the almost unfathomably complex nature of trying to assign a value to the actions of players in a soccer match. However, analysts at MIT have made interesting steps in basketball.

19.3.1 Expected Value of Possession

Cervone et al. (2014) present a major advancement in player ratings which represents a truly novel idea for how to use the new influx of player-tracking data available for team sports. They use player-tracking data from the National Basketball Association (NBA) to assign a point value to each movement during a possession by computing how many points the attacking team is expected to score by the end of the possession, a quantity they call *expected possession value*.

The idea can be adapted to soccer. Every time a player makes a decision to shoot, pass, or dribble the ball, there is an associated change in the likelihood of that possession ending in a goal. The instantaneous probability of the possession ending in a goal, calculated conditionally on the state of play (i.e. the positions [and speeds] of the players, and the ball, on the pitch), is the 'expected value of the possession'. By comparing the actual change in the expected value of the possession a player created, with that of the player's other options (and acknowledging the probabilities of each of the other options being successful), the player's decision making and ability to execute intended moves can be assessed.

In basketball a 'successful' possession can end with a 3-point score or a 1-point score. In soccer, there is only one goal awarded but, despite this simplifying feature of soccer over basketball, the computations are made many times more demanding as a consequence of the increase in the number of players on the pitch compared to that on the court in basketball. There are further complications if one is to implement the expected value of possession idea in soccer: off-the-ball runs and activity such as breaking free from defenders are a key part of a player's requirements in the team which are not recognised with reward; passes are often successful largely as a consequence of the receiver's skill and not the passer – how can points be divided up accordingly? Defensive actions in general are not monitored. Nevertheless, it may be possible to overcome these drawbacks and perhaps the use of expected value of possession will prove to be a major advancement in player ratings in soccer.

19.4 Predicting Future Performance and Regression to the Mean

The ideas discussed earlier concern rewarding past performance. Let us now consider the idea of predicting future performance. Consider a player who has reached the milestone of scoring 30 goals in one season. Investigating the players record, we learn that to score the 30 goals, the player achieved a well-above average goals per shot percentage of 30%. Would we expect the player to be able to repeat the feat (of both a high goals per shot ratio and hence a high number of goals) in the following season? As statisticians, we would probably expect that for all but the very top players, there might be *regression to the mean* and the player would score fewer goals in the following season. The principal reason for such intuition is the element of chance in converting a shot to a goal – some players may experience good luck whilst others will have some bad luck. As such, the player's goals per shot ratio might be expected to drop.

Predicting future performance of players is very much an exercise in separating the signal and the noise. This was the focus in McHale and Szczepanski (2013). The modelling

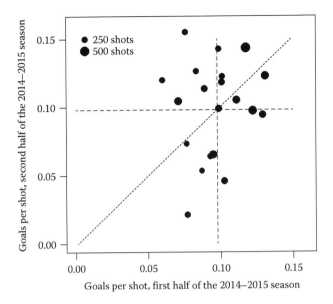

FIGURE 19.1
Goals per shot ratio for the second half of the 2014–2015 season versus the goals per shot ratio for the first half of
the 2014–2015 season. The horizontal and vertical lines show the mean ratio for all teams, and the diagonal dotted
line is the identity function.

approach they adopted was informed by ideas first discussed in relation to baseball in Efron
and Morris (1975) who used Stein's estimator to shrink observed batting averages and
subsequently showed a positive effect on predicted performance in subsequent seasons.
Continuing the theme, Albert (1992, 2006) used random effects models to analyse home
run hitters and estimated the ability of pitchers respectively. And related methods were
applied to analyse other aspects of baseball: Jensen et al. (2009) focussed on fielding performance, and Loughin and Bargen (2008) investigated pitcher and catcher influence on
base stealing.

To demonstrate the findings in McHale and Szczepanski (2013), consider the goals to
shots ratio at the team level. Figure 19.1 shows the goals to shots ratio for the second half
of the 2014–2015 versus the first half of the season. If there was a (strong) relationship, it
would be indicative of little randomness in goals per shot as a statistic. However, the figure
looks like noise – teams under-performing in the first half of the season are just as likely to
over-perform in the second half of the season. Checking this by fitting a linear regression
model confirms there is no relationship between the two as the coefficient on goals per shot
in the first half of the season is not statistically significant (far from it in fact).

How then can one deal with elements of chance when predicting future performance?
The general idea presented in McHale and Szczepanski (2013) and others (in baseball) is
to use random effects models. To estimate the ability of a player to convert shots to goals
we consider the outcome of each shot as a Bernoulli trial with the probability of success
depending on some covariates (e.g. distance from the goal, the team and the opposition)
and crucially the identity of the player. Letting the players themselves be a random effect
is akin to assuming that player abilities are random variables belonging to some probability distribution. The estimated value of the random effect for each player can be obtained
and it is this that serves as the estimate of the player's ability. The property that makes

this approach so attractive is that player strengths are 'shrunk' towards the mean and the amount of shrinkage is more if a player is observed only on a few occasions. In other words, for a player to attain a high rating, there must be a significant amount of evidence that this higher rating is deserved. The approach is both logically and intuitively appealing.

19.4.1 A Random Effects Model for Rating the Passing Ability of Players

In this section, we consider rating players for their ability to pass the ball. To do so, we adopt the random effects modelling framework described in the previous section. To estimate our model parameters, we use a rich data set provided by Prozone covering the 2012–2013 season of the English Premier League. The data come in two parts: (1) event data giving the x–y coordinates of all events (touch of the ball, pass, tackle, foul, shot and so on) occurring in a game, the players involved and the time of the event, and (2) player-tracking data giving the x–y coordinates of every player on the pitch, recorded at a frequency of 10 times per second.

The pitch dimensions are recorded such that the x-axis runs through the centre of each goal and is of mean length 103.25 m, whilst the y-axis runs across the halfway line and is of mean width 67.55 m.

In our data set there are 275,286 passes performed by 527 players with a completion rate of 85.71%. Table 19.1 shows the number of passes and the completion rate for each team in the league.

TABLE 19.1

Number of Passes and the Pass Completion Rate for Each Team during the 2012–2013 Season of the English Premier League, Sorted in Decreasing Order of Pass Completion Rate

Team	Number of Passes	Pass Completion Rate
Arsenal	18,493	0.90
Manchester City	18,030	0.89
Manchester United	18,053	0.89
Swansea City	17,637	0.88
Liverpool	18,286	0.88
Tottenham Hotspur	15,499	0.88
Chelsea	16,964	0.87
Fulham	14,727	0.87
Wigan Athletic	16,162	0.86
Everton	12,485	0.85
Newcastle United	12,538	0.85
Aston Villa	12,008	0.84
West Bromwich Albion	11,944	0.84
Queens Park Rangers	10,618	0.84
Southampton	13,587	0.83
Sunderland	9,849	0.83
Norwich City	10,171	0.82
West Ham United	9,869	0.81
Stoke City	8,980	0.79
Reading	9,386	0.78

The model we present is an enhancement on that described in Szczepanski and McHale (2015). The model estimated the probability of a pass being successful given several covariates generated from event data. The covariates are the *origin* and *destination* of the pass, the *time since the previous pass*, the *pass number* in the sequence of passes, *game time*, *player position* in the game, an indicator for whether the pass was a *headed pass* and a *home advantage* dummy. Random effects were used for each of the passer's team, the opposing team and the passer himself.

The model that we present here has access to player-tracking data, in addition to the event data used for the previously described model. As such we can use a different set of covariates to better account for the subtleties affecting whether a pass will be successful or not.

The list of covariates we derived, their definitions and their intended purposes are as follows:

1. *dist*: the distance (in metres) from the origin of the pass to the intended destination of the pass. It is expected that as pass distance increases, the probability of success will decrease.

2. *moe* (margin of error): the minimum angle between the line between the pass origin and intended destination and the opposition players (described in Figure 19.2). It is expected that as the margin of error increases, the probability of a pass being successful will increase.

3. *intendedX*: the intended x-coordinate of the pass destination.

4. *intendedY*: the intended y-coordinate of the pass destination.

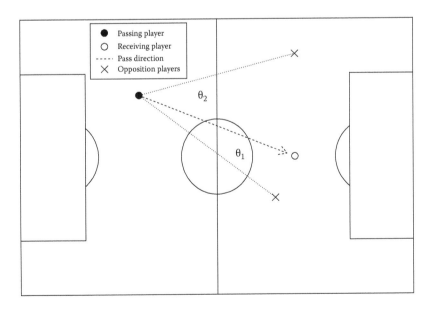

FIGURE 19.2
The plot depicts how we define margin of error, *moe*. The passer (solid black circle) is passing to the receiver (hollow circle) with two opposition players shown as the crosses. The passing player's margin of error is the minimum angle made by the vector from passer to receiver and the vector from passer to opposition player. In this example, it is θ_1.

5. *aveXoppTeam*: the average distance on the *x*-axis from the opposition goal of opposition team players. This variable is intended to capture whether the attacking team is on the counter-attack and thus has more room than would otherwise be the case.

6. *aveXpassersTeam*: the average distance on the *x*-axis from the opposition goal of the passing players teammates. In addition to *aveXoppTeam* mentioned earlier, this variable is intended to capture whether the attacking team is on the counter-attack and thus has more room than would otherwise be the case.

7. *passerPressure*: the number of opposition players within a 4 m radius circle of the passer.

8. *receiverPressure*: number of opposition players within a 4 m radius circle of the intended receiver.

9. *touches*: the number of touches the passing player has had. We expect that first touch passes are harder to complete.

10. *timeOnBall*: the time on the ball the passing player has had.

11. *home*: a dummy variable indicating that the player passing the ball is on the home team. There may be a positive effect on the probability of a pass being completed resulting from familiarity with the surroundings.

12. *passersTeam*: a factor for the passing player's team. This variable will account for the quality of the passer's teammates in being able to receive the ball.

13. *oppositionTeam*: a factor for the opposition team. This variable will account for the quality of the opposition team in stopping passes being completed.

The model we adopt to explain pass success is a Generalized Additive Mixed Model (GAMM), see Lin and Zhang (1999). A generalized additive mixed model is an extension of a generalised linear model in which the linear predictor is allowed to involve smooth functions of covariates as well as random effects. We use a smooth function, a tensor, to model the pass probability as a function of the *x*–*y* coordinates of the destination of the pass. Let the outcome of the the *i*th pass be o_i where $o_i = 1$ for a successful pass and 0 otherwise. We assume that the distribution of pass outcome follows a Bernoulli distribution with the probability of success represented by the inverse logit function of the linear predictor η_i:

$$(o_i|\eta_i) \sim \text{Bernoulli}\left(p_i\right) \tag{19.1}$$

where

$$p_i = \frac{\exp(\eta_i)}{1 + \exp(\eta_i)}. \tag{19.2}$$

We let η_i be a function of fixed effects, $\boldsymbol{\beta}$, and random effects, $\boldsymbol{b}$. The fixed effects correspond to a matrix, $\boldsymbol{W}$, of the variables listed earlier and the intercept. The random effects are given by a vector, $\boldsymbol{b}$, with the $K = 527$ elements representing the passing ability of players.

Within the GAMM framework, η_i is given by

$$\eta_i = \boldsymbol{W}_i\boldsymbol{\beta} + \boldsymbol{Z}_i\boldsymbol{b} + s(x_i, x_{i,end}, y_i, y_{i,end}). \tag{19.3}$$

where $\boldsymbol{Z}_i$ is a row of a design matrix selecting the elements of the random effects vector $\boldsymbol{b}$ corresponding to the player executing the *i*th pass. The variable *s* is a *tensor product smooth* we use to describe the origin and destination component of each pass. The tensor product smooth s allows the probability of a pass being successful to be a function of the x and y coordinates of the pass origin and destination.

For the random effects we assume

$$b \sim \mathcal{N}(0, \Sigma(\sigma)), \tag{19.4}$$

where $\Sigma(\sigma)$ is a skill covariance matrix with the elements on the diagonal equal to the player skill variance. This choice of random effect distribution reflects the belief that extremely good (and bad) players and teams are less common than *average* ones.

In this application, the values of the random effects for the players are the key parameters of interest since they can be interpreted as the passing ability of the players.

19.4.1.1 Results

Table 19.2 presents estimates of the parametric model terms contained in the vector β. As expected, the estimated coefficient on intended pass distance, *dist*, is negative so that as the distance increases the probability of the pass being successful decreases. Similarly as margin of error, *moe*, increases as the probability of the pass being successful increases. Regarding the passing player's team fixed effects, in this table Arsenal act as the reference category and, interestingly, the other top teams are not statistically significantly different. On the other hand, passes performed by lower rated teams such as Aston Villa, Stoke City and Reading are (statistically) significantly less likely to be completed than a pass by an Arsenal player. We expect this is a consequence of the quality of the receiving player. The estimated coefficients on the other variables are also as expected and, since the focus of this chapter is rating, we move on to using the random effects terms to obtain player ratings for passing.

The player ratings for passing ability from our model are given by the values of the random effects themselves. However, we can rescale the random effects so that the final ratings system is on a more intuitive scale. To do so, we use the model to predict the success probability of an 'average' difficulty pass for each of the $K = 527$ players. Thus, the ratings value for a player represents the probability of that player successfully completing a pass of average difficulty.

Table 19.3 shows the estimated top five players in each position. As with all rating systems, it is good to see the expected players present – those who have a reputation for being good at passing – but it is interesting to see a few unexpected names. For example, in the central midfield position, Paul Scholes is widely regarded as an excellent passer of the ball, whereas Fabian Delph, certainly during this season, did not have such a reputation. Since then Delph has made a move to one of the top teams in Europe, Manchester City.

19.5 Closing Remarks

In this chapter, we have presented our thoughts on the rationale for constructing a rating system for players in soccer. A key issue is whether the system is to be used to rate past performance or predict future performance. For predicting future performance, we describe the rationale for adopting random effects models. Such models are able to shrink estimated

TABLE 19.2

Model Estimates for the Fitted GAMM Model

Parameter	Estimate (s.e.)	z-Value
intercept	1.406 (0.097)	14.48
dist	−0.816 (0.007)	−112.90
moe	4.250 (0.067)	63.00
moe^2	−1.432 (0.045)	−31.54
aveXoppTeam	0.258 (0.043)	6.06
aveXpassersTeam	−0.196 (0.041)	−4.74
passerPressure	−0.098 (0.049)	−1.98
receiverPressure	−0.099 (0.051)	−1.94
touches	0.009 (0.011)	0.84
timeOnBall	0.274 (0.015)	18.04
timeOnBall^2	−0.030 (0.001)	−20.36
Aston Villa	−0.382 (0.135)	−2.84
Chelsea	−0.135 (0.137)	−0.99
Everton	−0.431 (0.141)	−3.05
Fulham	−0.204 (0.133)	−1.53
Liverpool	−0.098 (0.135)	−0.73
Manchester City	0.069 (0.136)	0.50
Manchester United	−0.060 (0.133)	−0.45
Newcastle United	−0.219 (0.131)	−1.68
Norwich City	−0.435 (0.140)	−3.11
Queens Park Rangers	−0.360 (0.130)	−2.77
Reading	−0.735 (0.132)	−5.58
Southampton	−0.264 (0.139)	−1.90
Stoke City	−0.622 (0.142)	−4.37
Sunderland	−0.503 (0.139)	−3.62
Swansea City	0.054 (0.138)	0.39
Tottenham Hotspur	−0.176 (0.134)	−1.31
West Bromwich Albion	−0.299 (0.137)	−2.18
West Ham United	−0.516 (0.138)	−3.74
Wigan Athletic	−0.248 (0.135)	−1.83

Note: We show the estimated values of the features derived from the Prozone data and the *passersTeam*, with Arsenal acting as the reference team for the latter parameter. Standard errors are shown in parentheses and z-values are displayed in the right-hand column.

abilities towards the mean and, critically, the amount of shrinkage depends on how much evidence there is of the player under-, or over-performing.

Models for rating players in soccer are very much in their infancy. There are many issues yet to be resolved, but perhaps the most difficult one will be to incorporate off-the-ball movement into a rating system. Another nontrivial problem is that of appropriately rewarding goalkeepers for their contribution to a team. They are almost playing a different sport (for starters they are permitted to use the hands!), but the complications arise from the fact that every saved shot they make is effectively a prevented goal. Using symmetry,

TABLE 19.3

Player Ratings: The Top Five Players by Position

Rating	Player	Pass Completion Rate	Number of Passes	Team	Position	Ability
1	Sung-Yueng Ki	0.959	1159	Swansea City	Centre Midfield	0.961
2	Paul Scholes	0.947	681	Manchester United	Centre Midfield	0.959
3	Javi Garcia	0.965	677	Manchester City	Centre Midfield	0.958
4	Fabian Delph	0.937	795	Aston Villa	Centre Midfield	0.957
5	Moussa Dembele	0.959	1308	Fulham	Centre Midfield	0.956
1	Shaun Maloney	0.931	1254	Wigan Athletic	Left Midfield	0.940
2	Damien Duff	0.931	533	Fulham	Left Midfield	0.931
3	Stephane Sessegnon	0.881	613	Sunderland	Left Midfield	0.928
4	Andreas Weimann	0.864	553	Aston Villa	Left Midfield	0.923
5	Eden Hazard	0.894	1136	Chelsea	Left Midfield	0.921
1	Seamus Coleman	0.845	595	Everton	Right Midfield	0.942
2	Barry Bannan	0.884	644	Aston Villa	Right Midfield	0.939
3	Samir Nasri	0.938	878	Manchester City	Right Midfield	0.933
4	Adam Johnson	0.849	689	Sunderland	Right Midfield	0.928
5	Antonio Valencia	0.902	716	Manchester United	Right Midfield	0.927
1	Javier Garrido	0.866	677	Norwich City	Left Back	0.954
2	Ashley Cole	0.918	918	Chelsea	Left Back	0.952
3	Joseph Bennett	0.893	563	Aston Villa	Left Back	0.951
4	Jan Vertonghen	0.926	1298	Tottenham Hotspur	Left Back	0.949
5	Leighton Baines	0.872	1137	Everton	Left Back	0.949
1	James Perch	0.932	618	Newcastle United	Right Back	0.957
2	Aaron Hughes	0.932	793	Fulham	Right Back	0.956
3	Sascha Riether	0.893	1181	Fulham	Right Back	0.956
4	Kyle Walker	0.896	1205	Tottenham Hotspur	Right Back	0.949
5	Joey O'brien	0.869	655	West Ham United	Right Back	0.947
1	Arouna Kone	0.899	673	Wigan Athletic	Attacker	0.922
2	Shinji Kagawa	0.916	675	Manchester United	Attacker	0.919
3	Clint Dempsey	0.915	586	Tottenham Hotspur	Attacker	0.912
4	Wayne Rooney	0.891	845	Manchester United	Attacker	0.912
5	Adam Lallana	0.899	666	Southampton	Attacker	0.911
1	Martin Skrtel	0.964	840	Liverpool	Centre Back	0.974
2	Per Mertesacker	0.965	1231	Arsenal	Centre Back	0.973
3	Ivan Ramis	0.929	622	Wigan Athletic	Centre Back	0.969
4	Daniel Agger	0.954	1367	Liverpool	Centre Back	0.969
5	Joleon Lescott	0.963	679	Manchester City	Centre Back	0.966

Note: In addition to each player's probability of completing an average difficulty pass, found under the column named 'Ability', the table also shows the player's pass completion percentage, the number of passes performed and the team he plays for.

one could argue that a save should be worth the same value as scoring a goal. However, a goalkeeper will make many more saves in the course of a match than the total number of scored goals. Does this mean a goalkeeper should be more highly rated? We would argue that they should not: many of the saves are easier for the goalkeeper to make than the goals are for the striker to score. How one deals with this properly is a difficult issue.

One area where, currently, there has been comparatively little research is in analysing (and ranking) players by their defensive capability. This is made difficult by the fact that defensive actions do not have an inherent number of points attached to them, making it a similar issue to determining how we award points to each goalkeeper for saving shots.

Some interesting work on analysing defensive performance has been done in basketball by Franks et al. (2015). They design a model to determine which defender is responsible for each attacker at any given time. This is followed by identifying how the defender affects the shot efficiency and shot frequency of the attacker in question. With some modifications a similar analysis could be performed for soccer and it will be interesting to see how defensive analysis evolves in the coming years.

Acknowledgement

We thank Prozone for allowing us to access their data.

References

Albert, J. (1992). A Bayesian analysis of a Poisson random effects model for home run hitters. *American Statistician*, 46, 246–253.

Albert, J. (2006). Pitching statistics, talent and luck, and the best strikeout seasons of all-time. *Journal of Quantitative Analysis in Sports*, 2, 1.

Cervone, D., D'Amour, A., Bornn, L. and Goldsberry, K. (2014). POINTWISE: Predicting points and valuing decisions in real time with NBA optical tracking data. *MIT Sloan Sports Analytics Conference* paper.

Efron, B. and Morris, C. (1975). Data analysis using Stein's estimator and its generalizations. *Journal of the American Statistical Association*, 70, 311–319.

Franks, A., Miller, A., Bornn, L. and Goldsberry, K. (2015). Characterizing the spatial structure of defensive skill in professional basketball. *Annals of Applied Statistics*, 9, 94–121.

Jensen, S. T., Shirley, K. E. and Wyner, A. J. (2009). Bayesball: A Bayesian hierarchical model for evaluating fielding in major league baseball. *Annals of Applied Statistics*, 3, 491–520.

Lin, X. and Zhang, D. (1999). Inference in generalized additive mixed models by using smoothing splines. *Journal of the Royal Statistical Society, Series B*, 61, 381–400.

Loughin, T. M. and Bargen, J. L. (2008). Assessing pitcher and catcher influences on base stealing in Major League Baseball. *Journal of Sports Science*, 26, 15–20.

McHale, I. G., Scarf, P. A. and Folker, D. E. (2012). On the development of a soccer player performance rating system for the English Premier League. *Interfaces*, 42, 339–351.

McHale, I. G. and Szczepanski, L. (2013). A mixed effects model for identifying goal scoring ability of footballers. *Journal of the Royal Statistical Society, Series A*, 397–417. DOI:10.1111/rssa.12015.

Szczepanski, L. and McHale, I. G. (2015). Beyond completion rate: Evaluating passing ability of footballers. *Journal of the Royal Statistical Society, Series A*, 513–533.

20

Effectiveness of In-Season Coach Dismissal

Lucas M. Besters, Jan C. van Ours, and Martin A. van Tuijl

CONTENTS

20.1 Introduction

Sports have become a fruitful field of research. First, studying sports is interesting in itself. Second, there are similarities between phenomena in sports and in other fields, such as labor economics (Kahn 2000, Szymanski 2003). One advantage of sports research is the wide availability and the relatively high frequency of data, such as match results in soccer. In this chapter, we will use such data and investigate the effect of an in-season coach dismissal of a professional soccer club on in-season performances of soccer teams. Given the resemblance between the position of a head-coach and a CEO (Ter Weel 2011), this question is interesting from a business perspective as well. Kuper and Szymanski (2010) indicate doubt about the impact of a soccer manager on team performance. Anderson and Sally (2013) argue that the influence of a soccer manager is nonnegligible. Most previous studies show that short-term performances do not improve as a result of a change in head-coach (Van Ours and Van Tuijl (2016) provide a recent overview of previous studies). Previous studies use various methods and data sets from different countries and different time periods. Koning (2003) shows that performances do not improve after the replacement of a head-coach. He uses data from the first tier of Dutch professional soccer ("Eredivisie"). Koning takes into account the heterogeneity of opponents prior to and posterior to a change of head-coach. Furthermore, Bruinshoofd and Ter Weel (2003) and Van Ours and Van Tuijl (2016), also studying the highest professional soccer league in the Netherlands, show that a sequence of bad performances is often followed by a recovery, irrespective of the replacement of the head-coach. This points at a so-called regression to the mean, that is, if a sequence of bad performances is random, better results will follow automatically although performance has not improved. Thus, the conclusions of the these studies are similar concerning the effectiveness of an in-season head-coach dismissal. However, the setup of the empirical analysis differs. And, although the authors all use data from the highest Dutch professional soccer league, their sample periods differ.

Professional soccer in Europe has experienced major developments in the past decades. Currently, Dutch professional soccer teams do not play an important role in European club competitions such as the Champions League and the Europa League. Competing with foreign clubs with a much looser budget constraint is virtually impossible. Nevertheless, the highest national league remains popular in the Netherlands, with high average attendance figures ranking sixth after the "Big Five" (England, France, Germany, Italy, and Spain). Professional soccer attracts much media attention (Internet, printed press, radio, and television). This has resulted in a mature soccer industry. According to PWC (2015), the value added of the Dutch soccer industry is about € 2.2 billion (approximately 0.34% of Dutch GDP in 2013), apart from substantial noneconomic benefits. Professional soccer clubs are comparable to small- and medium-sized enterprises but with a diverse group of stakeholders that might react in a rather emotional way. The general goal of professional soccer clubs is to maximize sportive performances. In case results are falling behind expectations, the management of a club might feel the pressure to act. Given the budget constraint and regulations that prohibit the hiring of new players outside the transfer window, replacing of the head-coach may be a strategy to boost the short-run performance. The question arises, however, whether this is an effective measure. We examine this question by using data from the highest Dutch professional soccer league for the seasons 2000/2001–2013/2014. We compare the three alternative methods mentioned earlier, demonstrating differences in methodology and investigating discrepancies in outcomes. Although the three methods use different identification strategies to establish the causal effect of coach replacement on team performance, the results are very much the same: in-season replacement of a coach does not affect performance.

The rest of this chapter is organized as follows. First, Section 20.2 briefly describes our data. Next, in Section 20.3, we discuss the different methods as well as the outcomes. Section 20.3.1 discusses Koning (2003), Section 20.3.2 deals with Bruinshoofd and Ter Weel (2003), while Section 20.3.3 is devoted to Van Ours and Van Tuijl (2016). Finally, Section 20.4 provides an evaluation of the different methods and the findings.

20.2 Data

Our analysis is based on matches from the highest professional Dutch soccer league for the seasons 2000/2001–2013/2014; the same data used by Van Ours and Van Tuijl (2016), viz. In the "Eredivisie," 18 teams compete for the Dutch championship according to a double round-robin format. For each of the 306 matches (34 rounds of 9 matches) per season, the home team, the away team, and the score by both teams at full time are recorded. Furthermore, the sample contains information about the head-coaches and, in case of a separation between club and coach, whether it was a dismissal ("forced") or a resignation ("quit"). We are especially interested in the effects of coach dismissals. However, in case of a resignation, it is not always clear that the separation was voluntary. In some cases, if no resignation would have occurred, the coach would have been fired. To account for the possibility that some resignations were in fact dismissals, we perform our analysis for both the coach dismissals and all coach replacements. The sequence number of the match in a particular season is recorded, as well as some coach characteristics (age, experience, tenure, and caps in the national team) and the rank of the opponent in the previous season.* For the replication of the analysis of Van Ours and Van Tuijl (2016), bookmaker data are included.† Table 20.1 provides an overview of the head-coach changes during the sample period. One should note that only the first change of a head-coach at a particular club within a season is taken into account. Sometimes, clubs appoint a caretaker after the exit of the former head-coach. The caretaker, in turn, is replaced by a new head-coach after some matches. In this chapter, we solely study the replacement of the initial head-coach. The subsequent performance is completely attributed to the "same" new head-coach. For this purpose, we will use the goal difference as the performance measure in all analyses. Following Van Ours and Van Tuijl (2016), we leave out all changes that took place before match 5 and after match 30, resulting in a total of 59 changes, including 42 dismissals. This consistent set of changes facilitates a comparison of the methods and the results.

As shown in Table 20.1, the number of coach dismissals in a season ranges from 1 in 2006/2007 to 6 in 2008/2009. The number of quits in a season ranges from 0 in the last two seasons to a maximum of 2 in some other seasons.

TABLE 20.1

Overview of Coach Changes during the Sample Period 2000/2001–2013/2014

Season	Dismissals by Club	Quits by Club	Dismissals	All
00/01	Graafschap, Fortuna Sittard, Sparta	AZ	3	4
01/02	Ajax, Fortuna Sittard, Roda JC	Vitesse	3	4
02/03	AZ, RBC, Vitesse	Groningen	3	4
03/04	Volendam, PEC Zwolle	ADO Den Haag, Willem II	2	4
04/05	Den Bosch, NEC, RBC	Ajax	3	4
05/06	NAC Breda, RBC, Willem II	Twente, NEC	3	5
06/07	RKC	ADO Den Haag, Roda JC	1	3
07/08	Heracles Almelo, Sparta, Willem II	Ajax, PSV	3	5
08/09	Graafschap, Utrecht, Feyenoord, PSV, Roda JC, Vitesse	Willem II	6	7
09/10	ADO Den Haag, AZ, Heerenveen, Willem II	NEC	4	5
10/11	VVV, Vitesse	Ajax	2	3
11/12	Graafschap, Twente, PSV	Utrecht, VVV	3	5
12/13	Twente, NAC		2	2
13/14	ADO Den Haag, AZ, Cambuur, Roda JC		4	4
Total			42	59

Note: The table provides information on the number of coach changes during the sample period, distinguishing between coach dismissals and quits. All represents the sum of the two. Since only the coach changes that happened after the 5th match and before the 30th match are considered, seven changes are left out of the analyses. These are: Feyenoord (06/07), ADO Den Haag (08/09), Ajax (08/09), Sparta (09/10), NAC Breda (10/11), Willem II (10/11), NEC (13/14).

---

* Teams that are promoted, and thus were not playing at the highest level in previous season, are given rank 18.
† Ninety-seven percent of the bookmaker data comes from William Hill, 2% from Ladbrokes and 1% from others. No data were available for two matches; see Van Ours and Van Tuijl (2016) for details.

20.3 Research Methods

As mentioned in Section 20.1, we use three different methods to address the effectiveness of an in-season head-coach dismissal empirically. These methods have been used before on different samples from the Dutch professional soccer league. This will be done in separate subsections by first describing the essence of the method, followed by some results. Section 20.3.1 presents the method of Koning (2003), who takes account of heterogeneity between opponents, but who does not use a specific control group. Section 20.3.2 discusses the method of Bruinshoofd and Ter Weel (2003). They construct a control group based on in-season short-term performances and apply a difference-in-differences approach. However, they do not apply any particular club fixed effect or season fixed effect. Section 20.3.3 deals with the method used by Van Ours and Van Tuijl (2016). They use bookmaker data to model expectations and use this to construct a control group for a difference-in-differences analysis in which club-season fixed effects are included.

20.3.1 Koning (2003)

Using match-specific data for the seasons 1993/1994–1997/1998 of the highest Dutch professional soccer league, Koning uses a rank model to investigate the effectiveness of an in-season coach dismissal. He argues that one of the main complications stems from the different conditions faced by the old and the new head-coach (which shows resemblance to a manager in a company). Quality differences of opponents seem to be the major factor. When analyzing the performances prior to and posterior to a change, therefore, one should take quality differences between adversaries into account. Koning uses a model similar to Clarke and Norman (1995), in which the goal difference of a match is related to home advantage, as well as to some quality indicators for the home team and the away team. The goal difference D_{ij} between home team i and away team j is defined as

$$D_{ij} = h_i + \theta_i - \theta_j + \varepsilon_{ij}, \tag{20.1}$$

where h_i denotes the home advantage (the expected goal difference in case of equal quality)* while θ_i and θ_j indicate the quality of the home team and the away team, respectively. The parameter ε_{ij} refers to a zero-mean error term with constant variance. In order to detect any effect of a coach change, two extensions are formulated:

$$h_i^n = h_i^o + k_i, \quad i \in F$$
$$\theta_i^n = \theta_i^o + \psi_i, \quad i \in F \tag{20.2}$$

in which n represents the new coach and o the old one. Furthermore, k_i is the change in home advantage, F is the set of teams that fired a coach during the season, and ψ_i the change in quality of the home team for the set of clubs that replaced their head-coach. Any effect of the replacement of a coach is then reflected in the values of k_i and ψ_i. This implicitly

---

* It is a well-known fact that home teams generally obtain better results than away teams. In our sample, the share of home victories is approximately 49%, the share of away victories is close to 28%, while the share of draws is roughly 23%.

assumes that both home advantage and the quality of the teams are fixed within a particular season in case no change of head-coach occurs. According to Koning, however, the composition of teams, and thus their quality, may change between seasons. Therefore, one should estimate the model for every season separately. While estimation can be done using dummies* one should impose the identification restriction that all quality indicators sum up to 0 ($\sum_i \theta_i = 0$). This yields the interpretation of a relative quality indicator, the value denoting the deviation from a hypothetical average team with a quality equal to 0.

Besides testing whether there is a positive effect of an in-season coach dismissal for individual teams (i.e., $k_i > 0$ and $\psi_i > 0$) it is also useful to test whether such an effect can be generalized over teams. This is done by imposing the constraint that all changes are equal across teams (i.e., $k_i = k$ and $\psi_i = \psi$).[†]

Table 20.2 provides some preliminary results. It presents the average number of goals scored and conceded for both the old and new head-coach. Furthermore, for both head-coaches, the average goal difference is presented. The fourth column shows that, in general, clubs that change their head-coach have a negative goal difference, indicating bad performances prior to the replacement. The goal difference for the new coach exhibits a considerable number of negative values. Nevertheless, these numbers all show an improvement compared to the average numbers of the old coach. The last column shows the difference-in-differences. It indicates that indeed positive results arise, when deducting the average score of the old coach from the average score of the new coach. This naïve approach suggests that in-season coach dismissals have a positive effect on performance. However, it does not take account of potential differences between opponents and the order of play. We can encounter these elements by using the model of Koning, as presented earlier. The model specification in which both extensions are estimated and both parameters are assumed to be constant across teams is of main interest. In his paper, Koning argues that if both are positive, an in-season effect might be present and he argues that this is the preferred model to be used.

Table 20.3 reports the parameter estimates of this extended model of Koning. In the first and second column, the estimates of the change in home advantage and team quality, respectively, are reported for all coach changes. The third and fourth columns show the same for coach dismissals. Only a few estimates are significantly different from zero, while it is never the case within a particular season that both estimates are significantly different from zero. Any potential positive effect of the replacement of a head-coach, which seems to present according to the findings in Table 20.2, disappears when taking account of home advantage and controlling for the heterogeneity in quality between teams (opponents). These analyses show that a naïve approach of simply looking at the short-term performance prior to and posterior to a coach replacement is not sufficient to draw conclusions on the efficacy of a coach dismissal. If anything, the findings suggest that there is no point in replacing the head-coach, since performances do not improve. However, we do not know how performances would have developed without the change in coach. In the following two sections, we will discuss two methods that attempt to deal with this topic by the construction of a counterfactual.

---

* In Clarke and Norman (1995), home advantage equals one for a home match and zero for an away match. Quality is equal to 1 if the team is playing at home, -1 if the team is playing away, and 0 otherwise. A similar method is used for the extensions of the model.

[†] Koning also makes a distinction between goals scored and goals conceded. He uses Poisson models to examine the effect of a coach change on those parameters. This does, however, not provide any further insight with respect to the effectiveness on an in-season coach dismissal. Thus, we will focus on goal difference only.

TABLE 20.2

Preliminary Analyses Method (Koning 2003)

Season	N	Old Scored	Old Conceded	Old Goal Difference	New Scored	New Conceded	New Goal Difference	New-Old Difference-in-Differences
All								
00/01	4	1.09	2.13	−1.04	1.25	1.77	−0.52	0.52
01/02	4	1.10	1.46	−0.36	1.35	1.33	0.02	0.38
02/03	4	1.16	2.06	−0.90	1.13	1.37	−0.24	0.66
03/04	4	0.95	1.96	−1.01	0.99	1.78	−0.80	0.21
04/05	4	1.17	1.72	−0.55	1.63	1.68	−0.05	0.50
05/06	5	0.98	1.68	−0.70	1.33	1.83	−0.49	0.21
06/07	3	0.93	1.79	−0.86	1.40	1.44	−0.04	0.82
07/08	5	1.78	1.59	0.19	1.70	1.42	0.28	0.09
08/09	7	1.15	1.57	−0.42	1.35	1.30	0.05	0.47
09/10	5	1.06	1.61	−0.55	1.47	1.67	−0.20	0.35
10/11	3	1.35	1.74	−0.39	1.50	1.58	−0.08	0.31
11/12	5	1.69	1.76	−0.07	1.88	1.70	0.18	0.25
12/13	2	1.11	1.57	−0.46	1.70	1.27	0.43	0.89
13/14	4	1.40	1.92	−0.52	1.38	1.66	−0.28	0.24
Total	59	1.22	1.74	−0.52	1.43	1.56	−0.13	0.39
Dismissals								
00/01	3	1.00	2.19	−1.19	1.25	1.86	−0.61	0.58
01/02	3	1.11	1.78	−0.67	1.30	1.32	−0.02	0.65
02/03	3	1.25	1.95	−0.70	1.24	1.51	−0.27	0.43
03/04	2	0.63	2.12	−1.49	0.82	1.91	−1.09	0.40
04/05	3	0.84	1.99	−1.15	1.44	1.88	−0.44	0.71
05/06	3	0.88	2.04	−1.16	1.26	2.31	−1.05	0.11
06/07	1	0.79	2.21	−1.42	1.10	1.45	−0.35	1.07
07/08	3	0.87	2.12	−1.25	1.43	1.64	−0.21	1.04
08/09	6	1.14	1.56	−0.42	1.46	1.20	0.26	0.68
09/10	4	1.08	1.65	−0.57	1.58	1.62	−0.04	0.53
10/11	2	0.97	2.11	−1.14	1.19	1.99	−0.80	0.34
11/12	3	1.98	1.64	0.34	2.09	1.53	0.56	0.22
12/13	2	1.11	1.57	−0.46	1.70	1.27	0.43	0.89
13/14	4	1.40	1.92	−0.52	1.38	1.66	−0.28	0.24
Total	42	1.11	1.87	−0.76	1.41	1.63	−0.22	0.54

Note: In the table, a distinction is made between all coach changes and dismissals. Furthermore, average values are given for the number of coach changes, the number of goals scored and conceded as well as the goal difference for both the old coach and new coach. The total for N is not an average, but the sum over all seasons. Due to the rounding of numbers, the values for Total might differ from the average of the values as given in a certain column. For each coach change, the whole season of the club at which the change took place is taken into account, consisting of 34 matches. The results of these 34 matches are split over the old and new coach depending on the time of the change.

TABLE 20.3

Parameter Estimates Extended Model (Koning 2003)

Season	All Changes		Dismissals	
	k	ψ	k	ψ
00/01	−0.729	0.692	−0.885	0.680
	(0.702)	(0.532)	(0.859)	(0.656)
01/02	1.333**	−0.490	1.276*	−0.098
	(0.606)	(0.495)	(0.716)	(0.577)
02/03	−0.155	0.792*	−0.145	0.566
	(0.584)	(0.431)	(0.707)	(0.457)
03/04	0.157	0.184	0.205	0.396
	(0.587)	(0.460)	(0.844)	(0.696)
04/05	0.134	0.697	−0.065	0.994
	(0.681)	(0.534)	(0.885)	(0.685)
05/06	−0.004	0.358	−0.146	0.263
	(0.449)	(0.361)	(0.585)	(0.478)
06/07	0.215	0.386	0.207	0.581
	(0.576)	(0.445)	(1.070)	(0.916)
07/08	0.155	0.206	0.411	0.920
	(0.547)	(0.515)	(0.667)	(0.580)
08/09	−0.0013	0.516	0.020	0.706**
	(0.482)	(0.330)	(0.519)	(0.330)
09/10	0.400	0.329	0.369	0.515
	(0.523)	(0.416)	(0.601)	(0.505)
10/11	1.276*	−0.400	1.060	−0.327
	(0.663)	(0.522)	(0.877)	(0.606)
11/12	−0.098	0.447	0.186	0.101
	(0.548)	(0.447)	(0.755)	(0.595)
12/13	−0.786	0.981	−0.786	0.981
	(0.857)	(0.650)	(0.857)	(0.650)
13/14	−0.208	0.406	−0.208	0.406
	(0.659)	(0.433)	(0.659)	(0.433)

Note: Observations per season 306; Robust standard errors in parentheses; **$p < 0.05$, *$p < 0.1$. The symbols have the interpretation as explained in the text.

20.3.2 Bruinshoofd and Ter Weel (2003)

Bruinshoofd and Ter Weel hypothesize that the replacement of a head-coach should lead to a short-term performance improvement because of the "shock effect". Therefore, they focus on in-season coach dismissals. They make a distinction between the effectiveness of a change, that is, whether the performance improves after a change, and the efficiency, that is, whether it was the change that caused any improvement. A before–after analysis is used to examine the effectiveness, while a difference-in-differences estimation is used to address whether a coach replacement was efficient. Their sample consists of all matches from the highest Dutch professional league in the seasons 1988/1989–1999/2000. The treatment group contains the actual coach changes, while the control group consists of cases in which performances follow a similar pattern (i.e., experiencing a performance dip) as those of the treatment group, but where no replacement takes place. Performances are measured in the following way:

- A moving average is computed based on the number of points earned during the previous four matches. This measure allows for short-term performances to vary. Thus, it captures any "shock effect".

- In order to evaluate this four-match point moving average, an "ordinary" performance level is defined as the average number of points in the current season. This is a constant measure, yet only observable after the season. The use of the current season prevents any potential bias resulting from buying and selling players to affect the results.

- Relative performance, at any moment during the season, is then defined as the four-match point moving average divided by the ordinary performance level. This performance measure has the same interpretation for all clubs.

Based on these performance measures, the following set of requirements is used to identify a performance dip and to construct the control group:

1. Relative performance has to be at most 1.1 four matches prior to a potential dip, indicating that performances are roughly average or worse at that moment.
2. Relative performance has to decline by at least 0.25 over the next four matches, indicating severe performance deterioration.
3. Relative performance has to be at most 0.65 after these four matches, indicating that performances are low compared to average.

In case all three requirements are met, a performance dip is identified. Performance dips have to be separated by at least four matches. Furthermore, for the construction of the control group, all performance dips that coincide with an actual change of head-coach are left out. A spell of four matches is necessary to obtain the performance indicator. Therefore, only the coach-changes after the 8th match and before the 31st match are considered. The first four-match performance moving average can only be constructed after four matches, while the next four matches are used to measure the decline. Thus, the first performance dip may only be identified after eight games.* A χ^2-test is used to check for uniformity of performance dips per season across teams, indicating uniformity. Bruinshoofd and Ter Weel argue that this allows them to group all data over teams and seasons. First, the effectiveness of a replacement is examined in a before–after analysis. Let $Y_{s,u}$ denote the performance level in case a coach change takes place after time t ($u > t$) and $Y_{s,-u}$ indicate the performance level in the case of a coach change before time t. Similarly, $Y_{c,u}$ and $Y_{c,-u}$ point at the performance levels after and before time t, respectively, in case a coach change did

---

* We will construct the control group in a slightly different way. In particular, we will only use the first performance dip occurring in a particular season, since we will only use the first coach-change in a season as well. Furthermore, we will use the coach-changes as discussed in Section 20.2.

not happen (i.e., the control group). Then $(Y_{s,u} - Y_{s,-u})$ and $(Y_{c,u} - Y_{c,-u})$ are the before–after estimators. Second, the efficiency of a replacement of the head-coach can be examined using the difference-in-differences estimator $((Y_{s,u} - Y_{s,-u}) - (Y_{c,u} - Y_{c,-u})).$*

Bruinshoofd and Ter Weel use the number of points as a performance indicator. For the sake of comparison of different methods and results, we use the goal difference as the performance measure. The construction of a four-match moving average, the "ordinary" performance level and the relative performance level, does not change. However, it is necessary to develop a new set of requirements for the construction of the control group, so that the performances of the control group prior to a performance dip follow a similar pattern as the performances of actual coach changes. The new set of requirements we propose are as follows:

1. Relative performance has to be at most 0.25 four matches prior to a potential dip.
2. Relative performance cannot increase over the next four matches (i.e., at most 0).
3. Relative performance has to be at most −0.1 after these four matches.

Although this set of requirements does not seem to be very strict, they result in a fairly good fit compared to the changes that happened. A total of 216 control groups are identified.

Figure 20.1 shows how the relative performance develops for all three categories of groups around the moment of an actual coach change or a performance dip. The figure

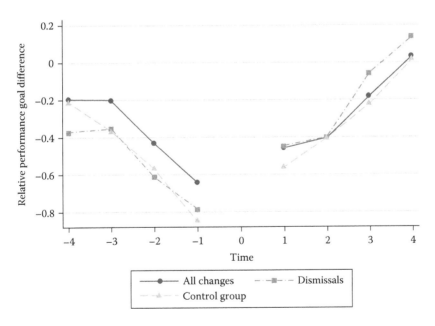

FIGURE 20.1
Relative performance based on goal difference. *Note:* The graph shows averages of the relative performance around the moment of an actual coach change or performance dip based on goal difference. Included are 59 observations for all changes, 42 dismissals and 216 control groups.

---

* Bruinshoofd and Ter Weel also distinguish between *ex post* successful and unsuccessful changes and use a similar method. We will ignore such a distinction.

shows a decline in performances prior to a change, while performances improve after the change. This indicates that a positive effect of a replacement of the head-coach might be present. This performance improvement is, however, also present for the control group after facing a performance dip. Both aspects have to be taken into account in order to address the overall effectiveness of an in-season replacement of the head-coach. This is done by estimation of the before–after differences and the difference-in-differences estimator. In Table 20.4, we present the relevant parameter estimates.

The first part of this table shows that performances significantly decline prior to the coach change (columns one and two) or a performance dip faced by the control group (column three). They tend to go up again afterward, though less significantly. This is in

TABLE 20.4

Parameter Estimates Method (Bruinshoofd and Ter Weel 2003)

	Performances				
	All Changes	**Dismissals**	**Control Group**	**All-Control**	**Dismiss-Control**
t_4	−0.194	−0.373***	−0.212***	0.018	−0.161
	(0.119)	(0.128)	(0.024)	(0.119)	(0.128)
t_3	−0.201*	−0.353***	−0.368***	0.167	0.015
	(0.106)	(0.117)	(0.049)	(0.106)	(0.117)
t_2	−0.430***	−0.615***	−0.569***	0.139	−0.046
	(0.118)	(0.125)	(0.051)	(0.118)	(0.125)
t_1	−0.644***	−0.790***	−0.850***	0.206**	0.060
	(0.101)	(0.118)	(0.040)	(0.101)	(0.118)
t1	−0.457***	−0.450***	−0.563***	0.106	0.113
	(0.098)	(0.120)	(0.050)	(0.098)	(0.120)
t2	−0.402***	−0.403***	−0.407***	0.005	0.004
	(0.101)	(0.123)	(0.060)	(0.101)	(0.123)
t3	−0.182*	−0.057	−0.223***	0.041	0.166
	(0.108)	(0.132)	(0.060)	(0.108)	(0.132)
t4	0.034	0.139	0.016	0.018	0.123
	(0.110)	(0.111)	(0.059)	(0.110)	(0.111)
Before–after and difference-in-differences estimation					
Before–After-1	0.186*	0.339***	0.287***	−0.101	0.052
	(0.095)	(0.110)	(0.043)	(0.095)	(0.110)
Before–After-2	0.028	0.212	0.162**	−0.134	0.050
	(0.131)	(0.144)	(0.074)	(0.131)	(0.144)
Before–After-3	0.020	0.296	0.145*	−0.125	0.151
	by: (0.155)	(0.179)	(0.082)	(0.155)	(0.179)
Before–After-4	0.229	0.512***	0.229***	−0.0002	0.283*
	by: (0.165)	(0.165)	(0.064)	(0.165)	(0.165)

Note: Observations 8568; Robust standard errors in parentheses; ***$p < 0.01$, **$p < 0.05$, *$p < 0.1$. In the first part average performance levels are presented in which t_X and tX indicate the number of match days prior or past time 0, which in turn indicates a performance dip or actual coach change. The Before–After-X indicates a comparison between the performances X before and X after time 0. The point estimates can be calculated by tX-t_x. These results are based on a total of 59 coach changes, 42 dismissals, and 216 control groups.

line with our observation in Figure 20.1. Furthermore, the positive values in the last two columns (four and five) suggest a (very) short-term gain in performances as a result of a coach change may be present, compared to the control group. However, the values are generally not significantly different from zero. The second part of the table confirms this finding. The first three columns show the before–after estimators. The results indicate an immediate improvement in performances (indicated by before–after-1), as well as some longer-term improvement (indicated by before–after-4). In terms of Bruinshoofd and Ter Weel, a change is (marginally) effective. The difference-in-differences estimations indicate, however, that they are not efficient. Only for the dismissed coaches and only when comparing four matches before and after the change, we find a significant positive effect. All other results are insignificant. Concerning efficiency, these results are in line with those of Bruinshoofd and Ter Weel. However, the authors even find significant negative coefficients, indicating it would even be better not to replace the head-coach. Our results here only suggest that it does not matter very much for the short-term performance.

A reason for the difference in results may be that the construction method of the control group is not exactly the same. Furthermore, a small difference in method already results in a rather large difference in the number of observations (e.g., by using a moving ordinary performance level instead of a constant one). Of course, the difference in outcome may also be caused by the difference in sample and the use of a different output measure. Many variants might be used as a check for the robustness of our results. We are mainly interested in the estimates of the difference-in-difference estimators. Therefore, we use two alternative sets of requirements for the construction of the control group, in which we allow the performances of the control group to decline somewhat more. The reestimated results are presented in Table 20.5. The results are rather similar to those presented in Table 20.4, although all significant results disappear when using alternative 2. This strengthens the

TABLE 20.5

Parameter Estimates Method Bruinshoofd and Ter Weel (2003) with Alternative Control Groups

	Alternative 1		Alternative 2	
Difference-in-Differences Estimation	**All-Control**	**Dismiss-Control**	**All-Control**	**Dismiss-Control**
Before–After-1	−0.100	0.053	−0.150	0.003
	(0.095)	(0.110)	(0.095)	(0.110)
Before–After-2	−0.065	0.119	−0.069	0.115
	(0.131)	(0.144)	(0.131)	(0.144)
Before–After-3	−0.061	0.215	−0.086	0.190
	(0.155)	(0.179)	(0.155)	(0.179)
Before–After-4	0.043	0.326*	−0.050	0.233
	(0.165)	(0.165)	(0.165)	(0.165)

Note: Observations 8568; Robust standard errors in parentheses; *$p < 0.1$. The Before–After-X indicates a comparison between the performances X before and X after time 0. The point estimates can be calculated by tX-t_x. Alternative 1 contains 211 control groups based on requirement 1: relative performance has to be at most 0.25, requirement 2: relative performance has to decline by at least 0.25, requirement 3: relative performance has to be at most −0.1. Alternative 2 contains 206 control groups based on requirement 1: relative performance has to be at most 0.25, requirement 2: relative performance has to decline by at least 0.25, requirement 3: relative performance has to be at most −0.5.

idea that the construction method of the counterfactual is an important issue. In the next section, we will examine a different approach, based on bookmaker data.

20.3.3 Van Ours and Van Tuijl (2016)

Using data from the highest Dutch professional soccer league for the seasons 2000/2001– 2013/2014, Van Ours and Van Tuijl (2016) investigate in-season coach dismissals in two steps. First, they consider the determinants of coach dismissals. Second, they use their findings of the first step to examine the effect of a replacement of the head-coach on in-season performances. Notably, they use bookmaker data to calculate expected performances.* The difference between the expected number of points and actual points earned in a particular match is called the "surprise". The sum of all surprises within a given season is the "cumulative surprise". To investigate the determinants of a replacement of the head-coach, an exponential hazard rate model is used to estimate the duration until a coach change. The results reveal that the cumulative surprise has a significantly negative effect on the probability of being dismissed. Thus, the coach of a team that performs worse than expected has a higher probability of being replaced than a coach of a team performing as (or even better than) expected. This finding is used to construct a control group, but first a naïve approach to address the effectiveness of replacement of the head-coach (without a control group) is formulated in which the following linear model is presented:

$$y_{ijk} = \eta_{ik} + r'_{ijk}\beta + \delta d_{ijk} + \varepsilon_{ijk}. \tag{20.3}$$

In this model, y_{ijk} represents the performance indicator of club i in match j of season k. The performance indicators include the number of points, whether or not a match was won and the goal difference measured as the goals scored by the home team minus the goals scored by the away team. The focus on in-season performances requires the inclusion of the term η_{ik} that represents club-season fixed effects. These fixed effects account for any (unobserved) quality of a team in a particular season. Without these fixed effects, the estimates will be biased. Furthermore, the vector r'_{ijk} represents potential determinants of performance, while d_{ijk} is a dummy indicating that the coach has been changed.[†] The vector β represents parameter estimates, ε_{ijk} is the error term, and δ is the estimate of the parameter measuring the effect of a coach change on performances. However, to interpret any significant effect as causal, a control group should be taken into account. As mentioned before, this control group, or counterfactual, is constructed by using the cumulative surprise obtained by using bookmaker data. For each coach change in a particular season for a particular club, a similar value of the cumulative surprise was sought (allowing a maximum difference of 0.5) for the same club but in a different season.[‡] Matching is thus based

---

* First, one calculates the "bookmaker premium" as the sum of the reciprocals of the odds. Next, one divides the reciprocal of the separate odds by this premium, in order to obtain the probabilities of a home win, a draw, or an away win, respectively. The expected number of points for the home (away) team is simply the probability of a home (away) win times three (points) plus the probability of a draw times one (point).

[†] As indicated before, coach changes before match 5 and after match 30 are left out of the analysis.

[‡] For some clubs, no proper counterfactual could be found, which might be due to the fact that the team only played one season in the league or because cumulative surprises deviate too much.

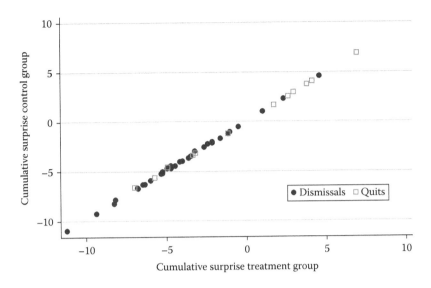

FIGURE 20.2
Scatterplot of matching procedure. *Note:* The graph represents the matching procedure based on cumulative surprise. Given that some coach changes did not have a counterfactual (e.g., because a club only played once in the league during the sample period or cumulative surprises deviated too much) the number of dismissals is 36. Adding to this a total of 12 quits results in 48 matched coach changes in total.

on proximity within the same club. In doing so, one controls for unobserved differences between clubs in dealing with the replacement of a coach. Furthermore, unobserved differences between the treatment and control group are controlled for by using club-season fixed effects.

Figure 20.2 shows the accuracy of the matching procedure, in which values of the cumulative surprise of the treatment group are matched to values of the control group in an exact way, represented by an almost straight line of dots.

By adding a dummy indicating the control group to the model as presented earlier, a difference-in-differences estimation is obtained. The first two columns of Table 20.6 present the results of the naïve approach. As discussed before, goal difference is the only performance variable to be used in the analyses. The results show that for both, all changes as well as the dismissals, a significant positive effect is found, suggesting a performance improvement after the replacement of the head-coach.

The next step is, thus, to examine whether this is the same for the control group. The third and fourth columns of Table 20.6 provide the relevant parameter estimates when a control group is considered. For both, all coach changes and for the dismissals, we find significant positive results, though the value as well as the significance is smaller than for the treatment group. An *F*-test for equality reveals that these differences are not significant, indicating that the performance improvement after the change of the head-coach is similar to the improvement of the short-term performances for a control group. This suggests that in-season performances would have improved irrespective of whether or not the replacement of the coach took place.

TABLE 20.6

Parameter Estimates Goal Differences Method (Van Ours and Van Tuijl 2016)

	Naïve Approach		Difference-in-Differences	
	All	Dismissals	All	Dismissals
Rank opponent	0.114***	0.118***	0.113***	0.118***
	(0.008)	(0.009)	(0.006)	(0.007)
Home	1.034***	1.095***	1.060***	1.066***
	(0.084)	(0.093)	(0.065)	(0.070)
Coach change	0.373***	0.484***	0.374***	0.458***
	(0.089)	(0.101)	(0.099)	(0.109)
Control group			0.273**	0.314**
			(0.117)	(0.129)
F-test equality treatment-control			0.43	0.73
Observations	2006	1428	3128	2346
Seasons	59	42	92	69
n-Treatment-Group			48	36
n-Control-Group			44	33

Note: Robust standard errors in parentheses; ***p < 0.01, **p < 0.05. Club-season fixed effects are used in all estimations. Rank opponent is the rank of the opponent in the previous season. Home is a dummy controlling for home advantage.

20.4 Evaluation and Discussion

In this chapter, we examine the effectiveness of an in-season coach dismissal in professional soccer using data from the highest Dutch professional league. Previous studies have shown that in many situations, it is ineffective to replace a head-coach, since performances would have improved anyway, irrespective of the replacement. These results are obtained by using different research methods and samples. We compare three methods using one sample period covering 14 seasons in the highest professional soccer league of the Netherlands. Although the methods are rather different in set up, they all result in the same conclusion, that is, it is ineffective to replace the head-coach after a sequence of bad results in order to try to improve the short term, in-season performance. This conclusion is in line with most of the previous findings. So, these different methods, taking into account different elements of the teams and performances, all result in this general finding that on average replacing a coach does not improve the performance of a soccer team. Of course, this does not rule out that some individual replacements do improve the performance while other replacements may lead to worse performances.

By way of concluding this chapter, we have a closer look at the (essential) elements of the methods used, making a distinction between heterogeneity between (the quality) of opponents; performance expectations; counterfactuals; matching method; within club-season analyses. Table 20.7 provides an overview. Looking at the heterogeneity of opponents, the interesting aspect of Koning (2003) is that his method controls for the order of play and uses a quality indicator that allows for relative differences between opponents, though keeping

TABLE 20.7

Overview of Methods

	Koning	Bruinshoofd and Ter Weel	Van Ours and Van Tuijl
Heterogeneity of opponents	Correcting for the order of play and using relative quality indicator	Correcting for the order of play	Correcting for the order of play and using crude quality indicator
Performance expectations	No	Based on in-season performances, only observable after the season	Based on exogenous in-season bookmaker data, observable any time
Counterfactuals or control groups	All clubs not replacing the head-coach	Based on performance patterns and a set of requirements	Based on cumulative surprises using bookmaker data
Matching method	No	No matching between individual observations of treatment and control	Individual matching between treatment and control based on nearest neighbor approach
Within club-season analyses	Within particular club and across clubs in a particular season	Across clubs across seasons	Using club-season fixed effects

them fixed during the season. Bruinshoofd and Ter Weel (2003) only marginally correct for the order of play by using a four-match time span, without using any further measure of quality.

Van Ours and Van Tuijl (2016) correct for the order of play and use a crude quality indicator based on the rank of the previous season, assuming that the quality difference between teams finishing after each other is similar at the top of the table as is at the bottom. With respect to performance expectations, we observe that Koning does not use them, while Bruinshoofd and Ter Weel use an expectation level based on some average at the end of the season (points per game in their case, we use goal difference per game). Van Ours and Van Tuijl use exogenously determined bookmaker data to construct a performance expectation and use it to calculate the cumulative surprise. The latter can be derived at any moment in time during the season. Thus, one does not need to know the final ranking, as is the case with Bruinshoofd and Ter Weel. Furthermore, using some form of performance expectations and by relating this to actual performances, enables comparison between clubs across seasons. Both Bruinshoofd and Ter Weel and Van Ours and Van Tuijl use it to construct a counterfactual or control group. On the one hand, Bruinshoofd and Ter Weel include all cases that fulfill some set of requirements and, thus, do not match individual cases across seasons. On the other hand, Van Ours and Van Tuijl only include the proximity case obtained from the same club in a different season. Any unobserved club- or season-specific element is controlled by using club-season fixed effects. Koning looks at performances within a particular club in a particular season. He does not relate that situation to some performances of another group or counterfactual outside of that season. Therefore, Koning does not need to use a matching procedure. His "control group" consists of all matches played by the clubs that did not replace their head-coach. Thus, this method is rather different from the other two methods. Therefore, we think that, in particular, further research should look for some way to combine the interesting element of heterogeneity of opponents used by Koning with some expectation based performance indicator and the construction of a control group.

References

Anderson, C. and D. Sally (2013): *The Numbers Game; Why Everything You Know about Soccer is Wrong*, New York: Viking/Penguin.

Bruinshoofd, A. and B. ter Weel (2003): Manager to go? Performance dips reconsidered with evidence from Dutch football, *European Journal of Operational Research*, 148, 233–246.

Clarke, S. R. and J. M. Norman (1995): Home ground advantage of individual clubs in English soccer, *The Statistician*, 4, 509–521.

Kahn, L. (2000): The sports business as a labor market laboratory, *Journal of Economic Perspectives*, 14, 75–94.

Koning, R. (2003): An econometric evaluation of the effect of firing a coach on team performance, *Applied Economics*, 35, 555–564.

Kuper, S. and S. Szymanski (2010): *Soccernomics*, London, U.K.: Harper Collins.

PWC (2015): De Kracht van het Voetbal: Sturen op de maarschappelijke impact van voetbal, [Online] www.pwc.nl (Accessed on October 28, 2015).

Szymanski, S. (2003): The assessment: The economics of sport, *Oxford Review of Economic Policy*, 19, 2003, 467–477.

Ter Weel, B. (2011): Does manager turnover improve firm performance? Evidence from Dutch soccer, *De Economist*, 159, 279–303.

Van Ours, J. C. and M. A. Van Tuijl (2016): In-season head-coach dismissals and the performance of professional football teams, *Economic Inquiry*, 54, 591–604.

21

Referee Bias in Football

Babatunde Buraimo, Dirk Semmelroth, and Rob Simmons

CONTENTS

21.1 Introduction

The possibility of referee bias occurring in football matches has exercised the attention of several scholars. Research is facilitated by the public availability of match-level data, including identity of referee and outcomes within games that fall under the discretion of referees. These include time keeping so referee bias might be revealed in the amounts of discretionary time added on at the end of games for stoppages during the match. Section 21.2 surveys the various studies of stoppage time. Referee bias may be revealed in terms of the awards of yellow and red cards as sanctions for player misdemeanours that fall outside the Laws of the Game. Section 21.3 surveys the literature on yellow and red cards. Section 21.4 summarises the strengths and weaknesses of existing statistical approaches with an emphasis on the capability of investigators to identify causal effects within their analyses. We also make some suggestions as to how future research on referee bias might proceed.

This chapter is primarily concerned with model specification choices and statistical methods so we will highlight the estimation procedures that have been followed. These are typically various forms of regression analysis, often just plain Ordinary Least Squares (OLS). An earlier influential and much-cited study by Nevill et al. (2002) performed an experiment in which two sets of referees viewed a videotape of a Premier League match under different conditions and were asked to nominate the award of free kicks. One set of referees viewed the replay with the sound of fan noise eliminated while the other group watched with sound retained. The latter group offered more decisions in favour of the home team. This study was influential since it revealed referee bias in the setting of a controlled experiment. But we will never know if the referees in that study were similarly prone to bias when they were officiating actual matches. External validity of the study is therefore in question and, unfortunately, researchers are unable to generate 'real' field experiments that

might usefully follow up on the study by Nevill et al. This problem of lack of causal identification will be a running theme throughout this chapter. However, we will try to show that
there does exist a substantial consensus on forms of referee bias, drawn from many football competitions around the world. This consensus is hard for football governing bodies
to ignore.

To aid comparison of the research reported here, Table 21.A.1 gives a simple, selective,
tabular guide to the various statistical models covered in the literature on referee bias.
The selection of papers was the same as in the economics-based survey on referee bias
by Dohmen and Sauermann (2016). Some more recent papers not covered by that paper
will be reported here in the main text.

21.2 Referee Bias through Stoppage Time

According to FIFA's Laws of the Game, the referee is the sole arbiter of time-keeping in a
football match. The referee blows a whistle to signal the beginning and end of each half
played in a match. In the current version of stoppage time, the referee indicates to an official on the touchline how much minimum stoppage time is to be added after 90 minutes
play has ensued. The fourth official then displays the referee's decision on the amount of
stoppage time to the crowd using an electronic board. Stoppage time is shown in discrete
form as a number of minutes. This is a minimum amount that takes into account various
stoppages such as injuries, substitutions, goal celebrations and deliberate time wasting by
players. Further delays during the announced stoppage time could lead to more time added
on by the referee.

Garicano et al. (2005) is an early and much-cited study of time added on at the end
of games. This covered Spain's La Liga over the 1994/1995 and 1998/1999 seasons. The
authors make the conjecture that referees 'internalise the preferences of the crowd' by systematically favouring the home team. That is, referees want to satisfy the crowd inside the
stadium, implicitly and tacitly.

The dependent variable was length of time added on in games where the score at
90 minute had one goal difference. Score difference is a dummy variable for whether the
home team was one goal ahead. The authors found that stoppage time falls if home team is
one goal ahead. The analysis does not consider games where home team is level. It could
be argued that *more* time is added on if home team is not winning in a close game. The
marginal effect of score difference on stoppage time varies from −1.76 to −1.88 [−1.80 in
most complete model with referee, home team fixed effects]. This marginal effect is significant at 1% level. Bias was not detected for half time extra time added on. The paper further
argued that if crowd is made up of more visitors then referee bias falls.

The complete model had score difference as the focus variable. Control variables for
potentially confounding factors were numbers of yellow and red cards, numbers of substitutions, year effects, home team budget (payroll), away team budget, league rank of
home team, difference in ranks of teams, home team and referee fixed effects. Sample size
(number of games) was 268, somewhat low because of the restriction to close games.

The study reported some interesting crowd effects on amount of stoppage time.
A one standard deviation rise in attendance raised bias by 20% but a higher
attendance-capacity ratio reduces the bias. In a more complete specification, score
difference retained its negative effect. Attendance interacted with score difference had

a negative coefficient. Attendance-capacity ratio interacted with score difference had a positive effect. An examination of referee fixed effects found no significant differences in magnitudes so referees were concluded to be equally biased.

This study generated several imitators from other Leagues. A relevant question is that of economic significance: In how many games where the home team was ahead did the away team get a goal in time added on? In how many games where the home team was behind at 90 minutes did the home team get a goal? If the home team was level how many times did it get a goal in time added on? Do the results change when referees add more time on?

More recently, Watanabe et al. (2015) analysed determinants (in seconds) of stoppage time in English Premier League close games, defined as those with one or two goals different, over the 2012/2013 and 2013/2014 seasons. Games with more than two goals differential were removed. Estimation was by Ordinary Least Squares with fixed effects for referees and teams (both home and away).

Goals scored, substitutions, yellow cards in regulation second half time and in stoppage time added significantly to time added on. Fouls in stoppage time added to extra time but fouls in regulation time did not. Red cards in regulation time added to extra time but not red cards in stoppage, most likely because these are rare events.

Referee experience was included but did not return a significant coefficient. Score difference at 90 minutes did not have a significant coefficient, contrary to Garicano et al., discussed earlier, and this discrepancy warrants further investigation. The authors have many control variables in their specification and it would have been useful to see correlation coefficients to help detect multicollinearity.

Quantifying the authors' results, they found that an increase in 10 seconds of second half injury time led to 6–7 seconds extra time added on for close games. The effects of substitutions and goals are much greater during stoppage time compared to regulation time (33 vs 8 seconds for subs and 30 vs 10–13 seconds for goals). This suggests that referees treated key events inconsistently with regard to stoppage time.

Butler and Butler (2015) is a notable working paper that also analyses stoppage time in the English Premier League, covering the 2009/2010 to 2012/2013 seasons. The paper's emphasis is on big club favouritism. The authors examine the effects of club size on stoppage time over and above attendance. The study identifies Arsenal, Chelsea, Manchester City, Manchester United and Tottenham as big clubs on basis of financial and sporting performances. Table 21.1 summarises the authors' main results.

TABLE 21.1

Results from Butler and Butler (2015)

Model	Variable	Impact of Variable on on Stoppage Time in Seconds
1	Home win	22.01
	Home loss	23.21
2	Big club at home losing	15.84
3	Small club at home drawing	−22.23
4	Big vs. small team losing	26.65
5	Small vs. big team winning	31.94
6	Small vs. small winning	22.70
	Small vs. small losing	31.88

Control variables include goal margin, substitutions, yellows, reds, serious injury, referee experience (not significant) and log attendance (not significant). Models include referee fixed effects but not team or year dummies. The authors do not include manager fixed effects even though the study is motivated by the alleged intimidatory behaviour towards referees by one notable manager, Sir Alex Ferguson of Manchester United. This yielded the catchphrase 'Fergie time', popular in the sports media. If the effect is manager-specific then it should disappear after he retired and was replaced by another head coach (David Moyes and then Louis van Gaal). The rows in Table 21.1 represent different specifications, and rows 2 and 4 show some evidence consistent with the presence of a big team bias. When the authors restrict attention to close games, the only significant coefficient is small vs. small with home team losing (16.36 seconds).

Dohmen (2008) used data on 3519 Bundesliga 1 games. He examined stoppage time and decisions to award goals and penalty kicks. The size of bias was dependent on crowd composition. Home bias fell when there were more supporters of the away team attending matches. Home team favouritism emerged as stronger when the match occurred without a running track. Hence, it appears that proximity of the crowd to the pitch (and therefore the referee) could be a relevant factor in award of stoppage time.

More stoppage time was awarded in close matches where the score margin was −1. Dohmen focussed on close games to replicate Garicano et al. (2005) for Germany. Supporters have weakened incentives to influence referees in decided matches where the match outcome is unlikely to change in stoppage time. The dependent variable was stoppage time and the focus variable, score difference, was again coded as one for home team ahead by one goal and zero for the away team ahead by one goal. Control variables were treatments for injury, substitutions, fouls, yellow cards, second yellow cards, red cards, season, relative strength and fixed effects for home and away teams and referees. The home team bias was smaller in the first half than in the second. Dohmen's point estimate was 22 seconds of extra stoppage time if the home team was behind by one goal. This was statistically significant in all specifications but less in magnitude than Garicano for Spain. The insignificant (zero) effect for first half stoppage time was in line with Garicano et al.

Dohmen also broke out the effects of crowd composition on stoppage time and Table 21.2 shows his results.

Here, score difference is +1 for if the home team was one goal ahead and −1 for one goal behind. An attractive team was one that appeared in the top third of a relative crowd composition index. This included well-known teams such as Bayern Munich and Borussia Dortmund. Results showed significant score difference effects for teams that are a long way apart in physical distance. They also show that the score difference effect was eliminated in stadia with running tracks. Without a running track, the score difference effect was −29.3. The removal of a running track by some teams offers potential for a natural experiment

TABLE 21.2

Dohmen's Results on Geographical Distance between Teams in Bundesliga

	Distance 150 km or Less	Distance More Than 150 km
Score difference	−4.20	−12.06***
Attractive team*score difference	−20.34*	3.60

Note: ***p < 0.01 and *p < 0.1.

design where the treated clubs are those that change stadium design and the control groups are those that retain current status.

Another study of stoppage time in German Bundesliga is Riedl et al. (2015). This covered the 2000/2001–2010/2011 seasons. They found 18 seconds more stoppage time if the home team was losing by one goal. The paper used the standard Garicano measure of score difference as +1 if leading by one goal, zero if behind by one goal. Control variables were added for yellow cards, red cards, 'incidents' and substitutions.

A novel feature of this paper is the idea of charity bias – referees could give both teams more opportunity to offset a handicap of one goal than to finish with win or loss. The study found that referees shortened injury time when the goal difference exceeded 1 (or fell below –1). Moreover, referees allowed 21 seconds more added time when a team gained the lead after the 79th minute of regulation time. The authors considered games where there was a draw between 46th and 79th minute and only one goal was conceded by a team between 80th and 90th minute (these games were coded one). If there was a draw for whole of the second half then the estimated stoppage time difference was 21 seconds shorter time added on [numbers of games coded were 316 for value 1 and 108 for value of 0]. In a second scenario, if a team was leading by at least one goal between minutes 46 and 79 but the other team equalised this was coded as one. A code of zero was entered if the score remained unchanged through the second half [number of games with code 1 was 252, 53 for 0]. This suggests that referees prefer games that are more clearly decided as they do not want to be seen to be determining the game outcome. The net effect of home team and charity biases was found to be zero. Since biases were opposite they did not contribute to home advantage.

Riedl et al. reported the actual implications of stoppage time for scorelines in their sample. Only 112 out of 3351 matches had outcomes changed by goals scored in stoppage time. That is, only 3.3% of games were affected this way. This arouses the suspicion that various studies of referee bias in discretionary stoppage time may actually have uncovered effects that are statistically significant yet economically and behaviourally unimportant. Despite the larger estimates of score difference effects found by Garicano et al. (2005) for Spain, it appears that later studies find rather low magnitudes of effects and, importantly following Reidl et al., the stoppage time effect may actually be benign in that match outcomes are not affected by the bias that is revealed. If the latter claim is true then football's governing bodies that are charged with monitoring referee performances may rest somewhat easier.

Scoppa (2008) analysed stoppage time in Italy's Serie A for 2003/2004 and 2004/2005 seasons. Injury time is a function of a home team behind variable, numbers of player substitutions, red cards, yellow cards, penalties, difference in league ranks, attendance, numbers of 'other stoppages', goals, a 2005 dummy, a dummy variable for presence of a running track and an interaction term for Track*home team behind.

Estimation was by OLS. Home team fixed effects were included but were not significant. In the fullest specification home behind had a coefficient of 0.489 and this was significant at 1%. Estimation only covered close games and home team behind is defined as one if the home team was behind by one goal or zero if the home team was ahead by one goal or drawing. The resulting sample size with these restrictions was 485.

The home team behind bias effect was much higher if there is no running track in stadium. The coefficient on home behind*track was −0.535 (significant at 5%). This implies that when a match was played in a stadium without a running track the home team benefited by 45 seconds extra injury time.

Scoppa's study considered teams suspected of connections with referees under the *Calciopoli* scandal. The variable for suspicious teams was called 'Suspected behind' and its coefficient was positive and significant. If these potentially corrupt teams were behind in score then the amounts of stoppage time were larger. Scoppa did not observe any shortening of games if the home team was ahead so the estimated significant effects were loaded on to more stoppage time added on if the home team was behind.

Rocha et al. (2013) analysed stoppage time in Brazil's top division over 2004–2008. The sample size was 1413 and this featured all matches, unusually for this literature. Referees were found to systematically add amounts of stoppage time when the home team was losing by one goal. However, the effects were small and it is useful to show a comparison of results:

- Sutter and Kocher (2004) 1.0 minutes
- Garicano et al. (2005) 1.8 minutes
- Dohmen (2008) 20 seconds
- Rocha et al. (2013) 11.3 seconds

Interestingly, the authors found that stoppage time in Brazil had an inverted U-shaped relationship with score margin. This was unconventionally defined as score difference at 85 minutes (not 90). Referee bias was found to be more important with lower visibility of the match where this is related to television live broadcasts. Matches without live TV coverage had a significant 12.1 fewer seconds of stoppage time when home teams were ahead in score in close games compared to televised games.

Similar to Reidl et al. (2015), although there were significant effects of score difference on added time, this did not translate into sizeable effects of added time on the probability of a goal being scored by the home team in injury time. The marginal effect was 0.034% from a probit model.

Yewell et al. (2014) studied referee bias and stoppage time in Major League Soccer in the 2011 and 2012 seasons, estimated separately. They estimated grouped data and partially adaptive models to deal with measurement error in time added on. For 2011, point estimates showed an increase in stoppage time of around 12–20 seconds for nationally televised matches. This could be due to TV renegotiation where the broadcaster prefers longer games (to facilitate more advertising revenue). For 2012, the home team got 33 seconds extra stoppage time when facing a one-goal deficit.

Reported stoppage time is interval-censored (i.e. discrete) so the authors used a grouped-data regression model. Data as recorded for the MLS show minimum time added on so do not deal with discretion during added time (injuries, substitutions, goals, cards, etc.). A relevant question is why use grouped data and not a count data model? The stoppage time data may be non-normal so the authors used a partially adaptive regression model. The 2011 results showed that national TV broadcasts were associated with more added time under each estimation (OLS, grouped data, partially adaptive). The 2012 results showed that if the home team was behind by one goal this had a positive effect of 26–29 seconds on time added on. However, this effect was not present in 2011. Results from 2012 were quite different to 2011 but results from alternative estimation procedures were quite similar.

The papers discussed thus far suggest that there is plenty of evidence pointing towards referee bias in discretionary time added on at the end of football matches. This evidence comes from several football leagues. But if this bias is present, does it persist over time? Perhaps bias might be reduced if it is publicised in the media. Perhaps professional football leagues might take an interest in the conduct of their referees and behavioural bias might be reduced. Rickman and Witt (2008) compared stoppage time before and after professionalization of referees in the English Premier League. This league switched referee compensation away from match fees (the predominant mode in most football leagues) towards an annual salary for a group of elite, selected referees. The research question is whether the professionalization of referees altered their behaviour in terms of stoppage time.

Rickman and Witt considered games that have one goal difference at 90 minutes. The authors compare the 1999/2000 season with the 2002/2003 season, before and after the change in referees' employment contracts. The Premier League was taken as a treatment group while the second tier (then branded as Division One, now known as the Championship) as a control group. The Premier League difference in added time was 0.410 and while the Division One difference in added time was 0.724. This gave a raw difference in difference was −0.314. To deal with confounding effects, the regression model had Premier League dummy, score difference, year, Premier League*score difference, Premier League*year and score difference*year as controls. In the fullest regression specification with referee and team fixed effects, PL*score difference*year had a marginally significant (at 10%) coefficient of 0.593. This equates to 35 extra seconds in Premier League after professionalization compared to before. This suggests that salaried, professional Premier League referees were inclined to lengthen games through more added time compared to the earlier match fee compensation system.

21.3 Home Advantage, Disciplinary Sanctions and Referee Bias

Several papers have tried to connect to referee bias the well-documented presence of home advantage recognising that referees can give unwarranted, favourable treatment towards home teams and against away teams. One mechanism by which this biased treatment might occur is through crowd pressure and several papers consider crowd variables as determinants of either referee outcomes or home advantage.

Boyko et al. (2007) is not directly about referee bias but links referees' decisions to home advantage. Their study covers English Premier League 1992/1993–2005/2006, covering 5244 matches. A broad-brush model was estimated where match outcome was a function of home advantage (home–away differentials), of home and away abilities and luck. Home advantage was broken out into differences in goals, yellow cards, red cards and penalties converted. The authors use referee fixed effects to test for effects of referees on home advantage. These are significant for: Goals scored by home team and goal difference but not totals or away goals; Yellow cards (difference, home team, away team, totals); Red cards (home team, away teams, totals but not difference); Penalties (difference home team, away teams, totals). Crowd density had significant effects on goal difference and home team penalties. Note that Boyko's results showing effects of referees on home advantage might depend on a small number of outlier referees.

Johnston (2008) replicated the Boyko et al. study for all 380 English Premier League games in 2006/2007. Home advantage was defined as goal difference in a given match, modelled as a function of mean home goals, mean away goals and attendance. Home advantage could have been modelled as count or ordered probit since it is discrete. The paper introduced referee fixed effects later on.

This paper did not have longitudinal data and applied OLS to just one season's cross-sectional data. The paper did not consider selection of referees into games and how this might relate to bias. Size of crowd does not equate to partisanship since smaller crowds can be more ferocious. The paper found no significant effects of referees on their measure of home advantage, thus contradicting Boyko et al.

Goumas (2014) also examined effects of crowd size on referee bias, this time concerning yellow cards in Champions League games (n = 422) and Europa League games (n = 950) over the 2009/2010 and 2010/2011 seasons. Since yellow cards is a count measure, Goumas used Poisson regression estimation. More precisely he uses a repeated measure regression analysis using a log-link Generalized Estimating Equation, essentially a log-link Poisson regression model. The estimates pool both Leagues but pooling restrictions are not rejected.

Data did not permit breaking cards down into categories: handball, dissent and foul play. The crowd variables were crowd size, crowd density and proximity (presence of a running track). To control for within-match attacking dominance, Goumas used shots at goal and corners for each team. Other controls were stage of competition, country, season and team ability but not referee fixed effects. As with Johnston (2008), this is analysis at the level of the whole game.

After controlling for within match dominance, referees issued 25% more cards to away teams in Champions' League and 10% more cards to away teams in the Europa League. The strongest crowd effect was for crowd density (0.043, significant at 1%) when other crowd factors and attacking plays were controlled for. In the most general specification, track and crowd size were not significant determinants of the away/home cards ratio though they were significant when entered separately.

Goumas found a strong positive association between home team bias and crowd density. At 25% stadium capacity, he found zero bias. At 50% capacity, there was an estimated 13% more yellow cards for the away team while at 100% capacity the estimates show 35% more yellow cards for the away team. A problem with this paper and others that use whole game indicators to assess home advantage and referee bias is the impossibility of identifying a causal channel from referee bias to home advantage. Does referee bias create home advantage, as some of the discussion in these papers implies, or does home advantage actually induce referee bias? A clearer identification strategy is needed and removal of a running track is one option to pursue here as a natural experiment.

Downward and Jones (2007) modelled effects of crowd size on referee bias defined in terms of home–away probability of receiving a yellow card in the FA Cup 1995/1996 to 2001/2002. This yielded 2622 observations on awards of yellow cards. The dependent variable was coded 1 if yellow card was for home team, 0 if it was for the away team. Estimation was by conditional logit. The paper did not offer a bivariate model nor did it control for within-game scorelines. Later papers addressed these limitations. The independent control variables were crowd size (negative coefficient) and crowd size squared (positive coefficient). Division difference and crowd-division difference interaction terms were added, each had a significant coefficient.

Page and Page (2010) analysed data from 15 competitions in England and Scotland plus internationals over 1994/1995 to 2006/2007. They asked whether referees differed in their effects on home advantage defined here as home minus away goals. If referees had no impact on home advantage then this should be the same for all referees. The basic model had home advantage as function of referee fixed effects, competition effects, attendances (entered non-linearly), differences in team quality and average team quality in a match.

The authors used a random effects approach with multi-level modelling. Individual and competition specific effects were treated as random effects. Page and Page found significant differences in home advantage between referees. Their modelling strategy was to first compare random effects and linear models with and without referee random effects. The linear model was rejected. They proceeded to include covariates and competition effects and again the linear model was rejected. In the fixed effects version of the model, the authors also found referee fixed effects to be significant.

Moreover, Page and Page found significant differences in attendance effects on home advantage across referees-without bias these interaction terms should have zero coefficients. These significant coefficients suggest differential referee responses according to crowd size.

The papers summarised so far in this section all suffer from an excess aggregation problem. The analysis is at match-level and does not facilitate identification of referee responses to crowd pressure separately from player responses. This relates to the issue of reverse causality between referee bias and home advantage. If there is inbuilt home advantage then players of away teams might try to offset this by using illegal aggression that manifests itself in more cards being awarded to away teams. This is a consequence of player actions rather than biased treatment afforded by the referees. Moreover, the literature on home advantage tends to have just a limited number of control variables and does not typically check results across different statistical estimation methods.

A move in the direction of greater econometric sophistication in the analysis of referee behaviour was offered by Dawson et al. (2007). They analysed the English Premier League over 1996/1997 to 2002/2003 seasons. Their dependent variable was disciplinary points rather than number of cards. Two points were imputed for a red card, one point for a yellow card. The article estimated disciplinary points for home and away teams simultaneously at the level of the whole match but did not control for within-game effects, for example, changing scoreline.

Dawson et al. fitted five estimations: Double Poisson, Bivariate Poisson, Zero inflated Poisson (preferred), Bivariate negative binomial and Zero inflated bivariate negative binomial. The authors tested for a home advantage effect – the tendency for away teams to get more disciplinary points than home team, after accounting for home advantage. They then tested for whether or not the average incidence of sanctions varied across referees. Further tests were for the extent to which away teams' greater relative discipline points varied across referees (referee consistency). Finally, the authors tested whether the average incidence of cards varied over time and in relation to crowd size (attendance). Control variables included match significance (for Championship, European qualification or relegation). This had a positive and significant effect on away team disciplinary points but not home team points. Managerial spells were used as a proxy for team quality (team payrolls are another possibility here). Season and referee fixed effects were included.

The authors found that referee fixed effects were significant so English Premier League referees were inconsistent. However, they did not control for selection of referees into games or referee careers, for example, tenure in the English Premier League. After controlling for home advantage there appeared to be a bias for home teams in terms of less disciplinary points. The authors rejected the hypothesis that the rate at which away teams got more discipline points than home team was invariant across referees.

There was evidence over the sample period that a rule change providing for automatic red card for tackle from behind temporarily led to a rise in discipline points in 1998/1999 but this was reversed in later seasons. Dawson et al. found that either match significance or attendance were significant but not both together. The model with match significance was preferred, in terms of fit, so they concluded that effects of gate attendance were not significantly different from zero.

The analysis of Dawson et al. (2007) was at match-level, although with greater econometric sophistication than most of the previous literature. In particular, home and away disciplinary sanctions were modelled as simultaneous, bivariate outcomes. The number and detail of control variables is impressive. Buraimo et al. (2010) went a stage further and presented a minute by minute bivariate probit model of home and away yellow cards for two leagues, English Premier League and Bundesliga. The minute-by-minute level of analysis facilitated controls for within-game events that might induce referee sanctions. These within-game confounding factors included scoreline and recent history of cards. The authors found evidence of referee bias. The Bundesliga had three teams that removed their running tracks either at the existing stadium or by relocation to a new stadium. These teams were Bayern Munich (which relocate), Hannover and Schalke 04 (which changed stadium design). The authors found that home teams with running tracks received more yellow and red cards than teams without. Hence, the running track removal offers a useful natural experiment design that helps identify referee bias more precisely than other studies where identification is less clear-cut.

The model of Buraimo et al. included minute and its square, 45th minute dummy, 90th minute dummy (since this includes stoppage time), home yellow card received in the last 3 minutes, away yellow last 3 minutes, home yellow card received prior to the last 3 minutes, away yellow card received prior to last 3 minutes, home second yellow cards, away second yellow cards, home red cards, away red cards, goal difference and its square, running track dummy (Bundesliga only), attendance, difference in bookmaker probability (for expected closeness of game). The authors performed a test for correlation between joint responses and found rho, the correlation coefficient between residuals, to be significantly different from zero.

The probability of a yellow card had an inverted U-shaped relationship with goal difference. For away teams this is shown in Table 21.3. It appears that turning points were quite different in the two leagues.

For home teams, the probability of a yellow card in a given minute was falling with goal difference at an increasing rate. If the running track was removed, home yellow card probability falls and away yellow card probability increased. The authors argued that it is

TABLE 21.3

Relationship between Goal Difference within Game and Probability of a Yellow Card

	Bundesliga	EPL
Goal difference	0.094	0.082
Goal difference squared	−0.021	−0.046
Turning point	2.24	0.89

possible to infer differential responses through players and referees. Their prediction is that if running track was removed, players will perform more aggressively so the probability of award of a yellow card rises. But the authors observed lower probability of yellow card to home teams so referee responses outweighed player responses. The conclusion was one of (indirect) evidence of favouritism under social pressure.

The authors then reconfigured matches as favourites versus underdogs, rather than home versus away. They found significant coefficients on a home underdog away dummy variable. This is further evidence of biased treatment of teams with regard to yellow and red cards.

Dawson and Dobson (2010) retained the match as unit of observation and analysed award of yellow and red cards in UEFA Cup and Champions' League 2002/2003 to 2006/2007. The focus of the paper is on nationality of referees. Referees in these European countries are drawn from neutral countries but the referee team is all from same country. There could be nationality bias as a result similar to the biases alleged for international figure skating and for Eurovision song contest. There were 1717 games in the sample with referees from 17 countries. Descriptive data revealed that Greek referees awarded more yellows to both home and away teams. Portuguese referees issued more yellow cards to away teams than to home teams. English referees were stricter than other nationalities in the award of red cards while Belgian referees were more lenient than others over player dismissals.

The authors used bivariate ordered probit for disciplinary points measured as 0, 1, 2, 3, 4 and 5+. A Lagrange ratio test rejected the null of independence of home and away card equations; similarly the rho statistic for correlation of residuals is positive and significant. Control variables were difference in team UEFA coefficient index, stage of competition, Champions League dummy, attendance (not significant), attendance/capacity (positive and significant), track (positive and significant for home team) and presence of a fence around the pitch. Referee nationality dummies were included and were significant, in line with the descriptive information. The authors concluded that referee bias remained in European competitions despite an explicit attempt to reduce this by using neutral referees.

A further step would be to break out referees into those that officiate in national team fixtures versus those that only officiate in club matches. Since the former have experience of different playing styles from different countries then they should *a priori* be less prone to bias.

Pope and Pope (2015) took an approach similar to Dawson and Dobson (2010) in covering Champions' League but focussed on own-nationality bias: Are referees more lenient to players of their own nationality? The analysis was at player-match level on this occasion and had controls for goals scored, shots on and off target, corners by home and away teams plus minutes played at individual level. There were 45,400 player-match observations. The mean foul difference was −0.06 where this refers to fouls suffered minus fouls committed. The authors estimated a coefficient of 0.220 on fouls difference for own-nationality in the fullest model specification with fixed effects for player, tournament stage, year, month, own team, opposing team, referee and match. When this was broken down into fouls suffered and fouls conceded it was fouls committed that had a negative, significant coefficient. Players of own-nationality were called up for fewer fouls than those of other nationalities whereas there was no difference in fouls suffered. Interestingly, the authors found no significant effects of own-nationality on offside calls and yellow or red cards issued.

TABLE 21.4

Key Results from Buraimo et al. (2012)

	Yellow Cards Home	Yellow Cards Away	Second Yellow and Red Cards Home	Second Yellow and Red Cards Away
Spain				
Track	0.115	−0.444	0.001	0.003
	(1.91)*	(8.34)***	(0.02)	(0.06)
Log	0.019	0.074	0.071	0.108
attendance	(0.48)	(1.87)*	(1.47)	(2.43)**
Champions League				
Track	−0.020	−0.05	0.148	−0.197
	(0.17)	(0.47)	(1.11)	(1.65)*
Log	0.133	0.126	−0.106	−0.026
attendance	(1.83)*	(1.81)*	(1.07)	(0.28)

Note: ***p < 0.01, ** p < 0.05, and *p < 0.1.

This paper benefited from the greater wealth of data now available. Previous papers were unable to model fouls while this paper has greater number of controls for game-level team outcomes in terms of shots and corners not just goals. This is where future research can usefully develop by utilising the richer array of within-game information that is gradually becoming available on web sites.

Buraimo et al. (2012) extended the analysis of Buraimo et al. (2010) to cover Spain's La Liga and the UEFA Champions' League. Similar to Buraimo et al. (2010), the authors exploited variation in running tracks within a minute-by-minute bivariate probit model of probability of award of yellow and red cards. Table 21.4 shows the key effects of presence of running track and crowd size on the probabilities of receiving a yellow or red card for home and away teams.

The data cover seasons 2003/2004 to 2006/2007. Control variables were as for the Buraimo et al. (2010). The models included referee and season dummies. The statistical significance of yellow cards is apparent for Spain where home teams get a higher probability of yellow card and away teams get a lower probability of yellow card when a running track is present in a Spanish stadium. For Champions League games, with neutral referees, these effects are mostly insignificant. There is weak evidence (significance at 10% only) that, in Champions' League games played with a running track, away teams have lower probability of a red card.

For the Champions' League, though, there is weak evidence that bigger crowds are associated with higher probability of yellow card for both teams so this is not consistent with home team bias. For Spain, there is a higher probability of an away team receiving a yellow card with a bigger crowd, imprecisely estimated, and a higher probability of an away team getting a red card as crowd size is greater (significant at 5%). Thus, elements of referee bias do appear to be revealed as documented by Dawson and Dobson (2010) for a similar sample period but a different statistical methodology.

Since yellow cards are quite frequent in top level football matches, with five or six in a typical game, the variation in awards is sufficient to allow for a reasonably full empirical model to yield several significant coefficients. Red cards are rare events, however. In a typical round of 10 weekend matches in England or Spain, only two to three red cards will be awarded in a given league. It is common for rounds of weekend matches to have no red card at all. Reilly and Witt (2013) examined the English Premier League for red cards over the 2003/2004 to 2007/2008 seasons using acquired OPTA data. They modelled probability of a red card for given player so the data were at player-match level. Estimations were by Chamberlain conditional logit with and without player fixed effects. Without fixed effects the models included player time-invariant covariates. Tests showed joint significance of player fixed effects. Significant control variables were fouls called and a yellow card dummy. It appeared that probability of dismissal varied positively with fouls called in a given game which appears intuitively plausible. In line with Dawson et al. (2007) and Buraimo et al. (2010), the results showed a negative and significant effect of home dummy with a coefficient of −0.30 in the fixed effects model. This shows a negative effect of home team on likelihood of a player receiving a red card and this is another possible source if referee bias.

Crowd trouble in Italy's Serie A has led on occasions to some teams playing some home games 'behind closed doors'. After serious violence at Catania versus Palermo in February 2007, at which game a policeman was stabbed and killed, a closed doors sanction was applied to teams that had serious stadium safety concerns. Excluding Catania, this sanction covered 21 out of 842 games in Serie A and B in the 2006/2007 season. The sanction led neatly to a simple natural experiment conducted by Pettersson-Lidbom and Priks (2010) since all affected teams reverted back to 'open doors' during the season once they received the relevant safety certificate from the authorities.

The authors' model had a standard difference-in-difference setup:

$$Y = \alpha + \beta \text{Home} + \gamma \text{Treatment} + \delta \text{Home}^*\text{Treatment} + v$$

where Y is an outcome measure for referee behaviour. The identifying assumption in this model is $E(v|\text{Treatment}^*\text{Home}) = 0$. This assumption was tested indirectly by showing that home-away differentials in team playing outcomes were invariant to whether the games were played behind open or closed doors. The difference-in-difference test is that $\delta = 0$. This null hypothesis was rejected in an OLS regression model where number of fouls was the dependent variable. The coefficient was 4.56 which translated to 23% more fouls being called for the away team relative to home team at the mean. If team fixed effects were excluded and referee effects included, the coefficient of treatment interaction on yellow cards was 0.68 and this was significant at the 5% level.

The sharpness and precision of the natural experiment in Pettersson-Lidbom and Priks (2010) is both a strength and a weakness. The authors were able to clearly identify treatment group (punished clubs with unsafe stadia) and control group (unpunished clubs with safe stadia). The common trends assumption for pre-event patterns in referee outcomes appeared to be valid for the two groups. The results emerge very easily and are intuitively plausible. But the results emerge from just 21 games in the treatment group and readers might question the external validity of inferences drawn from (fortunately) a very specific episode in Italian football. In the Champions' League, some clubs (e.g. Dynamo Kiev) have been forced to play matches behind closed doors due to racist behaviour of fans. Also, the Ukraine national team had to play a game behind closed doors for similar reasons. Recently (November 2015), Club Brugge was forced to play its Europa League match against Napoli behind closed doors due to a terror alert in nearby Brussels. These events might offer an opportunity for a follow-up study to verify the findings of Pettersson-Lidbom and Priks.

So far, there is no evidence that directly addresses the question of how referee performances affect fan enjoyment of football matches. This is inherently difficult as fans may well blame the referee for their team's defeat even when the referee had a good game by standards of peer review. Of particular interest is the extent to which referees make 'correct' calls. Here, there are both Type I and II errors to consider. The data needed to answer this important question would have to come from some official review body, although journalists' referee ratings in newspaper reports might be an acceptable proxy.

Dohmen (2008) obtained data on penalty kicks and goals awarded in the Bundesliga top division. 892 critical situations were reviewed by an independent assessment body using video evidence. These decisions were reviewed as correct, disputable or incorrect. There was no indication that referees awarded visitors fewer legitimate goals. But home teams were awarded more illegitimate goals than away teams. The preferential treatment effect was significant at the 10% level from a one-sided non parametric test. Also, home teams were awarded 28.7% of disputable penalties. Away teams were awarded 20.3% of disputable penalties. Home teams were significantly more likely to receive penalties when objectively they should be awarded and when penalty calls were disputable. Referees awarded fewer disputable and unjustified penalties when the crowd was separated from the pitch by a running track. This is consistent with the evidence of Buraimo et al. (2010) of more favourable treatment to home teams in the award of yellow cards when running tracks were removed in the Bundesliga. Unfortunately, it is difficult to persuade assessment bodies to part with useful information on peer review of referees' performances so Dohmen (2008) remains a rarity in this regard.

Fewer yellow and red cards given by referees might be indicators of better referee performance. A referee who issues many yellow and red cards is deemed to have 'lost control' and such a game will tend to be frustrating to watch for fans at the stadium and for TV viewers since the player misdemeanours and the referee's interventions will disrupt the flow of the game. This reduces the appeal of the spectacle. Buraimo et al. (2011) exploited the same natural experiment as Rickman and Witt (2008) in the English Premier League in 2000/2001 where referees switched from compensation by match fee to an annual salary. This was intended to make referees 'more professional' by offering them a career structure. Referees would be subject to more intensive peer review and more rigorous training sessions, often in partnership with English Premier League clubs. Similar to Rickman and Witt (2008), the authors had Tier 2 referees as a control group. Actually Premier League referees could officiate in Tier 2 games but Tier 2 referees could only officiate Premier League games if they were promoted to a central list of 20–24 elite referees. Descriptive data show that Premier League salaried referees gave out fewer yellow and red cards after 2000/2001 compared to non-salaried referees in Tier 2 before 2000/2001.

The regression model had number of first yellow cards as dependent variable where this was a proxy for referee performance. There were 11,169 games for analysis (10,904 in the fixed effects estimates). The focus variable is Tier 1*Post Salary where this is the difference-in-difference indicator for Premier League salaried referees. The model controls were Premier League dummy, dummy for Post Salary era, log attendance, score of game, score squared, derby, number of months left in a season, home points so far, away points so far and year dummies. Regressions were performed by OLS, by OLS with referee fixed effects and also by fixed effects Poisson estimation since cards can be considered as a count variable. Results for Poisson estimation were similar to OLS. With OLS, the coefficient of the difference-in-difference indicator in the yellow card estimates was −0.51 while with referee fixed effects the coefficient fell to −0.43 but was still statistically significant at 1%. In the second yellow/red card specification the coefficients on the focus difference-in-difference variable were −0.55 and −0.47 respectively. All control variables had significant coefficient of expected sign.

It could be argued that yellow cards are not a good proxy for referee performance since fewer cards issued simply indicates a more lenient referee possible to the extent of shirking or maybe vulnerability to social pressure. The authors had supplementary data on referee ratings issued by newspaper journalists for a subset of games. Controlling for referee and season fixed effects they found that first yellow cards were negatively associated with newspaper referee ratings. When red cards were added, the coefficient on this was significant as well as for first yellow card. Hence, the number of cards did appear to be correlated in the expected direction with one source of ratings of referee performance.

21.4 Conclusions

The literature on referee bias now offers several lessons to be taken forward in future research.

- Research on stoppage time added on by referees reveals that some bias is present but the size of effects tends to be small. With many applications already published we suspect that this literature has now run its course.

- Referees are found to be biased towards home teams in terms of cautions awarded. These biases are best revealed by estimations that use high frequency data; these enable researchers to control for confounding within-game influences such as earlier yellow and red cards and scoreline. Disturbingly for the stakeholders within football, evidence from international competitions with referees from neutral countries suggests that biases persist in this setting.

- Several studies have used natural experiments to investigate changes in referee behaviour, for example, to changes in playing rules or to employment contracts. This approach permits proper identification of effects and is to be encouraged in future work (Angrist and Pischke, 2015).

- An issue that requires further statistical investigation is referee corruption, especially match fixing. The analysis of Italy's *Calciopoli* episode by Boeri and Severgnini (2011) shows how referees may be induced, through career concerns, by club executives to give decisions in favour of specific clubs. Much concern has been expressed recently over allegations of match-fixing which often feature referees who are tempted by direct payments by bookmakers or clubs (Forrest, 2012). Football associations and tournament organisers are rightly concerned about match-fixing and it is important that the scientific community contributes to understanding sources and causes of match-fixing including the roles of referees.

21.A Appendix

TABLE 21.A.1

Summary of Literature on Referee Bias

Authors	Paper [League and Season]	N	Statistical Method	Control Variables [Dependent and Independent Variables]
Effect of Referee Bias on Stoppage Time				
Garicano et al. (2005)	*Primera Division, Spain (1994/1995 and 1998/1999)*	750	OLS (with six specifications).	The dependent variable *is the length of injury time in matches that ended with a one-goal difference. Controlling for player substitutions, year effect, Budget home/visitor, Rank home, the absolute value of the difference in ranks, yellow/red cards, team fixed effects, referee fixed effects. Further controls: game number, attendance, ration of attendance to capacity.*
Sutter und Kocher (2004)	*Bundesliga, Germany (2000/2001)*	306	A simple OLS-regression (testing three different dependent variables).	*The influence of the score margin on extra time: Extra time in second half Score margin +1/−1, Extra time in second half Score margin 0/−1, Extra time in first half Score margin +1/−1. Yellow cards, Red cards, Player substitution. Additional models controlling for (1) the influence of the relative strength of the home team vs. the visiting team (as measured by the rank in the previous round or at the end of the previous season, or alternatively measured by the budget of a team), (2) the stage of the season (by including a dummy for each round), or (3) for the number of spectators.*
Dohmen (2008)	*Bundesliga, Germany (1992/1993– 2003)*	3519	OLS	*Length of second-half stoppage time (in seconds) in matches in which one team was leading by one goal after regular time (Panel A) Length of first-half stoppage time (in seconds) in matches in which one team was leading by one goal after regular first-half playing time (Panel B) Home ahead, No. of treatments, substitutions, fouls, yellow cards, second yellow cards, red cards, season dummies, controls for relative strength, fixed effects for home team, visiting team, and referee. Further controls distance, attraction, attendance, ratio of attendance to capacity.*

(Continued)

TABLE 21.A.1 (*Continued*)

Summary of Literature on Referee Bias

Authors	Paper [*League and Season*]	N	Statistical Method	Control Variables [*Dependent and Independent Variables*]
Rickman and Witt (2008)	*Premier League, England (1999/2000 and 2002/2003)*	760	OLS. Similar to Garicano et al. (2005). Also study the effects of crowd composition, stadiums with and without track and the correctness of decisions to award penalties and goals.	*Length of injury time in games that ended with a 1 goal difference.* *Control variables: Score difference, year, interaction term score difference*year, yellow cards, red cards, player substitutions, referee fixed effects, team fixed effects.* *Controlling for turnover of clubs, table rankings of home teams, the absolute value of the difference in ranks, attendance, ratio of attendance to capacity, monthly dummies, controls for the effect of introducing professional referees.*
Scoppa (2008)	Are subjective evaluations biased by social factors or connections? An econometric analysis of soccer referee decisions [*Serie A and B, Italy (2003/2004–2004/2005)*].	686	A simple OLS regression with six specifications.	The dependent variable is *Injury time* [added at the end of the first or second half (in minutes)] . *Independent variables are: Home Behind, Player substitutions, Red cards, Yellow cards, Penalties, Difference in ranks, Attendance (1000s), Other Stoppages, Goals scored, Season 2005 dummy, Team fixed effects, home Interaction, Attendance, Budget difference, Interaction budget, Track, Track*Home Behind, team success.* *Moreover, they have tried to consider visiting team fixed effects (not reported) without appreciable modifications.*
Lucey and Power (2005)	*Serie A, Italy (2002/2003)*	283	OLS	*The dependent variable is the amount, in minutes, of added time played at the end of the second half.* *Controlling for yellow cards (second half), red cards (second half), goal difference, home/visiting team scoring during extra time, attendance and capacity, ranking variables.*
Lucey and Power (2005)	*Major League Soccer, USA (2003)*	159	OLS	*The dependent variable is the amount, in minutes, of added time played at the end of the second half.* *Controlling for yellow cards (second half), red cards (second half), goal difference, home/visiting team scoring during extra time, attendance and capacity, ranking variables.*
Rocha et al. (2013)	*Brazilian Football Championship (2004–2008)*	1413	OLS and probit models	*Dependent variable is the dummy variable extra time added at the end of match (in seconds).* *They included control variables that are related to the referee's choice of injury time, such as the number of yellow and red cards for both teams, the total number of substitutions in the game, time, stadium, home team, visiting team and referee dummies to control for unobservable factors affecting the dependent variable.*

(Continued)

TABLE 21.A.1 (*Continued*)

Summary of Literature on Referee Bias

Authors	Paper [League and Season]	N	Statistical Method	Control Variables [Dependent and Independent Variables]
Mendoza and Rosas (2013)	[Colombian Professional League (DIMAYOR) (2005–2010)]	720	OLS and Poisson regressions to estimate the effect of the score difference on the length of injury time added at the end of both the first and second halves of each game.	Length of injury time at the end of the first/second half. Controlling for number of player substitutions, the number of yellow and red cards, the occurrence of penalties or other unusual events during the game, fixed effects at the team and referee levels. They study how the size of the referee bias might depend on variables such as attendance, ranking difference, previous performance in the tournament, homicide rates in the city of the home team, as well as a measure of historical performance of each team.

Authors	Paper [League and Season]	Observations	Statistical Method	Control Variables [Dependent and Independent Variables]
Effect of Referee Bias in Other Decisions				
Boyko et al. (2007)	*Premier League, England (1992/1993–2006/2007)*	5,244	Ordinal regression model to determine whether the home advantage term systematically differs between referees.	Goal differential as the dependent variable. Controlling for the effect of home team ability, away team ability, crowd size, crowd density, between-season effects, and the particular referee for the match on several aspects of game outcome: the goal differential, match result (win/loss/draw), card differential, and penalty differential. Referee and season were treated as categorical variables, whereas the others were treated as continuous variables.
Johnston (2008)	*Premier League, England (2006/2007)*	380	OLS (four models used).	Home advantage Home goals, away goals, attendance (in 000s). In another regression the authors included besides referee effects.

(*Continued*)

TABLE 21.A.1 (*Continued*)

Summary of Literature on Referee Bias

Authors	Paper [League and Season]	Observations	Statistical Method	Control Variables [Dependent and Independent Variables]
Page and Page (2010a)	*English competitions (1994–2007).*	37,830; 872 referees	Random-effects and fixed-effects-models. Random effects approach uses multilevel modelling, which is a general approach to allow for random effects in the coefficients. For the fixed effects model they add dummies for each competition and referee.	*Differences in the home advantage by referee.* Controlling for several possible explanatory factors including *competitions, team quality, attendance, referee specific effects, home team effects.*
Dawson et al. (2007)	*English Premier League (1996–2003)*	2,660	Bivariate Poisson regression/ negative binomial distributions, defined by fitting the Frank copula. Test unconditional and conditional models.	*The total numbers of disciplinary 'points' incurred by the home and away teams in match denoted and calculated by awarding 1 point for a yellow card and 2 for a red card.* Controlling for *team behavior, individual referee effects, season effects, match attendance and live television broadcast.*
Buraimo et al. (2010)	*Premier League, England (2000/2001– 2005/2006)*	204,480	Minute-by-minute bivariate probit analysis of yellow and red cards issued in games over six seasons in the two leagues.	Dependent variable: *receiving a card in a given minute* Controlling for various **pre-game** (like *minute, goal difference, home/away yellow prior*) and **within-game variables** (like *track, derby, difference in bookmaker probability*). The models include sets of dummy variables for the fixed effects of referee, home teams, away teams and year.

(Continued)

TABLE 21.A.1 (*Continued*)

Summary of Literature on Referee Bias

Authors	Paper [*League and Season*]	Observations	Statistical Method	Control Variables [*Dependent and Independent Variables*]
Buraimo et al. (2010)	*Bundesliga, Germany (2000/2001– 2005/2006)*	159,210	Minute-by-minute bivariate probit analysis of yellow and red cards issued in games over six seasons in the two leagues.	Dependent variable: *receiving a card in a given minute* Controlling for various **pre-game** (*like minute, goal difference, home/away yellow prior*) and **within-game variables** (*e. g. track, derby, difference in bookmaker probability*). The models include sets of dummy variables for the fixed effects of referee, home teams, away teams and year.
Dawson and Dobson (2010)	*UEFA-tournaments, Europe (2002/2003– 2006/2007)*	1,717	Univariate and bivariate ordered probit models.	Dependent variable: *disciplinary points* Control variables are *crowd size, crowd density (attendance to capacity ratio), stadium architecture (i.e. presence of a running track and/or fencing), competition (Champions League or UEFA Cup), stage of competition, nationality of the referee, club nationality, league reputation (as measured by whether the team plays in one of the top five leagues).*
Buraimo et al. (2012)	*Primera Division, Spain (2003/2004 and 2006/2007)*	135,720	Within-game analysis: Bivariate probit models.	The dependent variables are pairs of binary indicators: The first pair is *the award of yellow cards to home and away teams in any given minute of the match.*A second pair of variables is constructed *for dismissals* Control for a large number of independent variables: *pregame (e.g. track, log attendance, derby) and within-game variables (e.g. goal difference, home/away yellow prior).* Dummy variables for *home teams, away teams, referees,* and *seasons* are included as control variables.
Buraimo et al. (2012)	[*UEFA's Champions League (2003/2004 and 2006/2007)*].	41,610	Within-game minute-level analysis: Bivariate probit models.	Dependent variables are pairs of binary indicators: The first pair is *the award of yellow cards to home and away teams in any given minute of the match.*A second pair of variables is constructed *for dismissals* Control for a large number of independent variables: *pregame (e.g. track, log attendance, derby) and within-game variables (e.g. goal difference, home/away yellow prior).* Dummy variables for *home teams, away teams, referees,* and *seasons* are included as control variables.

(*Continued*)

TABLE 21.A.1 (*Continued*)

Summary of Literature on Referee Bias

Authors	Paper [*League and Season*]	Observations	Statistical Method	Control Variables [*Dependent and Independent Variables*]
Downward and Jones (2007)	*FA Cup, England (1996/1997– 2001/2002)*	857	Binary logistic regression model.	Dependent variable, *home vs. away*, measured whether the first yellow card was awarded to a home-team or an away-team player. Independent variables are *crowd size* and its squared term, *crowdsq, a divisional difference variable Divdiff and an interaction term crowdiff.*
Reilly and Witt (2013)	*English Premiership League (2003/2004– 2007/2008)*	51,076 (1,162 players)	Chamberlain's (1980) conditional logit model incorporating player fixed effects; also a nonpanel logit model.	Dependent variable: *player dismissal (and player dismissal by fixture status*) Independent variables: *home fixture, yellow card, fouls called, 'veteran' player, number of prior games played, relative league position, 'derby' game, ln(attendance).* Tests of overall significance: *prob-value for referee effects, prob-value for fixed effects.*
Goumas (2014)	*Champions League and Europa League (2009/2010– 2010/2011)*	426 (CL) and 958 (Euro- League)	Regression-analyses, using log-link Poisson regression models.	*Match location and crowd factors on yellow cards issued by referees* *Crowd size, crowd density, track*
Pettersson-Lidbom and Priks (2010)	*Serie A and B, Italy (2006/2007)*	842	OLS Test whether home and away players are affected differently in games with and without spectators by estimating Equation 1 on a number of outcomes of players.	*Number of foals, number of yellow cards, number of red cards* *Referee bias effect, referee fixed effects, team fixed effects*

(Continued)

TABLE 21.A.1 (*Continued*)

Summary of Literature on Referee Bias

Authors	Paper [*League and Season*]	Observations	Statistical Method	Control Variables [*Dependent and Independent Variables*]
Nevill et al. (2002)	The influence of crowd noise and experience upon refereeing decisions in football [*laboratory study*].	40 referees	Binary logistic regression used to assess the effect of crowd noise and years of experience on referees' decisions.	The response variable was *categorical* (home foul vs. away foul vs. no foul vs. uncertain) The results in the figures illustrates the mean number of challenges for each of the four response options awarded by the noise and silent condition groups and the mean number of challenges for the remaining three response options awarded by the noise condition group, silent condition group and match referee (no foul and uncertain options collapsed into a single no foul option).
Boeri and Severgnini (2011)	*Serie A, Italy (2004/2005)*	1,444	Sequential probit model (four different specifications).	*Allocation of referees to matches (Table 7); referees' official evaluation (Table 8)* *A wide array of explanatory variables, controlling for the characteristics of referees, linesmen and matches.* *In particular, the first specification includes the usual controls for the career of the referee (less, intermediate and top) and interacts these controls with past involvement of the referees in matches for which there is hard evidence of match rigging.* *The second specification adds controls for the characteristics of the matches (e.g. the total number of fouls committed by the home and visitor teams, red and yellow cards depending on the nature of the foul, corners [corner] or the presence of a track in the stadium [track]).*
Nevill et al. (2013)	*Four professional English Leagues from 1946–1947 to 2010–2011 and Scottish Premier League from 1946–1947 to 1999–2000*	6,166	Multilevel model	*Home advantage (hi)* *Fixed explanatory variables, such as the random variance at each level: level 2 (between-team) variation, level 1 (within-team) variation, points per game or the examined leagues and seasons.*

References

Angrist, J. and J.-S. Pischke (2015). *Mastering 'Metrics'*. Princeton University Press, Princeton, NJ.

Boeri, T. and B. Severgnini (2011). Match rigging and the career concerns of referees. *Labour Economics* 18, 349–359.

Boyko, R., A. Boyko and M. Boyko (2007). Referee bias contributes to home advantage in English Premiership Football. *Journal of Sports Sciences 25*, 1184–1194.

Bryson,A., B. Buraimo and R. Simmons (2011). Do salaries improve worker performance? *Labour Economics 18*, 424–433.

Buraimo, B., D. Forrest and R. Simmons (2010). The twelfth man? Refereeing bias in English and German soccer. *Journal of the Royal Statistical Society: Series A (Statistics in Society) 173*, 431–449.

Buraimo, B., R. Simmons and M. Maciaszcyk (2012). Favoritism and referee bias in European soccer: Evidence form the Spanish League and the UEFA Champions' League. *Contemporary Economic Policy 30*, 329–343.

Butler, R. and D. Butler (2015). Favouritism in the allocation of additional time: The case of the English Premier League. Discussion paper, University College Cork, Cork, Ireland.

Dawson, P. and S. Dobson (2010). The influence of social pressure and nationality on individual decisions: Evidence from the behaviour of referees. *Journal of Economic Psychology 31*, 181–191.

Dawson, P., S. Dobson, J. Goddard and J. Wilson (2007). Are football referees really biased and inconsistent? Evidence on the incidence of disciplinary sanction in the English Premier League. *Journal of the Royal Statistical Society: Series A (Statistics in Society) 170*, 231–250.

Dohmen, T. (2008). The influence of social forces: Evidence from the behaviour of football referees. *Economic Inquiry 46*, 411–424.

Dohmen, T. and J. Sauermann (2016).Referee bias. *Journal of Economic Surveys*, 30, 679–695.

Downward, P. and M. Jones (2007). Effects of crowd size on referee decisions: Analysis of the FA Cup. *Journal of Sports Sciences 25*, 1541–1545.

Forrest, D. (2012). The threat to football from betting-related corruption. *International Journal of Sport Finance 7*, 99–116.

Garicano, L., I. Palacios-Huerta and C. Prendergast (2005). Favoritism under social pressure. *Review of Economics and Statistics 87*, 208–216.

Goumas, C. (2014). Home advantage and referee bias in European football. *European Journal of Sport Science 14*, S243–S249.

Johnston, R. (2008). On referee bias, crowd size and home advantage in the English soccer Premiership. *Journal of Sports Sciences 26*, 563–568.

Lucey, B. and D. Power (2005). Do soccer referees display home team favouritism? Discussion paper, Trinity College, Dublin, Ireland.

Mendoza, J. and A. Rosas (2013). Referee bias in professional soccer: Evidence from Colombia. Discussion paper, Pontifica Universidad Javeriana, Bogotá, Colombia.

Nevill, A., N. Balmer and A. Williams (2002). The influence of crowd noise and experience upon refereeing decisions in football. *Psychology of Sport and Exercise 3*, 261–272.

Nevill, A., T. Webb and A. Watts (2013). Improved training of football referees and the decline in home advantage post-WW2. *Psychology of Sport and Exercise 14*, 220–227.

Page, K. and L. Page (2010). Alone against the crowd: Individual differences in referees' ability to cope under pressure. *Journal of Economic Psychology 31*, 192–199.

Pettersson-Lidbom, P. and M. Priks (2010). Behaviour under pressure: Empty Italian stadiums and referee bias. *Economics Letters 108*, 212–214.

Pope, B. and N. Pope (2015). Own-nationality bias: Evidence from UEFA Champions League football referees. *Economic Inquiry 53*, 1292–1304.

Reilly, B. and R. Witt (2013). Red cards, referee home bias and social pressure: Evidence from English Premiership soccer. *Applied Economics Letters 20*, 710–714.

Rickman, N. and R. Witt (2008). Favouritism and financial incentives: A natural experiment. *Economica 75*, 296–309.

Riedl, D., B. Strauss, A. Heuer and O. Rubner (2015). Finale furioso: Referee-biased injury time and their effects on home advantage in football. *Journal of Sports Sciences 33*, 327–336.

Rocha, B., F. Sanchez, I. Souza and J. Carlos Domingos da Silva (2013). Does monitoring affect corruption? Career concerns and home bias in football refereeing. *Applied Economics Letters 20*, 728–731.

Scoppa, V. (2008). Are subjective evaluations biased by social factors or connections? An econometric analysis of referee decisions. *Empirical Economics 35*, 123–140.

Sutter, M. and M. Kocher (2004). Favoritism of agents – The case of referees' home bias. *Journal of Economic Psychology 25*, 461–469.

Watanabe, N., P. Wicker and J. Reuter (2015). Determinants of stoppage time awarded to teams in the English Premier League. *International Journal of Sport Finance 10*, 310–327.

Yewell, K., S. Caudill and F. Mixon, Jr. (2014). Referee bias and stoppage time in Major League Soccer: A partially adaptive approach. *Econometrics 2*, 1–11.

22

Golf Analytics: Developments in Performance Measurement and Handicapping

Mark Broadie and William J. Hurley

CONTENTS

22.1 Golf Performance Measurement in the ShotLink Era

In principle, the scientific analysis of golf should be similar to baseball. As both games progress, competitors move from one discrete state to the next. But whereas modeling and measurement are well developed in baseball, that effort is just in its infancy in golf. It is only recently that golf has begun to record the data in a way that enables some useful modeling and analysis. The PGA Tour's pioneering ShotLink™ system* records data for each shot a PGA Tour player takes.† Basically,

---

* See http://www.pgatour.com/stats/shotlinkintelligence.html.
† There are a number of "amateur" shotlink systems including *GameGolf* (https://www.gamegolf.com), *Arccos* (http://www.arccosgolf.com), *MyRoundPro* (TaylorMade, and http://www.myroundpro.com). These systems are based on GPS information, but a drawback is that commercial GPS isn't accurate enough for automatic recording of putts and can't distinguish fairway and rough when the ball is near the boundary of the two. Players need to manually input or adjust GPS data to create accurate shot records.

these systems record two critical items of information for each shot:

1. The *location* of the shot
2. The *condition* of the current location (tee, fairway, rough, recovery, sand, green)

The shot and hole locations allow the distance to the hole to be measured for each shot. At this writing, the PGA Tour's ShotLink system has recorded data on more than 15 million shots.

22.1.1 Problems with Traditional Golf Statistics

It is one thing to have data. It's quite another to make sense of it correctly. The traditional statistical measures in golf can sometimes be misleading. Take, for instance, statistics on driving. The PGA Tour measures a statistic called *total driving*. It's calculated as follows. All PGA Tour players are ranked on driving length (the average yards the ball is driven from the tee) and on driving accuracy (the percentage of time a drive comes to rest in the fairway). The total driving statistic is the rank of the sum of a golfer's ranks in these two categories. For example, for the 2015 season, Bubba Watson and Spencer Levin were tied for 44th with a total driving rank of 155. Watson ranked 2nd in distance and 153rd in accuracy; Levin ranked 139th in distance and 16th in accuracy. Is this evidence that Watson and Levin are comparable drivers?

Let's look at the primary statistics which determine the Distance and Accuracy ranks:

	Driving Distance (Yards)	Driving Accuracy (%)
Watson	315.2	56.55
Levin	283.7	69.09

On average, Watson drives the ball 30+ yards longer than Levin. Over an average of about 13 holes a round where players typically use a driver (on par-4 and par-5 holes), that's a big scoring advantage. On the other hand, Levin's nearly 13% point advantage in driving accuracy may not be that big a difference. Over a round, Levin would drive it in the fairway 10 times versus Watson's 8 times. Moreover, what if Watson's misses are often just off the fairway? There are two problems with the total driving statistic. First, there is no evidence that it is reasonable to give equal weight to distance and accuracy. Second, simply adding ranks is known to be flawed, because the difference in performance between players ranked 1 and 21 is much larger than the difference in performance between players ranked 71 and 91. We'll come back to this comparison once we define the strokes gained measure.

22.1.2 Strokes Gained

What gets around these problems is the *strokes gained* measure developed by Broadie (2008, 2012). Here is how it works. Suppose we could estimate a function $J(d, c)$ where d is the distance to the hole from the current location, c is the condition at that location (i.e., fairway, rough, etc.), and $J(d, c)$ is the average number of strokes a PGA Tour player would need to finish the hole. $J(d, c)$ is termed the *benchmark* for obvious reasons. Broadie (2012) has given the details of the estimation procedure for $J(d, c)$. Parameters of the $J(d, c)$ function are estimated with millions of ShotLink observations.

The benchmark is then used to define strokes gained. For shot i, suppose a golfer is at a distance d_i with condition c_i. Let us suppose that he ends up at a distance d_{i+1} and condition c_{i+1} after the shot. Then, we can define the *strokes gained* for shot i to be

$$g_i = J(d_i, c_i) - J(d_{i+1}, c_{i+1}) - 1.$$

The quantity $J(d_i, c_i) - J(d_{i+1}, c_{i+1})$ is the reduction in the number of strokes to complete the hole an average PGA Tour player would expend. And to do this, the player takes a stroke and hence 1 is subtracted.

For example, suppose a player is 185 yards from the hole in the fairway and hits an iron to 8 ft and then misses the putt (he has a 2 in. tap-in left after the miss). Let us suppose that the benchmark at 185 yards in the fairway is 3.1 shots and at 8 ft on the green it's 1.5 shots. On the approach shot, then, the strokes gained is

$$g_A = 3.1 - 1.5 - 1 = 0.6.$$

On the first putt, it's

$$g_{P1} = 1.5 - 1 - 1 = -0.5,$$

and on the second putt,

$$g_{P2} = 1 - 0 - 1 = 0.$$

All of this makes sense. The player hit a better-than-average approach shot into the green, ending up 8 ft from the hole. Hence the strokes gained on the approach, g_A, is positive. On the first putt, the benchmark is 1.5 suggesting that players make the putt half the time. Since our player missed the putt, his strokes gained is negative. Finally, for the tap-in, the average player will always make a 2 in. putt and hence the strokes gained is 0.

With this example, we can observe the general property that strokes gained observations are additive. Note that the player took 3 strokes to complete the hole from 185 yards and hence his overall strokes gained is $3.1 - 3 = 0.1$. And if we add the strokes gained attributed to each shot, we also get

$$g_A + g_{P1} + g_{P2} = 0.6 - 0.5 + 0 = 0.1.$$

Strokes gained has its antecedents in the work of Cochran and Stobbs (1968), Landsberger (1994), Riccio (1990), and Soley (1977). For instance, Cochran and Stobbs use a term similar to strokes gained but they do not compute strokes gained for a shot or suggest how to do it. The closest they come is to say that, on a par-5 hole, if the field's scoring average is 4.8, then a score of 5 is 0.2 strokes worse than the field average. Broadie (2008, 2012, 2014) was the first to set out the full theory and apply it extensively using the ShotLink database as well as amateur data observed through his Golfmetrics program. He is also responsible for making the connection between strokes gained and the dynamic programming principle (see Bellman, 1957; Broadie and Ko, 2009).

We now consider how effective strokes gained is (relative to the traditional statistics) with three examples.

Example 1. Let's return to the comparison of the driving ability of Bubba Watson and Spencer Levin. The Strokes Gained Driving and rank (based on Strokes Gained Driving) for each player for the 2015 season are shown below*:

---

* There are 200 PGA Tour players ranked with at least 30 rounds of ShotLink data.

	Stokes Gained Driving	Rank
Bubba Watson	1.29	1
Spencer Levin	0.15	78

Watson ranks first in driving on the PGA Tour; Levin ranks 78 and is a full stroke per round worse than Watson. Considering that Watson made $6 million more than Levin on tour in 2015, the firm conclusion is that the total driving statistic is not a good measure of driving ability.

Example 2. Consider the Greens in Regulation (GIR) and Putts per Round statistics. Golfer A and B play the same par-3 hole, say, 180 yards. Golfer A hits the green 60 ft from the hole and sinks the putt. Golfer B hits the green 2 ft from the hole and sinks the putt. Both had the same score, a birdie, both are credited with a GIR, and both took one putt. Can GIR and putts per round distinguish these two very different performances? No. Golfer A hit a mediocre shot relative to B but then followed up with a great putt. On the other hand, golfer B hit a great tee shot and then a below average putt. GIR and PPR can't distinguish this difference in performance, but strokes gained easily can.

Related to the issue of putting, interested readers are referred to Yousefi and Swartz (2013) who extend the Strokes Gained Putting measure to incorporate a spatial effect due to the greens and also to Hickman and Metz (2015).

Example 3. In Table 22.1, we present three traditional statistics for three PGA Tour players for 2013. Based on these statistics, we ask the reader to guess which player had the better season. The first author has put the question to many knowledgeable golfers and most pick either B or C. Few pick A. In Table 22.2, we reveal the identity of the players as well as the number of tournaments they played, their average strokes gained for the year, their PGA Tour earnings, and the number of tournaments they won. Golfer A—Tiger Woods—clearly had the best year with $8.5 million in earnings and five tournament wins. And this shows

TABLE 22.1

Selected Statistics for Three PGA Tour Players for the 2013 Season

	Golfer A	Golfer B	Golfer C
Fairways	62.5% (69)	66.3% (29)	67.8% (18)
Greens in Regulation	12.2 (24)	12.4 (10)	12.0 (37)
Putts per Round	28.85 (48)	29.10 (88)	28.68 (31)

TABLE 22.2

Strokes Gained and Other Performance Measures for the Three PGA Tour Players for the 2013 Season

	Golfer A	Golfer B	Golfer C
Name	Tiger Woods	Brendon de Jonge	Jordan Spieth
#Tournaments	16	30	23
Strokes Gained (Rank)	2.54 (1)	0.85 (37)	1.65 (12)
PGA Tour Earnings (Million)	$8.5	$1.8	$3.9
#Wins	5	0	1

up in strokes gained. Woods finished the year ranked first in strokes gained. Moreover, considering all three golfers, strokes gained correlates nicely (0.984) with dollar winnings. In contrast, the traditional statistics on fairways, greens in regulation, and putts couldn't identify one of the best seasons in PGA Tour history.

22.1.3 Applying Strokes Gained

There are two primary ways that strokes gained can be applied. One is to examine the various bits of conventional golf wisdom to determine whether they hold up to an analysis through the lens of strokes gained. The other is how a golfer might use strokes gained to improve his or her on-course strategy and determine the content of off-course practice. We discuss each in the following text.

22.1.4 The Relative Importance of Driving and Putting

Let's examine the conventional wisdom that a golfer "drives for show and putts for dough." This well-known aphorism suggests that the short game (putting, chipping, sand play, etc.) is more important than the long-game (driving and approaches). This view is not held by all golfers. For instance, Ben Hogan took the position that the drive was the most important shot in golf. Before strokes gained came along, the interpretation of the statistical evidence suggested the encomium was true (see Berry, 1999). A related issue is a consideration of what made Tiger Woods so dominant over his Golden Period. Generally, golf pundits and tour players take the position that Woods dominated because he was the best putter by a wide margin, an observation consistent with the received wisdom on the relative importance of the long and short games.

In his book, Broadie (2014) makes the case that the long game is the most important aspect of play at the professional level. This shows up in a strokes gained breakdown for top PGA professionals. In Table 22.3, we present a break-down of strokes gained for the professionals with the highest total strokes gained over the period 2004–2012. Total strokes

TABLE 22.3

A Strokes Gained Analysis of the 10 Top Professionals over the Period 2004–2012

	Strokes Gained per Round				
	Total	Driving	Approach	Short	Putting
Woods	2.8	0.6	1.3	0.3	0.6
Furyk	1.8	0.3	0.8	0.4	0.4
Donald	1.8	−0.1	0.7	0.5	0.7
Mickelson	1.7	0.4	0.7	0.5	0.1
Singh	1.6	0.6	0.7	0.4	−0.2
Els	1.4	0.4	0.8	0.3	−0.1
Garcia	1.4	0.5	0.7	0.2	−0.1
Stricker	1.3	−0.2	0.5	0.6	0.5
Scott	1.3	0.6	0.7	0.2	−0.1
Z. Johnson	1.2	0.3	0.4	0.1	0.4
Average	1.7	0.4	0.7	0.4	0.2
Fraction of Total (%)	100	20	45	20	15

gained is broken into four categories: driving, approach shots, short game, and putting. Let's first consider Woods. His total strokes gained is a whopping 2.8 strokes per round, a full stroke higher than Jim Furyk who ranked second. This means that, over the four rounds of a PGA Tour event, Woods was, on average, 4 strokes better than Furyk. More importantly, note that Woods's long game (strokes gained Driving and strokes gained on Approach shots) contributed $0.6+1.3 = 1.9$ strokes or about 68% of his total strokes gained. This is strong evidence that, while Woods definitely had a good short game, his long game contributed much more to his success. If you consider the averages for these top 10 players, their long games contributed 65% of the total strokes gained (1.1 of a total strokes gained of 1.9). So at least for the top PGA Tour professionals, it is patently untrue that they drive for show and putt for dough.

22.1.5 Using Strokes Gained to Determine Course Management Strategy

Ben Hogan was known to walk a course backwards in order to determine how best to play it. This, of course, is what the Bellman principle would suggest. For instance, where a drive should be placed in the fairway depends on what a golfer feels is the best avenue of approach to the green. There is no finer example of this than what Hogan did on the sixth hole at Carnoustie in the 1953 British Open. In 1953, Hogan won the Open, along with the Masters and US Open in what is considered to be the finest year ever by a professional golfer.

The sixth is a par 5 usually played into a prevailing wind which tends to magnify any movement left or right that a golfer puts on a shot. The short way to the green requires a drive to a patch of fairway about 20 yards wide between out-of-bounds on the left and two large bunkers on the right located in the center of the fairway. The safe play is to drive to the right of these bunkers away from the out-of-bounds. But the area to the left of the bunkers offers a much more direct and shorter path to the green.* In 1953, in all four rounds, Hogan drove his ball to this narrow, high-risk part of the fairway and birdied in each of the last two rounds. His play of the sixth hole is considered to be one of the primary reasons he won. Subsequently, the sixth hole was renamed "Hogan's Alley." But more importantly for this chapter, just about all shots in golf have a risk-return trade-off requiring a consideration of what lies ahead for strokes following the one a golfer is currently considering.

In the modern professional game, some golfers spend a considerable amount of time before a round even begins planning their approach to that day's round, hole by hole. For instance, LPGA Tour member Brooke Henderson spends upwards to an hour and half planning her round. Again, deciding on what strategy to use on a particular hole generally requires a selection along a risk–reward continuum working backwards from the green. Broadie (2014) outlines an approach to this problem using strokes gained. For the most part, the idea is to identify general principles that ought to govern such choice.

For instance, consider a 40 ft putt. Even the pros don't make many of these. But there is definitely a risk return trade-off with such a putt. There are two clear strategies. We'll term one the *Try-to-Make* (TTM) strategy and the other the *Lag* (L) strategy. With the TTM strategy, a golfer makes a strong effort to get the ball to the hole so that it has a chance to go in. On the other hand, with the L strategy, the golfer is trying to get the ball as close as

---

* To get a better idea of how this hole sets up, the interested reader is referred to this flyover: https://www. youtube.com/watch?v=XSioWl4u55U.

possible in order to have the best chance to make the second putt thus avoiding a 3-putt. Ian Poulter is notorious for using the TTM strategy. Very often he has comeback putts that are relatively long. Using the TTM strategy, we would expect the ball to be further from the hole after the first putt and therefore the chance of a 3-putt would be higher. Pelz (2000) and Broadie and Shin (2015) have studied this problem. Pelz suggests that, in general, the optimal target distance is about 1.5 ft beyond the hole, based solely on an analysis of the imperfections (i.e., the bumpiness) of greens, especially near the hole. The Broadie and Shin optimization model recommends a target that varies with the distance to the hole (the target distance decreasing as the putt distance increases), the slope and speed of the green, and the putt angle (i.e., whether the putt is uphill, downhill, or sidehill). Broadie and Shin analyze the ShotLink data and report that the implied putting targets used by PGA Tour players range from 1.5 to 4 ft beyond the hole and are therefore consistent with the TTM strategy on putts under 15 ft.

22.1.6 Using Strokes Gained to Guide Practice

Any golfer, amateur and professional, can improve his or her game with a program of focused, directed practice. Recently, the golf equipment manufacturers *Game Golf, Arccos,* and *TaylorMade* have come out with GPS-based shot-tracking systems for amateur golfers. Many of these systems now allow amateurs to see their strokes gained results. There are also a number of systems (apps) where an amateur can manually enter data to compute their strokes gained results. Because of the accuracy limitations of GPS, putting information and other "data clean-up" still needs to be done manually, but the technology is getting closer to automated shot data collection for weekend players. The issue is how these data can be used to guide practice. It seems reasonable that a player might want to spend relatively more time on his long-game but, of course, the best allocation of practice time will depend on the relative productivity of practice in these areas.

This is likely an area where research could make a significant improvement in the way teaching pros approach their students. Suppose a teaching professional had a strokes gained report on a student as well as a good working knowledge of the golfer's game. Based on experience, the teaching professional ought to be able to put a practice plan in place to improve the student's game. Over time, the professional and student could look at the way strokes gained measures change for the long and short games in order to track progress and make further refinements to practice.

It's instructive to examine data from the PGA Tour. We looked at PGA Tour players over the period 2003–2015 from one season to the next and considered those who were able to improve by at least 0.3 strokes gained per round. The results of that analysis are shown in Table 22.4. For those golfers whose weakest aspect of their games was driving, they managed to make the largest improvement (39%) in their driving. And this is true for each of the other categorizations (approaches, short game, and putting). These golfers, on average, improved in all aspects of their game but the biggest improvements came in the area where they were weakest to begin with.

22.1.7 Research Issues

In our view, there are a number of interesting research areas going forward. Here is a short list.

TABLE 22.4

The Areas Where PGA Tour Players Improved for Those Who Improved by
At Least 0.3 Strokes Gained per Round over the Period 2003–2015

Weakest	Breakdown of Improvement (%)			
	Drive	Approach	Short	Putting
Drive	39	34	4	23
Approach	10	55	9	25
Short	14	28	35	22
Putting	11	25	9	55

1. *Better Data*. It would be nice to have data on ball trajectories, wind, green speed and hardness, rough height, and wetness. These data would allow even more interesting analyses to be done.

2. *Expanded Data Sets*. It would be nice to get shot-tracking data for the British Open, the LPGA Tour, the European Tour, the Web.com Tour, Champions Tour, and amateurs at all levels. Data on the Ryder Cup and President's Cup competitions would make for interesting analyses as well.

3. *Analysis of Strategic Course Decisions*. As argued earlier, the point here is to determine principles that govern strategy choices on-course. Almost all shot selections involve a trade-off of risk and reward. The strokes gained methodology is extremely useful for such analyses.

4. *Analysis to Guide Practice*. Strokes gained data should make for more efficient practice if a golfer wishes to improve.

22.2 Handicap Research Issues

22.2.1 Background

The main goal of a handicap system is to allow golfers of differing skill levels to play against each other on a relatively even basis. In North America, handicap systems have been in place for several decades. Handicapping is a natural subject for statistical inquiry since golf scores have a significant random component—on successive days, the same golfer can shoot substantially different scores on the same course and in the same weather conditions.

The research on handicaps and fairness in North America begins with a series of papers by Scheid (1971, 1973, 1978, 1979). In Scheid (1971), he looked at a data set of 1000 rounds and used a resampling procedure to simulate a large number of pairwise competitions and found that with the handicap system in use at the time, lower handicapped players won significantly more often. In Scheid (1973), he argued that some golf courses are more difficult than others and that a handicap system ought to reflect those differences. This and work by others including Stroud and Riccio (1990) led to the "slope" system as we know it today. The slope system was approved for use by the United States Golf Association (USGA) beginning January 1, 1990 and is now in use by other major national golf associations.

Pollock (1977) presented a model which related the probability of a golfer winning a match to the differences in the handicaps of the players. His purpose was to examine the effects of different handicap scalings on the outcome of matches.

Kupper et al. (2001) examined the USGA handicap system, and with actual scoring data and a model, found that it favors lower handicapped golfers in a two-player net medal play* event. Lackritz (2011) had a similar finding. It should be noted that the USGA handicap system includes a factor termed a "bonus for excellence."[†] We will make it clear in the following text how this works, but it's an explicit factor in the calculation of a handicap that is *designed to give a benefit to lower handicap players*. Hence, the USGA system is not designed to be perfectly fair and so the results in Kupper et al. (2001) and Lackritz (2011) are not surprising.

In most handicap systems, a player's handicap is a single value determined as a function of a number of recent 18-hole scores. One of the difficulties in applying a single handicap number to different kinds of golf competitions is ranging from a two-player match play competition to a large field stroke play event. Bingham and Swartz (2000) argued that the current USGA system favors higher-handicapped golfers in large field stroke play events. They examined the probability that a particular player wins *given that both players are playing well*. They operationalize "playing well" by conditioning the probability a particular player plays at least k standard deviations below their average gross score. This approach makes sense to analyze a competition with many players where a player would have to play very well to get to the prize table.

In this regard, some researchers have questioned whether fairness is the right objective. McHale (2010) suggested a definition of fairness so that a better player will win a higher percentage of the time. He wrote: "surely it is fair that a better player wins more often than a player of lower ability, but that the probability of the better player winning is reduced so as to make the game more competitively balanced and interesting."

This research on fairness led Swartz (2009) to propose an interesting variation of the handicap system which the Royal Canadian Golf Association considered but rejected. We will have more to say about Swartz's ideas in a subsequent section.

There are many different golf competition formats (stroke play versus match play, individual play versus teams of two, scrambles, shambles, Pinehurst, better ball versus worst ball, etc.). The USGA has offered suggestions on how to handicap many of these. The interested reader is referred to Section 9 of the *USGA Handicap Manual*. Typically, the USGA recommends that a percentage of a competitor's handicap be used depending on the type of competition. Take, for example, the case of a "scramble," a popular competition where a team's players all hit a shot and the next shot is taken from the spot where their best previous shot finishes. The USGA has suggested percentages that range between 5% and 35%[‡] depending primarily on the number of players on the team (usually scrambles are for two or four-man teams). To us, the interesting aspect of Section 9 is the staggering number of golf competition types for which the USGA has been willing to make handicap suggestions. We also note that the typical adjustment is that players receive a fraction of their

---

* In a net medal play match, each of two golfers adjusts his or her gross 18-hole score by subtracting their respective handicaps to arrive at a net score and the golfer with the lowest net score wins the match. In contrast, net match play is a hole by hole competition where both golfers get to deduct strokes on the hardest holes as designated by the rating of the holes.
† See the USGS website at http://usga.org/Content.aspx?id=25472.
‡ In a scramble, the total handicap for a team is the sum of the individual handicaps of the players making up the team. A team's net score is then determined by subtracting a percentage of the team's total handicap from its gross score.

USGA handicap. Hence, a research issue is how well a "scaled" handicap works in the competitions it's designed for. A number of researchers have examined the handicap of group competitions (see Lewis, 2005; Pollard and Pollard, 2010; Tallis, 1994) but none have been able to recommend scaled handicaps that are fair for all competitions.

The basic difficulty is that a one-parameter handicap does not fit all competitions, and as we will argue in the following, there does not appear to be simple transformations of this handicap to make the various competitions fair.

In this regard, Hurley and Sauerbrei (2015) examined a model where USGA handicaps are used in team net best-ball matches. This is a match where two golfers form a team and play another two-man team. On each hole, a team's net best-ball is the lowest net score of the two players. A net best-ball match, then, is based on these team net scores. This is a very popular game among recreational weekend foursomes. Most golfers are of the view that the use of full handicaps makes such competition fair. But this is not the case as Hurley and Sauerbrei demonstrate. We will come back to their point in a moment. They also make the point that a simple scaling of USGA handicaps for net best-ball matches is not fair for all possible handicap combinations. As we see it, the portability of the basic handicap (one intended for medal play) to other kinds of competition is an open research issue.

Another issue is what to do in competitions where players do not have a handicap. The Callaway and Peoria systems* have been used successfully in these situations. However, we are not aware of any mainstream statistical research on these systems.

22.2.2 One Size Handicap Does Not Fit All Competitions

We first want to make the point that it's difficult to port a basic handicap, one designed for medal play, to different kinds of competition, and particularly team competitions. The following example is taken from Hurley and Sauerbrei (2015).

Suppose two golfers labeled Low and High are playing a one-hole match. The Low golfer is a good player and only gets a par or a birdie. On the other hand, the High golfer is only capable of a par, bogey, or double-bogey. Suppose the Low golfer gives a stroke to the Higher golfer. That is, for the purposes of determining who wins, the High golfer subtracts a stroke from the score he gets and then the two compare scores to see who wins.

Let us suppose that the probabilities of the possible results for each player are as specified below:

	Birdie	Par	Bogey	Double-Bogey
Low	p	$1-p$		
High		q_0	q_1	$1-q_0-q_1$

Once the High player subtracts a stroke from his score, he can get either a birdie, par, or bogey. With these probabilities defined, we can easily work out the probabilities that each wins:

$$\Pr(\text{Low Wins}) = pq_1 + 1 - q_0 - q_1, \text{ and}$$

$$\Pr(\text{High Wins}) = (1-p)q_0.$$

In order for this one-hole match to be fair, these two must be equal and this requires that

$$p = \frac{2q_0 + q_1 - 1}{q_0 + q_1}.$$

We suppose that this equality is satisfied and hence this match is fair.

Now consider a net best-ball team game where one team comprises two Low handicap golfers (the Low Team) and the other two High handicap golfers (the High Team). Again, these teams will play one hole. Critically, both High golfers will subtract one stroke from their individual scores for the purpose of calculating the High Team's net best-ball. Additionally, we assume independence of the individual scores.

Using the parameters p, q_0, and q_1, we can calculate the probabilities of various team outcomes net of the stroke that each player on the High Team gets:

	Birdie	Par	Bogey
Low	$P_{-1} = p^2 + 2p(1-p)$	$P_0 = (1-p)^2$	
High	$Q_{-1} = q_0(2-q_0)$	$Q_0 = q_1^2 + 2q_1(1-q_0-q_1)$	$Q_1 = (1-q_0-q_1)^2$

This gives the following win probabilities:

$$\Pr(\text{Low Team wins}) = P_{-1}(Q_0 + Q_1) + P_0 Q_1,$$

$$\Pr(\text{High Team wins}) = P_0 Q_{-1},$$

and the probability of a draw is $P_{-1}Q_{-1} + P_0 Q_0$.

Consider the following instance:

$$q_0 = 0.40, \quad q_1 = 0.40, \quad \text{and} \quad p = 0.25.$$

Under these assumptions, a match between a Low golfer and a High golfer where the High golfer gets 1 shot is fair since each golfer has a probability of 0.3 to win (the match is drawn with probability 0.4).

Now consider what happens in the team match where each High golfer gets a stroke. We have

$$\Pr(\text{Low Team wins}) = 0.18,$$

$$\Pr(\text{High Team wins}) = 0.36,$$

and the match is drawn with probability 0.46. *That is, the High Team's probability of winning is twice the Low Team's.* Hence, it does not follow that handicaps that make one-on-one matches fair also make team net best-ball matches fair.*

Hurley and Sauerbrei also demonstrate that there is no simple transformation of handicaps to make the net best-ball competition fair. Finding best handicap transformation rules where the competition is as fair as possible is an open research question.

---

* One of our colleagues, Jeff McGill (Queen's University), has suggested the idea of a random handicap. Applied to this example, suppose we arbitrarily gave one of the High players a handicap of 1; the other High player would receive a handicap of 1 stroke with probability h and 0 strokes with probability $1 - h$. We can show that there is an h^* that makes this competition fair.

22.2.3 The USGA Slope Handicap System

Scheid (1973) argued that the handicap system should reflect differences in course difficulty. One player can shoot a lower score than another because of a higher level of skill or because of playing an easier course. For example, suppose that players A and B have identical ability but A plays at a par 72 course which is much more difficult than the par 72 course B plays. Handicaps are based on a history of scores, so without any adjustment for course difficulty, A would have a higher handicap than B. If the two then played a match, say at A's home course, then A would get strokes and have a better chance of winning. Hence, there needs to be an adjustment for course difficulty.

The main idea of the slope system is to have an objective measure of the difficulty of a course. This is accomplished by having a group of people called course raters walk the course and apply a detailed set of formulas and procedures, which considers the length of holes, the prevalence of water, sand, trees, and other hazards, etc. The output of this procedure are two parameters for each course (and set of tees), the course rating and the slope rating. The course rating reflects the difficulty of the course for a highly skilled golfer, called a Scratch golfer, and the slope rating reflects the difficulty of the course for a less-skilled golfer, called a Bogey golfer.

An alternative approach, detailed in Broadie and Rendleman (2013), is to use only reported scores (together with player, course, and date identifiers), to simultaneously estimate player skill and course difficulty using a fixed effects regression approach. The output of this method is a set of player fixed effects which measure a player's skill on a "neutral" course and this could be the basis for a handicap system. Instead of using the method for the handicapping of amateur golfers, Broadie and Rendleman (2013) apply the method to the scores of professional golfers on different tours and demonstrate that the Official World Golf Ranking system exhibits significant bias.

Returning to the slope system, after courses are rated, the system proceeds with golfers submitting *adjusted scores*. An adjusted score is the *gross score* reduced by the number of strokes dictated by *equitable stroke control* (ESC). For instance, if a low handicap golfer gets a quadruple bogey on a hole, he or she must reduce their gross score by one stroke since the maximum he or she can take on any one hole is a triple bogey. ESC is in place so that "outlier" scores on a single hole do not unduly increase a player's handicap. It is also an effort to prevent "sandbagging" where a player shoots a low score for, say, 17 holes and wins a match, and then purposely shoots a high score on the last hole (which is inconsequential for the match) to preserve a higher handicap.

There are provisions for handicap calculation if fewer than 20 rounds have been played. But for ease of exposition, we'll explain the system when a golfer has played at least 20 rounds.

First, the adjusted score, X, is translated to a *differential*, D, according to this formula:

$$D = 113 \left(\frac{X - R}{S} \right)$$

where
   R is the *course rating*
   S is the course's *slope rating*

The course rating and slope are those of the course for which the score is being recorded.

The course rating is what a scratch golfer (a very good golfer) would shoot on average. The slope is another measure of difficulty for the course. If the course is harder than average, it would have a slope above 113. Easier courses have a slope lower than 113. Slopes and ratings differ between men and women and the particular tees that are played.

Coming back to the calculation, the next step takes the most recent 20 differentials and selects the best 10. Following Swartz (2009), suppose the last 20 differentials are

$$21.8 \quad 12.4 \quad 18.4 \quad 15.4 \quad 21.2 \quad 22.4 \quad 13.1 \quad 17.4 \quad 19.3 \quad 15.6$$
$$14.9 \quad 10.3 \quad 17.1 \quad 19.9 \quad 11.3 \quad 11.3 \quad 14.2 \quad 28.8 \quad 10.9 \quad 23.4\,\dot{}$$

These are then ordered:

$$10.3 \quad 10.9 \quad 11.3 \quad 11.3 \quad 12.4 \quad 13.1 \quad 14.2 \quad 14.9 \quad 15.4 \quad 15.6$$
$$17.1 \quad 17.4 \quad 18.4 \quad 19.3 \quad 19.9 \quad 21.2 \quad 21.8 \quad 22.4 \quad 23.4 \quad 28.8$$

Next, the *factor*, F, (sometimes called the *index*) is calculated by averaging the lowest 10 differentials and then multiplying this average by 0.96:*

$$F = 0.96 \left(\frac{10.3 + 10.9 + \cdots + 15.6}{10} \right) = 12.4$$

The *course handicap*, H, is calculated by adjusting the factor by the course slope, S:

$$H = F \left(\frac{S}{113} \right)$$

and then rounding to the nearest integer.

If two golfers are playing a *net medal play* competition, they will each subtract their course handicap from their gross score to arrive at a net score and the golfer with the lowest net score wins. If the two are playing *net match play*, the higher handicap golfer will be allotted a number of strokes equal to the difference in handicaps and these strokes will be taken on the course's most difficult holes according to the scorecard.

22.2.4 The Swartz Proposal

The golf research community has known for some time that the scores of relatively high handicappers exhibit more variation than those of low handicappers. Hence, it makes sense to consider a two-parameter handicap where there is a parameter for average score and one for the variation of this average score. Swartz (2009) has suggested such a system.

Swartz considers two cases. In one, termed casual play, he considers a net medal competition between two players. In the second, he considers the case of tournament play where a large number of players are competing for a relatively small set of prizes based on net medal score. Effectively, the second case is $N > 3$ players and he is trying to come up with a handicap that gives each player the same chance to win. To distinguish these two cases, we'll refer to the former as the *Medal2 handicap* and the later as the *MedalN handicap*; we'll refer to the respective competitions as the *Medal2* and *MedalN* competitions.

---

* The factor 0.96 is the "bonus for excellence" factor we referred to earlier.

Let's first look at the Medal2 proposal using the ordered differential data:

$$10.3 \quad 10.9 \quad 11.3 \quad 11.3 \quad 12.4 \quad 13.1 \quad 14.2 \quad 14.9 \quad 15.4 \quad 15.6$$
$$17.1 \quad 17.4 \quad 18.4 \quad 19.3 \quad 19.9 \quad 21.2 \quad 21.8 \quad 22.4 \quad 23.4 \quad 28.8$$

We are going to replace F with

$$
\begin{aligned}
\widehat{\mu} = {}& 0.0284\,(10.3) + 0.0365\,(10.9) + 0.0399\,(11.3) + 0.0424\,(11.3) \\
& + 0.0445\,(12.4) + 0.0463\,(13.1) + 0.0480\,(14.2) + 0.0496\,(14.9) \\
& + 0.0511\,(15.4) + 0.0526\,(15.6) + 0.0540\,(17.1) + 0.0554\,(17.4) \\
& + 0.0569\,(18.4) + 0.0584\,(19.3) + 0.0599\,(19.9) + 0.2762\,(21.2) \\
={}& 16.9
\end{aligned}
\tag{22.1}
$$

Swartz labels $\widehat{\mu}$ the *mean*.

First note that 16 differentials are used rather than 10. This makes sense statistically since, with more information, we ought to be able to measure a golfer's ability more accurately. Note that a different weight is put on each differential with the largest weight on the 16th highest differential. So where do these weights come from?

Swartz's development is based on the assumption that gross golf scores are normally distributed. A number of researchers have suggested that this is not the case. See, for example, Lackritz (2011). But, in our estimation, the departures from normality are not significant enough to affect the usefulness of Swartz's theory.

So assuming normality, the random variable

$$
113\left(\frac{Y - R}{S}\right)
\tag{22.2}
$$

(where Y is the gross score) is also normally distributed. Swartz then gives this normal random variable a mean μ and variance σ^2. In effect, μ is the golfer's average differential and hence it characterizes his or her average *ability*. The parameter σ^2 characterizes the golfer's *consistency*. We know that golf scores fluctuate round to round and it is the σ^2 parameter that characterizes this variation. Hence, we have that

$$
113\left(\frac{Y - R}{S}\right) \sim N(\mu, \sigma^2).
$$

Therefore, we can write that

$$
E\left[113\left(\frac{Y - R}{S}\right)\right] = \mu
$$

which simplifies to

$$
E(Y) = R + \frac{S}{113}\mu.
$$

Note that if $\mu = 0$, we have that $E(Y) = R$. That is, the golfer's average gross score is equal to the course rating and this is the definition of a scratch golfer. Hence, a scratch golfer has $\mu = 0$.

It turns out that $\widehat{\mu}$ is an estimator of μ with some special properties. Clearly, $\widehat{\mu}$ is linear and it can be shown that it is unbiased. Finally, among the class of linear estimators, $\widehat{\mu}$ has the lowest variance. Therefore, $\widehat{\mu}$ is a best linear unbiased estimator (BLUE). A loose argument for why $\widehat{\mu}$ is unbiased can be seen in the weights. Note that we have a right censored data set since we are ignoring the four highest differentials. Examining the weights in (22.1), note that the sum of the weights for the lowest eight differentials is less than the sum of the weights for next eight differentials. Hence, these higher differentials are given more weight and this pushes the mean to the right of where it would have had the differentials been given equal weights.

Swartz develops two interesting properties of the measure $\widehat{\mu}$. The first is that, with $\widehat{\mu}$, a golfer "shoots his handicap" half the time. To see this, note that a golfer's handicap is now given by

$$H_\mu = \widehat{\mu} \left(\frac{S}{113} \right).$$

A golfer will shoot his handicap when his net score is less than the course rating or when

$$Y - H_\mu < R$$

and this happens with probability

$$
\begin{aligned}
\Pr \left(Y - H_\mu < R \right) &= \Pr \left(Y - \widehat{\mu} \left(\frac{S}{113} \right) < R \right) \\
&\simeq \Pr \left(Y - \mu \left(\frac{S}{113} \right) < R \right) \\
&= \Pr \left(113(Y - R)/S < \mu \right) \\
&= 1/2
\end{aligned}
$$

since $113(Y - R)/S$ is a normal random variable with mean μ.

The second property is that, for two golfers of unequal ability, each has the same probability of winning. To see this, let's define two golfers with respective parameters (μ_1, σ_1^2) and (μ_2, σ_2^2). Let their respective gross scores be Y_1 and Y_2. Suppose, these players play on a course with rating R and slope S. Their respective handicaps on this course are

$$H_1 = \widehat{\mu}_1 \left(\frac{S}{113} \right)$$

and

$$H_2 = \widehat{\mu}_2 \left(\frac{S}{113} \right).$$

Then, the probability that player 1 wins is the probability his net score is lower than player 2's or

$$\Pr(Y_1 - H_1 < Y_2 - H_2) = \Pr\left[Y_1 - \hat{\mu}_1\left(\frac{S}{113}\right) < Y_2 - \hat{\mu}_2\left(\frac{S}{113}\right)\right]$$

$$\simeq \Pr\left[Y_1 - \mu_1\left(\frac{S}{113}\right) < Y_2 - \mu_2\left(\frac{S}{113}\right)\right]$$

$$= \Pr\left[Y_1 - \mu_1\left(\frac{S}{113}\right) - Y_2 + \mu_2\left(\frac{S}{113}\right) < 0\right]$$

Let's consider the random variable on the last line:

$$W = Y_1 - \mu_1\left(\frac{S}{113}\right) - Y_2 + \mu_2\left(\frac{S}{113}\right).$$

It's easy to show that $E(W) = 0$ and therefore W is normally distributed with a mean of 0. This gives

$$\Pr\left[Y_1 - \mu_1\left(\frac{S}{113}\right) - Y_2 + \mu_2\left(\frac{S}{113}\right) < 0\right] = \Pr(W < 0) = 1/2.$$

There is one other interesting issue with this system. Note that there is no mention of equitable stroke control. A golfer will be submitting his gross scores with no reduction for ESC. Two comments are in order. If gross scores are normally distributed, it is likely that gross scores adjusted for ESC are also normally distributed and the same analysis goes through. And second, if a golfer knowingly makes no adjustment of ESC when that is the law, he or she is likely cheating in a number of other ways. So requiring ESC is no guarantee that a handicap system using them is any more accurate that one not using them. ESC is at best a minor nuisance for a dyed-in-the-wool sandbagger.

Now, let's look at the MedalN proposal. What Swartz has in mind is a tournament with many players and he would like to specify a net score that gives each competitor the same chance of winning. To do this, he has to employ a second parameter, $\hat{\sigma}$, which he calls the *spread*. So now we will employ two parameters, the mean, $\hat{\mu}$, and the spread, $\hat{\sigma}$. As mentioned earlier, the mean measures a golfer's ability and the spread measures his or her consistency.

The mean is measured the same way as described earlier for the Medal2 competition. Using the same differential data as we did earlier, the spread is estimated with

$$\hat{\sigma} = -0.1511(10.3) - 0.1006\,(10.9) - 0.0792\,(11.3) - 0.0632\,(11.3)$$

$$- 0.0500\,(12.4) - 0.0384\,(13.1) - 0.0277\,(14.2) - 0.0178\,(14.9)$$

$$- 0.0082\,(15.4) + 0.0011\,(15.6) + 0.01030\,(17.1) + 0.0196\,(17.4)$$

$$+ 0.0291\,(18.4) + 0.03584\,(19.3) + 0.0492\,(19.9) + 0.3880\,(21.2)$$

$$= 4.86$$

where, again, the weights are specific to the case where the lowest 16 differentials are used. This spread is an estimator for the standard deviation of the normal distribution defined in (22.2) and it is also BLUE.

The proposal is to have a golfer's net score be determined by

$$N = R + \frac{452(Y - R)}{S\widehat{\sigma}} - \frac{4\widehat{\mu}}{\widehat{\sigma}}$$

(22.3)

where
  N is the net score
  Y is the gross score

This formula can be rewritten

$$N = R + \frac{4}{\widehat{\sigma}}\left[\frac{113(Y - R)}{S} - \widehat{\mu}\right].$$

For golfer i, we have

$$N_i = R + \frac{4}{\widehat{\sigma}_i}\left[\frac{113(Y_i - R)}{S} - \widehat{\mu}_i\right] \simeq R + \frac{4}{\sigma_i}\left[\frac{113(Y_i - R)}{S} - \mu_i\right].$$

N_i is normally distributed with mean $E(N_i) = 0$ and variance $Var(N_i) = 4^2$. Hence, every golfer's score follows the same normal distribution, and assuming independence, each will have the same probability of winning. Hence, the Swartz handicap system for a MedalN tournament is fair.

This system also has the nice property that, on average, half the golfers will shoot better than the course rating and half will shoot worse. We have not included a discussion of Swartz's treatment of ESC. Interested readers are referred to his paper for a full discussion.

Swartz tested his MedalN system on a fairly large data set in the following way. Over the years 1996–1999, a group of members of the Coloniale Golf Club in Beaumont, Alberta, recorded their full 18 hole scores. This resulted in 8000 rounds by 178 golfers whose abilities varied between plus factors to ones exceeding 30. To simulate a tournament, 99 rounds were drawn from the population of 8000 rounds and a hypothetical winner was determined. Swartz recorded whether the winner was from the top third (lowest handicaps) or the bottom third (highest handicaps). This procedure was then repeated 40,000 times. For a fair system, we would expect the winner to come from the top third 33% of the time and from the bottom third 33% of the time. Using the current handicap system, Swartz found that the top third won 27% of the time and the bottom third, 40% of the time. However, under the MedalN system, the top third won 29% of the time and the bottom third won 34% of the time. Hence, the Swartz system is a more equitable system.

We think the Swartz proposal has considerable merit. The empirical work he does suggests that the system is fairer than the current system, albeit the improvement is likely a second-order effect. Furthermore, the resulting handicap has the salutary property that a golfer's net score should be less than the course rating 50% of the time and that's a significant improvement over the current system where net scores should exceed the course rating most of the time. Most golfers have a difficult time figuring that out even when you explain that their handicaps are based on their 10 best scores.

But in the end, the question is whether this marginal improvement in fairness is worth the added calculation. In a world where computers handle the recording of scores and the handicap calculation, it seems to us that it would be no trouble to produce two handicaps: the standard handicap and the two-parameter handicap that Swartz proposes. Tournament committees could then decide which handicap to use for their particular competition. And for the weekend foursome playing a little money game based on handicaps, the two-parameter handicap would present another issue to bargain on the first tee, where, in the second author's opinion, most games are won and lost. In locales where handicaps are still calculated manually, the Swartz proposal may not be worth the added calculation.

Regarding handicap system "understandability" to the average golfer, the existing system is difficult to understand if not impossible. So the marginal increase in complexity of the two-parameter system is neither here nor there. That said, there is a black-box aspect to the Swartz system. Clearly, it doesn't boil down to a player subtracting a handicap from his or her score to arrive at net score (as the USGA system does). Rather the gross score must be fed into (22.3) to arrive at a net score. On the positive side, the Swartz system maintains the USGA slope system which is important.

The Swartz proposal was rejected by the Royal Canadian Golf Association (RCGA) primarily because Canadian handicaps would then not be portable to the competitions in the United States and other countries using the USGA system. It seems to us that the solution the RCGA might have considered would have been the two-handicap reporting we discussed earlier. Canadian players competing in international competitions requiring a USGA handicap would have one; and Canadian tournament committees (and players playing informal betting games) would have the flexibility to chose the handicap system they preferred.

22.2.5 Research Issues

To summarize, we feel the critical research issues for handicapping are as follows:

1. *How should a single-purpose handicap be ported to other kinds of competition?* It is clear that it is not easy to find simple handicap transformations for use in other kinds of competitions, particularly team competitions. One idea is to relax the fairness criterion. In our view, the focus should be on finding transformations that make such competitions as fair as possible.

2. *The course rating/slope system needs to be empirically tested.* To our knowledge, this system has not been subjected to any sort of empirical test of its accuracy.

3. *How should handicaps be adjusted when course conditions change.* In a conversation about what new golf equipment is doing to old courses, Tom Watson has remarked that a course can be protected by pin placements and green firmness. In our experience, this is certainly true. When greens are rolled and verticut and pins are put in difficult positions, scores can go up considerably. The interesting question is whether this has a differential effect on high and low handicappers. We think it does and that there should be an adjustment but clearly this is an empirical question and it needs to be studied.

In sum, work on handicapping is a fertile area of study where statisticians can make important contributions to practice.

References

Bellman, R. E. (1957). *Dynamic Programming*, Princeton University Press, Princeton, NJ.

Berry, S. M. (1999). Drive for show and putt for dough, *Chance*, 12(4), 50–55.

Bingham, D. R. and T. B. Swartz (2000). Equitable handicapping in golf, *The American Statistician*, 54(3), 170–177.

Broadie, M. (2008). Assessing golfer performance using golfmetrics, in: *Science and Golf V: Proceedings of the 2008 World Scientific Congress of Golf*, D. Crews and R. Lutz, eds., Energy in Motion Inc., Mesa, AZ, pp. 253–262.

Broadie, M. (2012). Assessing golfer performance on the PGA tour, *Interfaces*, 42(2), 146–165.

Broadie, M. (2014). *Every Shot Counts*, Gotham Books, New York.

Broadie, M. and S. Ko (2009). A simulation model to analyze the impact of distance and direction on golf scores, in: *Proceedings of the Winter Simulation Conference*, M. D. Rosetti, R. R. Hill, B. Johansson, A. Dunkin, and R. G. Ingalls, eds., The Society for Computer Simulation, Institute of Electrical and Electronics Engineers, Inc., Piscataway, NJ.

Broadie, M. and R. J. Rendleman (2013). Are the official world golf rankings biased? *Journal of Quantitative Analysis in Sports*, 9(2), 127–140.

Broadie, M. and D. Shin (2015). A simulation model for golf putting and its applications to optimal targeting strategy and attribution analysis, Working Paper, Graduate School of Business, Columbia University, New York.

Cochran, A. and J. Stobbs (1968). *The Search for the Perfect Swing*, Lippincott & Company, Philadelphia and New York.

Hickman, D. C. and N. E. Metz (2015). The impact of pressure on performance: Evidence from the PGA tour, *Journal of Economic Behavior and Organization*, 116, 319–330.

Hurley, W. J. and T. Sauerbrei (2015). Handicapping net best-ball team matches in golf, *Chance*, 28(2), 26–30.

Kupper, L. L., L. B. Hearne, S. L. Martin, and J. M. Griffin (2001). Is the USGA golf handicap system equitable? *Chance*, 14(1), 30–35.

Lackritz, J. R. (2011). A new analysis of the golf handicap system: Does the better player have an advantage? *Chance*, 24(2), 24–34.

Landsberger, L. M. (1994). A unified golf stroke value scale for quantitative stroke-by-stroke assessment, in: *Science and Golf II: Proceedings of the World Scientific Congress on Golf*, A. J. Cochran and M. R. Farrally, eds., E. & F.N. Spon, London, England, pp. 216–221.

Lewis, A. J. (2005). Handicapping in group and extended golf competitions, *IMA Journal of Management Mathematics*, 16, 151–160.

McHale, I. G. (2010). Assessing the fairness of the golf handicapping system in the UK, *Journal of Sports Sciences*, 28(10), 1033–1041.

Pelz, D. (2000). *Putting Bible*, Doubleday, New York.

Pollard, G. and G. Pollard (2010). Four ball best ball 1, *Journal of Sports Science and Medicine*, 9, 86–91.

Pollock, S. M. (1977). A model of the USGA handicap system and 'Fairness' of medal and match play, in: *Optimal Strategies in Sports*, S. P. Ladany and R. E. Machol, eds., Elsevier, Amsterdam, the Netherlands, pp. 141–150.

Riccio, L. (1990). Statistical analysis of the average golfer, in: *Science and Golf*, A. J. Cochran, ed., E. & F.N. Spon, London, England, pp. 153–158.

Scheid, F. (1971). You're not getting enough strokes, *Golf Digest*, June (reprinted in *The Best of Golf Digest*, 1975, 32–33).

Scheid, F. (1973). Does your handicap hold up on tougher courses? *Golf Digest* (October), 31–33.

Scheid, F. (1977). An evaluation of the handicap system of the United States Golf Association, in: *Optimal Strategies in Sports*, S. P. Ladany and R. E. Machol, eds., North Holland, Amsterdam, the Netherlands, pp. 151–155.

Scheid, F. (1978). The search for the perfect handicap, in: *78 Winter Simulation Conference*, Vol. 2, IEEE Computer Society Press, Los Alamitos, CA, pp. 889–896.

Scheid, F. (1979). Golf competition between individuals, in: *IEEE Winter Simulation Conference*, pp. 505–510.

Scheid, F. (1990). On the normality and independence of golf scores, with applications, in: *Science and Golf: Proceedings of the First World Scientific Congress of Golf*, A. J. Cochran, ed., E. & F.N. Spon, London, England, pp. 147–152.

Soley, C. (1977). How well should you putt? A search for a putting standard, *Soley Golf Bureau*.

Stroud, R. C. and L. J. Riccio (1990). Mathematical underpinnings of the slope handicap system, in: *Science and Golf*, A. J. Cochran, ed., E. & F.N. Spon, London, England, 1990, pp. 135–140.

Swartz, T. B. (2009). A new handicapping system for golf, *Journal of Quantitative Analysis in Sports*, 5(2), Article 9.

Tallis, G. M. (1994). A stochastic model for team golf competitions with applications to handicapping, *Australian & New Zealand Journal of Statistics*, 36(3), 257–269.

Yousefi, K. and T. Swartz (2013). Advanced putting metrics in golf, *Journal of Quantitative Analysis in Sport*, Issue 3, available at: http://www.degruyter.com/view/j/jqas.2013.9.issue-3/jqas-2013-0010/jqas-2013 -0010.xml?format=INT, accessed May 19, 2015.

23

Research Directions in Cricket

Tim B. Swartz

CONTENTS

23.1 Introduction

The handbook which you are reading is divided into six main sections covering the five major sports of baseball, basketball, hockey, soccer and football and a final section on other sports. Since this chapter on cricket resides in the final section, it is worth reflecting on what constitutes a major sport. In some ways, cricket may be considered a major sport with a long history dating back to the sixteenth century. Using various measures, cricket is typically regarded as the second most popular sport in the world after soccer (http://www.topendsports.com/world/lists/popular-sport/fans.htm). In terms of online media coverage, cricket ranks fourth worldwide after soccer, basketball and tennis (www.biggestglobalsports.com). In addition, the format of cricket known as Twenty20 is a 'big money' sport. Specifically, amongst professional leagues, the IPL (Indian Premier League) had the second highest average player salaries in 2014–2015 trailing only the National Basketball Association (Business Insider Sports 2015). However, in terms of advanced sports analytics, the sport of cricket appears to lag behind the major sports. This chapter attempts to provide the reader with an up-to-date survey of the quantitative work that has been done in relation to cricket and on the promising research opportunities that are available.

When discussing cricket analytics, it is important to distinguish between the three main formats of cricket. The traditional version of cricket and the 'highest' form of the sport known as *test cricket*. Test cricket matches can take up to 5 days to complete. Matches are played between two teams representing countries which are full members of the ICC (International Cricket Council). The 10 full-member cricketing nations of the ICC have connections to England through the expansion of the British Empire and consist of Australia, Bangladesh, England, India, New Zealand, Pakistan, South Africa, Sri Lanka, West Indies and Zimbabwe. In test cricket, each side has two *innings* (opportunities) for batting. Completion of an innings depends on the *dismissal* of the batting team (they are made "out") or the voluntary *declaration* by the batting team that their innings are complete or the time constraints for the match. Test matches often result in draws (roughly 25% of matches in recent years). Teams sometimes bat conservatively to attain a draw when they fear they are on the verge of losing; batting conservatively helps a team avoid dismissal and lengthens the duration of the innings. Hence test cricket tends to be a slow-moving game though full of strategy.

To accommodate the constraints of modern living and to infuse more excitement into cricket, the second form of cricket known as *one-day cricket* was introduced in the 1950s or 1960s depending on who is given credit for the origins of the game. However, there is no doubt that the format began in earnest in the 1970s with the first World Cup contested in 1975. One-day cricket is based on *limited overs* where each side has a single innings of batting. Batting is terminated by either dismissal or by completion of 50 overs (300 balls). In one-day international (ODI) cricket, draws are less common (1% compared to 26% in test cricket during the last 10 years). The style of play in one-day cricket is more aggressive with batsmen more willing to swing freely and score runs along with the associated increased risk of *wickets*. One-day cricket is seen by some as a brash upstart version of cricket, departing from tradition with various innovations including coloured uniforms, day/night games, etc.

The third form of cricket known as *Twenty20* or *T20* cricket is the most recent incarnation of cricket. Twenty20 cricket was introduced in 2003 in an English domestic tournament. It is also a form of limited-overs cricket and shares most of the characteristics of one-day cricket. The major difference is that Twenty20 cricket is based on 20 overs (120 balls) per side and therefore matches terminate in roughly 3 hours of playing time. As of 2016, there have been six World Cups in international Twenty20 cricket (2007, 2009, 2010, 2012, 2014 and 2016). Interest in Twenty20 has exploded with the introduction of professional leagues, most notably the IPL, which was formed in 2008, and the Australian Big Bash, which began in 2011. With fewer overs in which to lose wickets, Twenty20 cricket is even more aggressive than one-day cricket. In particular, batsmen score *4s* and *6s* at higher rates. It appears that this aspect of the game where the focus is primarily on scoring runs has lead to even greater excitement with perhaps less attention on in-game strategy.

Previously, the only overview of cricket analytics was provided by Clarke (1998). Much has changed since that time. Specifically, advances in computing have provided more tools to address large data sets. The structure of this chapter consists of sections that are mostly independent and focus on various topics of interest in cricket. The chapter concludes with some discussion on where cricket analytics may be heading.

23.2 Data

One of the great advantages of doing work in cricket analytics is the availability of reliable data. The primary source for data is the Cricinfo website (www.espncricinfo.com). In the match series archive, match information reaches back to the 1770s. In recent history, details on all international matches involving full ICC members have been recorded in the form of results, scorecards and match commentaries. There is also selective information on ICC Associate member's cricket, domestic cricket and women's cricket. The Cricinfo website also hosts a friendly search engine Statsguru, which offers a variety of options and provides immediate results.

On the other hand, one of the current limitations of cricket data is that it is typically presented in summary form. For example, although it is easy to find the number of runs that a batsman has scored in a particular match or a particular period of time, the breakdown of run scoring is not so easily accessible. For example, one may be interested in knowing how runs were accumulated by a batsman through the frequency of 1s, 2s, 3s, 4s and 6s at various stages (i.e. overs and wickets) of a match. We note that some researchers (e.g. Davis et al. 2015a) have developed parsers which scan match commentary logs to extract detailed ball-by-ball data. Detailed data has been helpful in analyzing the batting and bowling characteristics of individual players.

It now seems only a matter of time until more extensive cricket data will become widely available. Currently, the Hawk-Eye computer tracking system is in use in most major test, ODI and Twenty20 matches. Although its applications could be expanded, it is currently used for reviewing in-game umpiring decisions. For example, *leg before wicket* decisions may be appealed and challenged via Hawk-Eye technology. Hawk-Eye is also regularly used for providing graphics of ball locations and trajectories in live television broadcasts.

23.3 Resetting of Targets

For resetting targets in interrupted one-day cricket matches, the Duckworth–Lewis (D/L) method (Duckworth and Lewis 1998, 2004) supplanted the method of run-rates in the late 1990s and has since been adopted by all major cricketing boards. Given the popularity of cricket, it could be argued that the Duckworth–Lewis method has provided the greatest impact to the sporting world from a mathematical, statistical, and operational research perspective. What makes the acceptance of this method so remarkable is that the method is largely viewed by the public as a black-box procedure. The sporting world tends to like simple rules and simple statistics. From this point of view, the adoption of the Duckworth–Lewis method was a masterpiece in overcoming political hurdles.

The fundamental concept underlying this method is that of *resources*. In a one-day cricket match, the team batting first begins with 50 overs and 10 wickets at their disposal. They continue batting until either their overs are completed or 10 wickets have fallen. It is the combination of wickets and overs remaining in an innings that provides the capacity for scoring runs. The Duckworth–Lewis quantification of the combination of wickets and overs is known as resources. When innings are reduced due to an interruption, then targets are reset to 'fair' values based on the resources remaining. In one-day cricket, a team has 100% of its resources remaining at the beginning of its innings (50 overs and 0 wickets lost). When a team has used up all of its 50 overs, then 0% of its resources are remaining. Similarly, when 10 wickets are lost, a team has 0% of its resources remaining.

To get a sense of the use of the Duckworth–Lewis method, consider the abbreviated Duckworth–Lewis table presented in Table 23.1 based on the Standard Edition as taken

TABLE 23.1

Abbreviated Version of the Duckworth–Lewis Resource Table (2014–2015 Standard Edition)

Overs Available	Wickets Lost										
	0	1	2	3	4	5	6	7	8	9	10
50	100.0	93.4	85.1	74.9	62.7	49.0	34.9	22.0	11.9	4.7	0.0
40	89.3	84.2	77.8	69.6	59.5	47.6	34.6	22.0	11.9	4.7	0.0
30	75.1	71.8	67.3	61.6	54.1	44.7	33.6	21.8	11.9	4.7	0.0
20	56.6	54.8	52.4	49.1	44.6	38.6	30.8	21.2	11.9	4.7	0.0
10	32.1	31.6	30.8	29.8	28.3	26.1	22.8	17.9	11.4	4.7	0.0
5	17.2	17.0	16.8	16.5	16.1	15.4	14.3	12.5	9.4	4.6	0.0
1	3.6	3.6	3.6	3.6	3.6	3.5	3.5	3.4	3.2	2.5	0.0
0	0.0	0.0	0.0	0.0	0.0	0.0	0.0	0.0	0.0	0.0	0.0

Note: The table entries indicate the percentage of resources remaining in a match with the specified number of wickets lost and overs available.

from the 2014–2015 *ICC Playing Handbook* found at www.icc-cricket.com. Imagine that Team A is batting and scores 250 runs on completion of its innings. It then rains prior to the resumption of the second innings. When Team B comes to bat, they are only allotted 30 overs. Using the old method of run rates, Team B would need to score $250(30/50) + 1 = 151$ runs in order to win the match. The obvious problem with the run rate approach is that Team B can bat more aggressively since their 10 wickets are spread throughout 30 overs rather than 50 overs. Therefore the target of 151 runs is lower than what might be considered fair. However, using the Duckworth–Lewis approach, Table 23.1 indicates that with 30 overs remaining and zero wickets lost, Team B retains 75.1% of its resources. Therefore the D/L target is rounded up to $250(0.751) \to 188$ runs. The large difference between 151 and 188 runs indicates how unpalatable it was for matches to be determined by run rates. The *ICC Playing Handbook* describes other scenarios in which D/L can be used to reset targets in interrupted matches.

The Duckworth–Lewis methodology is a statistical approach based on an exponential decay model in a regression context. We note that in addition to the Standard Edition there is also a Professional Edition of this method that is used in international matches. Recently, Frank Duckworth and Tony Lewis have ceded the management of D/L to David Stern of Australia. In the tradition of Duckworth–Lewis, Stern (2016) has updated the approach to account for recent changes in scoring. The approach is now sometimes referred to as the Duckworth–Lewis–Stern (D/L/S) method.

It is worth stating that over the years, there have been a long list of competitors to the Duckworth–Lewis method (e.g. Clarke 1988, Christos 1998, Jayadevan 2002, Carter and Guthrie 2004, McHale and Asif 2013). However, it is the author's opinion that D/L/S will be difficult to supplant as it has passed the test of time and is well rooted in all levels of one-day cricket. Although competitors argue that their approach is better, the exercise of comparison seems futile. First, there is no gold standard to determine what is a 'fair' target. Second, most of the alternative approaches provide targets that differ only minimally from the D/L/S targets. Weatherall (2011) compares the adequacy of various methods for resetting targets in interrupted one-day cricket matches.

One area where there may be some legitimate concerns over D/L/S is the application of D/L/S in Twenty20 cricket (Perera and Swartz 2013). In Twenty20, the current solution for resetting targets involves scaling the resource table that was developed for one-day cricket. To be precise, refer again to Table 23.1 and note that 56.6% of a team's resources remain when 20 overs are available and zero wickets have been lost. Therefore, the modified resource table for Twenty20 is obtained from the original resource table by dividing each cell entry by 0.566. The modified Twenty20 table contains only the rows where the available overs do not exceed 20. To see that the mapping from the one-day format to Twenty20 is logically flawed, one need only consider the different stages at which powerplays occur in the two formats. The powerplay in Twenty20 occurs in the first 6 overs whereas the powerplay in one-day cricket occurs in the first 10 overs. Since run scoring rates are higher during powerplays, it is not reasonable to map the cells corresponding to 15–20 available overs in one-day cricket (a non-powerplay situation) to the cells corresponding to 15–20 available overs in Twenty20 cricket (a powerplay situation). However, at the end of the day, target resetting in Twenty20 is not a vital problem as match interruptions rarely occur. When the weather is bad, matches tend to be abandoned rather than shortened.

23.4 Match Simulation

When viewing a cricket match, the question that is continually on the minds of most supporters is how is their team doing. This is a mental exercise that is dynamic and requires the assessment of the current score, the overs remaining and the wickets lost. Formally, the scoring of runs is a stochastic process, and knowing the run scoring distribution would assist in prediction. Somewhat surprisingly, there has been little activity in the fundamental and related problem of developing match simulators. This is likely due to the complexity with which runs are scored in cricket.

Some of the earliest statistical publications concerning sport can be attributed to Elderton (1945) and Wood (1945) who proposed geometric distributions for individual runs scored in test cricket. Modelling test cricket seems particularly challenging as the run distribution depends on a team's objectives; sometimes teams are attempting to score runs at a high rate and sometimes teams are batting defensively.

Although likewise a difficult problem, there is more hope for modelling one-day and Twenty20 cricket. At least in these formats, the batting team is always attempting to score runs. However, the limited number of papers that have considered modelling/simulation in the one-day format tend to suffer from a lack of realism. For example, the papers may not have taken into account differences between batsmen, differences between bowlers, the stage of the match (i.e. overs completed and wickets taken), the home team advantage and modified tactics when chasing in the second innings.

In the author's opinion, the most comprehensive simulator to date which takes into account these situational effects is due to Davis et al. (2015a). Their simulator was developed for Twenty20 although the ideas are directly transferable to one-day cricket. We now provide a brief overview of their match simulator which has many applications, some of which are described in the following sections. One obvious application that has not been exploited is the calculation of in-match probabilities. In cricket, there are eight broadly defined outcomes that can occur when a batsman faces a bowled ball. These batting outcomes are listed below:

$$
\begin{aligned}
\text{Outcome } j = 0 &\equiv 0 \text{ runs scored} \\
\text{Outcome } j = 1 &\equiv 1 \text{ runs scored} \\
\text{Outcome } j = 2 &\equiv 2 \text{ runs scored} \\
\text{Outcome } j = 3 &\equiv 3 \text{ runs scored} \\
\text{Outcome } j = 4 &\equiv 4 \text{ runs scored} \\
\text{Outcome } j = 5 &\equiv 5 \text{ runs scored} \\
\text{Outcome } j = 6 &\equiv 6 \text{ runs scored} \\
\text{Outcome } j = 7 &\equiv \text{Dismissal}
\end{aligned}
\tag{23.1}
$$

In the list (23.1) of possible batting outcomes, *extras* such as *byes, leg byes, wide-balls* and *no balls* are excluded. In the simulator, extras are introduced by generating occurrences at the appropriate rates. Extras occur at the rate of 5.1% in Twenty20 cricket. The outcome $j = 3$ is not so common and the outcome $j = 5$ is rare, but both outcomes are retained to facilitate a straightforward notation.

According to the enumeration of the batting outcomes in (23.1), Davis et al. (2015a) suggested the following statistical model:

$$
(X_{iow0}, \ldots, X_{iow7}) \sim \text{multinomial}(m_{iow}; p_{iow0}, \ldots, p_{iow7})
\tag{23.2}
$$

where X_{iowj} is the number of occurrences of outcome j by the ith batsman during the oth over when w wickets have been taken. In (23.2), m_{iow} is the number of balls that batsman i has faced in the dataset corresponding to the oth over when w wickets have been taken. The dataset that they considered is special in the sense that it consists of detailed ball-by-ball data. The data were obtained using a proprietary parser, which was applied to the commentary logs of matches listed on the CricInfo website (www.espncricinfo.com).

The estimation of the multinomial parameters p_{iowj} in (23.2) is a high-dimensional and complex problem. The complexity is partly due to the sparsity of the data; there are many match situations (i.e., combinations of overs and wickets) where batsmen do not have batting outcomes. For example, bowlers typically bat near the end of the batting order and do not face situations when zero wickets have been taken.

To facilitate the estimation of the multinomial parameters, Davis et al. (2015a) introduced parametric simplifications and a hybrid estimation scheme using Markov chain Monte Carlo in an empirical Bayes setup. A key idea in their estimation procedure was a bridging framework where the multinomial probabilities in a given situation (i.e. overs and wickets lost) could be estimated reliably from a 'nearby' situation.

Given the estimation of the parameters in (23.2), first innings runs can be simulated for a specified batting lineup facing an average team. This is done by generating multinomial batting outcomes in (23.1) according to the laws of cricket. For example, when either 10 wickets are accumulated or the number of overs reaches 20, the first innings is terminated. Davis et al. (2015a) also provide modifications for batsmen facing specific bowlers (instead of average bowlers), they account for the home field advantage and they provide adjustments for second innings batting.

There may be several research avenues for improving the simulator. For example, additional covariates (such as the pitch condition) could be introduced. It might also be possible to augment the dataset for more reliable estimation. For example, can one-day cricket data be combined in some fashion with Twenty20 data? And as all cricket enthusiasts know, wickets are of great importance in determining the outcome of matches. For the simulator proposed in Davis et al. (2015a), it is essential that the wicket probabilities p_{iow7} in (23.2) are estimated reliably. Naturally, this is a difficult problem since the occurrence of wickets is rare for even well-established batsmen. Therefore, a future research direction is to consider ways to improve the estimation of batsman-specific wicket probabilities.

23.5 Evaluating Player Performance

Like all sports, cricket has statistics to measure player performance. However, this is an area where traditional statistics are lacking, and 'better' statistics have not yet been embraced or have yet to be developed. Perhaps it is the case that some of the advanced statistics are too difficult to calculate or are too complicated for the general public to comprehend.

The most commonly reported batting statistic is the *batting average*. The batting average (calculated separately in each of the three formats of cricket) is the total number of runs scored by a player divided by the total number of occasions where he has been dismissed. To see that the sole use of batting average is less than ideal, consider the pathological case of a player who has batted in 100 innings, has scored 100 runs and has been dismissed only once. The player would have a batting average of $100/1 = 100.0$ which would be the highest batting average ever attained! Yet such a player would be a detriment to his team since he scores runs at such a low rate. To get a better understanding of this batsman's performance, it is typical to simultaneously assess his *strike rate*, which is the average number of runs scored per 100 balls. Similar comments are applicable when discussing *bowling average* and *economy rate*, which are the commonly reported bowling statistics.

In this discussion, batting average fails to consider the number of balls faced and the strike rate fails to consider the number of dismissals. It is therefore clear that a single batting performance measure should at least take into account the following three important quantities: runs scored, balls faced, and the number of dismissals. Alternative statistics have been proposed that utilize all three quantities and are typically applicable in all formats of cricket. For example, Croucher (2000) proposed the *batting index*, which is defined as the product of the batting average and the strike rate. Although the batting index is simple to calculate, it is not readily interpretable. Using a similar rationale, van Staden (2009) proposed graphical displays to assess performance. Manage and Scariano (2013) considered principal component analysis to reduce six/four batting/bowling variables into a single variable using IPL data. Lemmer (2011) reviewed a number of related statistics and also proposed a statistic for individual matches that takes into account match conditions in limited overs cricket.

Rather than assessing runs scored relative to wickets lost or runs scored relative to overs consumed, Beaudoin and Swartz (2003) and Lewis (2005) considered a natural benchmark for batting in one-day cricket. They considered runs scored relative to resources consumed, where the Duckworth–Lewis concept of resources (see Section 23.3) is the standard quantity that measures the dual combination of wickets and overs. With bowlers, both papers explore symmetrical measures based on runs conceded relative to the resources used while bowling. There are at least two avenues where the general approach may be extended: (1) in Twenty20 cricket where the ideas translate directly and (2) the calculation of resources consumed by a player (during a match), which is a tedious exercise that would benefit from automation.

Getting away from formulae based on runs, wickets and overs, Davis et al. (2015b) addressed player evaluation using the simulator described in Section 23.4. The authors asked how does a player i contribute to winning in Twenty20 matches? Via the simulator they estimated the expected increase in the number of runs scored $E_s(i)$ when player i is inserted in the lineup of an average team. They also estimated the expected decrease in the number of runs conceded $E_c(i)$ when player i is inserted in the lineup of an average team. Then $E_s(i) - E_c(i)$ is the proposed measure of player worth as it speaks directly to the number of runs by which the player improves his team compared to an average player. It is seen that the very best players (both batsmen and bowlers) can improve an international Twenty20 side by 8–10 runs. The approach has great potential and flexibility as it is able to investigate combinations of players and hypothetical lineups. Along the same lines, Perera et al. (2016b) considered the insertion of players into lineups to assess their contribution to fielding. In this analysis, it was seen that the best Twenty20 fielders save on average roughly 1.2 runs per Twenty20 match compared to a typical fielder. In addition to the simulator, the fielding analysis used textual analysis and random forests. The first quantitative investigation of fielding was undertaken by Saikia et al. (2012). By contrast, their approach involved the subjective assessment of every fielding play to provide a weighted measure of fielding proficiency.

On the more technical side, Koulis et al. (2014) proposed a Bayesian hidden Markov model for assessing batting in one-day cricket. The basic modelling assumption is that a player's observed performance (runs scored) is based on his underlying form, and that his underlying form changes over time in accordance with a probability transition matrix with an unknown number of states. Through extensive computation, Koulis et al. (2014) estimate parameters associated with the underlying form to obtain measures of performance.

Two further topics related to player evaluation are worthy of mention. First, there is some discussion in the literature concerning batting *consistency* (Lemmer 2004, Borooah and Mangan 2010). The general feeling is that not only is a high batting average a measure of quality but it is preferable if the standard deviation of the runs scored across matches is small. Second, there exist commercial rating methods. The most prominent are the Reliance ICC Player Ratings that have gained stature primarily because they are endorsed by the ICC and appear on the ICC website. These ratings have been developed for batting and bowling in all three formats of cricket. On the downside, the ratings do not have straightforward interpretations as they are based on moving averages involving a somewhat arbitrary compilation of measurements. In addition, the precise details of the ICC ratings may be proprietary as they do not appear to be generally available.

An area of performance evaluation that has not been adequately investigated is performance under irregular conditions. For example, in the second innings of limited overs cricket when a team is chasing, batsmen become more aggressive and take more risks. Consequently, their dismissal rate increases. How should performance be assessed for batsmen and bowlers in these and other irregular situations? Some work in this direction has been considered by Akhtar et al. (2015) in the context of test cricket. They compare the contribution by players with match outcome probabilities using session-by-session performances.

23.6 Evaluating Team Strength

There are several complicating factors that come into play when assessing team strength in international cricket:

- Teams do not play often, and hence data are limited; for example, Sri Lanka played only 11 test matches in 2015 (cf. Major League Baseball where the regular season schedule consists of 162 games).
- There is a lack of balance in schedules; for example, New Zealand did not play any Twenty20 matches against Australia in 2015.
- There is no consistent dependent variable for winning margins which are reported in different scales; for example, a team may win by 25 runs or may win by 3 wickets (with 61 balls remaining).

These issues suggest that comparing cricketing sides is a difficult task. Accordingly, there has not been a lot of work done in the area and there may be opportunities for future research. One idea involves taking into account results from the three formats of cricket. It also strikes the author that the rating/ranking problem is similar to what exists in American college basketball where there are relatively few games and unbalanced schedules. In American college basketball, the topic of ranking is of great importance as tournament invitations are on the line.

The ratings/rankings that are widely referenced in cricket are commercial and are based on subjective combinations of performance measures. They do not offer interpretability nor are they readily convertible to probabilities of victory when one team plays another. Specifically, we point again to the ICC team ratings/rankings which are displayed prominently by the ICC, Cricinfo and the BBC. The rankings are available for each of test, one-day and Twenty20 cricket.

From an academic perspective, there has been little published on the evaluation of team strength. In de Silva et al. (2001), a Bayesian linear model is proposed for one-day cricket where the response variable is run differential. In matches where the team batting second wins, their runs are extrapolated according to the remaining Duckworth–Lewis resources (see Section 23.3). The linear model contains a strength parameter for each team, a home field advantage parameter and an error term where decreasing weight is given to matches occurring further back in time. The interpretation of the team strength parameter is that it is the expected number of runs that the team in question would defeat an average team on a neutral field.

Allsopp and Clarke (2004) proposed models that are very similar to de Silva et al. (2001). Some differences include the use of least squares estimation instead of a Bayesian analysis, decomposing the strength parameter into a batting and bowling parameter, the use of log transforms on the dependent variable and the consideration of first innings results in test matches to provide team ratings for test cricket.

Finally, the match simulator of Davis et al. (2015a) described in Section 23.4 provides a straightforward approach for assessing team strength in Twenty20 cricket. With simulated matches between two teams of interest based on specified lineups, one can calculate various probabilities of interest including the probability that a team defeats its opponent.

23.7 Optimal Lineups

Although fantasy leagues have been around for decades, their popularity has exploded in recent years. A fantasy league is a competition amongst individuals where participants 'select' players from a real sports league. A participant's fantasy team then consists of the players that the participant has selected. When a participant's players perform well, they accumulate points, and the fantasy team performs well. Although fantasy leagues have traditionally been an enjoyable past time amongst friends, there is now huge prize money involved through online sites such as DraftKings and FanDuel. Somehow these American enterprises have avoided being classified as gambling sites.

The selection of a fantasy team by a fantasy team owner should obviously depend on how fantasy teams accumulate points. Since fantasy leagues differ in how points are accumulated, there ought to be different objective functions for different fantasy leagues. It is also the author's opinion that a statistical consideration in building fantasy teams that has not been adequately considered is the probability distribution of the points associated with fantasy teams. In terms of winning or placing high in a fantasy league, a fantasy team owner should at least consider the expected number of points that his team accumulates, the variability of points associated with his fantasy team and the number of competitors in the fantasy league. Furthermore, expected fantasy points and point variability should not be straightforward calculations due to the correlations between players. Further complicating the selection problem are constraints imposed by fantasy leagues. For example, sometimes players are assigned a 'dollar value' and an owner has a fixed number of dollars to spend.

Clearly, there is considerable strategy involved in selecting fantasy teams. When huge prize money is involved, it is safe to say that the best strategies are kept confidential and are not published in academic journals. In cricket, Brettenny et al. (2012) is perhaps the only publication to directly address fantasy team selection.

We now transition to the real sporting problem of team selection; that is choosing the 11 cricketers which form a team. In cricket, team selection has traditionally been left to 'team selectors' whose methods are largely subjective. There are a number of papers that have been written on team selection algorithms including Lemmer (2013), Ahmed et al. (2013) and Kamble et al. (2011). The search algorithms used to select teams are varied and include genetic algorithms and binary integer programming. The approaches typically permit constraints on team selection. For example, sometimes a fixed number of pure batsmen, all-rounders and bowlers are imposed when forming a team. However, the feature which leads to the greatest differentiation in the approaches lies in the specification of player quality as expressed in the objective function. As we have seen in Section 23.5, there are various measures of performance and these measures are sometimes antagonistic in the context of team selection. For example, selecting a player with a high batting average who seldom makes out reduces the batting opportunities for other players.

In the papers previously mentioned, the problem of interest has been restricted to team selection; that is choosing the 11 cricketers to form a team. In the context of Twenty20 cricket, Perera et al. (2016a) extend the problem beyond team selection and also simultaneously consider the determination of the batting order and the bowling order amongst those selected. The objective function that they maximise is expected run differential, the expected number of runs by which one team defeats its opponent. Expected run differential is essentially a proxy for winning, and is therefore a sensible criterion for determining optimal lineups (i.e. team selection, batting order and bowling order). Their approach is again based on the use of the match simulator developed by Davis et al. (2015a) which is briefly described in Section 23.4. The problem of determining optimal lineups is computationally demanding, requiring a search over a vast combinatorial space. For example, given a selected team, there are $11! \approx 40$ million batting orders alone. Perera et al. (2016a) carry out the search through the implementation of a simulated annealing algorithm. The proposed approach has great potential for Twenty20 teams where non-standard yet effective lineups are sometimes revealed.

23.8 Tactics

One of the more challenging problems in sports analytics is the discovery and adoption of tactics to improve a team's chance of winning. With cricket being a slow-moving game with time for reflection and decision-making, it seems well suited for such discoveries.

In test cricket, a common scenario occurs when a team is batting in the third innings and has built a considerable lead. The question that faces this team (Team A) is whether they should voluntarily terminate their innings, and this is known as a declaration. The advantage of declaring early is that there will be more time remaining for Team A to dismiss its opponent and secure a win rather than a draw. The disadvantage is that the target is lower than what it would be if declaration had been delayed, and the opposition may therefore score sufficient runs to defeat Team A. The two relevant papers on the strategic problem of declaration are Scarf and Akhtar (2011) and Perera et al. (2014). Whereas the two approaches are completely different, there is some common ground in terms of the results. Specifically, both papers argue that it is often the case that teams do not declare sufficiently early. This cautiousness causes teams to draw more often and win less often. The misjudgement of declaring late is evident to most fans by the fact that declaring teams rarely lose. Perera et al. (2014) provide specific guidelines (Table 4 of their paper) whether teams should declare at various stages of matches (i.e. overs remaining) and under various circumstances (i.e. weak versus strong opposition, desire to win versus draw).

It is the author's opinion that in sport, the ability of teams to invoke strategy depends partially on the ability to retain possession. For example, the author suggests that in the following free-flowing games, strategy/possession is greatest in basketball followed by soccer followed by hockey. Cricket affords considerable strategy as there is an analogy between the ability to retain possession and the ability to vary the batting approach; both affect the pace and scoring in games. When a batsman is cautious, the dismissal probability decreases with the trade-off of a lower run-scoring rate. Since the level of cautiousness/aggressiveness is under the control of the batsman, this surely introduces strategic opportunities. The study of cautiousness/aggressiveness is an important problem in all three formats of cricket. In addition, training batsmen to have the optimal level of aggressiveness does not seem to be an easy task. In Davis et al. (2015a), it is suggested that teams that are trailing in the second innings of Twenty20 matches wait too long to increase their level of batting aggressiveness. In all formats of cricket, Gauriot and Page (2015) conclude that players and captains do not bat optimally from a team perspective, but sometimes focus on individual incentives at various stages of a match.

Scarf et al. (2011) use negative binomial distributions to model the runs scored in innings and the runs scored in partnerships during test matches. They then take the view that the batting team can control its run rate and its desired target during the third innings.

Under these assumptions, probabilities for winning, losing and drawing a match are estimated. Therefore teams can be tactical in their approach to batting by 'setting' the run rate
and target accordingly.

In a more technical paper, Preston and Thomas (2000) use survival analysis methodology and dynamic programming to investigate batting strategies in limited-overs cricket.
For both the first and second innings, they obtain optimal run rates which they compare
with actual run rates. Although there are a number of simplifying assumptions in their
approach, the paper contains some interesting general results. For example, Preston and
Thomas (2000) demonstrate that optimal first innings batting should follow the pattern of
an increasing run rate.

Silva et al. (2016) consider a number of radical tactics in Twenty20 cricket. One approach
is based on the premise that wickets are less important in Twenty20 cricket than in other
formats of cricket. From this premise, modification to batting and bowling orders are suggested. Another approach concerns a weak team facing a stronger team. The basic idea is
that the weak team may force increased variability in the match to improve their chance of
winning. This may be accomplished by varying the aggressiveness of batting.

In terms of future research ideas concerning tactics, the strategic selection of bowling
options has not been carefully explored. For example, it cannot be the case that all batsmen
are equally adept at handling the same type of bowled ball. The author is not aware of data
collected on individual batsmen consisting of the batting outcome, ball speed, ball action,
ball placement, etc. On the flip side, bowlers make choices in their deliveries. Is there a
clear understanding about their most effective bowled balls? And does the sequencing of
their bowling decisions matter? Another area where there seems to be opportunity is the
placement of fielders. The optimal placement of fielders would depend on the batsman,
the bowler and possibly the score.

23.9 Miscellaneous Topics

This section covers six miscellaneous topics in cricket that the author has found interesting.
As is typical in research, these topics are usually initiated by a question. And in sport, often
the question is one that challenges the status quo.

Topic 1: The home team advantage is ubiquitous in sport with various underlying causes
(Stefani 2008). The question in this case involves the assessment of the home team advantage in cricket. As was mentioned in the second bullet point of Section 23.6, the problem
is complicated by the fact that winning margins are expressed differently depending on
whether the team that bats first wins or loses the match. Also, in quantifying home team
advantage, some models propose a common home team advantage whereas other models permit a team-dependent home team advantage. In de Silva et al. (2001), the home field
advantage is estimated to be worth approximately 15.6 runs in ODI cricket (their model D).
This is corroborated by Clarke (1998), whose estimate for the common home field advantage in ODI cricket is 14 runs. In Davis et al. (2015a), a simple calculation of the home field
advantage in international Twenty20 cricket is given as 9.0 runs. Over a 10-year period in
test cricket, home teams have won, lost and drawn 46%, 28% and 26% of their matches,
respectively (Bettingexpert 2011).

Topic 2: Related to the home field advantage is the effect due to umpiring decisions. Sacheti et al. (2015) investigate leg before wicket (LBW) decisions based on 1000

test matches. Amongst their interesting findings is that there was previously a systematic bias by officials which favoured the home team. However, since the introduction of pairs of neutral umpires (in 2002), the bias has been virtually eliminated. The bias was estimated at approximately one extra LBW decision per innings in favour of the home team when both umpires had the same nationality as the home team. In a novel application, Manage et al. (2010) use receiver operating characteristic (ROC) curves to assess the quality of decisions by umpires in any format of cricket. The approach permits the comparison/ranking of umpires via simple ROC graphics and the corresponding area diagnostics. The ROC curves are based on the percentage of correct and incorrect umpiring decisions. These days, the *decision review system* (DRS) is a technology that is often utilized and enables retrospective judgements of on-field umpiring decisions. With the availability of DRS in test and ODI cricket, the accuracy of umpiring decisions is now at a very high level.

Topic 3: A cricket match begins with a coin toss where the team that wins the toss has the option between batting or fielding. The question in this case is what is the better decision. Pundits will tell you that the decision depends on your opponent, the pitch conditions, whether it is a day/night game, impending weather, etc. One of the interesting observations is that teams have greatly different perspectives on batting first versus batting second. de Silva and Swartz (1997) report that during 1990–1997, Australia chose to bat first 87% of the time in ODI matches upon winning the coin toss whereas Sri Lanka chose to bat first only 36% of the time. Furthermore, de Silva and Swartz (1997) demonstrate no appreciable advantage to winning the coin toss in ODI matches. Allsopp and Clarke (2004) corroborated this finding in both ODI and test cricket. On the other hand, McGinn (2013) argues that it is better to bat first in day/night games where the estimated advantage is roughly 0.2 runs per over (i.e. 10 runs in a 50-over match). An explanation is that the artificial lighting causes difficulties when batting at night. In addition, some blogs have suggested that particular teams are better at target-setting (batting first) than at chasing (fielding first). One wonders whether these team-specific results are due to multiple comparisons issues.

Topic 4: Much has been written in the media about the importance of partnerships and the special symbiosis that exists between certain players. However, is there any truth to this sentiment? Batting appears to be an individualistic activity and one would think that any top-level partner (who knows when to run and when not to run) may neither help not impede a particular batsman. Valero and Swartz (2012) investigate the importance of opening partnerships in test and one-day cricket. They arrive at the conclusion that synergies in opening partnerships may be considered a sporting myth.

Topic 5: In most sports, sporting bodies are reluctant to make frequent rule changes. When a rule change is made, a question of interest is how the change affects the game. This is particularly important for the sanctity of records in sport. One rule that has been tinkered with repeatedly is the powerplay in one-day cricket. Silva et al. (2015) investigated various powerplay rules in one-day cricket. Their approach was based on creating hypothetical parallel matches; that is matches where they impute scoring results into actual matches as if there had been no powerplay. They then compare the actual matches with the parallel matches. They conclude that powerplays increase run production but also increase wickets. It is interesting to note that the powerplay rule has changed again (2015), and has been simplified so that teams no longer determine the start of the powerplay.

Topic 6: One of the unique events in cricket is the annual IPL auction that determines the composition of IPL teams. The auction is different from drafts (see Chapter 16) as teams require different strategies to secure their best possible teams. The IPL auction has not been investigated in detail, and it is suggested that this topic has considerable research potential. Karnik (2010) considered the value of player characteristics using hedonic price models whereas Swartz (2011) argued that anomalies have occurred in previous auctions and that drafts may be preferable for dispersing talent.

23.10 Discussion

Hopefully, this chapter (1) has provided a nearly comprehensive picture of the important problems that have been addressed in cricket analytics and (2) will serve as a starting point for those interested in doing work in cricket analytics. It is anticipated that opportunities for research in cricket will flourish in the coming years as more data become available, especially via player and ball-tracking technologies.

What research problems are open to an individual who is interested in doing cricket analytics? One place to look is other sports. For example, baseball is a similar game to cricket and shares similar problems. However, unlike cricket, there has been a tremendous amount of quantitative research in the sport of baseball where SABR (the Society for American Baseball Research) boasts a membership of more than 6000 individuals, publishes articles and hosts meetings. One might also look to the sports gambling world as a benchmark for modelling. Whenever money is involved, predictive modelling tends to be as good as possible.

Although women's cricket has been around for a very long time, there are few (if any) quantitative academic articles on the sport. With media attention increasing, it appears that women's cricket is now on the rise. Investigations that distinguish features between the women's game and the men's game may provide various research opportunities.

As was argued in Section 23.1, cricket is one of the world's major sports. And what we have seen lately is that other major sports (e.g. baseball, soccer, football, hockey and basketball) are investing heavily in sports analytics. Forward-thinking teams have hired quantitative people who are looking at data to inform them on all aspects of their sport. To gain an edge, it seems that many topics are worthy of investigation, including nutrition, travel, strategy, injury prevention, player evaluation, salary constraints, drafting, etc. We therefore conclude the chapter with a provocative question. When will the gentleman's game of cricket begin to encourage and adopt modern analytics?

Acknowledgment

Swartz has been partially supported by grants from the Natural Sciences and Engineering Research Council of Canada. The author thanks Basil de Silva, Charith Karunarathna, Ananda Manage, Saman Muthukumarana, Harsha Perera, Lasantha Premarathna, Pulindu Ratnasekera and Rajitha Silva all of whom provided helpful advice on an earlier version of the chapter.

References

Ahmed, F., Deb, K., and Jindal, A. (2013). Multi-objective optimization and decision making approaches to cricket team selection. *Applied Soft Computing*, 13, 402–414.

Akhtar, S., Scarf, P.A., and Rasool, Z. (2015). Rating players in test match cricket. *Journal of the Operational Research Society*, 66, 684–695.

Allsopp, P.E. and Clarke, S.R. (2004). Rating teams and analysing outcomes in one-day and test cricket. *Journal of the Royal Statistical Society, Series A*, 167, 657–667.

Beaudoin, D. and Swartz, T.B. (2003). The best batsmen and bowlers in one-day cricket. *South African Statistical Journal*, 37, 203–222.

Bettingexpert. (2011) How often do home nations win in test cricket? http://www.bettingexpert.com/blog/what-is-home-field-worth-in-test-cricket, accessed December 20, 2015.

Borooah, V.K. and Mangan, J.E. (2010). The "Bradman Class": An exploration of some issues in the evaluation of batsmen for test matches, 1877–2006. *Journal of Quantitative Analysis in Sports*, 6, Article 14.

Brettenny, W.J., Friskin, D.G., Gonsalves, J.W., and Sharp, G.D. (2012). A multi-stage integer programming approach to fantasy team selection: A Twenty20 cricket study. *South African Journal for Research in Sport, Physical Education and Recreation*, 34, 13–28.

Business Insider Sports. (2015). The NBA is the highest-paying sports league in the world. http://www.businessinsider.com/sports-leagues-top-salaries-2015-5, accessed December 20, 2015.

Carter, M. and Guthrie, G. (2004). Cricket interruptus: Fairness and incentive in limited overs cricket matches. *Journal of the Operational Research Society*, 55, 822–829.

Christos, G.A. (1998). It's just not cricket. In *Mathematics and Computers in Sport*, N. de Mestre and K. Kumar (eds.), Bond University, Gold Coast, Queensland, Australia, pp. 181–188.

Clarke, S.R. (1988). Dynamic programming in one-day cricket—Optimal scoring rates. *Journal of the Operational Research Society*, 39, 331–337.

Clarke, S.R. (1998). Test statistics. In *Statistics in Sport*, J. Bennett (ed.), Arnold Applications of Statistics Series, Arnold, London, UK., pp. 83–103.

Croucher, J.S. (2000). Player ratings in one-day cricket. In *Mathematics and Computers in Sport*, G. Cohen and T. Langtry (eds.). Bond University, Gold Coast, Queensland, Australia, pp. 1–13.

Davis, J., Perera, H., and Swartz, T.B. (2015a). A simulator for Twenty20 cricket. *Australian and New Zealand Journal of Statistics*, 57, 55–71.

Davis, J., Perera, H., and Swartz, T.B. (2015b). Player evaluation in Twenty20 cricket. *Journal of Sports Analytics*, 1, 19–31.

de Silva, B.M., Pond, G.R., and Swartz, T.B. (2001). Estimating the magnitude of victory in one-day cricket. *Australian and New Zealand Journal of Statistics*, 43, 259–268.

de Silva, B.M. and Swartz, T.B. (1997). Winning the coin toss and the home team advantage in one-day international cricket matches. *New Zealand Statistician*, 32, 16–22.

Duckworth, F.C. and Lewis, A.J. (1998). A fair method for resetting targets in one-day cricket matches. *Journal of the Operational Research Society*, 49, 220–227.

Duckworth, F.C. and Lewis, A.J. (2004). A successful operational research intervention in one-day cricket. *Journal of the Operational Research Society*, 55, 749–759.

Elderton, W.E. (1945). Cricket scores and some skew correlation distributions. *Journal of the Royal Statistical Society, Series A*, 108, 1–11.

Gauriot, R. and Page, L. (2015). I take care of my own: A field study on how leadership handles conflict between individual and collective incentives. *American Economic Review: Papers and Proceedings*, 105, 414–419.

Jayadevan, D. (2002). A new method for the computation of target scores in interrupted, limited over cricket matches. *Current Science*, 83, 577–586.

Kamble, A.G., Venkata Rao, R., Kale, A.V., and Samant, S.P. (2011). Selection of cricket players using analytical hierarchy process. *International Journal of Sports Science and Engineering*, 5, 207–212.

Karnik, A. (2010). Valuing cricketers using hedonic price models. *Journal of Sports Economics*, 11, 456–469.

Koulis, T., Muthukumarana, S., and Briercliffe, C.D. (2014). A Bayesian stochastic model for batting performance evaluation in one-day cricket. *Journal of Quantitative Analysis in Sports*, 10, 1–13.

Lemmer, H.H. (2004). A measure for the batting performance of cricket players. *South African Journal for Research in Sport, Physical Education and Recreation*, 26, 55–64.

Lemmer, H.H. (2011). The single match approach to strike rate adjustments in batting performance measures in cricket. *Journal of Sports Science and Medicine*, 10, 630–634.

Lemmer, H.H. (2013). Team selection after a short cricket series. *European Journal of Sport Science*, 13, 200–206.

Lewis, A.J. (2005). Towards fairer measures of player performance in one-day cricket. *Journal of the Operational Research Society*, 56, 804–815.

Manage, A.B.W., Mallawaarachchi, K., and Wijekularathna, K. (2010). Receiver operating characteristics (ROC) curves for measuring the quality of decisions in cricket. *Journal of Quantitative Analysis in Sports*, 6, Article 8.

Manage, A.B.W. and Scariano, S.M. (2013). An introductory application of principal components to cricket data. *Journal of Statistics Education* [electronic journal], 21, Article available online at http://www.amstat.org/publications/jse/v21n3/scariano.pdf, accessed December 20, 2015.

McGinn, E. (2013). The effect of batting during the evening in cricket. *Journal of Quantitative Analysis in Sports*, 9, 141–150.

McHale, I.G. and Asif, M. (2013). A modified Duckworth-Lewis method for adjusting targets in interrupted limited overs cricket. *European Journal of Operational Research*, 225, 353–362.

Perera, H., Davis, J., and Swartz, T.B. (2016a). Optimal lineups in Twenty20 cricket. *Journal of Statistical Computation and Simulation*, 86, 2888–2900.

Perera, H., Davis, J., and Swartz, T.B. (2016b). Assessing the impact of fielding in Twenty20 cricket. *Journal of the Operational Research Society*, Under review.

Perera, H., Gill, P.S., and Swartz, T.B. (2014). Declaration guidelines in test cricket. *Journal of Quantitative Analysis in Sports*, 10, 15–26.

Perera, H. and Swartz, T.B. (2013). Resource estimation in Twenty20 cricket. *IMA Journal of Management Mathematics*, 24, 337–347.

Preston, I. and Thomas, J. (2000). Batting strategy in limited overs cricket. *The Statistician*, 49, 95–106.

Sacheti, A. and Gregory-Smith, I. (2015). Home bias in officiating: Evidence from international cricket. *Journal of the Royal Statistical Society, Series A*, 178, 741–755.

Saikia, H., Bhattacharjee, D., and Lemmer, H.H. (2012). A double weighted tool to measure the fielding performance in cricket. *International Journal of Sports Science and Coaching*, 7(4), Article 6.

Scarf, P.A. and Akhtar, S. (2011). An analysis of strategy in the first three innings in test cricket: Declaration and the follow-on. *Journal of the Operational Research Society*, 62, 1931–1940.

Scarf, P.A., Shi, X., and Akhtar, S. (2011). On the distribution of runs scored and batting strategy in test cricket. *Journal of the Royal Statistical Society, Series A*, 174, 471–497.

Silva, R., Manage, A.B.W., and Swartz, T.B. (2015). A study of the powerplay in one-day cricket. *European Journal of Operational Research*, 244, 931–938.

Silva, R., Perera, H., Davis, J., and Swartz, T.B. (2016). Tactics for Twenty20 cricket. *South African Statistical Journal*, To appear.

Stefani, R. (2008). Measurement and interpretation of home advantage. In *Statistical Thinking in Sports*, R. Koning and J.H. Albert (eds.), CRC: Boca Raton, FL, pp. 203–216.

Stern, S.E. (2016). The Duckworth-Lewis-Stern method: Extending the Duckworth-Lewis methodology to deal with modern scoring rates. *Journal of the Operational Research Society*, doi:10.1057/jors.2016.30.

Swartz, T.B. (2011). Drafts versus auctions in the Indian Premier League. *South African Statistical Journal*, 45, 249–272.

Valero, J. and Swartz, T.B. (2012). An investigation of synergy between batsmen in opening partnerships. *Sri Lankan Journal of Applied Statistics*, 13, 87–98.

van Staden, P.J. (2009). Comparison of cricketers' bowling and batting performance using graphical displays. *Current Science*, 96, 764–766.

Weatherall, D. (2011). Cross-validation of rain rules for interrupted one-day cricket matches. MSc thesis, Department of Mathematical Statistics and Actuarial Science, University of the Free State, Bloemfontein, South Africa.

Wood, G.H. (1945). Cricket scores and geometrical progression. *Journal of the Royal Statistical Society, Series A*, 108, 12–22.

24

Performance Development at the Olympic Games

Elmer Sterken

CONTENTS

24.1 Introduction

Competition and performance improvement are at the core of sports. The Olympic Motto *Citius, Altius, Fortius*, as proposed by the founder of the modern Olympic Games, Pierre de Coubertin, describes the ultimate intentions of sportsmen. Competition is a necessary condition for performance improvement. In this chapter, I review the improvement rates of the historical best performances in Track and Field, Rowing, Swimming, and Weightlifting at the Olympic Summer Games and Skating at the Olympic Winter Games. Indeed, performances improve as De Coubertin prescribed, but the rates of improvement vary substantially across events and over time. It is a fact that there are limits to improvement, although nobody knows the ultimate best performances. Athletic performances depend on the development of talents of sportsmen, the progress in knowledge of training, and technological progress in equipment and facilities.

Although the Olympic performance development curves look like the improvement curves of world records, they miss the property that changes are only one sided. Time series of world record improvements are typically lumpy. There is a rich variance between Olympic sports events in terms of rate of progress. The general rate of progress of performance improvement has decreased since the late 1980s, as compared to the three decades before (see Stefani, 2012). It is generally accepted that most sports events have undergone a strong professionalization, leading to a lower progress rate of the best performances and more competition. It is also shown that there is a gender gap, and that this gap might vary across sports, but that the gap has a continuous nature.

I review statistical methods to denote improving performance curves. In general, one can model the main flow of progress. Progress needs to be approached by nonlinear models: the S-shaped class of functions is attractive. These functions allow for a development phase, a professionalization time, and a maturation phase. Sometimes it seems that a final plateau is established. These properties of time series of best performances are hard to fit into a model. Progress can be modeled using the notion that in larger populations the probability that an extreme talented sports person will stand up and improve the best performance is larger.

I also discuss the so-called Gould (1996) hypothesis on the development of athletic performances. Gould predicts that due to professionalization, the rate of progress of performances will slow down as the sport matures. Moreover, the variance of top performances will also decrease, because the distribution of talent shifts more to the top-end side. I describe and explain inequality of performances of the eight best-ranked individual athletes in final events of both the Olympic Summer and Winter Games using a measure of inequality, the Gini coefficient. I observe a decrease of individual performance inequality across Olympic finals over time, which could be attributed to an increase in competition and so supports the Gould hypothesis.

In the next section I introduce a global description of performance improvement at the Olympic Games. I use some of the events, like the 100 m dash sprint, to illustrate progress. I compute average rates of progress for all events discussed. Next, I describe the methods to model progress. There are many technical articles that can help the model builder in finding a way: this section points at the directions to look for, and in Section 24.4 I discuss the development of competition at Olympic events. Through the Gould hypothesis, competition is related to the development of performance. I summarize and conclude in a final section.

24.2 Performance Improvement at the Olympic Games

Athletes perform at their best during the Olympic Games. This applies to both "interaction sports" and "non-interaction sports." And although it is obvious that there is a strong improvement of performance in interaction sports, it is typically hard to measure. In this section I describe the progress of athletic performance in some non-interaction sports at the Olympic Summer and Winter Games. In order to set the scene, Figure 24.1 shows the performance at the 100 m dash sprint for men and women. I normalized the data for 1928 (index = 100), the first edition of the Olympic Summer Games were women participating in Track and Field. Figure 24.1 shows that there is indeed progress and that the increases in performances can be lumpy, but that after 84 years there is a 10%–13% improvement. One can also see the exceptional performance of Florence Griffith Joyner at the 1988 Seoul Summer Olympic Games.

As can be seen, the progress for women is larger than for men since 1952. This has been observed in other studies like Whipp and Ward (1992) who concluded that the "gap is progressively closing" at the time of publication. But as Holden (2004) observes, it is really likely that there will be a plateau difference between the performances of the sexes and the level split can be explained by physical differences. This implies that the 1928 performance in the women's competition is typically less competitive than in the men's, allowing for catching up in the years after. I illustrate this finding in Figure 24.2: this figure contains the relative difference in percentages in the performances of men and women on the 100 m dash sprint, the 100 m freestyle swimming, the 500 m skating, the high jump, the marathon, and

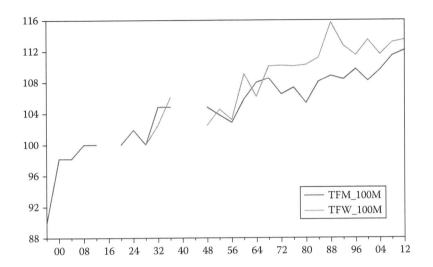

FIGURE 24.1
Performance index (1928 = 100) for the Olympic Gold performances in the 100 m dash sprint: 1896–2012 (TFM for men and TFW for women).

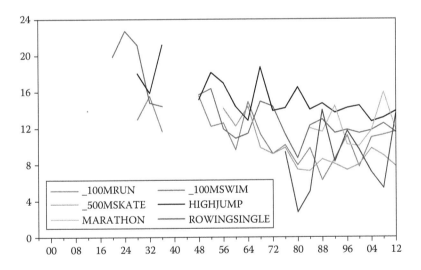

FIGURE 24.2
(See color insert.) Relative difference (in percentages) between Gold medal performances of women and men in the 100 m running, the 100 m freestyle swimming, the 500 m skating (data for 1994–2010 are plotted as 1996–2012), the high jump, the marathon, and the single sculls rowing.

the single sculls rowing. One can see that there is convergence to the 8%–13% difference region (see also Stefani, 2014).

Another way to look at the performance data is to plot the dynamic rate of change (in percentages) of the best performances. Figure 24.3 gives an illustration for the 100 m dash performance of the Gold medallists.

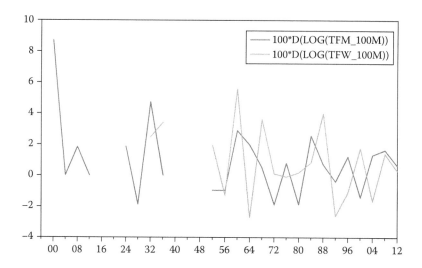

FIGURE 24.3
Rate of change of the Olympic Gold Medallists—100 m dash (TFM for men and TFW for women).

Both Figures 24.1 and 24.3 illustrate that since 1896 we can observe three subperiods: (1) the early development of the sports, (2) the "boost" in performances, and (3) maturation or saturation of performances. In order to illustrate this observation, I use the Olympic Track and Field tournament as an example. I make a distinction between running, jumping, and throwing events (see also Stefani, 2014). Because not all events are continued throughout the history of the Modern Olympic Summer Games, there is a problem with indexing (a fixed basis year cannot be selected, which troubles averaging). Therefore I plot Figure 24.4, like Figure 24.3, in terms of rate of progress. I exclude the pre–World War I editions of the Games, because these observations are by far larger and troubling the figure.

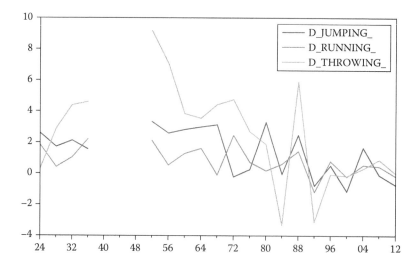

FIGURE 24.4
(See color insert.) Average rate of progress of the Olympic Gold performances in Jumping, Running, and Throwing (1924–2012, both men and women).

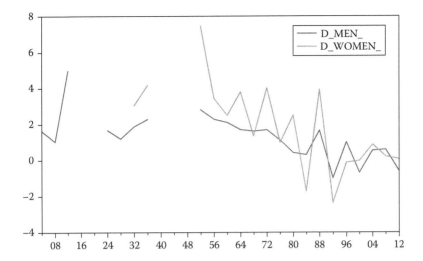

FIGURE 24.5
Average rate of progress of the Olympic Gold performances of men (blue) and women (red) in Track and Field (1904–2012).

One can observe the three phases as described. The rate of progress in throwing is very large in the years 1952–1980. After 1992 there is no significant progress observed. This is fully in line with the prediction made by Stephen Jay Gould (1996). Using theory from biology, Gould presented an evolutionary approach of the development of sports. Due to professionalization, the distribution of talented sportsmen will move to the upper end of the talent distribution. Pushing the limits is harder than ruling out the low performers in a professional sports market. So the rate of progress of the top performer will slow down, but also the variance of the top performances will become smaller. Denny (2008) labels this maturation of the competition as reaching a performance plateau. In general, and I will continue this argument in the next section, it seems that progress follows an S-shaped curve, allowing for a plateau at the end of the phase of development.

Figure 24.5 describes the rates of progress of performances of Olympic Gold medal-lists for both men and women in Track and Field. One can see that until 1980 the rate of progress of female performances has outnumbered the men's progress rate, but that there is convergence around 0 in recent years. One can also compute the compounded annual productivity growth rate of the different events in Track and Field. I give an overview in Table 24A.1. This table shows the index-years, the final 2012 index level, and the com-pounded annual improvement rates. It can be seen that in Track and Field the average rate of progress is 0.22 for both men and women. For running, the progress seems to be stronger for the longer distances (see also Dekerle et al., 2006).

24.2.1 Rowing, Skating, Swimming, and Weightlifting

Now we know the patterns in Track and Field, a natural question is whether this general picture applies to other sports as well. So I add data with similar information for rowing, skating, swimming, and weightlifting. These sports allow for a time series approach and provide some variation (including the Winter Games), but all are different from the track

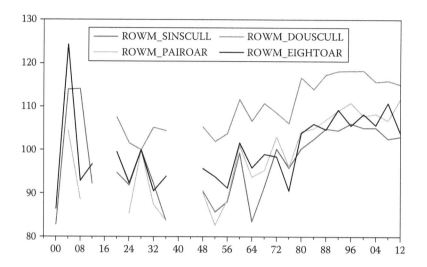

FIGURE 24.6
(See color insert.) Performance Index (1928 = 100) of the Olympic Gold performances of men's rowing events: Single Sculls, Double Sculls, Pair-Oared Shell without Coxswain, Eight-Oared Shell with Coxswain (1896–2012).

and field data in some respect. Note that cycling would have been a sport to present: there are measurement problems though for (discontinued) track events. I measured the annualized rate of progress for the 1000 m time trial for men (0.25) and the 500 m time trial for women (0.19), which are comparable to Track and Field. I summarize the results for the other sports in tables similar to Table 24A.1.

Starting with rowing events, one can observe in Figure 24.6 (for some of the men's events) that the rate of progress is typically smaller than the Track and Field (see also Stefani, 2014). As Table 24A.2 illustrates, the compounded growth rate for rowing is about half of the number in Track and Field. And although technological progress is relevant to rowing, the progress rates are modest, certainly compared to the next sports I discuss: speedskating (see also Haake (2009) for a discussion of technical progress in Olympic sports).

I illustrate another, slightly different, perspective for speedskating. Here one can observe a compounded annual growth rate for women and men of about 0.4 per year, almost double the equivalent figure in Track and Field. This can be explained by the capital intensity of speedskating and the rapid technological progress (indoor tracks, klap-skates, suits, see also Kuper and Sterken (2003)). From the 1950s on, it seems that progress is linear instead of a concave function. It seems that there might be a plateau once further technological innovation is discontinued (Figure 24.7).

Next, I illustrate a sport with a steady progress rate: swimming (see Pyne et al., 2004). One can observe that the progression in swimming is larger than in running and jumping in Track and Field, but indeed tends to be more constant than for other sports. There is a notable exception with respect to the role of swimming suites, which have been banned in recent years. There is discussion though whether these suites really have had an impact on the development of best performances.

And finally I illustrate progress in weightlifting. I give the results for the index of the men's competition in five weight classes. It is good to note that in 1976 all competitions excluded the "press" component pre-1972 events had, reducing the final scores by about

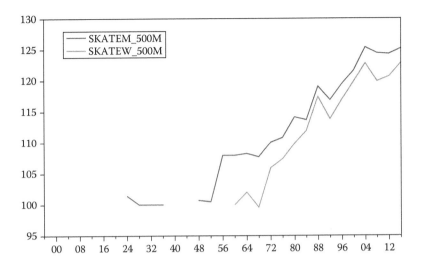

FIGURE 24.7
Performance Index (1928 = 100) of the Olympic Gold performances of the 500 m Speedskating for men and women (1896–2014).

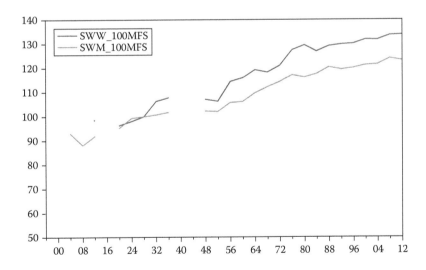

FIGURE 24.8
Performance Index (1928 = 100) of the Olympic Gold performances of the 100 m Freestyle Swimming events for men and women (1896–2012).

one-third. And after 1976, there have been several incidents with Performance-Enhancing Drugs. But also in weightlifting I observe slower growth rates in recent years than in the early Olympic years. The compounded growth rate for men up to 1976 was about 0.9, but only 0.4 after. For the women's events one can observe larger rates, indicating that the sport is still maturing (Figures 24.8 and 24.9).

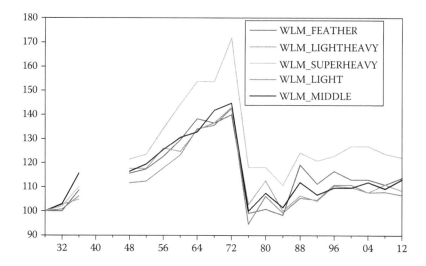

FIGURE 24.9
(See color insert.) Performance Index (1928 = 100) of the Olympic Gold performances of men's weightlifting events: Featherweight, Lightweight, Light heavyweight, Middleweight, and Super heavyweight (1928–2012).

24.3 Modeling Progress of Athletic Performance

Progress of human performances can best be measured in (controlled) sports events. Given the circumstances, population growth, training experience, and technical means athletes try to perform at the maximum, certainly with the honor of an Olympic medal waiting (see Haake et al., 2015). Therefore I can use the outcomes of world elite events as approximations of the maximum human performances.

The S-shaped historical curves found since the late nineteenth century up to the first fifteen years of the twenty-first century denote a likely pattern of performance development. It can be, though, that the shape of the S-curve differs across disciplines, as I illustrated in the previous section. After the starting phase one generally observes a steep progress of the top performances until a "plateau" is reached. The rate of progress on the plateau is typically smaller than in the years of development. The reaching of a plateau of course fascinates the sports statisticians. Is there an upper limit to performance or are there any invariant world records (see for instance Denny (2008), who compares running speed of dogs, horses, and human beings)?

There are two general arguments that allow even S-shaped curves or plateaus to move in time. The first argument is technological progress. It is evident that in some more capital-intensive sports, equipment and facilities allow either gradual or lumpy performance improvement (see Kuper and Sterken (2003) for an example in speedskating). The second argument is improvement in training. Knowledge about performance-enhancing training can lead to further professional specialization of athletes.

This has led Gould (1996) to apply theory of biology to the development of sports. In the early phases of development, professionalization is underdeveloped and the distribution of competing athletes is wide along a modest mean. Development leads to both an improvement of the extreme values of talent and skills, but more importantly to a more dense distribution toward the more talented athletes. This leads to the Gould-hypothesis that the more developed a sport is, the slower the rate of progress will be (in statistics the maximum of a distribution increases with the root of the total sample size), but also the lower the variance of top-sport performances.

Development of Olympic performances resembles the development of world records, but is not identical in a statistical sense. The development of world records is one sided: in terms of time used to complete an event, for instance, we only observe nonpositive changes. This leads to lumpy curves, which we observe less in the Olympic data.

Over the past years, multiple studies have made attempts to model the development of Olympic records, for instance, for running events (see for instance Pritchard and Pritchard, 1994). Chatterjee and Chatterjee (1982) analyze the development of the 100, 200, 400, and 800 m running at the Olympic Games. Stefani (1982, 1994, 2006) has published extensive work on predicting Olympic performances in Track and Field, Swimming, and Weightlifting. A famous example is the article by Tatem et al. (2004), who used a linear model for the 100 m sprint by women and men and concluded that women will outperform men in the year 2156.

Observing the time series of an Olympic event often reveals an inverted S-shaped pattern. At first competition is not intense. At the so-called inflection point, the rate of progress increases and more sportsmen participate. Professionalization starts and sports unions evolve. After the phase of rapid improvement saturation comes. It gets harder and harder to improve the best Olympic performance and only a highly talented individual is able to do so. For some sports events typical observations are available for some pieces of the curve. Piecewise linear techniques to model the development the performances is then popular. These linear approximations are computationally attractive, but miss the global view: the Tatem et al. (2004) article is an example. Of course linear approximations cannot be correct, especially not for long forecasting horizons. Wainer et al. (2000) analyze the development of progress in running and swimming and argue that women and men show similar development patterns of world records, but lag male equivalents.

Many studies are trying to get insight into the extreme limit of performance. A good example is Denny (2008), who compares the speed limits to running for dogs, horses, and humans. Denny uses extreme value theory combined with three underlying "core" models. The three different model versions are assumed: a linear trend, a logistic curve, and a model where the population size matters. The linear model works if a "plateau" can be verified. The logistic model describes the S-shaped version. Nevill and Whyte (2005) also use the logistic approach. Kuper and Sterken (2008) present a detailed overview of different types of functional forms that give an appeal to the S-shaped curve. A final choice is made in the analysis of running data for the Gompertz curve. The Gompertz curve is a relatively simple curve that allows for an asymmetric S shape; the implied limits can be derived easily.

In the remainder of this section, I describe which basic functional forms could be used to fit the performance progression of Olympic gold medalists. I will follow the Denny approach and suggest a three-step procedure:

1. Can we establish evidence of no trend: is there a plateau?
2. If not, what models can we use to denote the curvature?
3. If we know the basic shape of the development of performance progress, how can we apply extreme value statistics?

Once I reach a conclusion to this discussion, I proceed to the next section, where I include competition in the approach of performance development.

A plateau is defined as a multiyear time span without significant improvement of the performance (for instance velocity in running events). Denny (2008) suggests to pick a fixed window and to apply a backward recursive moving-window linear regression (say for 30 years). As long as there is no significant trend, the plateau is assumed to be valid. This implies that it is likely the progress function is at the ultimate side of the S-shaped function.

If there is curvature, Haake et al. (2014) give a description of functional forms applicable to model the (parts of the) S-shaped function. Kuper and Sterken (2008) discuss the selection of the functional form. Starting with the rejection of a simple linear approach, the next steps might be to opt for a polynomial (or spline) approach. As for all other models, the model choice should be based on the relative contribution of adding more parameters. A disadvantage of a higher-order polynomial is that there is no natural limit. Next, one might suggest a simple exponential model, maybe even in an extended form like in Blest (1996). The logistic curve and the double logistic curve are suitable candidates as well. The Gompertz curve has an attractive feature of having an asymmetric S-shaped form. Nevill and Whyte (2005) and Denny (2008) use a sigmoidal form.

The problem, however, is in identifying these different shapes of curves using so few data points. As an example I estimated the progress-index P (1928 $=$ 100) for the 100 m sprint for women and men. I compare the linear in time t model with a simple quadratic and the Gompertz curve. For men I found

$$P = 97.12 + 0.51t \qquad\qquad \text{Adjusted } R^2 = 0.84$$

$$P = 95.67 + 0.83t - 0.01t^2 \qquad \text{Adjusted } R^2 = 0.86$$

$$\log(P) = 4.72 - 0.17\exp(-0.07t) \qquad \text{Adjusted } R^2 = 0.86$$

For women I got

$$P = 96.92 + 0.62t \qquad\qquad \text{Adjusted } R^2 = 0.79$$

$$\log(P) = 4.81 - 0.29\exp(-0.05t) \qquad \text{Adjusted } R^2 = 0.80$$

Adding quadratic terms does not improve the fit of the model for women and the parameter estimate for the trend in the Gompertz curve is not significant at the 95% confidence interval. One can so observe that the nonlinear models are typically hard to identify and that (piece-wise) linear approximations are performing rather well. Therefore it can be concluded that fitting time series models of Olympic Gold Medal performances is troublesome and identification of the shape of the progress function is typically hard.

The main issue is whether Olympic best performances, like the incidental improvements of world records, offer sufficient statistical information to fit the nonlinear models. Although Olympic time series do not have the drawback to include one-sided changes only, the observations, certainly for some of the women's events, are rather sparse.

A final step one can take is after we have been able to establish the current position on the S-shaped progression function. Once we are on a plateau, there can still be improvement due to the occurrence of extremely talented athletes. The statistics of extreme occurrences shows that the distribution of the extreme improvements of best performances should conform to a Generalized Pareto Distribution (see Denny, 2008). I follow Denny (2008) and define the probability $P(V)$ that, for instance, a random running speed does not exceed V and can be defined by

$$P(V) = \exp\left\{-\left[1 + a\left(\frac{(V-b)}{c}\right)\right]^{(-1/a)}\right\},$$

where
  b is a location parameter
  $c > 0$ is a scale parameter

The scale parameter a can be nonnegative: in that case there is no defined limit. If $a < 0$ we have $V_{\max} = b - c/a$. Denny presents estimates of this maximum speed based on population statistics. The larger the population is, the higher the probability that there will be many subsamples that compete, and therefore the larger the probability that the best performance will be improved.

For the Olympic Games, there is hardly any evidence that population size is relevant to success anymore. There are many countries with large populations that hardly win any medal. It is more likely that competition will affect the record performances. In the next section I link indicators of competition with performance.

24.4 Concentration and Competition

Gould (1996) predicts that if professionalization of a sport increases, the rate of progress of the best performance will slow down and the variance of top athletic performances will decrease. This implies that the "market" for top performances becomes more competitive and the concentration of top athletes decreases. Sports competition is more interesting if the final result is rather unpredictable. A strong dispersion of sports results therefore might increase public interest. In particular, if the dispersion is an unbiased reflection of differences in quality, spectators are willing to accept and applaud clear performance differences. But inequality of results can also become predictable and especially in team sports this might lead to a lower attention. There are various initiatives to stimulate competitive balance principles (see Sanderson and Siegfried, 2003). For instance, sports organizations are interested in creating an optimal level playing field. Sometimes, equalizing principles like salary caps and first draft rights (like in the NBA) are used to create more equality among competing teams.

These principles can be applied to team sports, but are less applicable to individual sports. Sometimes, organizational incentives given to athletes can lead to uneven performances. Frick and Prinz (2007) present empirical evidence for professional marathon athletes, who are sensitive with respect to changes in prize money. But illegal stimuli can lead to unequal performances, like in recent times in professional cycling. So finding an interpretation of unequally distributed performances can find its origin in many aspects, ranging from differences in talent, in incentives, in training background, in (lack) of competition, or even illegal stimuli.

To measure concentration of results one can compute the market shares of participants in a competition, the so-called Herfindahl–Hirschmann index (the sum of squared market shares) and/or the Gini coefficient (Gini, 1912). Usually the Gini coefficient is used to measure inequality of income and or wealth in a population. For instance, for the United States, the wealth inequality exceeds the income inequality. The Gini coefficient is usually based on the Lorenz curve, which plots the proportion of variable A that is represented by the bottom y percent of the sample population. If the Gini coefficient measures 1, the distribution is unequal; for a Gini coefficient of 0, the distribution is fully equal. Deaton (1997) proposes to use the ratio to the mean of half of the pairs of the absolute deviations between the population elements. In a population $x(i)$, (with mean xk) of size N there are $N(N-1)/2$ distinct pairs, so the Gini coefficient is

$$G = \frac{1}{xk} N(N-1) \sum_{i>j} \sum_{j} |x(i) - x(j)|. \tag{24.1}$$

It is more convenient to write the Gini coefficient in the equivalent form GD, see Deaton (1997):

$$GD = \frac{N+1}{N-1} - \frac{2}{N(N-1)xk} \sum_{i=1}^{N} \rho(i)x(i). \tag{24.2}$$

where $\rho(i)$ is the rank of observation i in the x-distribution. If the Gini coefficient applies to income data, $xk(1 - GD)$ is a measure of welfare. The Deaton interpretation of the Gini coefficient satisfies the so-called transfer principle. Suppose that without changing the order of the distribution of the elements in the population, a fraction of a larger observation is transferred to a smaller observation, the measure should decrease (or not increase in a weaker form of the transfer principle).

I apply the Gini coefficient GD as a concentration measure to Olympic events. In general, the measures can be used in any competition that ranks the first eight outcomes by some variable. Whether this variable is time, distance, or ranking points is irrelevant. The GD can be used to measure inter-event comparisons: which event has experienced the lowest or highest inequality of results?

I use data of events held at both the Modern Summer and the Winter Olympic Games. Per event I collect the performances of the best eight athletes or the best eight teams of athletes if applicable. I also collect the total number of athletes N that competed in the event and the number of countries C that sent athletes to the specific event. If available, I also collect the ruling World Record WR performance (or in some cases the Best World Performance). Using the data I computed the following measures:

- The Deaton specification of the Gini coefficient GD: using the lower weights on the lower figures (so for the best completed times the lowest rank weights and for the largest heights or distances, e.g., the lowest rank weights)
- The relative deviation of the winning performance W to the world record performance WR (if available): $100 * (W/WR - 1)$

For the Summer Olympic Games I included Track and Field, Cycling, Rowing, Swimming, and Weightlifting for both men and women. In Track and Field I only use individual events (leaving out the relay-events, for instance). I include the 100, 200, 400, 800, 1500 m, 5k and 10k, marathon, 110 m hurdles for men (100 m hurdles for women), 400 m hurdles, 3k steeple, 20k walking, and 50k walking (the latter only for men); the high, long, and triple jump; pole vault; shot put; discus-, hammer-, and javelin throw (note that for the javelin throw the rules for the design of the javelin have changed over time, leading to a resetting of the world record performances). For Cycling, I included the 1k Track event for men and the 500 m for women, although these events are discontinued in 2008 (and I add the world cup data here). For Rowing, I included the Sculls, the Oared, and the Lightweight competitions. For Swimming I use the 50, 100, 200, 400, 800 m (women), and 1500 m (men) freestyle events, and the 100 and 200 m back- and breaststroke and butterfly events. In Speedskating I include the 500 m, 1k, 1500 m, 3k (women), 5k, and 10k (men) events.

In Table 24.1 I present the data, in averages over all editions of the Games, of the Gini coefficient of the first eight finishers *GD* and the gap of the gold medal performance to the world record or world best performance in percentages for the Track and Field events. One can observe that the men's 1500 m can be described as the event with the lowest historical inequality of the results. In general, all the running events seem to be more competitive than the field events. For women, the 5k, organized since 1996, has the lowest inequality measure *GD*, with the 1500 m (organized since 1972) being the second in a row. In general, as I will describe below, inequality of results is typically smaller for more recent events. The events for men have typically lower inequality measures than the women's events: I will illustrate this in a time series plot below. In terms of the gap of the winning performance to the world record the 400 m hurdles has the lowest value. Again, it seems that the field and road events have typically larger gaps than the track events.

In Table 24.2, I present the data for the Swimming events. Here I see that the 50 m freestyle and the breaststroke events have the lowest values of the inequality measures for the indoor events. The outdoor 10k event has a mass start character and has shown to have competitive finish procedures at the last two Olympic Summer Games editions (2008 and

TABLE 24.1

Performance Inequality Measures for Track and Field (in Percentages), Averaged over the Various Editions of the Summer Olympic Games

Men (1896–2012)	Gini	Gap to WR	Women (1928–2012)	Gini	Gap to WR
100 m	0.657	2.071	100 m	0.857	1.673
200 m	0.815	1.837	200 m	0.783	8.783
400 m	0.999	1.762	400 m	0.823	1.986
800 m	0.866	2.134	800 m	0.835	1.536
1500 m	0.441	2.002	1500 m	0.416	3.258
5k	0.543	2.198	5k	0.321	4.681
10k	0.655	1.591	10k	0.492	2.913
Marathon	1.215	2.589	Marathon	0.624	5.047
110 m hurdles	1.103	2.209	110 m hurdles	1.016	1.896
400 m hurdles	1.138	0.814	400 m hurdles	0.952	0.783
3k steeple	0.711	1.782	3k steeple	0.799	5.978
20k walk	0.804	3.62	20k walk	1.281	1.715
50k walk	1.213	1.895			
Triple jump	1.456	2.447	Triple jump	0.598	0.428
High jump	1.168	3.527	High jump	1.269	1.87
Long jump	1.775	4.139	Long jump	1.491	3.508
Pole vault	1.805	3.631	Pole vault	1.757	1.645
Discus throw	1.974	5.134	Discus throw	2.678	7.216
Shot put	2.205	5.07	Shot put	2.630	5.044
Hammer throw	2.280	4.758	Hammer throw	1.341	2.95
Javelin throw	2.231	5.948	Javelin throw	3.168	5.502

TABLE 24.2

Performance Inequality Measures for Swimming (in Percentages),
Averaged over the Various Editions of the Summer Olympic Games

Men (1908–2012)	Gini	Gap to WR	Women (1912–2012)	Gini	Gap to WR
50 m	0.579	0.940	50 m	0.683	0.865
100 m	0.764	1.026	100 m	1.12	0.194
200 m	0.739	0.171	200 m	0.733	0.546
400 m	0.869	1.216	400 m	1.064	0.651
1500 m	1.086	0.545	800 m	0.773	0.815
Backstroke			*Backstroke*		
100 m	1.050	0.957	100 m	0.882	1.005
200 m	0.893	0.751	200 m	0.900	0.497
Breaststroke			*Breaststroke*		
100 m	0.565	0.106	100 m	0.696	0.482
200 m	0.865	1.149	200 m	0.836	1.413
Butterfly			*Butterfly*		
100 m	0.686	0.551	100 m	0.913	0.846
200 m	0.731	0.458	200 m	0.855	1.076
Open water			*Open water*		
10k	0.135		10k	0.147	

2012). The gap to the world records in swimming is lower than in track and field events. This might partly be due to regulation issues, such as the "underwater starting rule" and the use of special swimming gear in more recent years. For Cycling I do not include a separate table, because I only have two events. For the mens 1k I find an average Gini coefficient of 0.736 (for 1928–2012) and an average gap to the world record of 2.505% (for 1948–2012); for the women's 500 m comparable data are 0.620 for the Gini coefficient and 0.280% for the average gap to the world record (over 2000–2012). These figures indicate that these Cycling events are relatively competitive.

Tables 24.3 and 24.4 present the Gini coefficients for the Rowing and Weightlifting events respectively. Table 24.4 illustrates that the Weightlifting events (especially for women) show a large spread in results. Next in Table 24.5, I present the performance inequality measures for the Speedskating events. One can observe that the men's events have smaller values for inequality and that the women's events have a smaller gap to the world record performance. The sprint events (500 m and 1k) seem to be the most competitive (Tables 24.3 through 24.5).

It is relevant to sketch the time series properties of the Gini coefficient. I take Track and Field for men and women as an example. Figure 24.10 plots the development of the Gini coefficient *GD*. It can be observed that the observed inequality for the men's events is smaller than for women's events. Overall, there is a decrease of inequality over time.

These could be due to the development of sports in terms of technology and training, but also to an increase in competition. I use some indicators, if available, to measure these determinants. First I use an indicator to measure technology and training progress, here measured by a variable labeled real gross domestic product per capita (GDPCAP) at the world level, as provided by the Growth and Development Center of the University of

TABLE 24.3

Performance Inequality Measures for Rowing (in Percentages), Averaged over the Various Editions of the Summer Olympic Games

	Gini
Men (1900–2012)	
Single Sculls	1.095
Double Sculls	0.888
Quadruple Sculls	0.699
Pair-Oared Shell	1.012
Four-Oared Shell	1.441
Eight-Oared Shell	0.796
Lightweight: Double Sculls	0.523
Lightweight: Four-Oared Shell	0.494
Women (1976–2012)	
Single Sculls	0.978
Double Sculls	0.975
Quadruple Sculls	0.834
Pair-Oared Shell	1.232
Eight-Oared Shell	0.815
Lightweight: Double Sculls	0.615

TABLE 24.4

Performance Inequality Measures for Weightlifting (in Percentages), Averaged over the Various Editions of the Summer Olympic Games

Men (1920–2012)	
Bantam	2.321
Featherweight	2.386
Lightweight	2.086
Middleweight	2.591
Light heavyweight	6.582
Middle heavyweight	2.060
Heavyweight	2.083
Super heavyweight	3.851
Women (2000–2012)	
Flyweight	3.737
Featherweight	4.309
Lightweight	2.369
Middleweight	4.685
Light heavyweight	3.236
Heavyweight	3.528
Super heavyweight	4.477

TABLE 24.5

Performance Inequality Measures for
Speedskating (in Percentages), Averaged over the
Various Editions of the Winter Olympic Games

	GD	Gap to WR
Men (1924–2014)		
500 m	0.381	2.117
1k	0.344	2.066
1500 m	0.514	3.200
5k	0.610	2.288
10k	0.857	1.355
Women (1960–2014)		
500 m	0.555	2.014
1k	0.546	2.306
1500 m	0.616	2.260
3k	0.747	1.395
5k	0.784	0.894

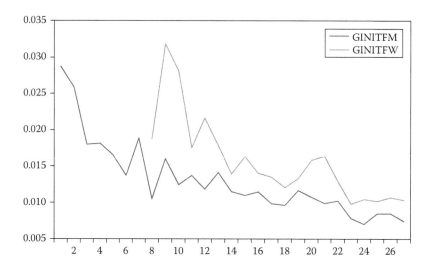

FIGURE 24.10
Average Gini coefficients for Track and Field Men (GINITFM) and Women (GINITFW).

Groningen (www.ggdc.net). This variable picks up the role of a time trend, explaining technological progress (see Haake, 2009), progress in training facilities, sports knowledge, etc. Next I try to measure competition in two ways. First, I measure the "broadness" of the sports, by including the number of countries C represented in the event (see Wallechinsky and Loucky, 2012). More competing countries could both lead to an increase or a decrease in inequality, depending on the available talents and skills. I also include the number of competitors per event N. An increase in the number of competitors should generally increase competition and decrease inequality of results. Next I include the performance of the winner of the gold medal W. This variable represents progress, but also indicates the incidence of an exceptional individual athlete (for instance, Jesse Owens or Usain Bolt in track and field). So an increase in W could possibly lead to more inequality of the results. If available, I include the relative deviation of the best performance W to the world record WR (or World Best Performance, as in the Marathon events): $GAP = 100 * (W/WR - 1)$. Larger deviations form the world record performances then would indicate less extreme performances of the Gold Medal Winner and lead to lower inequality. And finally I include for the Summer Olympic Games events the temperature, as indicated by Peiser and Reilly (2004). This variable is an indicator for the circumstances under which the athletes had to perform in competition.

I use a simple pooled linear model specification, using the various events in the sports branch as sectors and the various editions of the Olympic Games as time events (since I have a single event in Cycling, I use a simple linear regression for the men's 1k track competition; there are too few observations for the women's 500 m). I use event-specific dummy variables, but use GDP per capita as a proxy for the time dummy variables. I use the White-correction for heteroskedasticity and indicate the significance of the parameters with bold font. I present the estimated standard errors in the line below the parameter estimates.

Table 24.6 presents the overview of the estimation for men in the upper and for women in the lower panel. I observe that there is a trend effect in explaining the inequality of the results: the real per capita income GDPCAP increases over time and lowers the Gini coefficients. There is also an impact, on average, of an increase in the number of competitors per event on the inequality of the results: more competitors lead to a lower concentration. The number of countries that participate in the competition is a less relevant determinant of the Gini coefficient. If the performance of the gold medalist is better (or closer to the world record or the world best performance) the inequality of the competition outcomes tends to increase. And an increase in temperature affects the Track and Field and Weightlifting events. I observe that the model describes the Track and Field events better than the Rowing and/or Weightlifting events.

24.5 Summary and Conclusions

In this chapter, I review the development of performance in some non-interaction Olympic sports. Using examples from Track and Field, Cycling, Rowing, Speedskating, Swimming, and Weightlifting, it is illustrated that progress varies across sports, but is likely to follow an S-shaped pattern. After an early phase of development of the sport, professionalization gives a boost to performances. Once the professionalization comes to an end it is likely that a period of maturation starts, leading to a plateau of performances. Only individual excellent sportsmen are capable of rare improvements on the plateau.

Annual progress rates vary from 0.1 to 0.5. Performances in the 100 m dash in Track and Field are now about 12%–15% better than in 1928 (the benchmark year). There are no signs of a closure of the gap between the performances of women and men. In some sports, the progress rate for women exceeds the one for men for some decades, probably due to catching up.

I present an overview of the models one can use to describe progress. Although linear functions cannot be used in the long run, nonlinear functions, like the S-shaped curves,

TABLE 24.6

Estimation Results of the Pooled Model of the Historical Development of the Gini Coefficient

	Track and Field	Cycling	Rowing	Skating	Swimming	Weightlifting
Men						
Intercept	**112.23**	**202.63**	7.181	**162.58**	**145.27**	**375.07**
	22	46.16	117.29	13.96	22.72	68.08
Competitors	**−1.674**	**−3.767**	10.22	**−2.143**	**−1.033**	−6.918
	0.634	2.307	5.94	0.618	0.481	3.158
Countries	**2.196**	0.579	−9.876	0.077	0.792	−0.621
	0.811	2.083	6.004	0.987	0.59	5.077
GAP/W	**−4.297**	−4.584	0.166	−0.536	−0.448	0.552
	1.68	2.386	0.2	2.456	1.885	3.188
GDPCAP	**−3.604**	**0.192**	−0.925	**−1.626**	**−3.238**	**−4.648**
	0.743	0.725	0.517	0.187	0.486	1.401
Temperature	**3.636**	−2.011	0.94		0.82	**4.524**
	1.148	1.462	1.432		0.711	0.717
Adj. R^2	0.56	0.518	0.177	0.476	0.473	0.31
Women						
Intercept	**223.95**		168.61	**81.87**	**168.15**	−911.55
	29.98		92.32	16.4	25.03	805.57
Competitors	**−2.094**		0.047	**−1.461**	**−2.702**	−51.15
	1.077		2.96	0.475	0.701	33.6
Countries	**3.188**			**5.227**	**2.263**	24.72
	1.999			2.666	0.836	36.02
GAP/W	−0.292		−0.284	**−4.674**	**−11.302**	**21.75**
	0.488		0.051	0.695	4.769	8.941
GDPCAP	−3.588		−3.317	**−1.456**	**−3.827**	**74.36**
	1.914		1.254	0.119	0.735	33.01
Temperature	−0.313		4.531		**1.789**	**−22.4**
	0.883		3.105		0.634	8.051
Adj. R^2	0.645		0.313	0.467	0.52	0.667

Note: White-corrected standard errors; cross section fixed effects; significance at the 5%-confidence interval indicated by figures in bold.

are typically hard to identify with the sparse observations available. I also discussed the extreme value approach to rare improvements on the plateau.

Finally, I reviewed the Gould hypothesis that predicts convergence of top performances over time. Using historical data for both the Summer and Winter Olympic Games final events I analyze inequality of performance results. I show that over time generally results have become more competitive. I construct a simple model to explain inequality of results using the numbers of athletes, the number of countries represented, the gold medal performances, and real income per capita, and for the Summer Olympic Games the temperature as measures of circumstances as explanatory variables. I find that the increase in the number of competitors and the increase of real income per capita increases have led to lower inequality of results obtained in final events, maybe hinting at increases in competitiveness.

24.A Appendix Performance Improvement Tables per Sport

TABLE 24.A.1

Track and Field

	Index-Year	2012	Growth
Women			0.22
100 m	1928	113.5	0.15
100 m hurdles	1932	118.4	0.21
10k	1988	102.4	0.10
1500 m	1972	96.5	−0.09
200 m	1948	111.5	0.17
20k walk	2000	104.8	0.39
3000 steeple	2008	98.6	−0.36
400 m	1964	105.0	0.10
400 m hurdles	1984	103.6	0.13
5k	1996	99.5	−0.03
800 m	1928	117.7	0.19
Marathon	1984	101.2	0.04
Discus throw	1928	174.4	0.66
High jump	1928	128.9	0.30
Hammer throw	2000	109.9	0.79
Javelin throw	1932	159.2	0.58
Long jump	1948	125.0	0.35
Pole vault	2000	103.3	0.27
Shot put	1948	151.1	0.65
Triple jump	1996	97.7	−0.14
Men			0.22
100 m	1928	112.1	0.14
10k	1928	110.2	0.12
110 m hurdles	1928	114.6	0.16
1500 m	1928	108.9	0.10
200 m	1928	112.8	0.14
20k walk	1956	116.1	0.27
3000 m steeple	1928	112.7	0.14
400 m	1928	108.8	0.10
400 m hurdles	1928	112.1	0.14
50k walk	1932	134.3	0.37
5k	1928	106.9	0.08
800 m	1928	110.8	0.12
Marathon	1928	119.5	0.21
Discus throw	1928	144.3	0.44

(Continued)

TABLE 24.A.1 (*Continued*)

Track and Field

	Index-Year	2012	Growth
High jump	1928	122.7	0.24
Hammer throw	1928	156.8	0.54
Javelin throw	1928	127.0	0.28
Long jump	1928	107.5	0.09
Pole vault	1928	142.1	0.42
Shot put	1928	137.9	0.38
Triple jump	1928	117.1	0.19

TABLE 24.A.2

Rowing

	Index-Year	2012	Growth
Women			0.12
Single sculls	1976	103.4	0.09
Double sculls	1976	107.3	0.20
Quadruple sculls	1988	96.1	−0.17
Paired oar	1976	107.3	0.20
Eight oar	1976	113.1	0.34
Lightweight double	1996	100.8	0.05
Men			0.08
Single sculls	1928	103.1	0.04
Double sculls	1928	115.1	0.17
Quadruple sculls	1976	109.6	0.25
Paired oar	1928	111.7	0.13
Four oar	1928	108.1	0.09
Eight oar	1928	104.0	0.05
Lightweight double	1996	96.4	−0.23
Lightweight four	1996	101.8	0.11

TABLE 24.A.3

Swimming

	Index-Year	2012	Growth
Women			0.37
50 m freestyle	1988	106.0	0.24
100 m freestyle	1928	134.0	0.35
200 m freestyle	1968	114.9	0.32
400 m freestyle	1928	142.0	0.42
800 m freestyle	1968	114.0	0.30
10k freestyle	2008	101.6	0.39
100 m back	1928	140.6	0.41
200 m back	1968	116.7	0.35
100 m breast	1968	115.8	0.33
200 m breast	1928	138.0	0.38
100 m butterfly	1956	126.8	0.43
200 m butterfly	1968	116.6	0.35
Men			0.32
50 m freestyle	1988	103.7	0.15
100 m freestyle	1928	123.3	0.25
200 m freestyle	1968	111.7	0.25
400 m freestyle	1928	137.0	0.38
1500 m freestyle	1928	136.8	0.37
10k freestyle	2008	101.8	0.44
100 m back	1928	130.8	0.32
200 m back	1964	114.9	0.29
100 m breast	1968	115.8	0.33
200 m breast	1928	132.6	0.34
100 m butterfly	1968	109.2	0.20
200 m butterfly	1956	123.3	0.37

Note: For weightlifting I have a structural break in 1976, where the competition skipped the 'press' component and only left the 'snatch' and 'jerk'.

TABLE 24.A.4

Weightlifting

	Index-Year	1972	Growth	Index-Year	2012	Growth
Men			0.88			0.26
Bantam	1948	122.8	0.86	1976	111.6	0.31
Feather	1928	140.0	0.77	1976	114.7	0.38
Light	1928	142.6	0.81	1976	112.8	0.33
Middle	1928	144.8	0.84	1976	113.1	0.34
Lightheavy	1928	143.0	0.82	1976	105.5	0.15
Middleheavy	1952	118.0	0.83	1976	109.3	0.25
Heavy	1972	100.0		1976	107.0	0.19
Superheavy	1928	171.8	1.24	1976	103.4	0.09
Women						0.68
Flyweight				2000	110.8	0.86
Feather				2000	100.4	0.04
Light				2000	110.6	0.84
Middle				2000	101.0	0.09
Lightheavy				2000	107.6	0.61
Heavy				2000	118.8	1.44
Superheavy				2000	111.0	0.87

TABLE 24.A.5

Skating

	Index-Year	2014	Growth
Women			0.41
500 m	1960	122.9	0.40
1k	1960	127.1	0.46
1500 m	1960	127.9	0.47
3k	1960	130.8	0.52
5k	1988	105.5	0.22
Men			0.38
500 m	1928	125.2	0.27
1k	1976	116.0	0.41
1500 m	1928	134.4	0.35
5k	1928	143.1	0.43
10k	1928	146.1	0.45

References

Blest, D. C. (1996), Lower bounds for athletic performances, *The Statistician*, 45, 243–253.

Chatterjee, S. and S. Chatterjee (1982), New lamps for old: An exploratory analysis of running times in Olympic Games, *Applied Statistics*, 31, 14–22.

Deaton, A. (1997), *Analysis of Household Surveys*, Johns Hopkins University Press, Baltimore.

Dekerle, J., X. Nesi and H. Carter (2006), The distance–time relationship over a century of running Olympic performances: A limit on the critical speed concept, *Journal of Sports Sciences*, 24, 1213–1221.

Denny, M. W. (2008), Limits to running speed in dogs, horses and humans, *The Journal of Experimental Biology*, 211, 3836–3849.

Frick, B. and J. Prinz (2007), Pay and performance in professional road racing: The case of city marathons, *International Journal of Sport Performance*, 2, 25–35.

Gini, C. (1912), *Variabilita e mutabilita, Contributo allo Studio delle Distribuzioni e delle Relazioni Statistiche*, C. Cuppini, Bologna, Italy; Reprinted in: *Memorie di metodologica statistica* (E. Pizetti and T. Salvemini, eds.), Libreria Eredi Virgilio Veschi, Rome, Italy.

Gould, S.J. (1996), *Full House: The Spread of Excellence from Plato to Darwin*, Three Rivers, New York.

Haake, S. J. (2009), The impact of technology on sporting performances in Olympic sports, *Journal of Sports Sciences*, 27, 1421–1431.

Haake, S. J., L. I. Foster and D. M. James (2014), An improvement index to quantify the evolution of performance in running, *Journal of Sports Sciences*, 32(7), 611–622.

Haake, S. J., D. M. James and L. I. Foster (2015), An improvement index to quantify the evolution of performance in field events, *Journal of Sports Sciences*, 33(3), 255–267.

Holden, C. (2004), An everlasting gender gap? *Science*, 305, 639–640.

Kuper, G. H. and E. Sterken (2003), Endurance in speedskating: The development of world records, *European Journal of Operational Research*, 148(2), 293–301.

Kuper, G. H. and E. Sterken (2008), Modelling the development of world records in running, in: J. Albert and R. H. Koning, eds., *Statistical Thinking in Sports*, Chapman & Hall/CRC, Boca Raton, FL, pp. 7–31.

Nevill, A. M. and G. Whyte (2005), Are there limits to running world records? *Medicine and Science in Sports and Exercise*, 37, 1785–1788.

Peiser, B. and T. Reilly (2004), Environmental factors in the summer Olympics in historical perspective, *Journal of Sports Sciences*, 22, 981–1002.

Pritchard, W. G. and J. K. Pritchard (1994), Mathematical models of running, *American Scientist*, 82(6), 546–553.

Pyne, D. B., C. B. Trewin and W. G. Hopkins (2004), Progression and variability of competitive performance of Olympic Swimmers, *Journal of Sports Sciences*, 22, 613–620.

Sanderson, A. R. and J. J. Siegfried (2003), Thinking about competitive balance, *Journal of Sports Economics*, 4, 255–277.

Stefani, R. T. (1982), Olympic winning performances: Trends and predictions, *Olympic Review*, 176, 357–364.

Stefani, R. T. (1994), Athletics, swimming and weightlifting from Barcelona to Atlanta, *Olympic Review*, 326, 598–603.

Stefani, R. T. (2006), The relative power output and relative lean body mass of World and Olympic male and female champions with implications for gender equity, *Journal of Sports Sciences*, 24(12), 1329–1339.

Stefani, R. T. (4–8 April 2012), Overcoming the doping legacy: Can London's winners outperform the drugs of 1988? *Significance*, 9(2), 4–8.

Stefani, R. T. (2014), The power-to-weight relationships and efficiency improvements of Olympic Champions in Athletics, Swimming and Rowing, *International Journal of Sport Science and Coaching*, 9(2), 271–285.

Tatem, A. J., C. A. Guerra, P. M. Atkinson and S. I. Hay (2004), Momentous sprint at the 2156 Olympics? *Nature*, 431, 525.

Wainer, H., C. Njue, and S. Palmer (2000), Assessing time trends in swimming & running, *Chance*, 13(1), 10–15.

Wallechinsky, D. and J. Loucky (2012), *The Complete Book of the Olympics: 2012 Edition*, Aurum, London, UK.

Whipp, B. J. and S. A. Ward (1992), Will women soon outrun men? *Nature*, 355, 25.

Index

Milton Keynes UK
Ingram Content Group UK Ltd.
UKHW052025071024
449327UK00027B/2425